FRONTIERS IN PROTEIN CHEMISTRY

DEVELOPMENTS IN BIOCHEMISTRY

FRONTIERS IN PROTEIN CHEMISTRY

Proceedings of the Conference on Protein Chemistry, University of Hawaii, Honolulu, Hawaii, U.S.A., July 2-6, 1979

Editors:

TEH-YUNG LIU, Ph.D.
Director, Biochemistry Branch, Division of Bacterial Products, Bureau of Biologics, Food and Drug Administration, Bethesda, Maryland, U.S.A.

GUNJI MAMIYA, M.D.
Professor, Department of Biochemistry, National Defense Medical College, Tokorozawa, Japan

KERRY T. YASUNOBU, Ph.D.
Professor and Chairman, Department of Biochemistry-Biophysics, University of Hawaii, Honolulu, Hawaii, U.S.A.

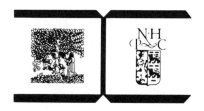

ELSEVIER/NORTH-HOLLAND
NEW YORK • AMSTERDAM • OXFORD

Published by:

Elsevier North Holland, Inc.
52 Vanderbilt Avenue, New York, New York 10017

Sole distributors outside USA and Canada:

Elsevier/North-Holland Biomedical Press
335 Jan van Galenstraat, P.O. Box 211
Amsterdam, The Netherlands

Library of Congress Cataloging in Publication Data

Conference on Protein Chemistry, University of Hawaii, 1979.
    Frontiers in protein chemistry.
    (Developments in biochemistry; v. 10)

    Includes index.
    1. Proteins—Congresses. 2. Biological chemistry—Congresses.
    I. Liu, Teh-Yung, 1932-   II. Mamiya, Gunji. III. Yasunobu, Kerry Tsuyoshi,
    1925- IV. Title. [DNLM: 1. Proteins—Analysis—Congresses.
    W1 DE997VG v. 10 / QU55 C749f 1979]
QP551.C717   1979   574.19'245   80-14454
ISBN 0-444-00414-9

Manufactured in the United States of America

Contents

SECTION III - PROTEASES AND OTHER HYDROLYTIC ENZYMES
Chairmen: R.L. Heinrikson, K. Narita

SECTION IV - OXYDASES, DEHYDROGENASES AND HEMOGLOBIN
Chairman: T. Matsuda

Preface

A group of physical biochemists, bioorganic chemists, protein chemists and enzymologists met under balmy weather conditions and blue skies in the middle of the Pacific Ocean to discuss research related to proteins. The meetings were held in the Asian Room of the East-West Center adjacent to the University of Hawaii in a miniature United Nations conference room. The meeting was held on July 2-5, 1979 with a break on July 4th for a tour of the islands and a welcome opportunity not only to enjoy the beautiful flora and fauna of Honolulu, but to allow the participants an opportunity to discuss science.

This was the first of a series of planned meetings to discuss topics of research which fall into the category of "Frontiers in Protein Chemistry." Although the meeting was planned to take advantage of the scientists from Asia who were on their way to the International Biochemistry meeting in Toronto, Canada, it turned out that the two meetings conflicted with one another. Plans are underway to hold the next meeting in 1982 when the International Biochemistry Congress will be held in Perth, Australia.

Nevertheless, the organizers of this symposium felt that the meeting was quite successful and a number of extremely interesting findings were presented at this meeting.

The organizers of this symposium are very greatful to the sponsors of this symposium, especially to John E. Fogarty International Center, Acting Chancellor Howard McKaughan of the University of Hawaii, Mr. James McCoy of the East-West Center and the Department of Biochemistry and Biophysics at the University of Hawaii, who took care of the meeting arrangements. Finally, we wish to thank all the participants of the meeting. To all of these people, we wish them a fond aloha.

Teh-yung Liu

Gunji Mamiya

Kerry T. Yasunobu

Participants

Lyman Armes
Department of Biochemistry-Biophysics
University Of Hawaii School of Medicine
Honolulu, Hawaii 96822

Ralph Bradshaw
Department of Biological Chemistry
Washington University School of
   Medicine
St. Louis, Missouri 63110

Antonio Camargo
Protein Chemistry Laboratory
Department of Pharmacology
Faculty of Medicine of Ribeirão Preto
Ribeirão Preto, S.P., Brazil

Sunney I. Chan
A.A. Noyes Laboratory of Chemical Physics
California Institute of Technology
Pasadena, California

Jack S. Cohen
Developmental Pharmacology Branch
National Institute of Child Health and
   Human Development
Bethesda, Maryland

William Egan
Bureau of Biologics
Food and Drug Administration
Bethesda, Maryland

Nobutaka Fujii
Faculty of Pharmaceutical Sciences
Kyoto University
Kyoto, Japan

Richard Guillory
Department of Biochemistry and Biophysics
University of Hawaii
Honolulu, Hawaii

William Habig
Bureau of Biologics
Food and Drug Administration
Bethesda, Maryland

Mitsuru Haniu
Department of Biochemistry-Biophysics
University of Hawaii
Honolulu, Hawaii

Robert Heinrikson
Department of Biochemistry
University of Chicago
Chicago, Illinois

Seikoh Horiuchi
Department of Biochemistry
Kumamoto University Medical School
Kumamoto, Japan

Sadaaki Iwanaga
Department of Biology
Faculty of Science
Kyushu University
Fukuoka 812, Japan

Lynn Jelinski
National Institute of Dental Research
National Institutes of Health
Laboratory of Biochemistry
Bethesda, Maryland

P.G. Katsoyanis
Department of Biochemistry
Mt. Sinai School of Medicine
City University of New York
New York, New York

T.-Y. Liu
Division of Bacterial Products
Bureau of Biologics
Food and Drug Administration
Bethesda, Maryland

Tung-bin Lo
Institute of Biological Chemistry
Academia Sinica
Taipei, Taiwan (ROC)

Gunji Mamiya
Department of Biochemistry
National Defense Medical College
Tokorozawa, Saitama, Japan

John Markley
Biochemistry Division
Department of Chemistry
Purdue University
West Lafayette, Indiana

Genji Matsuda
Department of Biochemistry
Nagasaki University School of
   Medicine
Nagasaki, Japan

Shigeru Matsukawa
Department of Biochemistry
Kanazawa University School of
   Medicine
Kanazawa, Japan

Masayuki Miyawaki
The Department of Biochemistry
Kumamoto University Medical School
Kumamoto, Japan

Yoshimasa Morino
Department of Biochemistry
Kumamoto University Medical School
Kumamoto, Japan

Howard Morris
Department of Biochemistry
Imperial College of Science and
   Technology
London, United Kingdom

Takashi Murachi
Department of Clinical Science
Kyoto University Faculty of Medicine
Kyoto, Japan

Shin Nakamura
Primate Research Institute
Kyoto University
Aichi 484 Japan

Kozo Narita
Institute for Protein Research
Osaka University
Osaka, Japan

Robert Newcomb
Department of Biochemistry-Biophysics
University of Hawaii School of Medicine
Honolulu, Hawaii

Eduardo Oliveria
Instituto de Ciências Biológicas
Universidade Federal de Minas Gerais
Belo Horizonte, Brazil

Paul Schmidt
Laboratory of Protein Studies
Oklahoma Medical Research Foundation
Oklahoma City, Oklahoma

Robert Seid
Division of Bacterial Products
Bureau of Biologics
Food and Drug Administration
Bethesda, Maryland

Tsunejiro Sekita
Clinical Chemistry Laboratory
Kawasaki Municipal Hospital
Kawasaki, Japan

Nobuo Tamiya
Department of Chemistry
Tohoku University
Aobayama, Sendai, Japan

Masaru Tanaka
Department of Biochemistry-Biophysics
University of Hawaii
Honolulu, Hawaii

Sumio Tanase
The Department of Biochemistry
Kumamoto University Medical School
Kumamoto, Japan

Jordan Tang
Department of Biochemistry and
   Molecular Biology
University of Oklahoma Health
   Services Center
Oklahoma City, Oklahoma

Akio Tomoda
Department of Biochemistry
Kanazawa University School of
   Medicine
Kanazawa, Japan

Joseph Villafranca
Department of Chemistry
The Pennsylvania State University
University Park, Pennsylvania

Marguerite Volini
Department of Biochemistry and
   Biophysics
University of Hawaii
Honolulu, Hawaii

Haruaki Yajima
Faculty of Pharmaceutical Sciences
Kyoto University
Kyoto, Japan

Kerry Yasunobu
Department of Biochemistry and
   Biophysics
University of Hawaii
Honolulu, Hawaii

Yoshimasa Yoneyama
Department of Biochemistry
Kanazawa University School of Medicine
Kanazawa, Japan

Hussein Zeidan
Department of Biochemistry
  and Biophysics
University of Hawaii School of
  Medicine
Honolulu, Hawaii

FRONTIERS IN PROTEIN CHEMISTRY

I
Physicochemical Studies

Published 1980 by Elsevier North Holland, Inc.
Liu/Mamiya/Yasunobu, eds. Frontiers in Protein Chemistry

CONFORMATION, MECHANISM AND PEPTIDE EXCHANGE OF ^{13}C-ENRICHED RIBONUCLEASE S STUDIED BY ^{13}C NUCLEAR MAGNETIC RESONANCE SPECTROSCOPY

JACK S. COHEN, CHIEN-HUA NIU[*], SHUJI MATSUURA[+], AND HEISABURO SHINDO[*]
Developmental Pharmacology Branch and [+]Endocrinology and Reproduction
Research Branch, National Institute of Child Health and Human Development,
National Institutes of Health, Bethesda, Maryland 20205

ABSTRACT

Enzymatically active RNase S' complexes were prepared from selectively 90% ^{13}C-enriched synthetic (1-15) peptides and RNase S-protein (residues 21-124). The ^{13}C signals of residues Ala 5 (u.l.), His 12 (C^{ε}), Met 13 (C^{ε}) and Asp 14 (C^{γ}) were monitored as a function of pH, temperature and in the presence of enzyme inhibitors. The titration curves of His 12 (pK_a = 5.7) and Asp 14 (pK_a = 2.4) exhibited no common pK_a values and indicated that Asp 14 is hydrogen bonded in the pH range 3-8. Also, Asp 14 is unaffected by inhibitors, while His 12 is significantly affected, although its pK_a value was not changed by the substrate analog UpcA. These results are related to the acid pH ($pK \approx 4$) conformational transition of RNase and to its mechanism of action. The pH and temperature denaturation of RNase S were monitored at each ^{13}C-enriched site, of which only Met 13 exhibited a different mid-point from the overall transition, indicating easier unfolding of this residue. The exchange of ^{13}C-enriched peptide with RNase S and of unenriched S-peptide with ^{13}C-enriched RNase S' were monitored as a function of time by the ^{13}C NMR method. The dissociation constant of the RNase S' complex of (1-15) peptide was found to be (0.2 ± 0.8) x 10^{-7} M at pH 5.2, 4°C, and the value of the constant for the native RNase S (containing S-peptide, residues 1-20) was 4 x 10^{-9} M. By following the relative areas of the ^{13}C enriched peptide and RNase S' complex resonances at different temperatures the difference in the enthalpy of binding for the (1-15) and (1-20) peptides was calculated to be 1.7 kcal/mole. We have shown that peptide-protein interactions can be monitored quantitatively in solution by the use of nondisturbing stable isotope probes.

A preliminary account of some of this work was presented at the American Society of Biological Chemists' Meeting, Atlanta, Georgia, June, 1978.
* Present addresses: Chien-Nua Niu, Laboratory of Biochemical Pharmacology; Heisaburo Shindo, Laboratory of Chemical Physics; both at National Institute of Arthritis Metabolism and Digestive Diseases, National Institute of Health, Bethesda, Maryland 20205.

4

INTRODUCTION

The subtle relationship between protein conformation and enzymatic mechanism is not entirely revealed by static protein structures. In order to investigate this relationship in solution for ribonuclease with a minimally disturbing probe we have utilized the combination of stable isotopic enrichment with ^{13}C and observation by ^{13}C nuclear magnetic resonance (NMR) spectroscopy.[1]

The ribonuclease system has certain advantages in that relatively large quantities of amino terminal peptides, RNase (1-15), could be synthesized with selected ^{13}C-enriched amino acids inserted into them (Fig. 1). Then fully active enzyme could be obtained by noncovalent complexation of the synthetic peptides with RNase S-protein (residues 21-124).[2] We have been able to resolve the individual ^{13}C resonances of the resulting semi-synthetic RNase S' complexes.[1] In the present work we describe the effects of changing pH and of adding competitive inhibitors of the enzyme. Together with our earlier studies utilizing ^{1}H and ^{31}P NMR,[3-5] these results are analyzed in terms of aspects of the

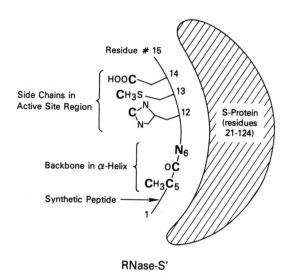

RNase-S'

Fig. 1. Schematic representation of the sites enriched in [13]C (bold letters) in residues 5, 12, 13 and 14 in the RNase (1-15) peptide, and their relationship to the RNase S-protein in forming the enzymatically active RNase S' complex.[1] (One site, residue 6, was also enriched in ^{15}N, although this was not studied in the present work.)

solution conformation and the mechanism of action of ribonuclease. In addition, we have utilized the stable isotope as a probe to quantitate the exchange of the peptides with the binding site on the S-protein.

EXPERIMENTAL PROCEDURES

The materials and experimental procedures carried out in these studies were the same as described previously.[1]

The semi-synthetic selectively ^{13}C-enriched RNase S' complexes used were the same ones prepared previously. The inhibitors P_i and the 5'-phosphonate ester analog of the natural substrate UpA, which is abbreviated to UpcA (this material was a gift of Syntex Laboratories, Inc.), were added in increasing concentrations. The actual molar ratios of inhibitors used is given in the figure legends. The concentration of P_i was determined in duplicate by the phosphomolybdate method. Protein concentration was determined using an extinction coefficient of 0.695 for a 1 mg/ml solution of RNase S.[6] For NMR measurements RNase S' concentrations were ca. 3 mM. ^{13}C and ^{1}H NMR spectra were obtained as described previously at 67.89 and 220 MHz respectively. The sample temperature was 20 ± 1°C in both cases unless otherwise indicated.

RNase S' complexes for exchange experiments were prepared in sodium acetate buffer (0.06 M, pH 5.2) by dialysis and stored in ice until the exchanging peptide was added. The samples were then quickly transferred to the NMR spectrometer and spectra were determined over short time intervals using an automatic program, usually at 4°C. Spectra were obtained with gated decoupling to avoid differential NOE values.

RESULTS

The ^{13}C NMR titration curves of His 12 C^ϵ and Asp 14 C^γ of the semisynthetic His 12 C^ϵ/Asp 14 C^γ RNase S' complex alone[1] and in the presence of P_i and UpcA are shown in Figs. 2 and 3. The pK_a values derived from curve-fitting the data are given in Table 1. Several results can be summarized; a) increasing concentrations of P_i increases the pK_a value of His 12; b) the presence of P_i results in the observation of a lower pKa inflection for His 12; c) UpcA does not change the pK_a of His 12 itself, but does result in an additional lower pK_a inflection; d) neither P_i nor UpcA significantly affect the ^{13}C NMR titration curve of Asp 14; e) there is no correspondence between the pK_a's observed for His 12 and those observed for Asp 14, indicating that they each experience different ionization processes. In addition, no effect of P_i (0.15 M) was observed on the [13]C^ϵ resonance of Met [13]C^ϵ RNase S' at pH 5.7.

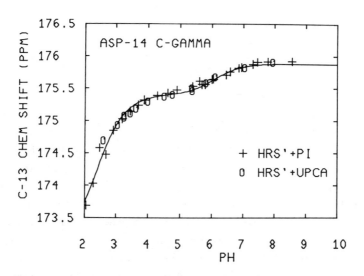

Fig. 2. ^{13}C NMR titration curves of [γ—^{13}C] Asp 14 RNase S' in 0.1 M NaCl at 20°, and in the presence of 5 x 10^{-3} M P$_i$ (+) and 10^{-2} M UpcA (0). In view of the absence of any significant effect of the inhibitors on this titration curve all the data, including that for the protein alone, and with excess P$_i$ (Table 1) were fitted together to provide the curve in this figure.

The signal intensity of the ^{13}C enriched RNase S' complex was monitored as a function of time until equilibrium was established both on the addition of S-peptide and on the addition of enriched (1-15) peptide to native RNase S (Fig. 4). The equations used for curve fitting are given below. The value of K$_d$ obtained was (0.2 ± 0.8) x 10^{-7} M. Other values are given in the Discussion Section. The ^{13}C signals of the enriched peptide and complex were also monitored at equilibrium in the presence of an approximately equimolar ratio (1:1.2) of S-peptide (Fig. 5). The Van't Hoff plot (Fig. 6) gave a difference in the enthalpies of interaction of the two peptides at the S-protein binding site of 1.6 ± 0.4 kcal/mole.

DISCUSSION

Conformation

The amount of structural information on ribonuclease, its derivatives and inhibitor complexes is extensive.[7-9] NMR studies have provided a wealth of detailed information, especially on the pH-dependence of histidine residues,[4,10] and effects of inhibitors utilizing both ^{1}H and ^{31}P NMR.[3,11,12] A conformational

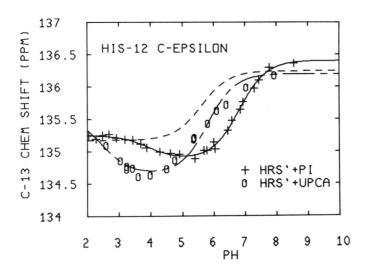

Fig. 3. ^{13}C NMR titration curves of $[\varepsilon - ^{13}C]$ His 12 RNase S' (---) in 0.1 M NaCl at 20°, and in the presence of 5 x 10^{-3} M P$_i$ (+) and 10^{-2} M UpcA (0). Inhibitor to protein molar ratios were 1.6 and 3.1 respectively.

TABLE 1

IONIZATION CONSTANTS FOR His12($^{13}C^{\varepsilon}$)/Asp14($^{13}C^{\gamma}$) RNase (1-15) S' COMPLEX[a]

	His12(C^{ε})		Asp14(C^{γ})	
	pK$_1$	pK$_2$	pK$_1$	pK$_2$
Complex[b]				
+ 5.0 x 10^{-3} M P$_i$	3.9 ± 0.1	5.68 ± 0.03	2.41 ± 0.06[c]	6.07 ± 0.09[c]
+ 1.9 x 10^{-2} M P$_i$	3.9 ± 0.3	6.82 ± 0.04		
+ UpcA	∿ 2.5	5.8 ± 0.10		

[a]Derived from best fits to an equation for single (pK$_1$) or double (pK$_2$) transitions.

[b]In the presence of 1.5 x 10^{-4} M P$_i$ (molar ratio HRS' : P$_i$ = 42:1).

[c]These values derived from fits including data in the presence of inhibitors (P$_i$, UpcA).

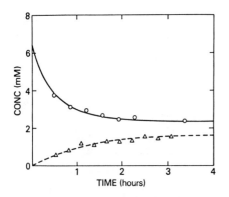

Fig. 4. Plot of concentration of RNase S', derived from the area of the ^{13}C signal of the ^{13}C-enriched complex, as a function of time; 0, for the reaction of RS* and S-peptide; Δ, for the reaction RS and S*. The lines are theoretical curves from converged best fits (see text).

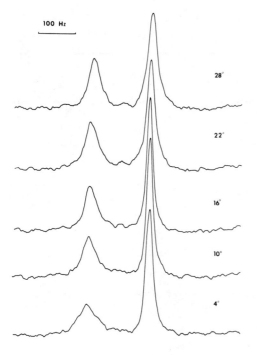

Fig. 5. Temperature dependence at equilibrium of the ^{13}C signals of [ε - ^{13}C] His 12 (1-15) peptide and its RS* complex in the presence of S-peptide (molar ratio 1:1.2). Spectra were obtained with gated decoupling without NOE.

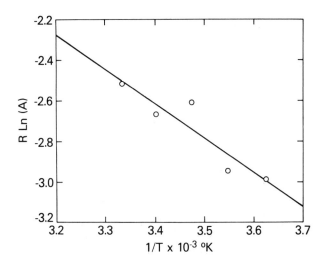

Fig. 6. Van't Hoff plot of data derived from Fig. 5, for temperature dependence of exchange of S-peptide and (1-15) peptide with RNase S-protein site, where A is the ratio of the dissociation constants for the S-peptide and the (1-15) peptide (A = K_d'/K_d).

transition in the acid pH region has been detected for RNase by several techniques.[10-16] We have shown that this transition with a pK $\simeq$ 4 affects three of the histidine residues of RNase, including His 48 and the two active site histidines 12 and 119.[4] In view of the proximity of Asp 14 to His 48 we previously suggested that the carboxyl group of Asp 14 was responsible for this acid transition.[4] Our current results (Fig. 2) enable us to eliminate Asp 14 as the interacting group, since its ^{13}C NMR titration curve exhibits no pK_a transition in this pH range in the RNase S' complex (Table 1).

In Table 2 we have surveyed the interactions of all carboxyl groups in RNase S from the X-ray structure.[17] We presume that one of these must be the origin of the acid-conformational transition. It is clear that only Asp 14 and Asp 121 are adjacent to histidine residues, and that Asp 121 is the only carboxyl residue within hydrogen-bonding distance of an active site histidine residue. We are now carrying out studies on ^{13}C-enriched carboxyl-terminal peptides[18,19] to determine whether Asp 121 is the source of this acid transition in ribonuclease. The observation of a carboxyl group at the active site of RNase T_1,[20,21] prompts us to suppose that this acid-transition may have functional significance, particularly since it also appears to affect the kinetics of the enzymatic reaction.[22]

TABLE 2

HYDROGEN BONDS FORMED BY CARBOXYL GROUPS IN RIBONUCLEASE S[a]

Carboxyl Residue	Hydrogen Bond To	Closest Approach of H-Bonding Atom (A)
Glu 2	Arg 10 side chain	2.8
Glu 9	None	-
Asp 14	Tyr 25 side chain	2.6
	His 48 side chain	3.3
Asp 38	Arg 39 side chain	3.0
Glu 49	None	-
Asp 53	None	-
Asp 83	Thr 45 side chain	2.6
Glu 86	Ser 90 side chain	2.4
Glu 111	None	-
Asp 121	Lys 66 backbone	3.1
	His 119 side chain	3.1
Val 124	None	-

[a]From AMSOM;[17] X-ray structure of RNase S plus UpcA inhibitor from Richards and Wyckoff.[7] The distances are for closest approach of N or O atoms excluding H. In some cases, such as imidazole rings, it is not possible to distinguish between C and N atoms; in these cases the closest approach is also quoted.

Mechanism

We have previously published a mechanism of action of ribonuclease[3] which conforms to the basic push-pull type mechanism.[23] No phosphate pseudo-rotation[24,25] was presumed to occur, but rather an on-line, direct axial displacement of the 5'-oxygen leaving group from the penta-coordinate intermediate,[26] consistent with other findings.[9,27-29] This mechanism is shown schematically in Fig. 7.

One complication relating to this mechanism since its original conception[3] is that the assignments of the active site histidine C^{ε}-H NMR resonances utilized in the original work[30] have been reversed.[5,31-33] Not only have we now definitely confirmed this corrected assignment by the observation of the ^{1}H NMR spectra of [ε-^{13}C] His 12 RNase (1-15) S' complex,[1] but we are also able in the present work to measure the pK_a of His 12 directly by ^{13}C NMR (Fig. 3). Thus, we are able to include with confidence the pK_a values for the active site histidine residues in the proposed mechanism (Fig. 7). In addition, in the

FIG. 7

A. RNase – UpA Complex

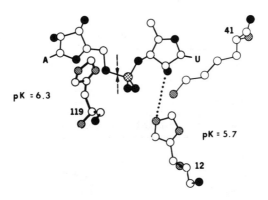

B. Cleavage Step

    Penta-coordinate Phosphorus Transition State

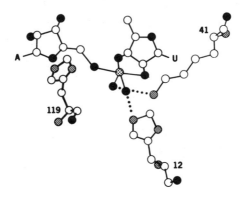

C. Products

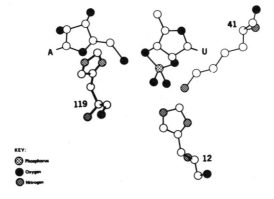

Fig. 7. A. Proposed structure of the RNase-UpA Michaelis complex based on the X-ray crystallographic and NMR data from the RNase-UpcA complex. The side chain of histidine residue 12, and the ribose-phosphate-ribose portion of UpA, are roughly in the plane of the page, while lysine residue 41 is behind the plane of the page. The side chain of histidine residue 119 is above the plane of the page with its side chain projecting upward. The dotted line indicates that histidine 12 may act as a base to abstract a proton from the 2'-hydroxyl group. (See text for further discussion.) The pK_a values of histidine residues 12 and 119 are shown adjacent to them. B. Proposed structure of the transition state for cleavage of UpA by RNase in which the phosphorus atom is penta-coordinate. The location of the phosphorus atom moves so that oxygen atoms are brought into contact with both histidine 12 and lysine 41 side-chains. C. Proposed structure of the RNase-product complex following cleavage of UpA. (Water molecules and hydrogen atoms are omitted from this figure for clarity.)

present results we see no evidence for a proposed interaction of the side chain of His 119 with that of His 12,[10,34] and in fact, further evidence from the binding of EDTA to RNase, which affects only the pK_a of His 12,[35] also supports this lack of histidine-histidine interaction.[4] As noted earlier[5] this reversal of the assignment of the histidine C^ϵ-H resonances does not basically alter our earlier conclusions,[3] which were that,

a) evidence from the effects of mononucleotides are inadequate to provide a mechanism of action, due to the significant effects of the second negative charge on the phosphate group;[11,12,36]

b) a phosphate diester substrate-analog is required (in this case UpcA);

c) in the penta-coordinate intermediate P-O groups are within contact of histidine 12 and lysine 41 side chains.

These conclusions were derived from the fact that no significant effects of UpcA on either the His 12 or His 119 ^{1}H NMR titration curves were recorded in the previous work.[3] We have been able to confirm this for His 12 in the present work, since UpcA produced no change in the His 12 pK_a value (Table 1). That this is not fortuitous is shown by the consistent increase in the His 12 pK_a value on the addition of P_i (Table 1 and reference 13) and by the alteration in the shape of the His 12 ^{13}C NMR titration curve in the presence of UpcA. This lack of a pK_a change indicates that His 12 is interacting with a neutral group, which we presumed to be the 2'-hydroxyl[3] as others have also indicated.[9]

Since the pK_a of His 12 is 5.7, it cannot be protonated in the pH range where RNase manifests its maximal enzymatic activity. Hence, it may be presumed to act as a base to abstract a proton from the 2'OH group (Fig. 7). The low pK_a value of His 12 and the downfield shift of its [13]C^ϵ resonance in the complex[1] are consistent with a positively charged environment for this residue, presumably due to the proximity of the side chain of Lys 41.

Peptide Exchange

Kinetic determinations of peptide exchange with protein binding sites can be made by few techniques. Measurements of enzyme activity,[37-39] tritium exchange[40] and antibody interactions[41] have been used to quantitate the rates of peptide-protein exchange. We now describe the use of the ^{13}C-enriched synthetic RNase (1-15) peptides to follow peptide exchange in the RNase S system. Using the stable isotopic enrichment enabled us to follow the kinetics of the exchange between labeled and unlabeled species at equilibrium, and avoided any problems connected with the self-association of RNase S-protein.[42]

Since the natural S-peptide is 20 amino acids long and the synthetic peptide is 15 amino acids in length, a preference for the former is expected for the peptide binding site on the S-protein. In effect, we can perform competition experiments between S-peptide (1-20) and (1-15) peptide. The reaction is,

$$RS^* \underset{k_2}{\overset{k_1}{\rightleftharpoons}} R + S^* \tag{1}$$

$$RS \underset{k_4}{\overset{k_3}{\rightleftharpoons}} R + S \tag{2}$$

$$K_d = k_1/k_2 \equiv \frac{[R] \cdot [S^*]}{[RS^*]} \tag{3}$$

$$K_d' = k_3/k_4 \equiv \frac{[R] \cdot [S]}{[RS]} \tag{4}$$

Where K_d is the dissociation constant of the RNase S' complex containing the ^{13}C enriched (1-15) peptide (S^*), and K_d' is the corresponding value for the natural (1-20) peptide (S). By monitoring the ^{13}C resonance of the enriched complex RS^* on the addition of natural (1-20) peptide, a value for K_d was determined. The data of signal intensity as a function of time was fitted (Fig. 4) with the equations,

$$\frac{d [RS^*]}{dt} = - k_1 [RS^*] + k_2 [R][S^*] \tag{5}$$

$$\frac{d [R]}{dt} = k_1 [RS^*] + k_3 [RS] - k_2 [R][S^*] - k_4 [R][S] \tag{6}$$

Assuming an equal rate for k_2 and k_4 these equations can be solved for k_1 and k_2.

In order to evaluate K_d the equation for exchange at equilibrium must also be solved, i.e., when $K_d << 1$ and $K_d' << 1$,

$$(1 + Af_\infty)[R]_o - (1 + Af_\infty)[RS^*]_\infty - [S]_o = 0 \tag{7}$$

where $A = K_d'/K_d$, and $f_\infty = [S^*]_\infty/[RS^*]_\infty$. Under the experimental conditions used (4°C, pH 5.4) the value of A was found to be 0.20. From this a value of k_1 of $(4.2 \pm 0.4) \times 10^{-4}$ sec^{-1} was evaluated and a value of K_d of $(0.2 \pm 0.8) \times 10^{-7}$ M was determined. This gives a value of $K_d' = 4 \times 10^{-9}$ M, which is consistent with the values determined by Schreier and Baldwin[40] and others.[37-39] It is known that K_d is sensitive to both pH[42] and temperature[38] as well as the presence of substrate when the enzymatic activity is used to quantitate K_d.[43]

Another system was examined by monitoring the resonance of the ^{13}C signal of natural RNase S upon addition of ^{13}C-enriched (1-15) peptide (Fig. 4). A consistent value of $A = 0.18$ was determined for this equilibrium, and a similar rate k_1 of $(3.9 \pm 0.2) \times 10^{-4}$ sec^{-1} was obtained by fitting using the value of K_d previously determined. The absence of a convergent fit results primarily from the interdependence of k_1, k_2 and K_d. Similarly, an experiment in which ^{13}C-enriched (1-15) peptide (S^*) was added to RNase S' complex prepared from S-protein and synthetic unenriched (1-15) peptide, failed to give a convergent fit; although from the equilibrium a value of $A = 1.04$ was obtained, which is very close to the expected value of unity.

We also studied the effect of increasing temperature on the ratio of the ^{13}C-enriched peptide and complex signals at equilibrium in the presence of S-peptide (Fig. 5). For reactions (1) and (2) we cannot measure an absolute value of the enthalpy of binding, but rather a difference,

$$\frac{d(R\ln A)}{d(1/T)} = -(\Delta H^o_* - \Delta H^o) = -\delta\Delta H^o \tag{8}$$

where, ΔH^o_* is the enthalpy of binding for the (1-15) peptide and ΔH^o for the (1-20) peptide. The relative areas of the peaks are plotted in a Van't Hoff plot in Fig. 6, giving a value of $\delta\Delta H^o$ of 1.7 ± 0.4 kcal/mole. This is of the order that might be expected between two such similarly binding peptides,[2] and of course, is a fraction of the value of -24 kcal/mole for the overall interaction of (1-20) peptide determined calorimetrically.[38] The difference in enthalpy implies that residues (16-20) contribute little to the binding, and the value determined may reflect a hydrogen-bond for Ser 16 to His 48 in the complex.

CONCLUSION

Unique information on the solution conformation, mechanism of action, exchange kinetics and thermodynamics have been determined for the (1-15) and (1-20) amino terminal peptides bound to RNase S-protein using selective ^{13}C enrichment and ^{13}C NMR observation.

ACKNOWLEDGMENTS

We thank Gordon Jones of Syntex, Inc. for a generous gift of UpcA, and Elmer Leininger for typing this manuscript.

REFERENCES

1. Niu, C., Matsuura, S., Shindo, H. and Cohen, J.S. (1979) J. Biol. Chem., 254, 3788-3796.
2. Potts, J.T., Jr., Young, D.M. and Anfinsen, C.B. (1963) J. Biol. Chem., 238, 2593-2594.
3. Griffin, J.H., Schechter, A.N. and Cohen, J.S. (1973) Ann. N.Y. Acad. Sci., 222, 693-708.
4. Cohen, J.S. and Shindo, H. (1975) J. Biol. Chem., 250, 8874-8881.
5. Shindo, H., Hayes, M.B. and Cohen, J.S. (1976) J. Biol. Chem., 251, 2644-2647.
6. Sherwood, L.M. and Potts, J.T., Jr. (1965) J. Biol. Chem., 240, 3799-3805.
7. Richards, F.M. and Wyckoff, H.W. (1971) in The Enzymes, Boyer, P.D., ed., Vol. 4, Academic Press, New York, pp. 647-806.
8. Carlisle, C.H., Palmer, R.A., et al. (1974) J. Mol. Biol., 85, 1-18.
9. Wodak, S.Y., Liu, M.Y. and Wyckoff, H.W. (1977) J. Mol. Biol., 116, 855-875.
10. Markley, J.L. and Finkenstadt, W.R. (1975) Biochemistry, 14, 3562-2566.
11. Haar, W., Thompson, J.C., et al. (1973) Europ. J. Biochem., 40, 259-266.
12. Haar, W., Maurer, W. and Ruterjans, H. (1974) Europ. J. Biochem., 44, 201-211.
13. Cohen, J.S., Griffin, J.H. and Schechter, A.N. (1973) J. Biol. Chem., 248, 4305-4310.
14. French, T.C. and Hammes, G.G. (1965) J. Biol. Chem., 87, 4669-4673.
15. Donovan, J.W. (1965) Biochemistry, 4, 823-829.
16. Markley, J.L. (1975) Biochemistry, 14, 3554-3561.
17. Feldmann, R. (1977) Atlas of Macromolecular Structure on Microfiche (AMSOM) Tracor-Jitco, Rockville, Maryland.
18. Liu, M.C. (1970) J. Biol. Chem., 245, 6726-6731.
19. Hayashi, R., Moore, S. and Merrifield, R.B. (1973) J. Biol. Chem., 248, 3889-3892.
20. Arata, Y., Kimura, S., et al. (1976) Biochem. Biophys. Res. Comm., 73, 133-140.
21. Walz, F.G., Jr. (1977) Biochemistry, 16, 4568-4571.
22. Walker, E.J., Ralston, G.B. and Darvey, I.G. (1976) Biochem. J., 153, 329-337.
23. Findlay, D., Herries, D.C., et al. (1962) Biochem. J., 85, 152-153.
24. Westheimer, F.H. (1968) Accts. Chem. Res., 1, 70-78.
25. Usher, D.A., Richardson, D.I., Jr., and Oakenfull, D.G. (1970) J. Amer. Chem. Soc., 92, 4699-4711.
26. Roberts, G.C.K., Dennis, E.A., et al. (1969) Proc. Natl. Acad. Sci., U.S.A., 62, 1151-1158.
27. Richards, F.M. and Wyckoff, H.W. (1973) in Atlas of Protein Structures, Phillips, D.C. and Richards, F.M., eds., Vol. 1, Oxford University Press.

28. Usher, D.A., Richardson, D.I., Jr., and Eckstein, F. (1970) Nature, 228, 663-665.
29. Usher, D.A., Erenreich, E.S. and Eckstein, F. (1972) Proc. Natl. Acad. Sci. U.S.A., 69, 115-118.
30. Meadows, D.H., Jardetzky, O., et al. (1968) Proc. Natl. Acad. Sci. U.S.A., 60, 766-772.
31. Markley, J.L. (1975) Biochemistry, 14, 3546-3554.
32. Patel, D.J., Canuel, L.A. and Bovey, F.A. (1975) Biopolymers, 14, 987-997.
33. Bradbury, J.H. and Teh, J.S. (1975) Chem. Comm., 936-937.
34. Ruterjans, H. and Witzel, H. (1969) Europ. J. Biochem., 9, 118-127.
35. Brauer, M. and Benz, F.W. (1978) Biochim. Biophys. Acta, 533, 186-194.
36. Meadows, D.H., Roberts, G.C.K. and Jardetzky, O. (1969) J. Mol. Biol., 45, 491-511.
37. Richards, F.M. and Vithayathil, P.J. (1959) J. Biol. Chem., 234, 1459-1465.
38. Hearn, R.P., Richards, F.M., et al. (1971) Biochemistry, 10, 806-817.
39. Levit, S. and Berger, A. (1976) J. Biol. Chem., 251, 1333-1339.
40. Schreier, A.A. and Baldwin, R.L. (1976) J. Mol. Biol., 105, 409-426.
41. Sachs, D.H., Schechter, A.N., et al. (1972) Proc. Natl. Acad. Sci. U.S.A., 69, 3790-3794.
42. Richards, F.M. and Logue, A.D. (1962) J. Biol. Chem., 237, 3693-3697.
43. Woodfin, B.M. and Massey, V. (1968) J. Biol. Chem. 243, 889-892.

NMR AND EPR STUDIES OF THE ACTIVE SITE OF GLUTAMINE SYNTHETASE

JOSEPH J. VILLAFRANCA
Department of Chemistry, The Pennsylvania State University, University Park,
Pennsylvania 16802.

ABSTRACT

A variety of biophysical techniques have been used to study glutamine synthetase from $E.$ $coli$.   Electron paramagnetic resonance experiments of enzyme-
bound Mn^{2+} were used to follow changes in the metal ion environment in response
to substrate and inhibitor binding. Novel electron paramagnetic resonance (EPR)
experiments were conducted to establish the distance between the two metal ion
binding sites per subunit of this dodecameric enzyme. These experiments utilized substitution inert Cr^{3+}-nucleotides as substrate analogues for the metal-
nucleotide site of glutamine synthetase and revealed that the metal ion sites
move closer as substrates or inhibitors bind to the enzyme. A combination of
nuclear magnetic resonance (NMR), EPR and fluorescence energy transfer experiments led to the determination of the spatial relationship between the catalytic
site and the regulatory covalent adenylyl site. In $E.$ $coli$, adenylylation of a
specific tyrosyl residue on each subunit provides a major pathway for regulation of glutamine biosynthesis. Experiments with the diamagnetic ATP analog
$Co(III)(NH_3)_4ATP$ were conducted to map the distances between the two metal ion
sites and the ATP site. These studies along with ^{18}O-^{16}O exchange studies
provide insite into the structure and function of catalysis and regulation of
this crucial allosteric enzyme.

Glutamine synthetase is an essential enzyme in the assimilation of NH_3 in
bacteria. The reaction that glutamine synthetase catalyzes is the following:

$$L\text{-glutamate} + ATP + NH_3 \xrightleftharpoons[\quad]{2M^{2+}} L\text{-glutamine} + ADP + P_i$$

Divalent cations are required for activity (Mg^{2+}, Mn^{2+}, Co^{2+}) and highly
purified homogeneous preparations of glutamine synthetase are available from
many bacterial sources including $Escherichia$ $coli$[1,2] and $Salmonella$ $typhimur$-
ium.[3,4]

Glutamine synthetases of bacterial origin are dodecameric with molecular
weights in the range of $\sim$600,000.

Recognition of the central role for glutamine synthetase in nitrogen metabolism has produced considerable interest in all aspects of catalysis and regulation of this key enzyme. The amide nitrogen of glutamine is involved in synthesis of precursors of protein (tryptophan, histidine, asparagine, etc.) and DNA biosynthesis (CTP, AMP, GMP) as well as other important metabolic compounds including several enzyme cofactors (folate, NAD). The regulation of glutamine synthetase activity in E. coli has been reviewed.[1,2,5] The synthesis of the enzyme is repressed when E. coli are grown on media high in ammonia, whereas, growth conditions on limiting ammonia produce a 20-fold derepression of the enzyme. Additionally, feedback inhibition by multiple end products of glutamine metabolism (Ala, Gly, His, Trp, CTP, AMP, glucosamine-6-P) also regulate the activity of existing glutamine synthetase molecules. Finally, E. coli have developed an elegant method to control enzyme activity that occurs by covalent modification of each subunit. In this latter reaction, a single tyrosyl residue per subunit is adenylylated to produce a stable 5'-adenylyl-O-tyrosyl derivative. Recent NMR and fluorescence data will be reviewed concerning the nature of this adenylyl site and its spatial relationship to the catalytic site. The enzymes responsible for the adenylylation comprise a "cascade system" for amplifying the activation or inactivation of glutamine synthetase molecules.[5]

To establish the structure of the active site components of glutamine synthetase, experimental techniques for determining distances must be employed. Techniques that are available for these studies are X-ray crystallography, EPR, NMR and fluorescence energy transfer. The X-ray structure of the E. coli enzyme is currently being determined by Dr. David Eisenberg of the University of California at Los Angeles.[6] My laboratory and those of Dr. Earl Stadtman at the National Institutes of Health are presently employing the latter three techniques to determine the three dimensional relationships among metal ions, substrates and allosteric effector molecules.

The outline that follows will deal with recent studies on the enzyme from E. coli. Specifically, 1) EPR studies to probe the n_1 and n_2 metal ion sites of glutamine synthetase to follow geometric changes in the metal ion environment, 2) metal-metal distances between the n_1 and n_2 sites established by EPR, 3) distances between the covalent adenylyl regulatory site and the catalytic site, 4) NMR data on the spatial relationships of the substrates at the catalytic site, and 5) kinetic studies of ^{18}O exchange from P_i catalyzed by the enzyme.

EPR studies of conformational changes at the n_1 metal ion site

As discussed earlier, the enzymic reaction catalyzed by glutamine synthetase requires the presence of divalent metal ions. Extensive work has been conducted on the binding of Mn^{2+} to the enzyme isolated from E. coli.[2,7-11] Two important types of sites, each with different affinities for Mn^{2+}, exist per dodecamer: 12 sites (n_1) of high affinity, responsible for inducing a change from a relaxed metal ion free protein to a conformationally tightened, catalytically active protein; 12 sites (n_2) of moderate affinity, involved in active site activation via a metal-ATP complex. The state of adenylylation and pH value alter the metal ion specificity and affinities. Adenylylated subunits bind Mn^{2+} tightly resulting in catalytically active enzyme whereas unadenylylated enzyme is specifically activated for catalysis by Mg^{2+}. Unadenylylated enzyme binds Mn^{2+} with about equal affinity as does fully adenylylated enzyme ($K_1 = 5 \times 10^{-7}$ M) at the n_1 sites. The affinity of Mn^{2+} for the n_2 sites is about two orders of magnitude weaker than for the n_1 sites but the presence of glutamine tightens the Mn^{2+} binding at n_2 to about 5×10^{-7} M for the fully adenylylated enzyme.[12] Our NMR and EPR results show that the "tight" metal ion site is likely involved in orienting the γ-carboxylate of glutamate and the second metal ion site (metal-ATP site) is close to the "tight" n_1 site. Since the binding constants of Mn^{2+} to the tight and weak metal ion sites of unadenylylated enzyme are 5.0×10^{-7} and 4.5×10^{-5} M, respectively, the tight site can be selectively populated under conditions where [enzyme] > [Mn^{2+}]. Figure 1 shows EPR spectra obtained with a solution of 0.79 mM enzyme subunit concentration and 0.7 mM Mn^{2+} concentration.[13] This spectrum represents Mn^{2+} bound only at the "tight" sites with no free Mn^{2+} present and shows that bound Mn^{2+} is in a relatively isotropic environment, i.e., the zero field splitting is small.

L-glutamate

I

"Tetrahedral Intermediate"

II

L-methionine-(S) -sulfoximine phosphate

III

Our studies of the substrates, glutamate (I) and ATP and of substrate analogues, AMP-P-(CH_2)-P, and methionine sulfoximine, reveal interactions between

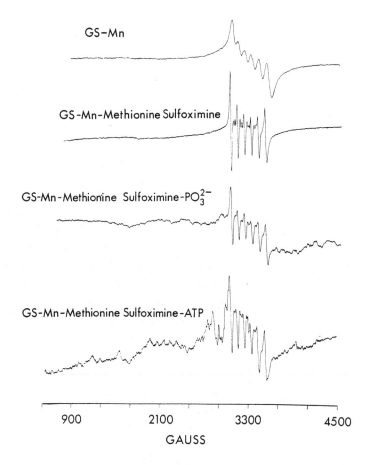

GS–Mn

GS-Mn-Methionine Sulfoximine

GS-Mn-Methionine Sulfoximine-PO$_3^{2-}$

GS-Mn-Methionine Sulfoximine-ATP

900   2100   3300   4500

GAUSS

Fig. 1. EPR spectra at x-band for some complexes of glutamine synthetase. Experimental conditions are given in Villafranca et al.[13]

both substrate sites and both metal ion sites. Others have shown that the irreversible inhibition of glutamine synthetase in the presence of L-methionine (S)-sulfoximine and ATP was due to formation of the sulfoximine phosphate (III). The tetrahedral geometry at the sulfur atom of the sulfoximine was suggested by modeling studies[14] and chemical identification[15] to be a "mimic" of the "active structure" of the adduct of γ-glutamyl phosphate and ammonia (II). Data in our laboratory provide spectroscopic evidence that methionine sulfoximine and the sulfoximine phosphate are transition-state analogs of the glutamine synthetase reaction.[13,16]

In the presence of a saturating concentration of L-methionine (SR)-sulfoximine, a dramatic change in the line width and resolution of the EPR spectral intensities of Mn^{2+} bound at n_1 are noted. The line widths change from about 30 G peak to peak to about 12 G (middle spectrum, Fig. 1). Two observations can be made about this spectrum: (1) the narrow line widths are symptomatic of Mn^{2+} ions that are occluded from solvent bombardment and this is supported by NMR data of solvent water relaxation in enzyme solutions containing saturating concentrations of L-methionine (SR)-sulfoximine; (2) the bound Mn^{2+} is in a nearly cubic environment based on the EPR spectrum that shows six -1/2 to 1/2 transitions and no apparent transitions at higher or lower field strengths. Thus, L-methionine (SR)-sulfoximine may have induced a distortion of the metal ion environment as a result of displacement of solvent. Alternatively, close proximity of sulfoximine to the Mn^{2+} ion (~ 5 Å) may have immobilized a water molecule in the coordination sphere producing the observed distortion of the Mn^{2+} ion.

When 3.3 mM L-methionine (SR)-sulfoximine phosphate is present in a solution of enzyme-Mn^{2+}, the resultant spectrum is quite different from that detected for the unphosphorylated sulfoximine. Additional fine structure splittings are noted around 3300 G in addition to poorly resolved resonances at about 2400 G. This indicates that the sulfoximine phosphate binds differently to the enzyme and induces different changes in the Mn^{2+} spectra than does either the sulfoximine itself or the methionine sulfoximine phosphate formed *in situ* (below). The addition of 10 mM $MgCl_2$ and 10 mM ADP to an enzyme solution containing the sulfoximine phosphate did not produce significant spectral changes from those shown in Fig. 1.

The spectrum at the bottom of Fig. 1 is obtained when Mg-ATP is added to a solution of enzyme-Mn^{2+}-L-methionine (SR)-sulfoximine. The solution is allowed to incubate for 10 min at 25°C, cooled to 1°C, and the spectrum taken. This spectrum is quite reproducible if Mg^{2+} and ATP are added in a 1:1 ratio at a two- to five-fold molar excess over enzyme. Higher concentrations produced no additional changes. No detectable changes are seen in the spectrum of enzyme-Mn^{2+}-L-methionine (SR)-sulfoximine with a 40-fold excess of either Mg^{2+} or ATP alone. This experiment demonstrates that the Mn^{2+} is locked into the tight metal ion site by the sulfoximine and not displaced or perturbed by Mg^{2+} or ATP alone. Thus, the presence of Mg-ATP produces a dramatic change in the spectrum of bound Mn^{2+} when this complex is bound at the metal ion-nucleotide site. This spectrum could result from the *in situ* formation of the sulfoximine phosphate and result in stabilization of an enzyme conformation and Mn^{2+} distortion unobtainable with the separate addition of sulfoximine phosphate and ADP.

Nuclear magnetic resonance (NMR) studies

   Analysis of NMR titration data with sulfoximine in the presence of ATP gave a dissociation constant of 0.8 μM.[13,17] The K_I value from kinetic determinations of Wedler and Horn[18] for L-methionine (SR)-sulfoximine is 1.5 μM with inhibition being linearly competitive with respect to L-glutamate. These kinetic and spectroscopic approaches provide the first direct measurement of the equilibrium binding constant for L-methionine (SR)-sulfoximine to the Mn-enzyme. Interestingly, the conversion of the δ-atom from trigonal (carbon for glutamate) to tetrahedral (sulfur for sulfoximine) symmetry results in a decrease in the binding constant by two orders of magnitude (from 3 mM to 30 μM comparing L-glutamate to the L-sulfoximine. The binding of ADP and P_i to the sulfoximine-enzyme complex tightens the binding approximately another ten fold, whereas the phosphorylation of the inhibitor by ATP lowers the binding constant by only about five fold. This suggests that the tight binding of the sulfoximine is due primarily to tetrahedral geometry at the δ-position and conformational effects (synergism of binding) by bound nucleotide rather than to formation of a phosphorylated intermediate.

   Other published EPR data[13] show that L-glutamate produces a small axial distortion of the environment of enzyme-Mn(II) with unresolved fine structure and no obvious change in line width. Addition of MgATP results in a diminution of all spectral intensities and the anisotropic spectrum shows poorly resolved fine splitting. The appearance of these additional sets of transitions indicated that the environment of Mn^{2+} at the tight site is changed when metal nucleotide is bound. NH_4^+ produces additional subtle changes in the EPR spectrum of Mn^{2+}.

   The EPR data presented above provide a model of the active site in which Mn^{2+} at the "tight" n_1 metal ion site may be involved in orienting the substrate L-glutamate and substrate analogue inhibitors. This orientation of the metal ion site is influenced by addition of the substrates Mg-ATP and NH_4^+ since progressive changes in the EPR spectra are observed.

Metal-metal distances between the n_1 and n_2 sites

   Recent data from this laboratory[19,20] gave the first distance determinations between the two metal ion sites of glutamine synthetase. These experiments represent a breakthrough in that many enzymes require two or more metal ions for catalysis and our work provides the groundwork for the design of rational experimentation by others to determine metal-metal distances on enzymes. In our experiments, the interaction between two enzyme-bound paramagnetic metal ions was monitored by EPR and the results were used for calculations of metal-metal distances employing the Leigh theory.[21] The procedure involved binding of Mn^{2+} at one site and Cr^{3+}-ATP, Cr^{3+}-ADP or Co^{2+}-nucleotides at the metal-nucleotide site. These studies represented the first estimates of metal-metal distances on an enzyme by this EPR technique.

First it was established by kinetic methods that the substitution-inert Cr^{3+}-nucleotides, CrADP and CrATP, were inhibitors of unadenylylated *Escherichia coli* glutamine synthetase. Both complexes were linear, competetive inhibitors versus MgATP in the biosynthetic assay. The K_I values were 9.6 $\pm$ 0.6 µM and 25 $\pm$ 1 µM for CrATP and CrADP, respectively. Addition of a saturating amount of CrATP produces a 60% decrease in the EPR spectrum of enzyme-bound Mn^{2+}. This electronic spin-spin interaction between Mn^{2+} and Cr^{3+} was analyzed at both 9 and 35 GHz. A distance of 7.1 Å between Mn and Cr was obtained by analysis of these EPR data. Only the data at 35 GHz can unambiguously be interpreted as being due to dipolar relaxation and thus used to calculate distances. Titration experiments with CrATP were conducted by following the decrease in the EPR spectral amplitude of enzyme-Mn^{2+} and a K_D value of 0.30 $\pm$ 0.04 mM was calculated. Figure 2 shows a typical titration and the narrow lowest field line of the bound Mn^{2+} sextet. These experiments were also conducted in the presence of glutamate, glutamine, or methionine sulfoximine. The K_D and metal-metal distances were 0.28 mM and 5.2 Å, 0.055 mM and 5.9 Å, and 0.20 mM and 6.8 Å, respectively. Substrates or inhibitors were thus shown to move the n_1 and n_2 metal ion sites closer together. These data correlate well with other data that demonstrate induced conformational changes in the enzyme and synergism in substrate binding.[22]

Experiments were conducted with CrADP that also show synergistic interaction between substrate sites. These Mn to Cr distances were 5.9, 5.2 and 4.8 Å with the CrADP, CrADP plus P_i, and CrADP, P_i plus glutamine complexes, respectively. Data with substitution labile Co(II)-nucleotides were also obtained. Titrations with both Co(II)-ADP and Co(II)-ATP in the presence of the inhibitor methionine sulfoximine produced a diminution of the EPR spectrum of enzyme-bound Mn(II). The Co to Mn distances were 6.5 and 5.2 Å for the Co-ADP and Co-ATP complexes, respectively.

These data demonstrate a novel application of dipolar electron-electron relaxation between enzyme-bound metal ions for the determination of distances between the n_1 and n_2 metal ion sites of *E. coli* glutamine synthetase. The conclusion is that the two metal ions are in close enough proximity to both be involved in substrate orientation, binding and catalysis.

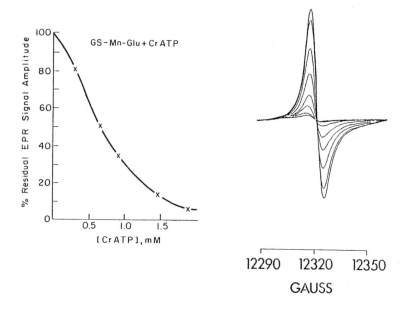

Fig. 2. EPR spectrum and titration curve showing the effect of Cr(III)-ATP on the spectrum of the glutamine synthetase-Mn(II)-glutamate complex. Only the highest field line of the sextet is shown. Conditions given in Villafranca et al.[19] and Balakrishnan and Villafranca.[20]

Distances between the adenylyl site and the metal ion sites

For the first time, three physical methods, vis., NMR, EPR, and fluorescence energy transfer were used to determine the spatial relationship between well-defined sites on an enzyme. With glutamine synthetase these three techniques were used to establish relationships between the adenylyl, catalytic (n_2) and the divalent metal ion activating sites (n_1). Glutamine synthetase of low adenylylation state ($E_{\overline{1.0}}$) was enzymatically adenylylated with either [2-^{13}C] ATP, ε-ATP(1-N^6-etheno-ATP) or 6 amino-TEMPO-ATP.[23,24] Since the adenylylation reaction is a specific enzymatic process and since the metal ion sites are well established there is little ambiguity in the design of these experiments.

Paramagnetic effects on the $1/T_1$ and $1/T_2$ relaxation rates of ^{13}C and ^{31}P nuclei of adenylylated glutamine synthetase ([2-^{13}C]AMP-GS) were measured when the two enzyme bound Mg^{2+} ions were replaced by Mn^{2+}.[23] Also, the paramagnetic effects of Co^{2+} were measured on the ^{31}P relaxation rates of adenylylated enzyme. Distances between these metal ions and the ^{13}C and ^{31}P nuclei are summarized in Fig. 3.

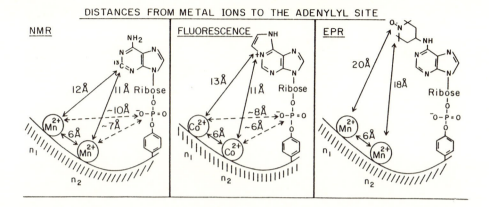

Fig. 3. Summary of ^{13}C, ^{31}P-NMR data, fluorescence energy transfer data and EPR data on various derivatives at the covalent adenylyl site on glutamine synthetase.

Figure 3 also depicts the distances calculated from fluorescence quenching experiments by enzyme bound Co^{2+} on the ε-adenosine moiety of ε-ATP adenylylated glutamine synthetase (ε-AMP-GS). Additional data were gathered by an EPR method that measured the decrease in EPR amplitude of the nitroxide spin labeled adenylylated enzyme (TEMPO-AMP-GS) due to enzyme bound Mn^{2+}.[24] Distances between Mn^{2+} and the N$\pm$O of the spin label are also in Fig. 3.

Thus, we have determined the distances between the adenylyl moiety and the two divalent metal ion binding sites on glutamine synthetase by ^{13}C and ^{31}P NMR, spin labelled EPR and fluorescence energy transfer methods. The results obtained from each method are in good agreement. The data show that the adenylyl regulatory site is close to the catalytic site (12-20 Å). Additional data on the rotational correlation time of the adenyl derivatives reveal that the adenylyl site is located on the surface of the enzyme.

Kinetic studies of ^{18}O exchange from P_i

Studies were initiated to determine how the state of adenylylation, pH, metal ion (Mn^{2+} or Mg^{2+} or Co^{2+}) content, or presence of feedback inhibitors, changes individual rate constants in the overall reaction catalyzed by glutamine synthetase. The initial approach that we used involved following the "wash out" of ^{18}O from ^{18}O-enriched P_i during the net reverse reaction, i.e., formation of ATP, NH_3 and glutamate from ADP, $[^{18}O]P_i$ and glutamine.

Recently several methods of analysis of the $^{18}O/^{16}O$ content among the five species of inorganic phosphate, i.e., $P^{16}O_4$, $P^{16}O_3{}^{18}O_1$, $P^{16}O_2{}^{18}O_2$, $P^{16}O_1{}^{18}O_3$, $P^{18}O_4$ have been reported. These include 1) mass spectral analysis of both the tris-(trimethylsilyl) phosphate ester[25] and the trimethylphosphate esters[26] and 2) a direct ^{31}P-NMR analysis that relies on the isotopic shift differences exerted by ^{18}O and ^{16}O in the five species listed above.[27] We have applied the ^{31}P(^{18}O/^{16}O)-NMR method to our investigation. Our studies were designed to complement those of Stokes and Boyer[28] who demonstrated that oxygen exchange from [^{18}O]P_i to L-glutamine (adenylylated enzyme, Mn^{2+}-activated) occurred at a rate five times faster than net formation of L-glutamate. Our experiments also supplied information on the rotational freedom or restriction of P_i on the enzyme and the relative rates of P_i release compared to enzymatic interconversion steps involving ^{18}O/^{16}O exchange.

The results of our studies indicate that there is rapid exchange of ^{18}O out of [^{18}O]P_i compared to net formation of ATP with the Mg^{2+}-activated unadenylylated enzyme. Thus, under similar reaction conditions both Mn^{2+}-activated adenylylated subunits and Mg^{2+}-activated unadenylylated subunits catalyze exchange reactions with nearly the same relative rate ratios.

Figure 4A shows the ^{31}P NMR spectrum of 77.6 atom % [^{18}O]P_i. The percent of total of the five species agrees with that predicted by a binomial distribution of the given atom % ^{18}O. Glutamine synthetase was then added to the reaction mixture and aliquots were withdrawn at various time intervals. Figure 4B shows the ^{31}P NMR spectrum of a sample after 2 hrs of the reaction at 37°C. The ^{18}O content of the sample dropped to 58.9 atom % and the ^{18}O/^{16}O distribution is still binomial. Figure 4C shows the spectrum after 18 hrs reaction. The distribution is binomial and ^{18}O content is 47.5 atom %.

The percentage distribution of ^{18}O/^{16}O species was calculated for samples withdrawn at various time intervals and the data are plotted in Fig. 5. The curves drawn in the figure are from a model that assumes random removal of ^{18}O from any of the P_i species and that the bound P_i has rotational freedom. The rate constant used to generate the curves for loss or production of an individual species is 410 ± 40 min^{-1}.

The % distribution of ^{18}O in P_i was obtained from the individual peaks by a curve digitizer or by measuring the intensities of the individual peaks. The exchange of ^{18}O out of P_i was analyzed by a computer program which fits the ^{18}O distribution with time to the model

$$P^{18}O_4 \xrightarrow{k} P^{18}O_3{}^{16}O_1 \xrightarrow{3/4K} P^{18}O_2{}^{16}O_2 \xrightarrow{1/2k} P^{18}O_1{}^{16}O_3 \xrightarrow{1/4k} P^{16}O_4$$

An interactive chemical reaction simulator program was used as supplied by

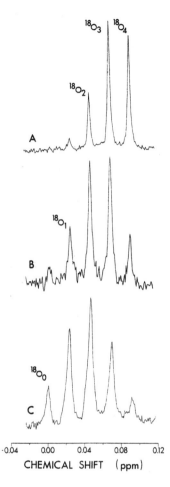

Fig. 4. $^{18}O-^{16}O$ exchange in inorganic phosphate catalyzed by glutamine synthetase. Experimental conditions given in Balakrishnan et al.[30]

Dr. S. J. Benkovic. The algorithm used involves calculation of the reaction time course by means of Taylor series expansions of the differential equations describing the changes in concentration of the chemical species in the system.

In a separate experiment, the initial velocity of ATP production was measured under identical reaction conditions by assaying with hexokinase and glucose-6-phosphate dehydrogenase, and was found to be 62 ± 4 min^{-1}. Thus, ^{18}O exchange out of P$_i$ is ~ 7 times faster than the overall back reaction.

Our data were fit to a scheme for the overall kinetic reaction (see below) originally reported by Rhee and Chock.[29] The analysis led to a model in which partition of intermediate chemical species in the reaction involving P$_i$ and γ-glutamyl-P could be detected by this technique.[30] Further experiments will show

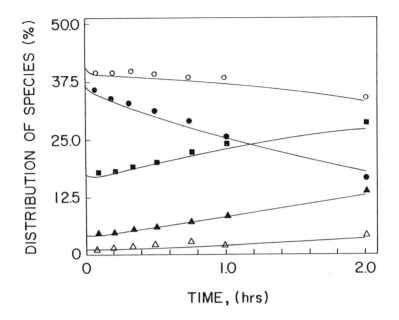

Fig. 5. Time course of the $^{18}O-^{16}O$ exchange catalyzed by glutamine synthetase. Experimental conditions given in Balakrishnan et al.[30]

whether different metal ions, feedback modifiers or pH alter the partitioning of these enzyme-bound intermediates.

$$Mg \cdot E_{1.7} \underset{k_{-1}}{\overset{k_1}{\rightleftharpoons}} Mg \cdot E_{1.7} \cdot ATP \cdot Glu \cdot NH_3 \quad (I)$$

$$k_5 \left\Vert\, k_{-5} \right. \qquad\qquad\qquad\qquad k_{-2} \left\Vert\, k_2 \right.$$

$$Mg \cdot E_{1.7} \cdot ADP \cdot P_i \cdot Gln \underset{k_4}{\overset{k_{-4}}{\rightleftharpoons}} I_2 \underset{k_3}{\overset{k_{-3}}{\rightleftharpoons}} I_1$$

As mentioned before, glutamine synthetase from *Escherichia coli* has two metal ion binding sites per subunit, the n_1 site and the n_2, metal-nucleotide substrate site. With either Mg(II) or Mn(II) bound at n_1, two substitution-inert analogs of metal-ATP, vis., Co(III)(NH$_3$)$_4$ATP and Cr(III)-ATP, were used to map distances between the metal-substrate and n_1 metal ions sites of unadenylylated glutamine synthetase (GS). Thus, the GS-Mn(II)-Co(III)(NH$_3$)$_4$ATP enzyme complex was studied by ^{31}P-NMR and $1/T_1$ and $1/T_2$ relaxation rate data were used to

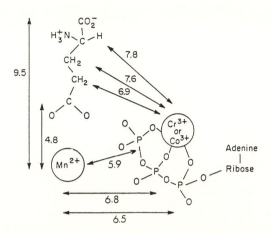

Fig. 6. Spatial relationships between the metal ion and substrate sites on glutamine synthetase from *E. coli*.

calculate the Mn(II) to γ-P (5.9 Å) and $\alpha\beta$-P (6.5-6.8 Å) distances. The GS-Mg(II)-Cr(III)ATP-L-glutamate complex was studied by ^{1}H-NMR. The Cr(III) to ^{1}H distances obtained from $1/T_1$ data with glutamate were 6.9 Å (γ-CH$_2$), 7.6 Å (β-CH$_2$) and 7.8 Å (α-CH). These data demonstrate the utility and versatility of using substitution inert analogs to study the topography of substrate binding sites of metal ion-nucleotide requiring enzymes. The data also establish the proximity of the n_1 and n_2 sites and the glutamate and ATP sites. These distance data are summarized in Fig. 6.

ACKNOWLEDGMENT

This work was supported by the U.S. Public Health Service (GM23529) and the National Science Foundation (PCM78-07845). J. J. Villafranca is an Established Investigator of the American Heart Association.

REFERENCES

1. Stadtman, E.R., and Ginsburg, A. (1975) The Enzymes, Vol. X, Academic Press, New York, pp. 755-812.
2. Ginsburg, A. (1972) Adv. Protein Chem., 27, 1-79.
3. DeLeo, A.B. and Magasanik, B. (1975) J. Bacteriol., 121, 313.
4. Balakrishnan, M.S., Villafranca, J.J. and Brenchley, J.E. (1977) Arch. Biochem. Biophys., 181, 603-615.
5. Ginsburg, A. and Stadtman, E.R. (1973) in The Enzymes of Glutamine Metabolism, Prusiner, S., and Stadtman, E.R., eds., Academic Press, New York, p. 44.
6. Kabsch, W., Kabsch, H. and Eisenberg, D. (1976) J. Mol. Biol., 100, 283-291.
7. Hunt, J.B., Smyrniotis, P.Z., et al. (1975) Arch. Biochem. Biophys., 166, 102-124.

8. Villafranca, J.J. and Wedler, F.C.   (1974) Biochemistry, 13, 3286-3291.
9. Villafranca, J.J., et al.   (1976) Biochemistry, 15, 536-544.
10. Shapiro, B.M. and Ginsburg, A. (1968) Biochemistry, 7, 2153.
11. Segal, A. and Stadtman, E.R.   (1972) Arch. Biochem. Biophys., 152, 356.
12. Shrake, A., et al.   (1977) Biochemistry, 16, 4372-4381.
13. Villafranca, J.J., et al.   (1976) Biochemistry, 15, 544-553.
14. Gass, J.D., and Meister, A.   (1970) Biochemistry, 9, 1380-1390.
15. Manning, J.M. et al.   (1969) Biochemistry, 8, 2681-2685.
16. Villafranca, J.J., et al.   (1975) Biochem. Biophys. Res. Commun., 66, 1003-1010.
17. Villafranca, J.J. (1977) in Biomolecular Structure and Function, Agris, P.F., ed., Academic Press, New York, p. 353-362.
18. Wedler, F.C. and Horn, B.R. (1976) J. Biol. Chem., 251, 7530.
19. Villafranca, J.J., et al.   (1977) Biochem. Biophys. Res. Commun., 75, 464-471.
20. Balakrishnan, M.S. and Villafranca, J.J. (1978) Biochemistry, 17, 3531-3538.
21. Leigh, J.S. (1970) J. Chem. Phys., 52, 2608-2612.
22. Rhee, S.G., and Chock, P.B.   (1976) Biochemistry, 15, 1755-1760.
23. Villafranca, J.J., et al.   (1978) Proc. Natl. Acad. Sci. USA, 75, 1255-1259.
24. Chock, P.B., et al.   (1979) in NMR and Biochemistry, Opella, S., Lu, P., eds., Marcel Dekker, New York, 405-418.
25. Eargle, D.H., Jr., et al.   (1977) Anal. Biochem., 81, 186.
26. Midelfort, C.F. and Rose, I.A. (1976) J. Biol. Chem., 251, 5881.
27. Cohn, M., and Hu, A. (1978) Proc. Natl. Acad. Sci. USA, 75, 200.
28. Stokes, B.O. and Boyer, P.D. (1976) J. Biol. Chem., 251, 5558.
29. Rhee, S.G. and Chock, P.B. (1976) Proc. Natl. Acad. Sci. USA, 73, 476.
30. Balakrishnan, M.S., et al.   (1978) Biochem. Biophys. Res. Commun., 85, 991.

NUCLEAR MAGNETIC RESONANCE STUDIES OF SERINE PROTEINASES

JOHN L. MARKLEY, DARROW E. NEVES,[+] WILLIAM M. WESTLER, IGNACIO B. IBAÑEZ,
MICHAEL A. PORUBCAN[*], AND MARY WELCH BAILLARGEON
Biochemistry Division, Department of Chemistry, Purdue University,
West Lafayette, Indiana 47907

ABSTRACT

The properties of the histidyl residues of bovine trypsinogen, porcine trypsinogen and trypsin, bovine chymotrypsinogen and chymotrypsin A_α, and α-lytic
proteinase were studied by ^{1}H nuclear magnetic resonance (NMR) spectroscopy.
The diisopropylphosphoryl-derivatives of these proteins were investigated by ^{1}H
NMR and ^{31}P NMR spectroscopy. Complexes between several of these proteins and
soybean trypsin inhibitor or bovine pancreatic trypsin inhibitor were examined
by ^{1}H NMR and ^{13}C NMR spectroscopy. Tryptic catalysis of a ^{13}C labeled ester
substrate was investigated by ^{13}C NMR at subzero temperatures.

The data do not support the "charge relay hypothesis" which predicts that the
pK_a' of aspartate-102 is higher than that of histidine-57. The pK_a' of histidine-
57 is different in the various proteins studied, ranging from 5.0 to 7.7. In
all cases where a pK_a' could be inferred for aspartate-102, it was lower than the
pK_a' assigned to histidine-57.

The properties of the catalytic residues are different in zymogens and
enzymes. The pK_a' of histidine-57 is higher in zymogens than in corresponding
enzymes.

The pK_a' of histidine-57 is higher in diisopropylphosphoryl-enzymes than in
native enzymes. The NMR data for diisopropylphosphoryl-zymogens are consistent
with recent X-ray crystallographic results.

Histidine-57 does not titrate in intact (proteinase)-(protein proteinase
inhibitor) complexes. Evidence is presented suggesting that the histidine-57
ring is positively charged in these complexes.

[90% 1-^{13}C]L-arginine was incorporated into the reactive site (residue 63) of
soybean trypsin inhibitor. ^{13}C NMR spectroscopy of the trypsin complex revealed
that the labeled carbon is trigonal. It is concluded that the complex is a noncovalent species.

In the normal mechanism proposed for serine proteinases, histidine-57 is
protonated only in intermediates where the P1 carbonyl is tetrahedral. Comparison of the ^{1}H NMR and ^{13}C NMR results indicated that (proteinase)-(protein

*
Present address: Squibb Institute for Medical Research, Princeton, New Jersey
08540; [+]Central Data Processing, 100 N. Senate Street, Indianianapolis, Indiana
46223.

proteinase inhibitor) complexes lie off the normal catalytic pathway.

A tetrahedral species as identified by ^{13}C NMR spectroscopy was trapped in subzero temperature studies of the hydrolysis of Nα-acetyl-[90% 1-^{13}C]L-arginine methyl ester by trypsin in 70/30 (v/v) methanol/water at -60°C.

INTRODUCTION

For the past few years we have been using NMR spectroscopy to investigate the active sites of serine proteinases and their mechanism of interaction with small proteins that are specific inhibitors. We summarize here our cumulative results and compare them with other work in this field. (For a more extensive, but less recent, review see reference 1.)

Serine proteinases comprise a ubiquitous class of enzymes whose function is to hydrolyze peptide bonds. They differ greatly in their specificities but have a number of uniform characteristics. We have investigated several serine pro-teinases that are related evolutionarily to chymotrypsin: chymotrypsin itself from bovine pancreas, trypsin from bovine and porcine pancreas, and α-lytic pro-teinase from the bacterium, *Lysobacter enzymogenes*. We have also studied the "inactive" zymogen precursors of the pancreatic enzymes listed above. These proteins have well-characterized sequence homologies;[2,3] and their three-dimensional structures have been obtained by X-ray crystallography.[4-12] In addition, the large number of chemical and enzymatic studies of these proteins make them perhaps the best studied class of enzyme. In our NMR studies, we have attempted to address unresolved questions in this field and to provide in-formation complementary to X-ray and chemical results.

Catalytic mechanism. The reaction sequence of serine proteinases is under-stood in detail. It has been postulated to consist of the symmetrical series of steps shown in Fig. 1.[13] The key role of serine-105 as a nucleophile in the attack of substrates was elucidated by studies of inhibitors such as diisopropyl phosphoro fluoridate.[14-16] The identity of the reactive serine was established by enzymatic cleavage and isolation of the organic phosphorus labeled peptide.[17] The catalytic function of the critical serine was confirmed by trapping an acetyl enzyme intermediate in substrate hydrolysis.[18] Enzymatic degradation of the acetyl enzyme established that serine-195[19] in the chymotrypsinogen se-quence[20-22] is acylated. Chemical modification by photooxidation[23,24] and by acylation with site-specific reagents[25,26] demonstrated that a histidine is also required for enzyme activity. A critical histidine residue had been sus-pected on the basis of pH-activity profiles that reveal that serine proteinases are activated with a pK_a' around 7.[27]

The first high-resolution X-ray structure of a serine proteinase, that of

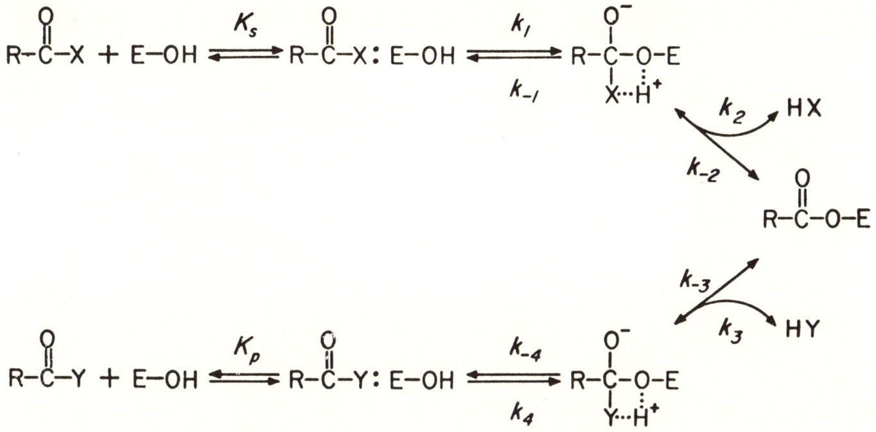

$$R-\overset{\overset{O}{\|}}{C}-X + E-OH \underset{}{\overset{K_s}{\rightleftharpoons}} R-\overset{\overset{O}{\|}}{C}-X : E-OH \underset{k_{-1}}{\overset{k_1}{\rightleftharpoons}} R-\overset{\overset{O^-}{\|}}{\underset{\overset{|}{X\cdots\overset{\cdot\cdot}{H}^+}}{C}}-O-E$$

$$\underset{k_{-2}}{\overset{k_2}{\rightleftharpoons}} \; HX$$

$$R-\overset{\overset{O}{\|}}{C}-O-E$$

$$\underset{k_{-3}}{\overset{k_3}{\rightleftharpoons}} \; HY$$

$$R-\overset{\overset{O}{\|}}{C}-Y + E-OH \underset{}{\overset{K_p}{\rightleftharpoons}} R-\overset{\overset{O}{\|}}{C}-Y : E-OH \underset{k_4}{\overset{k_{-4}}{\rightleftharpoons}} R-\overset{\overset{O^-}{\|}}{\underset{\overset{|}{Y\cdots\overset{\cdot\cdot}{H}^+}}{C}}-O-E$$

Fig. 1. Reaction sequence postulated for serine proteinases (from reference 13).

tosyl chymotrypsin,[4] demonstrated that the side chains of histidine-57 and serine-195 are adjacent in the active site and further that the side chain carboxyl of residue-102 is hydrogen bonded to the imidazole of histidine-57.[28] Blow and co-workers[28] correctly predicted that the original sequences of chymotrypsin and trypsin contained errors and that residue 102 is an aspartate rather than an asparagine. Discovery of the orientation of these three residues led to the important "charge relay" hypothesis concerning their function. As originally formulated,[28] the charge relay was envisioned as a mechanism for the generation of an alkoxide ion at serine-195, which would explain its high reactivity. The triad of catalytic side chains came to be referred to as the "charge relay system."

The charge relay hypothesis generated a large number of theoretical and experimental studies that tested the ideas involved. The existence of an alkoxide ion in solution is unlikely because of the extremely high pK_a' of a serine hydroxyl. The original hypothesis was later modified to include a concerted transfer of two protons, which accompanies attack on the substrate by serine-195.[29] Here the essential feature is the relay of charge from the histidine imidazole to the aspartate. Proponents of this hypothesis have emphasized that proton transfer from histidine-57 to aspartate-102 is important for catalysis because it provides a significant decrease in the free energy of the transition state.[29,30] The idea is that the free energy of a neutral histidine-aspartate is lower than that of a histidine-aspartate cation-anion pair, given their environment in the enzyme. A corollary of the charge-relay hypothesis is that the pK_a' of aspartate-102 is less than that of histidine-57; in other words, the microenvironment of these side chains in the protein leads to a reversal of the pK_a' values they normally have in aqueous solution. This prediction was supported by molecular orbital calculations.[31-34]

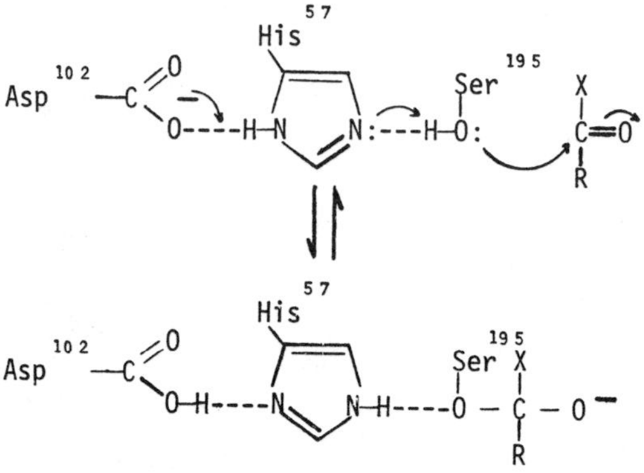

The protonation scheme for a hydrogen bonded imidazole-carboxyl couple is shown in Fig. 2. According to the charge relay hypothesis, species (HE) rather than species $(EH^{-,+})$ should predominate after the first protonation step. Since NMR spectroscopy is well suited for determining the protonation state of imidazole groups in proteins[35] it provides various means for testing this hypothesis.

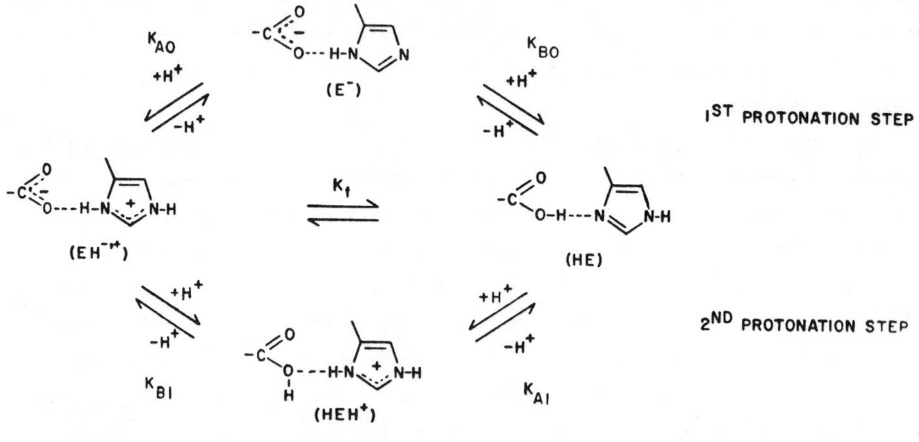

Fig. 2. Protonation scheme for the active site residues histidine-57 and aspartate-102 of serine proteinases (from reference 49).

Zymogen activation

Many mammalian serine proteinases are produced by cells as inactive precursors called pre-enzymes or zymogens. The zymogens are converted to activated enzymes by limited proteolysis near the amino terminal. The activation peptide may be lost as in the case of trypsin or may be retained by a disulfide linkage as in the case of chymotrypsin A_π.

One of the goals in the serine proteinase field has been to determine how cleavage of a single peptide bond can result in a million-fold increase in catalytic activity. The picture provided by the first comparison of X-ray structures of a zymogen, chymotrypsinogen,[5,36] and its activated enzyme, chymotrypsin,[4] indicated that the zymogen lacks both a developed binding pocket and "oxanion hole." One of the hydrogen-bonding groups (glycine-193 N-H) that stabilizes the oxanion transition state in the enzyme is improperly positioned in the zymogen. Similar changes have been reported to accompany activation of trypsinogen to trypsin.[10,11,37] The primary event in the zymogen-to-enzyme conversion in proteinases related to chymotrypsin is the cleavage of a single peptide bond, generating a new amino terminal with an IleVal- or ValVal- sequence.[38] The new positively-charged amino terminus forms an ionic interaction with the carboxylate of aspartate-194 adjacent to the active site. This interaction triggers a reorganization of hydrogen bonds, leading to a conformational change and the generation of increased activity. The above X-ray studies indicated no significant differences in the positioning of the catalytic triad (serine-195, histidine-57, aspartate-102) between enzymes and zymogens. This view has been modified recently since studies of refined X-ray structures reveal that the hydrogen bond between the serine-195 O^γ and the histidine-57 $N^{\epsilon 2}$ is strained or more probably absent in enzymes but present in zymogens.[39,40] It was hoped that NMR spectroscopy could detect some of the subtle changes that accompany activation of the pancreatic zymogens.

The bacterial serine proteinase, α-lytic proteinase, on the other hand, does not contain the normal IleVal- or ValVal- amino terminal.[41] The critical interaction with aspartate-194 is provided instead by the side chain of arginine -138.[3] No zymogen has been found for α-lytic proteinase or related bacterial serine proteinases.

Inhibitor complexes

Naturally occurring small protein inhibitors of serine proteinase such as bovine pancreatic trypsin inhibitor (BPTI),[42] and soybean trypsin inhibitor (STI), form very tight 1:1 complexes with the proteinases that they specifically inhibit. A substoichiometric amount of the proteinase, however, usually catalyzes the cleavage of the peptide bond at the reactive site of the inhibitor.[43] Hence, these inhibitors function as substrates for which one of the kinetic steps in the forward catalytic mechanism (Fig. 1) is very slow. It is necessary to know the stage at which the reaction is halted in order to understand the mechanism of inhibition. The initial X-ray crystallographic studies of the BPTI-trypsin complex[44] and the STI-trypsin complex[45] indicated that the reaction was frozen as the first tetrahedral adduct, normally the species of highest energy in the hydrolysis of peptide bonds. Subsequent analysis of refined X-ray data revealed that the BPTI-trypsin complex cannot be a tetrahedral adduct because the serine-195[1] O^γ - reactive site C' bond length is too long (2.6 Å)[46] for a covalent bond (1.4 Å). On the other hand, the refined X-ray results indicated that the reactive site carboxyl is pyramidalized.[46-48] Because of the strong dependence of ^{13}C NMR chemical shifts on bond order, NMR spectroscopy was expected to offer a sensitive approach to evaluating the hybridization of the reactive site carbonyl.

NMR APPROACHES

A wide variety of methods have been used in NMR studies of serine proteinases. The C^ε ring proton of histidine generally can be resolved in ^1H NMR spectra of small proteins.[35] We have used this approach to investigate the histidyl residues of porcine trypsin,[49] porcine and bovine trypsinogen,[50] bovine chymotrypsin and chymotrypsinogen,[51] and α-lytic proteinase.[52] The enzyme α-lytic proteinase contains only one histidyl residue (histidine-57). The other proteinases studied contain more than one histidine (Table 1) and required assignment of a particular peak to histidine-57.

Robillard and Shulman[55-57] resolved a low-field N-H ^1H NMR peak in solutions of serine proteinases in ^1H_2O which they assigned to the proton hydrogen bonded between histidine-57 and aspartate-102. We have reproduced some of their results and have used the low field peak to characterize (proteinase) ÷ (proteinase inhibitor) complexes.[58]

In elegant experiments, the single histidine of α-lytic proteinase has been labeled biosynthetically with ^{13}C in the C^ε position for ^{13}C NMR studies[29,59] and with ^{15}N in the N^δ and N^ε positions for ^{15}N NMR spectroscopy.[60]

It has been suggested that diisopropylphosphoryl derivatives of serine proteinases resemble the tetrahedral transition state since the phosphate is tetrahedral and the phosphate oxygen occupies the "oxyanion hole."[7,13] These derivatives may be studied by ^{31}P NMR spectroscopy.[61,62] We have carried out parallel

TABLE 1

POSITIONS OF HISTIDINE RESIDUES (CHYMOTRYPSINOGEN NUMBERING
SYSTEM) IN THE SERINE PROTEINASES INVESTIGATED BY NMR SPECTROSCOPY

Proteinase	Histidines in positions				Reference
α-lytic proteinase[a]		57			41
bovine chymotrypsin(ogen) A[b]	40	57			20-22
bovine trypsin(ogen)[c]	40	57		91	53
porcine trypsin(ogen)[d]	40	57	71	91	54

^{31}P and ^{1}H NMR investigations of these derivatives to determine how the modification affects the pK$_a'$ values of active site groups.[63]

Sealock and Laskowski[64] developed an enzymatic procedure for removing and
reinserting the P1 reactive site amino acid residue in soybean trypsin inhibitor (residue-63). This provides a means for isotopically labeling this critical
residue in the inhibitor. The carbonyl of the P1 residue of STI,[65,66] the C$^\zeta$ of
the P1 arginine of STI (M.W. Baillargeon, M. Laskowski, Jr., D.E. Neves, M.A.
Porubcan, R.E. Santini and J.L. Markley, Biochemistry, submitted), and the carbonyl of the P1 residue of BPTI (R. Richarz, H. Tschesche, and K. Wüthrich, personal communication) have been enriched with ^{13}C by this technique for ^{13}C NMR
studies.

Recently, we made use of a ^{13}C labeled ester substrate of trypsin in a low
temperature experiment in which it appears by ^{13}C NMR that a tetrahedral species
is trapped.[67] Subzero temperature NMR spectroscopy may prove exceedingly useful for investigations of enzyme mechanisms.

NMR STUDIES OF FREE ENZYMES

α-lytic proteinase

We recently discovered that α-lytic proteinase can exist in two conformational
forms in solution as determined by ^{1}H NMR spectroscopy (W.M. Westler and J.L.
Markley, unpublished results). The two forms were detected through a doubling
of the peak assigned to the histidine-57 C$^\varepsilon$- H (Fig. 3). The "fresh" form of
α-lytic protinease is obtained by dialyzing the enzyme against water, lyophil-
izing, and dissolving in 0.2 M KCl (in either [1]H$_2$O or [2]H$_2$O) at low pH (between

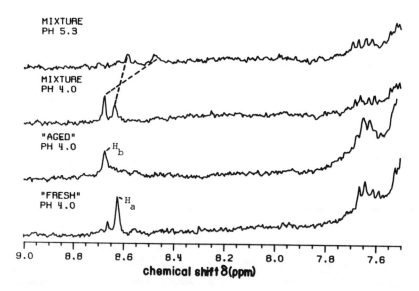

Fig. 3. 360 MHz ^{1}H NMR spectra of 1 mM α-lytic proteinase in ^{2}H$_2$O with 0.2 M
KCl at various pH* values showing peaks from the "fresh" and "aged" forms (W.M.
Westler and J.L. Markley, unpublished results).

pH 3-5). Upon incubation at pH 8.0 in 0.2 M KCl (in either ^{1}H$_2$O or ^{2}H$_2$O), α-
lytic proteinase converts to the "aged" form with a half-time of about 2.5 hr.
The "aged" form is stable at high or low pH over a long period of time, but it
can be reconverted to the "fresh" form by dialysis and lyophilization.

Both ^{1}H and ^{13}C NMR spectra reveal a large number of differences between the
two conformational forms in addition to the change in histidine-57. The only
other perturbation that can be assigned at present affects the position of a
methionine methyl resonance: 0.03 ppm downfield in the "aged" form compared to
the "fresh" form.

The pK$_a'$ of histidine-57 is 0.6 pH units lower in the "aged" form compared to
the "fresh" form as determined by pH titration studies of the ^{1}H NMR peaks
(Fig. 4). The transition affecting the chemical shift of histidine-57 in each
conformational state is assigned to protonation of the side chain imidazole.
The rationale for this is the rather normal chemical shift range for the titration curves.[35] In neither species is there evidence for a low pH inflection
in the titration curve. Studies of chymotrypsin and trypsin discussed below
suggest that the chemical shift of histidine-57 is perturbed by titration of
the adjacent aspartate-102. Hence, we conclude that in α-lytic proteinase, the
pK$_a'$ of aspartate-102 is below 3, in the region inaccessible to NMR studies
because of precipitation of the protein.

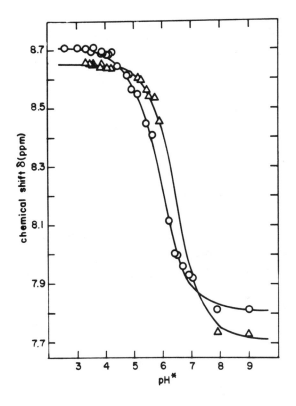

Fig. 4. Plots of the pH dependence of the ^{1}H NMR peaks
assigned to the histidine-57
C^ε -H of "fresh" (Δ) and "aged"
(o) α-lytic proteinase. Sample
spectra are shown in Fig. 3
(W.M. Westler and J.L. Markley,
unpublished results).

TABLE 2

HISTIDINE-57 OF α-LYTIC PROTEINASE: COMPARISON OF NMR STUDIES

Method	pK$_a'$ values	Comments	Reference
^{13}C NMR	3.3 6.8	Three species at pH 3.3	29
^{15}N NMR	7.0		60
^{1}H NMR	5.7 ("aged")	Two conformational species "aged" and "fresh," can be prepared and distinguished by ^{1}H NMR.	W.M. Westler and J.L. Markley, unpublished
^{1}H NMR	6.6 ("fresh")		

The results of these ^{1}H NMR experiments are compared with other NMR studies
of α-lytic proteinase in Table 2. Our conclusions are more in line with those
of Bachovchin and Roberts[60] who assigned a pK_a' of about 7.0 to histidine-57 than
those of Hunkapiller and co-workers[29] who ascribed a pK_a' of around 3.3 to
histidine-57. The interpretation of the latter experiment has been criti-
cized[56,68] on the grounds that the ^{13}C NMR data could also support a histidine-
57 pK_a' of 6.8. Hunkapiller et al.[29] reported the presence of three species at
low pH, one of which was ascribed to denatured enzyme; the other two may correspond to the "fresh" and "aged" forms detected by ^{1}H NMR spectroscopy (Fig. 3).

Chymotrypsin

Bovine chymotrypsinogen and chymotrypsin have histidyl residues in positions
40 and 57 (Table 1). ^{1}H NMR peaks corresponding to the C^{ε}-H of each residue
were resolved with chymotrypsinogen A and chymotrypsin A_α (Fig. 5). The pH
dependence of the chemical shifts of these peaks was used to construct titration
curves (Fig. 6). The two peaks of chymotrypsinogen A were assigned to histidines 40 and 57 on the basis of chemical modification with iPr_2P-F; the two
peaks of chymotrypsin A_α were assigned by reference to changes when the enzyme
binds bovine pancreatic trypsin inhibitor.[51] The ^{1}H NMR titration study indicated that the chemical shift and pK_a' values of both histidine residues (40 and
57) are affected by zymogen activation. The low pH perturbations present in the
histidine titration curves are assigned to perturbations by neighboring groups.
In the case of histidine-57, the perturbation is caused most probably by the
adjacent carboxyl of aspartate-102 and may be used to estimate its pK_a'. The
results of the NMR titration studies are summarized in Table 3. The experiments indicate that the pK_a' values histidine-57 and aspartate-102 both change
upon conversion of zymogen to enzyme. The pK_a' values for histidine-57 are in
rough agreement with those derived from ^{1}H NMR titration studies in [1]H_2O of the
low field N-H peak assigned to the proton hydrogen bonded between histidine-57
and aspartate-102 (pK_a' value of 7.5 for histidine-57 in both chymotrypsinogen
and chymotrypsin).[55-57] The chemical shift of the N-H peak, however, appears
not to be sensitive to the differences in the environment of histidine-57 in the
zymogen and enzyme; and the low pH transitions were not observed.[55-57]

Trypsin

^{1}H NMR studies of bovine trypsin have proved frustrating; histidyl peaks
have not been resolved, probably because of structural heterogeneity of the
enzyme.[69] On the other hand, all four histodine C^{ε}-H peaks of porcine typsin
and trypsinogen (Table 1) have been resolved, and a peak has been assigned to
histidine-57 in the enzyme[49] and zymogen.[50] Typical spectra are shown in Fig. 7.

Figure 5

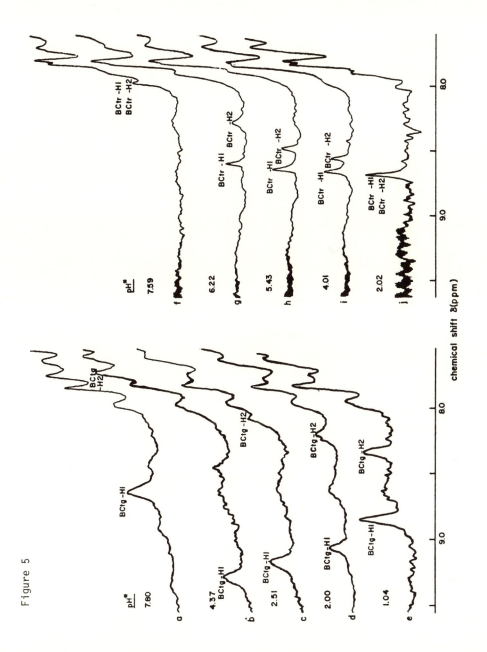

chemical shift δ(ppm)

Fig. 5.   250 MHz correlation ^{1}H NMR spectra of the histidyl $C^{\varepsilon 1}$-H regions of heat preexchanged bovine chymotrypsinogen A (spectra a-e) and bovine chymotrypsin A_α prepared from the preexchanged zymogen (spectra f-j) at selected pH* values shown in the figure: 25 mg of protein/0.5 mL of 0.5 M KCl in ^{2}H$_2$O, 31°C. Peaks BCtg-H1 and BCtg-H2 of chymotrypsinogen A are assigned to histidines 57 and 40, respectively; peaks BCtr-H1 and BCtr-H2 of chymotrypsin A_α are assigned to histidines 40 and 57, respectively (from reference 49).

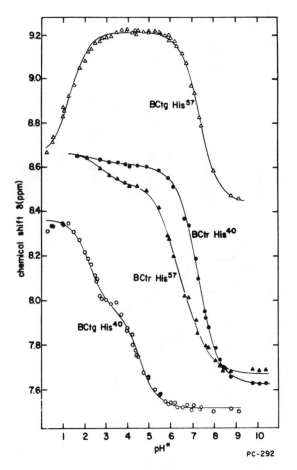

Fig. 6.   Current ^{1}H NMR titration curves for the histidine residues of chymotrypsinogen A and chymotrypsin A based on data obtained at 360 MHz: 25 mg of protein/0.5 mL of 0.5 M KCl in ^{2}H$_2$O, 31°C (I.B. Ibañez and J.L. Markley, unpublished results).

Peak H4 of trypsin was assigned to histidine-57 by three independent experiments: BPTI binding,[49] modification of histidine-57 by tosyl L-lysine chloromethyl ketone,[69] and modification by iPr$_2$-P-F.[63]   Peak H1 of porcine trypsinogen was assigned to histidine-57 by BPTI binding and iPr$_2$-P-F modification.[50]

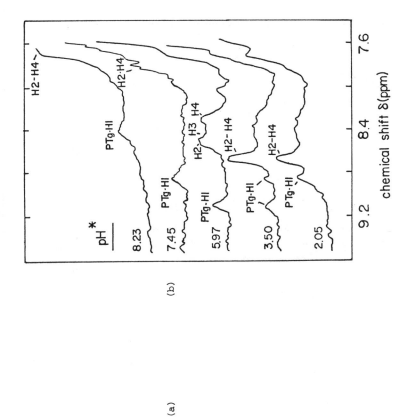

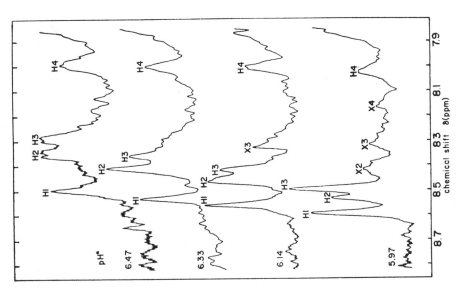

Fig. 7. 250 MHz correlation ^{1}H NMR spectra of the histidyl C$^\varepsilon$-H regions of heat preexchanged (a) porcine trypsin and (b) porcine trypsinogen at various pH* values indicated. Peak H4 of trypsin is assigned to histidine-57.[49,50] Peak H1 of trypsinogen is assigned to histidine-57.[51] Samples contained 2 mM protein in 0.5 m KCl in ^{2}H$_2$O.

44

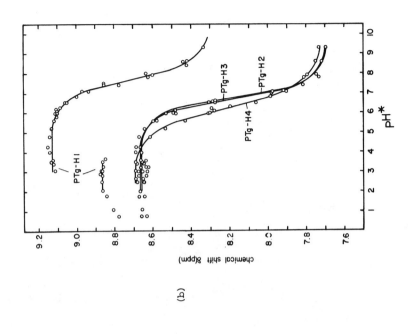

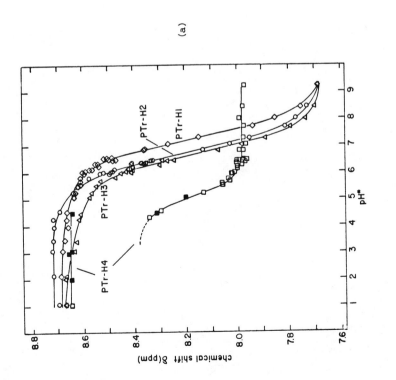

Fig. 8. Titration curves derived from the pH dependence of 250 MHz ^{1}H NMR spectra of (a) porcine trypsin (peak PTr-H4 is assigned to histidine-57) and (b) porcine trypsinogen (peak PTg-H1 is assigned to histidine-57). Sample spectra are shown in Fig. 7. [49,50,51]

TABLE 3

COMPARISON OF THE ^{1}H NMR TITRATION BEHAVIOR OF PEAKS ASSIGNED TO HISTIDINE-57 C^ϵ-H IN SERINE PROTEINASES[a]

	First protonation			Second protonation	
	δ high pH (ppm from DSS)	pK'_1	$\Delta\delta_1$ (ppm)	pK'_2	$\Delta\delta_2$ (ppm)
bovine chymo-trypsinogen A[b]	8.44	7.3	0.77	1.4	− 0.56
bovine trypsinogen[c]	8.36	7.7	0.68	1.8	− 0.43
porcine trypsinogen[c]	8.31	7.7	0.81	3.4[g]	− 0.28
bovine chymo-trypsin A_α[d]	7.92	6.8	0.79	3.2	0.14
porcine trypsin[e]	7.94	5.0	0.41	4.5[g]	0.28
α-lytic proteinase[f]					
form a "fresh"	7.71	6.5	0.94		
form b "aged"	7.81	5.9	0.90	[h]	

[a]All data were obtained at 31°C in ^2H_2O.

[b]From reference 51.

[c]From reference 50.

[d]I.B. Ibañez and J.L. Markley, unpublished

[e]From reference 49.

[f]W.M. Westler and J.L. Markley, unpublished data.

[g]This transition appears to be cooperative in hydrogen ion; hence the value is a pH_{mid} rather than a pK'_a.

[h]The pK' of Asp[102] is probably less than 3.

Titration curves for the histidyl peaks (Fig. 8) show that the peaks assigned to histidine-57 undergo low pH transitions. As in chymotrypsinogen/chymotrypsin, these are attributed to aspartate-102. The titration data for trypsinogen and trypsin are summarized in Table 3. Note that the low pH transition present in trypsinogen and trypsin is slow on the NMR time scale (discontinuous) rather than rapid (continuous) as in chymotrypsinogen/chymotrypsin. The pK'_a and environment of histidine-57 are clearly different in trypsinogen and trypsin.[50]

DIISOPROPYLPHOSPHORYL DERIVATIVES

Diisopropyl phosphorofluoridate (iPr$_2$P-F), the classic inhibitor of serine proteinases, reacts exclusively with serine-195 to form an inactive (diisopropyl-phosphoryl) serine derivative.[17] It is known that the iPr$_2$P-derivative can undergo two reactions: the iPr$_2$P group can be removed in a "reactivation" reaction to generate active enzyme;[70] or an "aging" reaction can take place by which the tertiary phosphate ester is converted to a secondary phosphate ester with the loss of an isopropyl group.[71-72] "Aged" enzyme cannot be "reactivated." iPr$_2$P-derivatives of chymotrypsin[73] and trypsin (R.M. Stroud, personal communication) have proved unstable over the long periods of time required for X-ray crystallography. More recently, it was found that iPr$_2$P-F reacts with zymogens[74,75] in the same way as enzymes, leading to loss of their low level of catalytic activity.[75,76]

We initially used iPr$_2$P-derivatives to make assignments of the histidine-57 C$^\varepsilon$-H peaks;[50-51] however, the changes that could be observed in these derivatives prompted a more detailed study.[63] We found from ^{1}H NMR spectra that the pK$_a'$ of histidine-57 and the pK$_a'$ attributed to aspartate-102 are altered in the iPr$_2$P-derivatives. The pK$_a'$ values of the derivatives are reflected also in the ^{31}P resonance of the iPr$_2$P group. Paralleled ^{1}H and ^{31}P NMR titration experiments were carried out with seven iPr$_2$P-derivatives. Typical spectra and titration curves are shown for iPr$_2$P-α-lytic proteinase (Figs. 9 and 11) and iPr$_2$P-trypsin (Figs. 10 and 12). Table 4 summarizes pK$_a'$ values for all derivatives studied.

COMPLEXES BETWEEN PROTEINASES AND NATURAL INHIBITORS

^{1}H NMR studies

Histidine C$^\varepsilon$-H peaks have been resolved in ^{1}H NMR spectra of BPTI-porcine trypsin,[49] BPTI-chymotrypsin,[51] and BPTI-porcine (and bovine) trypsinogen complexes.[50] The results in all cases were similar: the peak assigned to histidine-57 C$^\varepsilon$-H fails to titrate in the pH range where the complex is intact whereas other histidine C$^\varepsilon$-H resonances of the proteinases (BPTI contains no histidine) titrate normally. Examples are shown in Fig. 13. The same conclusion was reached by Robillard and Shulman in their investigation of the low field N-H peak of BPTI-chymotrypsin.[56] The X-ray structures of BPTI-trypsin[44] and BPTI-trypsinogen[78] demonstrate that the imidazole ring of histidine-57 is shielded from solvent.

It is critical for an understanding of the mechanism of inhibition to know the charge state of the histidine-57 imidazole in these complexes. Robillard

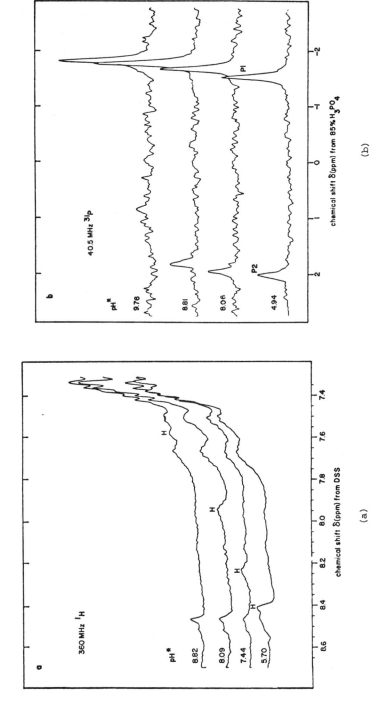

Fig. 9. (a) 360 MHz ^{1}H NMR spectra, using a Carr-Purcell-Meiboom-Gill-pulse sequence with a total sequence length of 6 ms to remove unexchanged broad N-H peaks,[77] and (b) 40.5 MHz ^{31}P NMR spectra of (diisopropylphosphoryl)α-lytic proteinase at 30°C. Samples contained 2-4 mM protein in 0.4 M KCl in ^{2}H$_2$O. Peak H in the ^{1}H NMR spectra is assigned to the C^ε-H of histidine-57. Peak P1 in the ^{31}P NMR spectra is assigned to (diisopropylphosphoryl)serine-195; P2 appears in aged samples and is unassigned.[63]

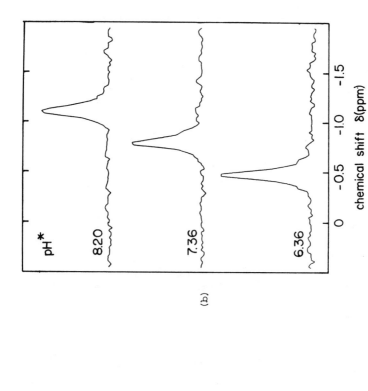

(a)

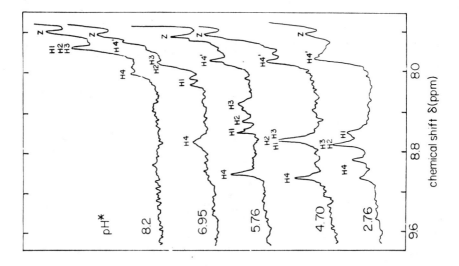

(b)

Fig. 10. (a) 250 MHz ^{1}H NMR spectra of 2 mM solutions of (diisopropylphosphoryl)porcine trypsin in 0.5 M KCl in ^{2}H$_2$O, 30°C.[69] Peak H4
is assigned to histidine-57 C$^\epsilon$-H; peak Z is a nontitrating peak attributed to a deshielded aromatic group. (b) 40.4 MHz ^{31}P NMR spectra of 1 mM solutions of (diisopropylphosphoryl)porcine trypsin in
0.5 M KCl in ^{2}H$_2$O, 30°C.[63]

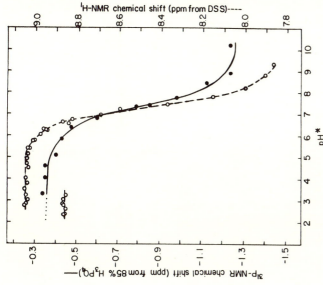

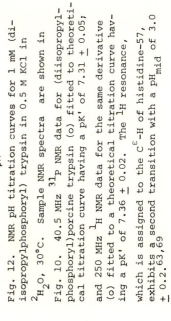

Fig. 12. NMR pH titration curves for 1 mM (di-isopropylphosphoryl) trypsin in 0.5 M KCl in ^2H_2O, 30°C. Sample NMR spectra are shown in Fig. 10. 40.5 MHz ^{31}P NMR data for (diisopropyl-phosphoryl)porcine trypsin (o) fitted to theoreti-cal titration curve having a pK' of 7.31 ± 0.05; and 250 MHz ^1H NMR data for the same derivative (o) fitted to a theoretical titration curve hav-ing a pK' of 7.36 ± 0.02. The ^1H resonance, which is assigned to the C$^\epsilon$-H of histidine-57, exhibits a second transition with a pH$_{mid}$ of 3.0 $\pm$ 0.2.63,69

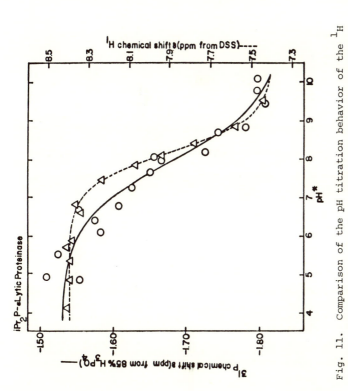

Fig. 11. Comparison of the pH titration behavior of the ^1H NMR peak assigned to the C$^\epsilon$-H of histidine-57 (($\triangle$) and the ^{31}P NMR peak assigned to the (diisopropylphosphoryl)serine-195 (o) of (diisopropylphosphoryl)α-lytic proteinase. Sample spectra are shown in Fig. 9. The computer fitted titration curves yield pK' values of 8.16 ± 0.03 (-----) and 7.91 ± 0.2 (——).63

TABLE 4

THE EFFECT OF INHIBITION BY DIISOPROPYL PHOSPHOROFLUORIDATE (iPr_2P-F)
ON THE pK' VALUE OF HISTIDINE-57 IN VARIOUS ENZYMES AND ZYMOGENS

Species	pK' value of histidine-57		pH_{mid} of low pH transition attributed to aspartate-102	
	native	iPr_2-derivative[a]	native	iPr_2P-derivative[a]
α-lytic proteinase	5.9[b]	8.0	--[d]	--[d]
bovine chymotrypsin A	6.8[c]	7.5	3.2	4.3
bovine trypsin	--[d]	7.7	--[d]	--[d]
porcine trypsin	5.0[e]	7.3	4.5	--[d]
bovine chymotrypsinogen A	7.3[f]	7.6	1.4	3.2
bovine trypsinogen	7.7[g]	8.0	1.8	4.3
porcine trypsinogen	7.7[g]	7.5	3.4	4.3

[a] Average of ^{31}P NMR and ^{1}H NMR data where available.
[b] W.M. Westler and J.L. Markley, unpublished; "aged" form.
[c] I.B. Ibañez and J.L. Markley, unpublished. The native pK' is for chymotrypsin A_α; the iPr_2-derivative pK' is for chymotrypsin A_δ.
[d] Not measured.
[e] Reference 49.
[f] Reference 51
[g] Reference 50.
[h] Conditions: 30°; in $^{2}H_2O$; solutions contained 0.5 M KCl, except native and inhibited α-lytic proteinase solutions which contained 0.2 M KCl.

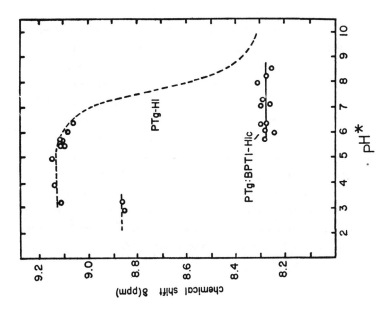

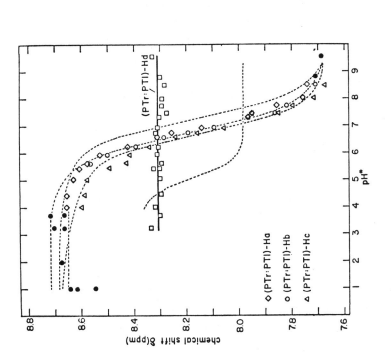

Fig. 13. The pH dependence of the histidyl C^ε-H peaks in complexes. 250 MHz ^{1}H NMR spectra, 0.2 mM protein in 0.5 M KCl in ^{2}H$_2$O, 31°C. (a) Porcine trypsin-bovine pancreatic trypsin inhibitor complex; (peak (PTr:PTI)-Hd is assigned to histidine-57 in the complex; dashed lines are the titration curves for free porcine trypsin.[49] (b) Porcine trypsinogen-bovine pancreatic trypsin inhibitor complex; peak PTg-BPTI-Hlc is assigned to histidine-57 in the complex; dashed lines are the titration curves of free porcine trypsinogen.[50]

and Shulman[57] inferred from the chemical shift of the low field N-H peak that histidine-57 is uncharged in the BPTI-chymotrypsin A_δ complex. We have reproduced this spectrum and have found that the chemical shift of the low field N-H peak in various complexes depends more on the nature of the proteinase than the inhibitor (i.e., the chemical shifts of STI and BPTI complexes with a given proteinase are similar).[58] We have argued that, since the chemical shift of the N-H is highly dependent on the strength of the hydrogen bond, the position of this peak may not be a reliable indicator of the charge state of histidine-57 in the complex. We have preferred to rely on the chemical shift of the C^ϵ-H peak which predicts that histidine-57 is positively charged in the complexes.[49-51] Our conclusion that histidine-57 is positively charged could be in error if we failed to account for environmental influences on the C^ϵ-H chemical shift that greatly deshielded it in the complexes. However, we had apparent succes in explaining the chemical shift difference of the C^ϵ-H of histidine-57 in trypsin and trypsinogen[50] by reference to changes in its magnetic environment as predicted by the X-ray structures.[10,79]

^{13}C NMR studies

There has been considerable speculation about the mechanism of interaction between inhibitors like BPTI or STI and proteinases. An acyl enzyme complex has been considered as well as noncovalent structures.[43] The initial X-ray data for BPTI-trypsin[44] and STI-trypsin[45] were interpreted as evidence for a tetrahedral adduct with a covalent bond between the P1 carbonyl of the inhibitor and the serine-195 O^γ. Upon refinement of the X-ray data for BPTI-trypsin it was concluded that the interaction is not covalent but that the P1 carbonyl is tetrahedrally distorted.[46]

Sealock and Laskowski[64] developed an enzymatic procedure for removing the P1 (reactive site) residue of STI and replacing it with another amino acid. We recently adapted this semisynthetic approach for the production of STI labeled with ^{13}C in the reactive site in amounts sufficient for ^{13}C NMR spectroscopy (Fig. 14). The chemical shift of the P1 carbonyl of STI ([1-^{13}C]Arg-63) is 173.8 ± 0.1 ppm. When the STI-trypsin complex is made, the P1 carbonyl resonance shifts to 174.0 ± 0.1 ppm.[65]

The chemical shift of an arginyl carbonyl in a simple peptide is around 173.4 ppm.[80] In the STI-trypsin complex this chemical shift could be perturbed by three possible effects: (1) Covalent tetrahedral bond formation between the arginine-63 carbonyl of the inhibitor and the serine-195 O^γ of the proteinase should shift the resonance upfield by 30-55 ppm. (2) Hydrogen bonding of the arginine-63 carbonyl in the oxyanion hole could shift the resonance downfield

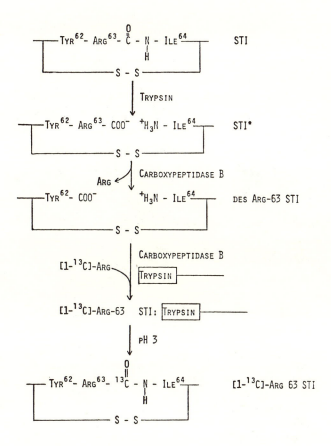

Fig. 14. Flow chart illustrating the enzymatic removal of residue 63 of soybean trypsin inhibitor followed by enzymatic resynthesis and incorporation of ^{13}C labeled arginine (M.W. Baillargeron, M. Laskowski, Jr., D.E. Neves, M.A. Porubcan, R.E. Santini, and J.L. Markley, Biochemistry, submitted.

by a maximum of 5 ppm.[81]   (3) Out-of-plane distortion of the arginine-63 carbonyl could shift the resonance downfield by a maximum of 2 ppm.[82]

Since the P1 carbonyl resonance of STI moves 0.9 ppm downfield upon complex formation, covalent tetrahedral bond formation is ruled out.[65] Similar ^{13}C NMR results were obtained for the STI-trypsin complex by Hunkapiller et al.[66] and for the BPTI-trypsin complex by Richarz et al. (personal communication). Hunkapiller et al. observed doubling of the carbonyl peak in the STI trypsin complex and attributed this to an equilibrium between a noncovalent trigonal complex and an acyl-enzyme species.[66] We found only a single peak for the complex (Fig. 15) and suggest that the second peak of Hunkapiller et al. corresponds to uncomplexed

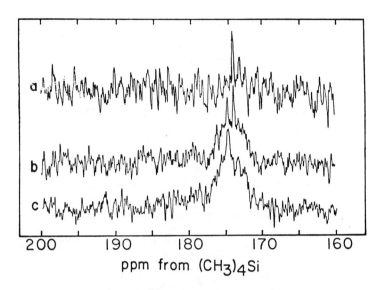

ppm from (CH₃)₄Si

Fig. 15. The carbonyl region of ^{13}C NMR spectra at 90 MHz. (a) Soybean trypsin inhibitor (STI) in which residue-63 has been labeled with 90% enriched [1-^{13}C]L-arginine by means of enzymatic semisynthesis (see Fig. 14). (b) Following addition of half stoichiometric amount of unlabeled porcine trypsin. (c) Following addition of a slight excess of unlabeled porcine trypsin. The sharp peaks at 173.8 ppm and 174.7 ppm are assigned to the reactive site Pl carbonyl (of arginine-63) in free STI and in the STI-trypsin complex, respectively. The broad hump of peaks corresponds to natural abundance ^{13}C carbonyl resonances of the proteins. (M.W. Baillargeon, M. Laskowski, Jr., D.E. Neves, M.A. Porubcan, R.E. Santini, and J.L. Markley, Biochemistry, submitted).

STI (M.W. Baillargeon, M. Laskowski, Jr., D.E. Neves, M.A. Porubcan, R.E. Santini, and J.L. Markley, Biochemistry, submitted). An acyl-enzyme complex would have a chemical shift similar to that observed; however, this structure can be eliminated on the grounds that anhydrotrypsin forms a species that appears identical with authentic complex by kinetic[83] and X-ray analysis.[47]

CONCLUSIONS

   The pK_a' values of active site residues derived from NMR studies carried out in this laboratory are collected in Table 3. A few generalizations emerge. First of all, the data do not support the "charge relay hypothesis" which predicts that the pK_a' of aspartate-102 is higher than that of histidine-57.[28-34,59] The experimental values demonstrate that the pK_a' of histidine-57 is higher than that of aspartate-102 in all zymogens and enzymes.[49-52] The validity of the "charge relay hypothesis" has been questioned recently since upon refinement of the X-ray structures, there appears to be no hydrogen bond between histidine-57 and serine-195.[40]

Second, the properties of the catalytic residues are different in zymogens and enzymes.[50,51] The pK_a' of histidine-57 is higher in zymogens than in corresponding enzymes. This result is also in agreement with the recent X-ray refinements[40] which show that the hydrogen bond between histidine-57 and serine-195 is present in zymogens but not in enzymes. Zymogens are catalytically less active than their respective enzymes because of lowered efficiency of the binding and catalytic apparatus.[5,8,11,36] Since the pK_a' of histidine-57 is higher in zymogens than in enzymes, a part of this inefficiency at physiological pH must be attributed to fractional protonation of histidine-57 which renders it catalytically inactive.

Third, the pK_a' of histidine-57 is higher in iPr_2P-enzymes than in native enzymes.[50,51,63] This result is important insofar as iPr_2P-enzymes resemble the transition state.[7,13] The lower pK_a' of histidine-57 in free enzymes ensures that more of the enzyme is in the catalytically active, deprotonated form. After formation of the Michaelis complex and as the transition state is approached, the pK_a' of histidine-57 is raised. This makes the imidazole a more efficient base for accepting the hydroxyl proton of serine-195. In agreement with Reeck et al.,[62] we find that the ^{31}P resonance of iPr_2P-zymogens is shielded compared to that of the corresponding iPr_2P-enzymes. Recent experiments of Stroud and co-workers show that the position of the phosphate is very different in crystals iPr_2P-F inhibited trypsinogen as compared to iPr_2P-F inhibited trypsin. The X-ray results indicate that there are three hydrogen bonds donated to the phosphate in iPrP-enzymes (the backbone NH's of glycine-193 and serine-195 and the N^ϵ-H of histidine-57) but no hydrogen bonds to the phosphate in iPr_2P-zymogens (R.M. Stroud, personal communication). Since hydrogen bonds donated to the phosphate should result in a downfield shift, the ^{31}P NMR data[63] are consistent with such a structural change. The ^{31}P NMR titration shift which accompanies the high pH transition in iPr_2P-trypsin is 2.4 times as large as that in iPr_2P-trypsinogen, suggesting that the phosphate is farther from the imidazole of histidine-57 in iPr_2P-trypsinogen than in iPr_2P-trypsin. This result is consistent with the X-ray data which indicate a hydrogen bond between histidine-57 and the phosphate in the enzyme derivative but not the zymogen derivative (R.M. Stroud, personal communication). ^1H NMR results with iPr_2P-derivatives[63] also are in agreement with displacement of histidine-57 away from aspartate-102 in iPr_2P-trypsinogen (R.M. Stroud, personal communication) as compared with native trypsinogen.[9-11] The ^1H NMR chemical shift of the C^ϵ-H of histidine-57 is altered substantially when trypsinogen is converted to iPr_2-trypsinogen; and the magnitude of the low pH transition attributed to protonation of Asp^{102} is attenuated by a factor of almost four.[50] In addition, the pK' attributed to aspartate-102 is more normal in iPr_2P-bovine trypsinogen (4.3) than in bovine trypsinogen native (1.8),[50] in agreement with increased solvent accessibility of aspartate-102 in the iPr_2P-derivative (R.M. Stroud, personal communication). Similar changes were observed with iPr_2P-derivatives of porcine trypsinogen[50] and bovine chymotrypsinogen.[51]

The pK_a' of about 7 obtained from steady state kinetics generally has been attributed to protonation of the catalytic triad in the free enzyme. The pK_a' obtained from the pH dependence of k_{cat}/K_m for a variety of substrates with bovine chymotrypsin at 25°C and 0.1 M ionic strength is 6.80 ± 0.03 with only one exception.[84] The pK_a' value of histidine-57 in chymotrypsin A_α derived from ^{1}H NMR data in ^2H_2O and 30°C and 0.5 M ionic strength is 6.5 (Table 3). [Note that this pK_a' is higher than the value (6.1) reported previously.[51] The earlier studies[51] were carried out at 250 MHz; the titration curves can be traced more accurately at 360 MHz (I.B. Ibañez and J.L. Markley, unpublished)]. The very low pK_a' (5.0) obtained for histidine-57 of porcine trypsin[49] is puzzling. We recently repeated the titration experiment at 360 MHz (D.E. Neves and J.L. Markley, unpublished) with results similar to those at 250 MHz. At 360 MHz the histidine-57 C^ε-H resonance could not be followed between pH 3 and 5 (apparently because of exchange broadening), but the data points that could be obtained at 360 MHz superimposed with those previously obtained at 250 MHz.

A striking area of agreement between the NMR and X-ray data concerns the BPTI-trypsinogen complex. The chemical shift of the histidine-57 C^ε-H is the same in BPTI-trypsin and BPTI-trypsinogen complexes whereas the chemical shift is very different in free trypsin and free trypsinogen.[50] The X-ray results for BPTI complexes with bovine trypsinogen and trypsin indicate that binding of BPTI induces a conformational change in trypsinogen leading to an active site structure that resembles that of trypsin.[8,9,85]

Since inhibitors like STI and BPTI serve as trypsin substrates, one can write the mechanism for them shown in Fig. 16. The charge state of histidine-57 and the hybridization of the P1 reactive site C-1 carbon are shown for each species. As discussed above, the ^{13}C NMR studies of complexes[65,66] (and R. Richarz, H. Tschesche, and K. Wüthrich, personal communication) are in agreement that the P1 carbonyl is trigonal. Other considerations appear to rule out trigonal species, A and EI*.[47,82] However, the ^{1}H NMR data for complexes have been interpreted to indicate that histidine-57 is positively charged.[49-51] A positively charged histidine-57 is consistent with a tetrahedral complex[44,45] but

Species	His[57]	P1 Residue C-1
E + I	0	TRIGONAL
⇅		
EI	0	TRIGONAL
⇅		
T_1	+	TETRAHEDRAL
⇅		
A	0	TRIGONAL
⇅		
T_2	+	TETRAHEDRAL
⇅		
EI*	0	TRIGONAL
⇅		
E + I*	0	TRIGONAL

Fig. 16. The charge state of histidine-57 and the hybridization of the P1 residue C^1 in the hypothetical steps of the reaction between proteinase (E) and inhibitor (I). Modified inhibitor is designated I*; T_1 and T_2 represent the two tetrahedral states; and A represents the acyl enzyme intermediate.

this structure has been abandoned on X-ray[46] and NMR[65,66] grounds. As pointed out above, correct assignment of charge of the imidazole depends on identifying all significant environmental contributions to the ^{1}H NMR chemical shift of the histidine C^ε-H. We have used the X-ray structure of the BPTI-trypsin complex to estimate these, and conclude that histidine-57 is positively charged. If this is true, the complex probably is a noncovalent species which differs from EI in that it has picked up a proton. Protonation of histidine-57 would prevent the catalytic reaction from proceeding to the next species (T_1) in the catalytic mechanism and would help explain why the catalytic reaction is stopped at this stage.

One goal of NMR spectroscopy has been to probe intermediates in proteolytic reactions.[86-88] A successful approach to trapping intermediates has been the use of cryosolvents and low temperature,[89,90] but NMR spectroscopy has been little used[91] to characterize such trapped intermediates probably owing to technical difficulties. In recent low-temperature experiments, we resolved a signal which appears to be due to a trapped tetrahedral species.[67]

Porcine trypsin was dissolved in a methanol/water cryosolvent 70/30 (v/v). The solution was 1 mM in trypsin and 20 mM in cacodylate buffer at neutral pH. After cooling to -60°C, the substrate, N-α-acetyl [90% 1-^{13}C]L-arginine methyl ester, was added. The final concentration of enzyme and substrate were

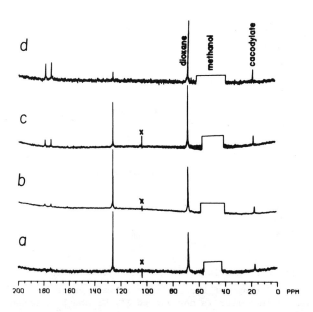

Fig. 17. Low temperature experiments: sample contained 1 mM trypsin, 1 mM Nα-
acetyl-[90% 1-^{13}C]L-arginine methyl ester, 20 mM cacodylate buffer pH 7, dissolved in 70/30 (v/v) methanol/water. Dioxane was used as an internal standard
and assigned a chemical shift of 67.8 ppm. The intensity between 40 and 60 ppm
is due to methanol and was reduced to permit on-scale representation of the
spectra. (a) -60°C; initial spectrum after adding the substrate, Nα-acetyl-
[90% 1-^{13}C]L-arginine methyl ester; (b) -60°C; 1 hour after substrate addition;
(c) -48°C; 6 hours after substrate addition; (d) -5°C; 8 hours after substrate
addition. (D.E. Neves and J.L. Markley, unpublished spectra.) The signals
marked with X's are artifacts resulting from quadrature detection.

approximately equimolar. ^{13}C NMR spectra were obtained at 37 MHz. The initial
spectrum revealed a single peak with chemical shift 126.0 ppm downfield from
$(CH_3)_4Si$ (Fig. 17a). After one hour at -60°C, a ^{13}C NMR spectrum revealed two
additional peaks of small intensity with chemical shifts 174.4 and 178.9 ppm
(Fig. 17b). The temperature was raised in increments to a final temperature
of -5°C. ^{13}C NMR spectra revealed that as the temperature increased, the
intensities of the downfield peaks increased with concomitant decreased intensity of the upfield peak (Fig. 17 c,d). The rates of increase in intensities
of the downfield peaks were larger as the temperature was raised. Note added
in proof: The signal at 126.0 ppm has a T_1 shorter and a T_2 longer than predicted by theory for a protein-bound species. Additional experiments are
required to determine the identity of this species.

REFERENCES

1. Markley, J.L. (1979) In NMR in Biology, Shulman, R.G., ed., Academic Press, New York, N.Y., 397-461.
2. DeHaen, C., Neurath, H., and Teller, D.C. (1975) J. Mol. Biol., 92, 225-259.
3. James, M.N.G., Delbaere, L.T.J., and Brayer, G.D. (1978) Can. J. Biochem., 56, 396-402.
4. Matthews, B.W., Sigler, P.B., et al. (1967) Nature, 214, 652-656.
5. Freer, S.T., Kraut, J., et al. (1970) Biochemistry, 9, 1997-2009.
6. Tulinsky, A., Vandlen, R.L., et al. (1973) Biochemistry, 12, 4185-4192.
7. Stroud, R.M., Kay, L.M., and Dickerson, R.E. (1974) J. Mol. Biol., 83, 185-208.
8. Bode, W., Fehlhammer, H., and Huber, R. (1976) J. Mol. Biol., 106, 325-335.
9. Bode, W., Schwager, P., and Huber, R. (1979) In Proteolysis and Physiological Regulation, Academic Press, New York, N.Y., 43-76.
10. Fehlhammer, H., Bode, W., and Huber, R. (1977) J. Mol. Biol., 111, 415-438.
11. Kossiakoff, A.A., et al. (1977) Biochemistry, 16, 654-664.
12. Delbaere, L.T.J., Brayer, G.D., James, M.N.G. (1979) Nature, 279, 165-168.
13. Kraut, J. (1977) Annu. Rev. Biochem., 46, 331-358.
14. Jansen, E.F., Nutting, M.D.F., Balls, A.K. (1949) J. Biol. Chem., 179, 201-204.
15. Schaffer, N.K., May, S.C., Jr., and Summerson, W.H. (1953) J. Biol. Chem., 202, 67-76.
16. Cohen, J.A., Oosterbaan, R.A., et al. (1955) Faraday Discuss. Chem. Soc., 20, 114-119.
17. Oosterbaan, R.A., Kunst, P., and Cohen, J.A. (1955) Biochim. Biophys. Acta, 16, 299-300.
18. Oosterbaan, R.A., and van Adrichem, M.E. (1958) Biochim. Biophys. Acta., 27, 423-245.
19. The chymotrypsinogen numbering system is used for all proteinases discussed here.
20. Hartley, B.S. (1964) Nature, 201, 1284-1287.
21. Hartley, B.S., and Kauffman, D.L. (1966) Biochem. J. 101, 229-231.
22. Meloun, B., Kluh, I., et al. (1966) Biochim. Biophys. Acta., 130, 543-546.
23. Weil, L., James, S., and Buchert, A.R. (1953) Arch. Biochem. Biophys., 46, 266-278.
24. Ray, W.J., Jr., and Koshland, D.E., Jr. (1960) Brookhaven Symp. Biol., 13, 135-150.
25. Schoellman, G., and Shaw, E. (1962) Biochem. Biophys. Res. Commun., 7, 36-40.
26. Ong, E.B., Shaw, E., and Schoellman, G. (1964) J. Am. Chem. Soc., 86, 1271-1272.
27. Hess, G.P. (1971) Enzymes, 3, 213-248.
28. Blow, D.M., Birktoft, J.J., and Hartley, B.S. (1969) Nature, 221, 337-340.
29. Hunkapiller, M.W., et al. (1973) Biochemistry, 12, 4732-4743.
30. Koeppe, R.E., II, and Stroud, R.M. (1976) Biochemistry, 15, 3450-3458.
31. Amidon, G.L. (1974) J. Theor. Biol., 46, 101-109.
32. Scheiner, S., Kleier, D.A., and Lipscomb, W.N. (1975) Proc. Nat. Acad. Sci. USA, 72, 2606-2610.
33. Kitayama, H.P., and Fukutome, H. (1976) J. Theor. Biol., 60, 1-18.
34. Beppu, Y., and Yomosa, S. (1977) Japan. J. Phys. Soc., 42, 1694-1700.
35. Markley, J.L. (1975) Acc. Chem. Res., 8, 70-80.
36. Robertus, J.D., Kraut, J., et al. (1972) Biochemistry, 11, 4293-4303.
37. Bode, W., Schwager, P., and Huber, R. (1978) J. Mol. Biol., 118, 99-112.
38. Zwilling, R., Neurath, H., et al. (1975) FEBS Lett., 60, 247-249.
39. Birktoft, J.J., Kraut, J., and Freer, S.T. (1976) Biochemistry, 15, 4481-4485.
40. Matthews, D.A., Alden, R.A., et al. (1977) J. Biol. Chem., 252, 8875-8883.

41. Olson, M.O.J., Nagabhushan, N., et al. (1970) Nature, 228, 438-442.
42. Abbreviations used are: BPTI, bovine pancreatic trypsin inhibitor; STI, soybean trypsin inhibitor; iPr$_2$P-, diisopropylphosphoryl-; pH*, uncorrected pH meter reading of a ^{2}H$_2$O solution using a glass electrode calibrated with ^{1}H$_2$O buffers.
43. Laskowski, M., Jr., and Sealock, R.W. (1971) Enzymes, 3, 375-473.
44. Rühlmann, A., et al. (1973) J. Mol. Biol., 77, 417-436.
45. Sweet, R.M., et al. (1974) Biochemistry, 13, 4212-4228.
46. Huber, R., Kukla, D., et al. (1974) J. Mol. Biol., 89, 73-101.
47. Huber, R., Bode, W., et al. (1975) Biophys. Struct. Mech., 1, 189-201.
48. Huber, R., and Bode, W. (1978) Acc. Chem. Res., 11, 114-122.
49. Markley, J.L., and Porubcan, M.A. (1976) J. Mol. Biol., 102, 487-509.
50. Porubcan, M.A., Neves, D.E., et al. (1978) Biochemistry, 17, 4640-4647.
51. Markley, J.L., and Ibañez, I.B. (1978) Biochemistry, 17, 4627-4640.
52. Westler, W.M., and Markley, J.L. (1978) Fed. Proc., Fed. Am. Soc. Exp. Biol., 38, 1795.
53. Walsh, K.A., and Neurath, H. (1964) Proc. Natl. Acad. Sci. USA, 52, 884-889.
54. Hermodson, M.A., Erisson, L.H., et al. (1973) Biochemistry, 12, 3146-3153.
55. Robillard, G., and Shulman, R.G. (1972) J. Mol. Biol., 71, 507-511.
56. Robillard, G., and Shulman, R.G. (1974) J. Mol. Biol., 86, 519-540.
57. Robillard, G., and Shulman, R.G. (1974) J. Mol. Biol., 86, 541-558.
58. Markley, J.L. (1978) Biochemistry, 17, 4648-4656.
59. Hunkapiller, M.W., Smallcombe, S.H., and Richards, J.H. (1975) Org. Magn. Reson., 7, 262-265.
60. Bachovchin, W.W., and Roberts, J.D. (1978) J. Am. Chem. Soc., 100, 8041-8047.
61. Gorenstein, D.G., and Findlay, J.B. (1976) Biochem. Biophys. Res. Commun., 72, 640-645.
62. Reeck, G.R., Nelson, T.B., et al. (1977) Biochem. Biophys. Res. Commun., 74, 643-649.
63. Porubcan, M.A., et al. (1979) Biochemistry, 18, 4108-4116.
64. Sealock, R.W., and Laskowski, M., Jr. (1969) Biochemistry, 9, 3703-3710.
65. Neves, D.E., Markley, J.L., et al. (1979) Fed. Proc., 38, 474.
66. Hunkapiller, M.W., et al. (1979) Biochem. Biophys. Res. Commun., 87, 25-31.
67. Neves, D.E., and Markley, J.L. (1979) Int. Symp. Magn. Reson. in Chem. Biol. and Physics, Argonne, Illinois, Abstract, p. 39.
68. Egan, W., Shindo, H., and Cohen, J.S. (1977) Ann. Rev. Biophys. Bioeng., 6, 408, 1977.
69. Porubcan, M.A. (1978) Ph.D. Thesis, Purdue University, West Lafayette, Indiana.
70. Cunningham, L.W., Jr., and Neurath, H. (1953) Biochem. Biophys. Acta., 11, 310.
71. Berends, F., Posthumus, C.H., et al. (1959) Biochim. Biophys. Acta., 34, 576-578.
72. Lee, W., and Turnbull, H.J. (1961) Experientia, 17, 360-361.
73. Blow, D.M. (1969) Biochem. J., 112, 261-268.
74. Morgan, P.H., Robinson, N.C., et al. (1972) Proc. Nat. Acad. Sci. USA, 69, 3312-3316.
75. Robinson, N.C. Neurath, H., and Walsh, K.A. (1973) Biochemistry, 12, 420-426.
76. Kay, J., and Kassell, B. (1971) J. Biol. Chem., 246, 6661-6665.
77. Campbell, I.D., Dobson, C.M., et al. (1975) FEBS Lett., 57, 96-99.
78. Bode, W., Schwager, P., and Huber. (1978) J. Mol. Biol., 118, 99-112.
79. Bode, W., and Schwager, P. (1975) J. Mol. Biol., 98, 693-717.
80. Richarz, R., and Wüthrich, K. (1978) Biopolymers, 17, 2133-2141.
81. Llinas, M., Wilson, D.M., and Klein, M.P. (1977) J. Am. Chem. Soc., 99, 6846-6850.

82. Grathwohl, C., Tun-Kyi, A., et al. (1975) Helv. Chim. Acta., 58, 415-423.
83. Ako, H., Foster, R.J., and Ryan, C.D. (1974) Biochemistry, 13, 132-139.
84. Fersht, A.R. (1977) Enzyme Structure and Mechanism, W.H. Freeman and Co., San Francisco, 146-147.
85. Bode, W., and Huber, R. (1976) FEBS Lett., 68, 231-235.
86. Robillard, G., Shaw, E., and Shulman, R.G. (1974) Proc. Natl. Acad. Sci. USA, 71, 2623-2626.
87. Lowe, G., and Nurse, D. (1977) J. Chem. Soc. Chem. Commun., 815-817.
88. Niu, C-H., et al. (1977) J. Am. Chem. Soc., 99, 3161-3162.
89. Douzou, P. (1977) Cryobiochemistry, Academic Press, New York.
90. Fink, A.L. (1977) Accts. Chem. Res., 10, 233-239.
91. Ghisla, S., Hastings, J.W., et al. (1978) Proc. Natl. Acad. Sci. USA, 75, 5860-5863.
92. Supported by National Institutes of Health Grants GM 19907 and RR 01077 (to the Purdue University Biochemical Magnetic Resonance Laboratory).

Published 1980 by Elsevier North Holland, Inc.
Liu/Mamiya/Yasunobu, eds. Frontiers in Protein Chemistry

PROTON MAGNETIC RESONANCE STUDIES OF PEPSINOGEN

AND PEPSIN-INHIBITOR INTERACTIONS

PAUL G. SCHMIDT
Laboratory of Protein Studies, Oklahoma Medical Research Foundation, and the
Department of Biochemistry and Molecular Biology, University of Oklahoma Health
Sciences Center, Oklahoma City, Oklahoma 73104

ABSTRACT

Pig pepsinogen in ^2H_2O displays 3 His C-ε_1 resonances in proton NMR spectra

after peptide NH protons have been exchanged for ^2H. All three histidines

titrate with apparent pKa values near 9, suggesting that each is involved in

interactions with other protein groups. When pepsinogen is activated by drop-

ping the pH to 4 there is an immediate change in the chemical shifts of numer-

ous peaks suggesting a conformational rearrangement, with little evidence for

completely displaced amino acids. Over a period of 14 hrs. peaks appear which

are due to released peptides. Pepstatin, a potent peptide inhibitor, induces

numerous chemical shift changes in the pepsin spectrum. The exchange rate

of pepstatin is slow, with a lifetime, τ_{ex}, greater than 0.1 sec. Pepstatin

binding stoichiometry is 1:1. Several analogs of pepstatin also produce sig-

nificant spectral changes. Those analogs containing statine (4-amino-3-hydroxy-

6-methyl heptanoic acid) with stereochemistry (3S,4S) give rise to a protein-

analog spectrum which resembles closely the one induced by pepstatin. When

the stereochemistry is the non-natural, (3R,4S), the spectrum is affected much

less and in a different way, confirming the key role of the hydroxyl group of

statine and suggesting that the diastereomeric peptides bind with different

orientations at the active site.

INTRODUCTION

Pepsin and the other acid proteases have finally "come of age"[1] since now

sequences[2,3] and an atomic resolution crystal structure[4] are available. Rapid

advances in understanding of the enzyme should follow since spectroscopic and

other methods applied to solution structure and mechanism studies turn out to

be far more compelling when interpreted in light of known primary and tertiary

structure. Nuclear magnetic resonance[1] data, in particular, benefit from ref-

erence to a crystal structure. Now that an acid protease with high homology to

pepsin (penicillopepsin from the mold *Penicillium janthinellum* has a known

structure,[4] NMR is an attractive method for studies of the enzyme.

Our work has focused initially on pig pepsin, since it and its zymogen are readily available and most previous work has been done on the porcine molecules. Pepsin is not necessarily an ideal candidate for NMR, however. The enzyme has a relatively high molecular weight (34,000 d for pepsin and 40,000 d for pepsinogen) and a multidude of overlapping aromatic residues. To date only 2 papers have reported the use of NMR for studies of pepsin. Hunkapiller and Richards used a fluorinated substrate for probing active site residue pKâ's and the catalytic mechanism.[5] Roberts and coworkers published an ^{15}N spectrum of alkaline denatured pepsin.[6] No attempt was made to interpret the result.

We report here a proton NMR investigation of pepsin and pepsinogen. The zymogen contains 3 histidines, 2 of which are part of an activation peptide (the amino-terminal 44 amino acids) cleaved off during conversion of pepsinogen to pepsin in acid. The pH dependence of pepsinogen His proton peaks is rather unique and suggests a structural role for these residues. When the zymogen is activated there are significant conformational changes. These changes show up clearly in NMR difference spectra where release of the activation peptide(s) can be followed.

Pepsin's catalytic mechanism centers around two aspartate side chain carboxyl residues, but there is not yet agreement on their roles.[7] Specificity of the enzyme is not precise but hydrophobic residues on either side of the scissile bond are highly favored.[8] The peptide substrate binding site is extended, and the catalytic rate constant (k_{cat}) may depend rather strongly on the nature of groups bound to subsites.[7]

Amidst a somewhat nebulous picture of substrate—enzyme interactions one exception looms large. A peptide called pepstatin was discovered by Umezawa and colleagues in cultures of Actinomycetes and was found to be an incredibly potent inhibitor of pepsin.[9] The molecule is Isovaleryl-Val-Val-Sta-Ala-Sta[10] where Sta is statine, 4-amino-3-hydroxy-6-methyl heptanoic acid.[11] Recent binding studies using a radiolabeled pepstatin analog[12] gave a K_D of 4.5×10^{-11}! We have used pepstatin for NMR binding experiments with pepsin. Extensive conformational changes are detected in difference spectra, and work with several analogs of pepstatin, including optical isomers, has helped define key interactions.

MATERIALS AND METHODS

Enzyme. Pig pepsinogen was of the highest specific activity commercially available from both Worthington Biochemical Corp. and Sigma Chemical Co. Specific activities and NMR spectra of these preparations were essentially the same except for the presence of a peak apparently due to Tris in the Worthington sample.

Peptide NH protons in pepsinogen were exchanged for ^{2}H by incubation of the enzyme at pH 9.3 and 40° for 2 hr in 50 mM borate buffer. The mixture was then dialyzed against D_2O. Pepsinogen undergoes a reversible, partial denaturation at alkaline pH,[13] a situation highly favorable to exchange of interior peptide amide protons in globular proteins.[14] Acid activation and assay of pepsinogen were done by the acid-denatured hemoglobin hydrolysis protocol.[15]

Porcine pepsin was obtained from Worthington for some experiments. However, most work reported here involved pepsin made from activation of pepsinogen. In preparing pepsin from pepsinogen, the protocol of Rajogopalan et al.[16] was followed. Deuterium-exchanged pepsinogen was incubated at pH 2 and the result-ing mixture applied to a column of Sephadex C-25. Pepsin was eluted at 4° with an acetate buffer, pH 4.4. The pooled pepsin fractions were dialyzed against H_2O in the cold and then lyophilized. NMR spectra of pepsin obtained from pepsinogen showed almost no detectable peptide fragments, in contrast to com-mercial pepsin which contains autolysis products and which usually has fairly prominent, sharp peaks for small peptides. The pepsinogen generated pepsin also had the highest possible specific activity as measured by the hemoglobin assay.[15]

Peptides. Pepstatin a was from Sigma Chemical Co. It appeared > 95% pure based on NMR spectra in CH_3OH-d_4 and $(CH_3)_2CO-d_6$. Several protected peptides containing statine were synthesized by Professor Dan Rich of the University of Wisconsin,[17] and kindly supplied to us.

NMR measurements. Proton NMR spectra were run at 270 MHz with a Bruker Instruments magnet and probe (5 mm), Nicolet 1180 data system and home built, quadrature rf transmitter-receiver system. 90° pulses were generally used with wait times between scans sufficient to allow at least 90% recovery of all pro-tein peaks. Difference spectra were produced through use of the standard Nico-let software. Appropriate corrections were made for concentration changes due to addition of reagents.

Chemical shifts were referenced to the omnipresent HDO peak and are expres-sed relative to DSS (2,2-dimethyl-2-silapentane-5-sulfonate). A separate cali-bration established the HDO-DSS chemical shift as a function of temperature. Most spectra reported here were run at the ambient probe temperature of 28° where HDO is at 4.77 ppm from DSS.

Protein solutions for NMR were generally 0.5 mM or 1 mM. Higher concentra-tions of pepsin showed evidence of aggregation particularly at pH* values of 4 or less. Much lower concentrations required extended time-averaging and since pepsin is a proteolytic enzyme there was the problem of autolysis. Ample signal-to-noise ratios were obtained in 10 min or less of time averaging, a short enough period to avoid detectable proteolysis.

Buffers. Several buffers were needed to cover the range pH 3 to 8 for pepsin and pepsinogen. We sought ones without carbon-bound protons for benefit of the NMR. Oxalate served nicely from pH 2 to 4.5, pyrophosphate and phosphate from 4.5 to 7.5; borate was used near pH 9. In several cases no buffer was used; there was no detectable difference in NMR spectra for either pepsin or pepsinogen in comparing a sample at the same pH with or without buffer. Addition of 2 mM EDTA did not change the protein spectra suggesting that paramagnetic metal ions are not a problem.

pH values in D_2O solutions were measured on a Radiometer model 26 using a thin Ingold combination electrode standardized in H_2O buffers. Reported values are uncorrected for the deuterum glass electrode isotope effect and are denoted pH*.

RESULTS AND DISCUSSION

Pepsinogen amide proton exchange. When pig pepsinogen is dissolved in D_2O it yields an NMR spectrum still containing a great number of peptide NH resonances. These exchange only very slowly with ^2H and are still present weeks later for solutions near neutral pH. Undoubtedly pepsinogen, like other protease zymogens and active enzymes including pepsin[4] has a tightly organized secondary structure that effectively screens interior groups from rapid solvent exchange.

Figure 1 compares pepsinogen before and after the amide protons were deliberately solvent exchanged in D_2O by bringing the pH* to 9.3 and incubating at 40° for 2 hrs. The solution was returned to neutrality, dialyzed against D_2O, and lyophilized. Potential pepsin activity of pepsinogen was unaltered by the exchange process, and the NMR spectrum, except for loss of peptide resonances was virtually identical before and after. Exchange of amide protons allows unambiguous observation of the histidine ring protons of pepsinogen as can be seen in the inset (Fig. 1b).

Dynamics of basic amino acid side chains. A striking characteristic of the pepsinogen amino acid sequence[17,18] is the presence of almost all its basic groups (13 out of 17) in the amino terminal 44 residues. This piece, called the activation peptide, is eventually lost as pepsinogen undergoes conversion to pepsin. An attractive model for the pepsinogen structure has at least some of the Arg, Lys and His side chains interacting with carboxylates of the main pepsin chain.

The pepsinogen spectrum contains a relatively narrow resonance at 3.0 ppm corresponding to the ε-CH_2 of Lys. This peak appears in all of the various pepsinogen preparations we have used and does not decrease in amplitude after

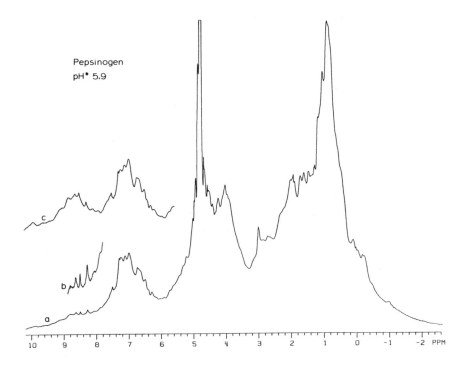

Fig. 1. (a) 0.5 mM porcine pepsinogen in ^2H_2O, pH* 5.9. The sample had been pre-exchanged with ^2H_2O to eliminate most peptide NH resonances. 512 scans; 3 sec pulse repetition time, 16 K points in the time domain, 2 Hz broadening on the free induction decay. (b) High vertical gain readout to show His C-ε_1 protons. (c) Same as (a) before exchange of amide NH protons.

dialysis. With an area of 5-6 protons it represents approximately 3 Lys groups out of a total of 10 in the zymogen. No corresponding narrow peak appears for the δ-CH$_2$ group of Arg which, in small peptides, is found at 3.2 ppm.

The width across the Lys multiplet is 20 Hz, only 5-10% greater than the width of the 3 lysine ε-CH$_2$ peaks in the isolated dodecapeptide from pepsinogen's amino terminus. Although its multiplicity obscures the actual line-width, such a narrow peak still implies a high degree of mobility for the side chains of 3 Lys residues in pepsinogen. If the -CH$_2$ group were reorienting slowly relative to the overall protein its linewidth would be at least 75 Hz, based on an expected rotational correlation time for pepsinogen of $3 \pm 1 \times 10^{-8}$ sec.[19] Spin echo measurements by the Carr-Purcell method confirm the long T_2 value implied by a narrow line and also show that all other peaks in the region around 3 ppm relax much faster (data to be published). We conclude that all of the Arg side chains and 7 of 10 Lys have significant interaction with other groups on pepsinogen.

It is interesting to note that Rimon and Perlmann[20] found that 3 Lys residues could be carbamylated with KCNO without loss of potential pepsin activity and with no change in ORD parameters of pepsinogen. Further reaction of Lys led to loss of activity and spectral changes. We do not know whether or not 3 *specific* residues were modified in the carbamylation reaction, or if they are the ones in the spectrum, but the parallel with the NMR data is striking.

Histidine pH titration. After the peptide NH's are exchanged for ^{2}H three relatively narrow peaks appear between 8 and 9 ppm in the pepsinogen spectrum (Fig. 1b). We assign these to the C-ε_1 protons of the 3 His residues, two of which are in the activation peptide at positions 29 and 31 and one in the pepsin chain at pepsinogen position 97 (pepsin 53). We have not made complete individual assignments of the His peaks but the farthest upfield resonance is one of the activation peptide groups, a conclusion based on its behavior during pepsinogen conversion to pepsin.

Over the pH range 4.5 to 8 these peaks change only a few hundredths of a ppm. They finally titrate between pH 8 and 10 (Fig. 2). Above pH 8.2 the peaks broaden and become difficult to follow. It is well known that pepsinogen undergoes reversible conformational changes at pH values above 8.5.[21,13] At least 2 transitions can be distinguished fluorometrically[22] one of which occurs with a midpoint near pH 9.4 and another close to 10. It appears from the NMR data that the histidines are titrating approximately with the first transition. With such a high pH required before any of the 3 His rings begin to lose a proton it seems probable that each group is involved in some sort of stabilizing structure. We suggest that histidines 29 and 31 provide 2 of the anchors of the activation peptide-pepsin interaction.

The His group common to both pepsinogen and pepsin must also be involved in structure in pepsinogen. When we titrated *pepsin* over the pH range 4 to 7.5 we found that the chemical shift of the single C-ε_1 His proton was 8.4 ± 0.01 ppm from pH 4 to 6.5. Above pH 6.5 pepsin denatures and the peak moves upfield. Lack of any shift change from pH 4 to 6.5 is consistent with the environment of the homologous His ring in the penicillopepsin crystal structure[4] where it is apparently hydrogen bonded to 2 backbone peptide groups.

Pepsinogen Activation. Creation of pepsin from pepsinogen superficially involves clipping off 44 amino acids from the zymogen amino terminus when the pH is lowered. But it is clear from numerous studies that the process does not merely involve hydrolysis of the 44-45 peptide bond. U.V. difference spectra gives evidence for some sort of conformational change (or perhaps solvent

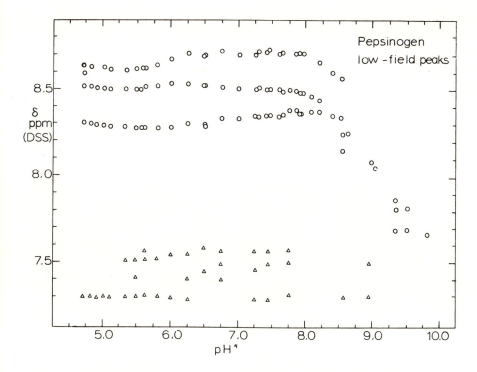

Fig. 2. pH dependence of lowfield carbon-bound proton peaks in pepsinogen. Below pH 4.7 the sample rapidly activates to give pepsin activity. Above pH 9.5 the molecule denatures.

exposure) involving tryptophan when the pH of a pepsinogen solution is lowered.[23,24] This structural rearrangement is at least partially reversible if the solution is returned to neutrality. Marciniszyn et al. found evidence for several intermediate structures in pepsinogen activation, including one where the active site is accessible but the activation peptide is still attached and another where the peptide lies in the site prior to being hydrolyzed.[25]

We have approached the question of structure changes in pepsinogen activation by monitoring the proton NMR spectrum after the pH was dropped to 4 to initiate the reaction. The results are presented as *difference spectra* in Fig. 3. The bottom spectrum is the difference between pepsinogen 2 hrs after it was activated at pH 4 and before it was activated (pH 5.9). The 2 hr spectrum was very similar to one taken immediately after activation. The first difference spectrum shows fairly extensive changes in chemical shifts, as evidenced by "dispersion-like" peaks in the aromatic region and near 0-1 ppm in the methyl region. These shifts suggest that conformational rearrangements have taken

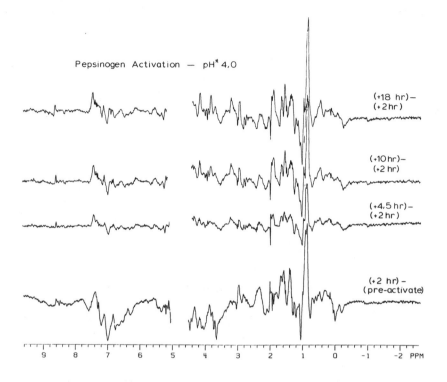

Fig. 3. NMR spectral changes in pepsinogen associated with activation to pH 4. The lower plot is a difference spectrum where a pepsinogen spectrum at pH 5.9 was subtracted from the spectrum obtained 2 hrs after the pH was lowered to 4.0. The other difference spectra were produced by subtracting the 2 hr spectrum from those obtained at later times as indicated. Each spectrum required 20 min of time averaging.

place within the time required for a spectrum to be acquired ($\sim$ 10 min), leading to displacement of groups relative to neighboring anisotropic centers such as aromatic rings. Some relatively narrow resonances appear as positive peaks in the difference spectrum which seem not to have counterparts in the negative direction. An example is a peak at 3 ppm which we assign to Lys ε-CH_2. These narrow, positive peaks might result from new, independent motion of parts of the protein after activation. One prospect would be release of the activation peptide, residues 1-44. But not nearly enough peaks are found in the first difference spectrum. Furthermore, several resonances are unique markers for the activation peptide, including His 29 and 31 and Tyr 37. The His region is somewhat altered initially, but a peak remains at 8.27 ppm, just as in unactivated pepsinogen, which is assigned to either His 29 or 31. No discernible peaks for mobile Tyr are found.

Over time there are further changes and these are seen in the other difference spectra of Fig. 3 where the 2 hr spectrum has been subtracted from scans taken at later times. By subtracting the early post-activation spectrum the major conformational changes do not obscure more subtle events which take place later. As the activation mixture incubates, the His peak at 8.27 ppm and increased intensity is found at 8.6 ppm, the chemical shift expected for His $C-\varepsilon_1$ protons of the uncomplexed amino acid. A doublet grows in near 6.9 ppm which we assign to one of the Tyr peaks. The peak at -1 ppm that serves as a trademark for pepsinogen also disappears over time (but not in the first activation events). The Lys $\varepsilon-CH_2$ continues to grow with time (not surprisingly since there are 9 Lys residues in the activation peptide). Our interpretation of these data involves a modest change in protein conformation as an immediate consequence of lowering the pH to 4. Release of the activation peptide proceeds much more slowly and, in particular, the residues in the region 29-37 stay with the protein for hours. Peptide 1-16 is thought to be the first one released.[26] It contains 3 Lys, 2 Arg, 4 Leu and 3 Val among others. The NMR does *not* provide evidence for a *free* peptide of this composition immediately after activation, but the peptide is a good inhibitor of pepsin ($K_i = 0.25 \times 10^{-6}$ M)[27] so it should be bound anyway.

Pepstatin binding to pepsin. A potent inhibitor of pepsin was discovered by Umezawa and coworkers[9] in cultures of Actinomycetes. This compound, named pepstatin, turned out to be isovaleryl-Val-Val-Sta-Ala-Sta where Sta is statine (4-amino-3-hydroxy-6-methyl heptanoic acid). The extremely hydrophobic nature of the pepstatin side chains along with the unique structure of statine undoubtedly contribute to its strong interaction with pepsin. A recent binding study using an ^{125}I labelled pepstatin derivative gave a dissociation constant of 4.5×10^{-11} M.[12]

Rich has explored pepstatin binding by synethesizing a number of analogs and testing their inhibition of pepsin catalysis.[28] Data for some of the analogs are collected in Table 1. The carboxy-terminal statine is far less important than Sta[4] for strong binding. Furthermore the hydroxyl of statine is a key functional group; when it is absent, the inhibition is 18-fold weaker.[17] When the stereochemistry is changed from (3S,4S) to (3R,4S) for the hydroxy and amino carbons the binding is 2 orders of magnitude weaker. In collaboration with Dr. Rich we have begun to explore NMR binding studies of pepsin with pepstatin and analogs in order to define more precisely the inhibitor-protein interactions.

A spectrum of pepsin in D_2O is shown in Fig. 4. The single histidine $C-\varepsilon_1$ proton is visible at 8.5 ppm, especially in pepsin produced from pre-^2H_2O exchanged pepsinogen whereby the large amide proton background is greatly

TABLE 1

PEPSTATIN ANALOG INHIBITION CONSTANTS[*]

	Inhibitor	K_I
Pepstatin	Ival-Val-Val-Sta-Ala-Sta	10^{-10} M
	Boc-Val (3S,4S) Sta-Ala-NH-Iamyl	10^{-9}
	Boc-Val (3R,4S) Sta-Ala-NH-Iamyl	2×10^{-6}
	Boc-Sta (3S,4S) Ala-NH-Iamyl	2×10^{-9}
	Boc-Sta (3R,4S) Ala-NH-Iamyl	2×10^{-6}

[*] From reference 27 and D. Rich, private communication.

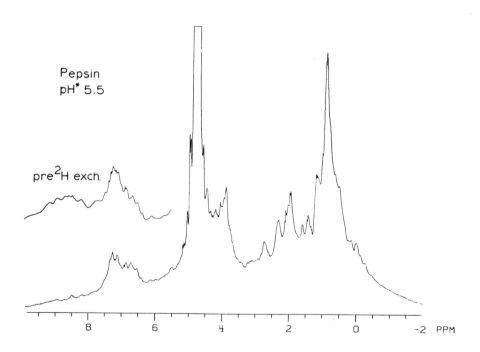

Fig. 4. Pepsin spectrum. The inset shows the aromatic region of a sample without ^{2}H exchange. The main spectrum is of a sample generated from pepsinogen that had been pre-exchanged with ^2H_2O. 0.5 mM pepsin in ^2H_2O, 28°; 256 scans; 3 sec pulse period; 2 Hz line broadening.

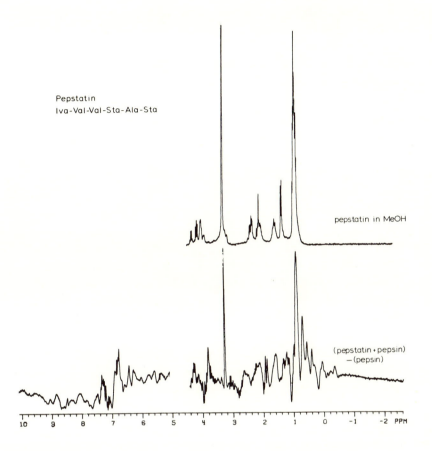

Fig. 5. Pepstatin effects on pepsin spectrum. The top spectrum is pepstatin in
deuterated methanol. The bottom plot is a difference spectrum between 0.5 mM
pepsin complexed with 0.5 mM pepstatin and uncomplexed pepsin. Each spectrum
required 256 scans.

suppressed. No peaks for Lys or Arg are resolved, suggesting their involvement
in interactions with other protein residues.

When pepstatin (10 mM in MeOH-d$_4$) is added in aliquots to a 0.5 mM pepsin
solution at pH* 4.5 the inhibitor binds and produces a marked change in the NMR
spectrum. Figure 5 shows the *difference spectrum* between 0.5 mM pepsin plus
0.5 mM pepstatin and pepsin alone. There are changes in chemical shifts of
numerous pepsin methyl (or possibly methylene) groups as shown by the dispersion
peaks between 0 and 1 ppm. Near 2 ppm it appears that one or possibly 2 sing-
lets are shifted by pepstatin binding. We tentatively assign those to methionine methyls. The peak at 3.3 ppm is the residual signal from MeOH-d$_4$. The

aromatic region shows evidence of a great number of aromatic protons having been displaced relative to other aromatic rings. Since pepstatin has no aromatic groups of its own the changes in the region 5 to 9 ppm probably arise solely from changes in the pepsin conformation. The histidine $C-\varepsilon_1$ proton at 8.48 ppm in pepsin is shifted only slightly downfield but it shows up clearly in a difference spectrum because of the very narrow linewidth.

Overall it appears that most peaks in the difference spectrum have not been greatly shifted, but that there are a large number of protons contributing. Part of the efficacy of pepstatin binding may lie in the involvement of a sizeable fraction of the protein in small, energetically favorable, conformational accommodations. The small shift of histidine probably reflects the fact that this group is about 20 Å from the carboxyl groups at the active site and experiences only indirect effects of pepstatin binding.

The difference spectra generated by less than 1:1 ratios of pepstatin:pepsin are simply lower amplitude versions of the 1:1 complex. No peaks are exchange broadened. This implies that the peptide is in slow exchange on the NMR time scale; a lower limit of $\tau_{ex} \gg 0.1$ sec can be set for the exchange lifetime, based on the His $C-\varepsilon_1$ proton shift of only 8 Hz. It is not surprising that pepstatin exchanges slowly with pepsin considering its extremely tight binding constant. As aliquots of pepstatin are added to pepsin, the difference spectrum increases in amplitude up to a 1:1 molar ratio. Further addition of pepstatin results in precipitation of the very hydrophobic peptide and no further change in the difference spectrum, implying a 1:1 stoichiometry.

Pepstatin analogs. The synthetic, protected peptide BOC-Val-Sta-Ala-isoamyl amide closely resembles pepstatin in that it has 5 of 6 hydrophobic side chains and the sequence Val-Sta-Ala. When this peptide, containing the (3S,4S) statine isomer, is added to a pepsin solution the resulting difference spectrum (Fig. 6) is remarkably similar to that produced by pepstatin. Particularly in the aromatic region from 5.2 to 7.5 ppm the difference spectra correspond closely. (Non-D_2O exchanged pepsin was used for these spectra so lost N-H resonances appear in the region 7 to 10 ppm.) In the methyl region the tert-butyl peak is shifted downfield from 1.50 to 1.65 ppm, possibly due to an average position of the 9 methyl protons edge-on to a ring. Other protons of the peptide can not be picked out so readily due to a strong interaction with the active site and corresponding broad lines for most peaks.

When the same peptide, but this time containing the (3R,4S) Sta stereoisomer, is added to a pepsin a very different picture emerges. The bottom spectrum of Fig. 6 shows the result. It is clear that the changes in structure accompanying (3R,4S) binding are different and much less extensive than with (3S,4S). Note

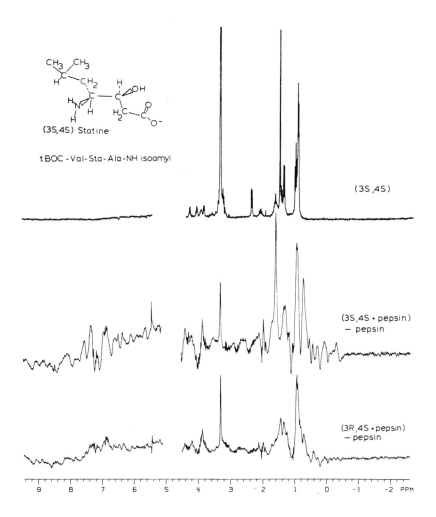

Fig. 6. Effects of BOC-Val-Sta-Ala-isoamyl amide on pepsin spectrum. The inset shows the structure of statine with the optical configuration of (3S,4S). The top spectrum is BOC-Val-Sta-Ala-isoamyl amide with (3S,4S) statine. The middle plot is the difference spectrum for that compound plus pepsin minus pepsin. At the bottom is the difference spectrum for the peptide with statine in the (3R, 4S) configuration.

that now the tert-butyl peak has shifted *upfield* slightly on binding and is

noticeably broadened. (The spectrum of free (3R,4S) is not shown for the sake

of space-saving; it is similar to that of 3S,4S) shown at the top of Fig. 6).

These difference spectra confirm that the statine hydroxyl group plays a key

role in binding of pepstatin and its analogs,[28] as implied by the weaker K_i for

(3R,4S). In particular the NMR shows that the peptide is forced to reorient

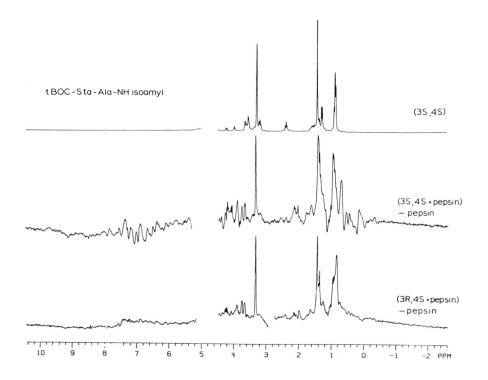

Fig. 7. BOC-Sta-Ala-isoamyl amide effect on pepsin NMR spectrum. Experimental
details as in Fig. 6.

when the hydroxyl points in the wrong direction. We submit that the difference
in binding constant is not merely due to loss of a hydrogen bond (for example)
but involves a different interaction of the rest of the peptide with its site.

The peptide BOC-Sta-Ala-isoamyl amide lacks the Val residue of the previous
analog. This time the difference spectrum (Fig. 7) of the (3S,4S) isomer does
not so closely resemble that produced by pepstatin. But significant changes
are found throughout the aromatic region and in the 0-1 ppm methyl region. The
BOC methyls are not shifted, in contrast to BOC-Val-Sta-Ala-isoamyl amide.
Still, the difference spectrum generated by the (3S,4S) steroisomer is at least
an order of magnitude greater than that with (3R,4S).

NMR difference spectra produced by pepstatin analogs can be compared to the
kinetic inhibition constants measured for them (Table 1). By and large the
magnitude of the difference spectrum correlates with the value of K_I. However
pepstatin, with an order of magnitude stronger binding constant gives essentially
the same difference spectrum as BOC-Val-Sta-Ala-isoamyl amide. The extra

hydrophobic side chain on pepstatin may contribute to binding free energy but be uninvolved in the key interactions leading to a protein conformational re-arrangement.

ACKNOWLEDGMENTS

I thank Professor Daniel Rich for generously supplying the pepstatin analogs and Drs. Jordan Tang and Jean Hartsuck for many helpful discussions. Jeffrey Savidge provided expert technical assistance. This work was supported by Research Grant GM 25703 from the National Institutes of Health.

REFERENCES

1. Tang, J.   (1977) Nature, 266, 119-120
2. Sepulveda, P., et al.   (1975) J. Biol. Chem., 250, 5082-5088.
3. Tang, J., et al.   (1973) Proc. Natl. Acad. Sci. U.S.A., 70, 3437-3439.
4. Hsu, I.-N, et al.   (1977) Nature, 266, 140-145.
5. Hunkapiller, M.W. and Richards, J.H. (1972) Biochemistry, 11, 2829-2839.
6. Gust, D., Moon, R.B. and Roberts, J.D. (1975) Proc. Natl. Acad. Sci. USA, 72, 4694-4700.
7. Fruton, J.S.   (1977) in Acid Proteases. Structure, Function and Biology, Tang, J., ed., Plenum Press, New York, pp. 131-140.
8. Trout, G.E. and Fruton, J.S. (1969) Biochemistry, 8, 4183-4190.
9. Umezawa, H., Aoyagi, T., Morishima, H., et al.   (1970) J. Antibiotics, 23, 259-260.
10. Morishima, H., Takita, T., et al.   (1970) J. Antibiotics, 23, 263-265.
11. Rich, D.H., et al.   (1978) J. Org. Chem., 43, 3624-3636.
12. Wortman, R.J. and Burkitt, D.W. (1979) Arch. Biochim. Biophys., 194, 157-164.
13. Perlmann, G.E. (1963) J. Mol. Biol., 6, 452-464.
14. Markley, J.L. and Porubcan, M.A. (1976) J. Mol. Biol., 102, 487-509.
15. Anson, M.L. and Mirsky, A.E. (1932) J. Gen. Physiol. 16, 59-63.
16. Rajagopalan, T.G., et al.   (1966) J. Biol. Chem., 241, 4940-4950.
17. Rich, D.H., et al.   (1977) Biochem. Biophys. Res. Commun., 74, 762-767.
18. Ong, E.B., and Perlmann, G.E. (1968) J. Biol. Chem., 243, 6104-6109.
19. Fratalli, V., Steiner, R.F. and Edelhoch, H. (1965) J. Biol. Chem., 240, 112-121.
20. Rimon, S. and Perlmann, G.E. (1968) J. Biol. Chem., 243, 3566-3572.
21. McPhie, P. (1975) Biochemistry, 14, 5253-5256.
22. Steiner, R.F., et al.   (1965) J. Biol. Chem., 240, 122-127.
23. McPhie, P. (1972) J. Biol. Chem., 247, 4277-4281.
24. Al-Janabi, J., Hartsuck, J.A. and Tang, J. (1972) J. Biol. Chem., 247, 4628-4632.
25. Marciniszyn, J., Jr., et al.   (1976) J. Biol. Chem., 251, 7095-7102.
26. Dunn, B.M., et al.   (1978) J. Biol. Chem., 253, 7269-7275.

Published 1980 by Elsevier North Holland, Inc.
Liu/Mamiya/Yasunobu, eds. Frontiers in Protein Chemistry

[13]C NUCLEAR MAGNETIC RESONANCE STUDIES OF HEMOGLOBIN S AGGREGATION

J.W.H. SUTHERLAND[*+], A.N. SCHECHTER[*], D.A. TORCHIA[#], AND W. EGAN[**]
*National Institute of Arthritis, Metabolism, and Digestive Diseases, #National Institute of Dental Research, National Institutes of Health, Bethesda, Maryland 20205 and **Bureau of Biologics, Food and Drug Administration, Bethesda, Maryland 20205

SUMMARY

Scalar and dipolar decoupled ^{13}C nuclear magnetic resonance techniques have been used to estimate the fraction of aggregated deoxy hemoglobin S in both cell-free solutions and in intact erythrocytes. At 37°C, approximately 30% of the hemoglobin in a 28 $gm \cdot dL^{-1}$ solution and 60% of a 36 $gm \cdot dL^{-1}$ solution is aggregated; in whole, deoxygenated sickle cells, approximately 75% of the hemoglobin is aggregated. The rigid nature of the backbone of these aggregated molecules is evidenced by residual chemical shift anisotropy broadening of carbonyl and aromatic carbon atoms and the efficiency with which the α carbons of the backbone are cross-polarized. The spectrum of the nonaggregated hemoglobin is virtually unaffected by the presence of polymer, suggesting that the system behaves essentially as a two-phase system. The dependence of the extent of aggregation on temperature for the 28 $gm \cdot dL^{-1}$ solution and for the intact erythrocytes was investigated; at 4°C, no aggregate was evidenced for the 28 $gm \cdot dL^{-1}$ solution whereas no reversal of aggregation was evidenced for the erythrocyte, even after 24 hours at 4°C.

INTRODUCTION

The pathophysiology of sickle cell disease is associated with the intracellular aggregation of the individual hemoglobin S (HbS) molecules that occurs on deoxygenation.[1] It is presumably this aggregation that causes the normally biconcave, discoid erythrocyte to assume the characteristic sickle (or holly) shape from whence the disease derives its name. At a specified temperature and above a certain critical concentration, HbS aggregation occurs upon deoxygenation.[2] The molecular basis for the decreased solubility of HbS relative to adult hemoglobin HbA is the substitution of valine for glutamic acid at the 6 position of the β chains of the Hb tetramer.[3]

While considerable progress has been made toward an understanding of the physical chemistry of HbS aggregation in cell-free solutions, relatively little

---

[+]Present address: Kodak Park Research Laboratories, Rochester, New York 14550

progress has been made in this regard with the intact erythrocytes. This talk describes a ^{13}C nuclear magnetic resonance (NMR) technique for quantifying the fraction of aggregated HbS molecules in both cell-free systems and intact erythrocytes.

The ^{13}C NMR linewidths associated with isotropically mobile ($\tau_r < ca.$ 10^{-6} sec, where τ_r is the rotational correlation time) molecules are generally less than $ca.$ 100 Hz. On going from a mobile liquid phase to a motionally restricted microcrystalline solid phase, a number of changes are observed in the ^{13}C NMR spectrum, the most readily apparent being a line broadening, the severity of which is strongly dependent on the degree of rotational reorientational restriction.[4] The line-broadening derives primarily from two effects--static dipolar interactions and chemical shift anisotropy. (Under conditions of rapid isotropic rotational reorientation, static dipolar and chemical shift anisotropy contributions to the line width are averaged to zero.)

The line broadening of a ^{13}C NMR spectrum due to static dipolar interactions with protons can be removed by high power (also termed "dipolar") decoupling.[5] Just as scalar ^{13}C--^{1}H couplings (on the order of 200 Hz) are removed by low power irradiation ($\gamma H_2/2\pi \sim$ 3 kHz) at the proton resonance frequency, dipolar ^{13}C--^{1}H couplings (on the order of 40 kHz) are removed by high power ($\gamma H_2/2\pi \sim$ 60 kHz) irradiation at the proton resonance frequency. Scalar couplings are removed by dipolar decoupling. ^{1}H dipolar decoupling does not remove line broadening contributions from chemical shift anisotropy, nor does it remove line broadening contributions from non-^{1}H static dipolar interactions, e.g., ^{14}N--^{13}C dipolar couplings.

Due to the relatively long spin lattice relaxation times (T_1s) of carbon atoms in solid samples, it is often difficult to observe totally unsaturated signal intensities in their ^{13}C spectra obtained by the dipolar decoupling technique. This difficulty may be somewhat circumvented by employing the cross-polarization technique of Hartmann and Hahn,[5,6] in which magnetization is transferred from the abundant ^{1}H to the isotopically dilute ^{13}C spins. As protons have shorter T_1s in general, the efficiency of acquiring a ^{13}C spectrum of a solid is greatly increased. Several difficulties, however, attend the cross-polarization technique; namely, the degree of magnetization transfer can vary for different carbon atoms in the sample. This makes quantitative measurements difficult without first "calibrating" the molecular system. This difficulty is, however, more than offset by the ease with which the cross-polarization technique allows obtaining a ^{13}C spectrum of a solid. In contrast to the dipolar decoupling technique which furnishes a spectrum of both mobile and immobile carbon atoms, the cross-polarization technique furnishes a ^{13}C spectrum

only of immobilized carbon atoms and this is especially the case when a temperature inversion pulse sequence is employed.[5,7]

In a sample containing both rotationally free and rotationally immobilized
molecules, three distinct experiments are envisaged: a scalar decoupled experiment, a dipolar decoupled experiment, and a cross-polarization experiment. The
first allows selective observation of rotationally free molecules, the second
allows observation of both free and immobilized molecules, and the third allows
selective observation of immobilized molecules.

More than 90% of the carbon atoms of an erythrocyte derive from hemoglobin
molecules.[8] This facilitates ^{13}C NMR experiments with erythrocytes in that
the subtraction of nonhemoglobin deriving carbon signals is not required. We
will now report the results of the double resonance experiments with hemoglobin
in solution and in erythrocytes.

RESULTS AND DISCUSSION

The ^{1}H scalar decoupled ^{13}C NMR spectrum of oxy and deoxy HbA (as a 28 gm.
gm·dL^{-1} solution in water at 37°C) is shown in Fig. 1. Three signal areas are
readily distinguished: carbonyl carbons at 15-25 ppm, aromatic carbons at 60-75
ppm, and aliphatic carbons at 120-200 ppm. The aliphatic region may be further
divided into the backbone α carbon region between 120 and 150 ppm and the remaining side-chain aliphatic region between 150 and 200 ppm. The dipolar
decoupled spectrum of deoxy HbA (Fig. 1) was indistinguishable from the scalar
decoupled spectrum; this is most easily visualized by the lack of observable
resonances in the difference spectrum (DD-SD) shown in the figure. It was not
possible to obtain a cross-polarization spectrum with deoxy HbA (see Fig. 1).
Dipolar and scalar decoupled spectra of oxy HbA were found to be identical and,
here again, no cross-polarization spectrum was obtainable.

The ^{1}H scalar decoupled ^{13}C NMR spectrum of a 28 gm·dL^{-1} solution of oxy HbS
at 37°C is shown in Fig. 2. It is indistinguishable from the spectrum obtained
with oxy HbA. The scalar decoupled spectrum of HbS alters, however, following
deoxygenation (Fig. 2). The spectral change is most apparent in the aliphatic
region, being observable as a decrease in integrated absorption intensity. This
decrease is clearly manifested in the difference spectrum (DD-SD). A cross-
polarization spectrum was obtained with deoxy HbS, consistent with the presence
of immobilized HbS molecules and a change in absorption intensity between scalar
decoupled and dipolar decoupled spectra. Similar, although markedly more pronounced results were obtained with HbS erythrocytes; see Fig. 3. The spectrum
of HbA erythrocytes did not change on going from the oxy to the deoxy state and
was identical to the spectrum obtained with oxy HbS erythrocytes.

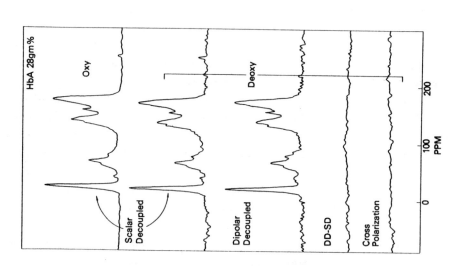

Fig. 1. Comparison of spectra of a 28g/dl preparation of hemoglobin A at 37°C; (a) oxygenated sample, 90°- t pulse sequence, t = 1s, scalar decoupled, 32768 transients accumulated; (b) deoxygenated sample, 90°- t pulse sequence, t = 1s, scalar decoupled, 4096 transients accumulated; (c) deoxygenated sample, 90°- t pulse sequence, t = 1s, dipolar decoupled, 2048 transients accumulated; (d) spectrum (c) minus spectrum (b); (e) deoxygenated sample, proton enhanced, 1 ms Hartmann-Hahn matched contact, 2s repetition time, dipolar decoupled, 2048 transients accumulated. Spectra have been normalized to compensate for differences in numbers of accumulations. Digital line broadenings of 20 Hz, in (a), (b) and (c), and 50 Hz in (d) and (e), were employed to enhance sensitivity. The chemical shift scale is relative to external CS_2.

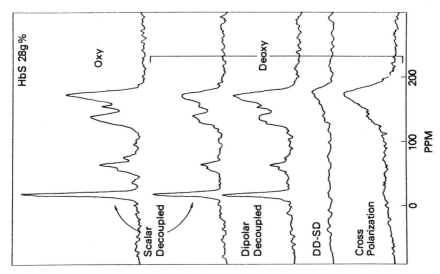

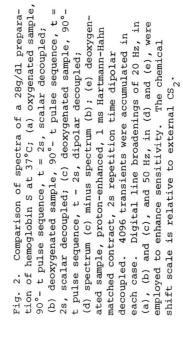

Fig. 2. Comparison of spectra of a 28g/dl preparation of hemoglobin S at 37°C; (a) oxygenated sample,
90°- t pulse sequence, t = 2s, scalar decoupled;
(b) deoxygenated sample, 90°- t pulse sequence, t =
2s, scalar decoupled; (c) deoxygenated sample, 90°-
t pulse sequence, t - 2s, dipolar decoupled;
(d) spectrum (c) minus spectrum (b); (e) deoxygenated sample, proton-enhanced, 1 ms Hartmann-Hahn
matched contract, 2s repetition time dipolar
decoupled. 4096 transients were accumulated in
each case. Digital line broadenings of 20 Hz, in
(a), (b) and (c), and 50 Hz, in (d) and (e), were
employed to enhance sensitivity. The chemical
shift scale is relative to external CS_2.

84

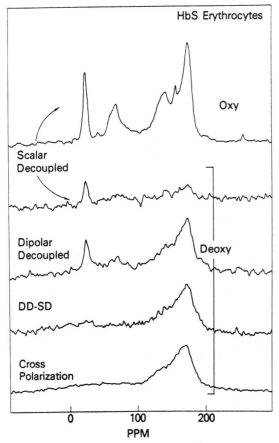

Fig. 3. Comparison of spectra of sickle erythrocytes at 37°C; (a) oxygenated sample, 90°- t pulse sequence, t = 1s, scalar decoupled, 8192 transients accumulated; (b) deoxygenated sample, 90°- t pulse sequence, t = 10s, scalar decoupled, 2048 transients accumulated; (c) deoxygenated sample, 90°- t pulse sequence, t = 2s, dipolar decoupled, 2048 transients accumulated; (d) spectrum (c) minus spectrum (b); (e) deoxygenated sample, proton-enhanced, 1 ms Hartmann-Hahn matched contact, 2s repetition time, dipolar decoupled, 8192 transients accumulated. Spectra have been normalized to compensate for differences in numbers of accumulations. Digital line broadenings of 20 Hz, in (a), (b), and (c), and 50 Hz, in (d) and (e), were employed to enhance sensitivity. The chemical scale is relative to external CS_2.

The mass fraction, f, of immobilized deoxy HbS can be determined by two ^{13}C NMR methods: subtraction of the scalar decoupled deoxy HbS spectrum from the scalar decoupled oxy HbS spectrum (Method A) or subtraction of the scalar decoupled deoxy HbS spectrum from the dipolar decoupled HbS spectrum (Method B). The application of either method requires the meeting of certain conditions. With Method A, it is necessary to know (i) the amount of immobilized hemoglobin present in the oxy form and (ii) the effect of the paramagnetism associated with the deoxy state of hemoglobin on its carbon signal intensity (the oxy form of hemoglobin in diamagnetic). For Method B, it is additionally necessary to know the effect of paramagnetism on the ^{13}C spectrum of immobilized deoxy HbS. Additionally, both methods require that unsaturated signals be observed or, alternatively, that the degree of saturation be known.

The intensities of the signals in the aliphatic region of the spectrum of scalar decoupled oxy and deoxy HbS were unaltered when the interval between 90° pulses was lengthened from 2 to 10 seconds; hence the aliphatic carbon magnetizations had recovered to their equilibrium value within two seconds. Spin lattice relaxation times for the aliphatic carbon atoms in solid deoxy HbS were determined by an inversion recovery technique, recently described by Torchia.[9] From these T_1 measurements, it was estimated that in solid deoxy HbS approximately 85% of the aliphatic carbon magnetizations had returned to equilibrium within the 2-second period between 90° pulses; there is thus a slight amount of signal saturation associated with these resonances when measured by a dipolar decoupling technique. As established above, full intensity is observed for the mobile fraction.

Since rotationally restricted hemoglobin was not detectable in oxy HbS solutions either by dipolar decoupling or cross-polarization methods (Fig. 2), and the absolute signal intensities of HbA solutions were unaltered on going from the diamagnetic oxy to the paramagnetic deoxy states, it may be concluded that Method A should provide a reliable estimate for the mass fraction of immobilized hemoglobin. With reference to Method B, it was unfortunately not possible to precisely determine the extent of signal loss due to paramagnetic line broadening; however, as the intensities of the aliphatic regions of the scalar decoupled oxy HbS and dipolar decoupled deoxy HbS are nearly identical, this loss should not exceed 10%.

Determinations of $\underline{f}$ by Methods A and B are presented in Table 1. As expected, Method A offers the slightly higher value. In accord with previously published studies,[10] f is larger for a 36 gm·dL^{-1} than for a 28 gm·dL^{-1} solution. Interestingly, although the ratio of f values determined for the two solutions by ultracentrifuge[2] and by NMR are the same, the NMR method yields the lower absolute value for f; the cause of this discrepancy is not apparent.

The amount of immobilized hemoglobin in deoxy HbS erythrocytes (f = 0.75) is greater than found for a comparable concentration of cell-free deoxy HbS (f = 0.5). Although such results are possible, the erythrocyte result must be regarded as too preliminary for detailed comment.

The above experiments were carried out at 37°C. On lowering the temperature of the 28 gm·dL^{-1} solution to 4°C, the intensity of the scalar decoupled spectrum increased. Moreover, at 4°C, it was no longer possible to obtain either an enhancement of dipolar decoupling or a cross-polarization spectrum, evidencing that little, if any, aggregated material was present at this temperature. Determining cross-polarization and dipolar decoupled spectra after warming to 37°C, demonstrated the reversibility of the aggregation.

TABLE 1

FRACTION OF AGGREGATED DEOXY HEMOGLOBIN S IN VARIOUS SAMPLES AT 37°C

Sample	Method	Fraction Polymerized[b]
28 g·dL^{-1}	(SD'– SD)[a]	0.40
	(DD – SD)	0.35
37 g·dL^{-1}	(SD'– SD)	0.60
	(DD – SD)	0.45
HbS Erythrocyte	(DD – SD)	0.75

[a]The method represented as (SD'– SD) involves subtracting the scalar decoupled deoxy HbS spectrum from the scalar decoupled oxy HbS spectrum; the method represented as (DD – SD) involves subtracting the scalar decoupled spectrum of deoxy HbS from the dipolar decoupled spectrum of deoxy HbS. Both methods are discussed in the text.

[b]The fraction of polymerized material is rounded off to the nearest 0.05; the error is probably $\pm$ 0.05.

Similar results were not obtained with deoxy HbS erythrocytes. Lowering the temperature to 4°C produced no change in spectral properties relative to those observed at 37°C. Thus, it was possible to obtain spectral enhancements by dipolar decoupling and a cross-polarization spectrum was readily obtainable. Keeping the sample at 4°C for 24 hrs did not alter these spectral characteristics.

Using backbone T_1 measurements for nonaggregated hemoglobin at 37°C, and assuming isotropic reorientation, a rotational correlation time of *ca.* 3 x 10^{-8} sec is obtained. The broad signal observed for aggregated deoxy HbS for the carbonyl and aromatic carbons in the cross-polarization spectrum shows that chemical shift anisotropy and ^{13}C--^{14}N dipolar interactions are not averaged out by molecular motions. The correlation time of the backbone carbons in aggregated hemoglobin must therefore exceed 10^{-4} seconds. This is a change in rotational correlation time that is *at least* four orders of magnitude; very likely it is larger.

The line shape of the nonaggregated hemoglobin was not influenced, at least to any great extent, by the presence of aggregated material, indicating that, at least to a first approximation, the monomer and polymer exist as separate phases, a model previously proposed by Minton.[10]

CONCLUSION

Nuclear magnetic double resonance techniques have been shown to provide accurate, nondestructive estimates of the fraction of aggregated hemoglobin S both in cell-free solutions and in intact erythrocytes. The technique is readily extendable to studies of various factors that affect HbS aggregation, such as O_2 concentration, Hb concentration, temperature, and inhibitors.

REFERENCES

1. Dean, J. and Schechter, A.N. (1978) New Engl. J. Med., 299, 756-763.
2. Ross, P.D., Hofrichter, J., and Eaton, W.A. (1977) J. Mol. Biol., 112, 111-134.
3. Ingram, V.M. (1961) Hemoglobin and Its Abnormalities, C.C. Thomas Publishers, Springfield, Illinois.
4. Slichter, C.P. (1978) Principles of Magnetic Resonance, 2d Ed., Springer Verlag, New York, pp 137-251 and references cited therein.
5. Torchia, D.A. and VanderHart, D.L. (1979) in Topics in Carbon-13 NMR Spectroscopy, Levy, G., ed., Harper and Row, New York, Vol. 3, in press.
6. Hartmann, S.R. and Hahn, E.L. (1962) Phys. Rev., 128, 2042-2053.
7. Stejskal, E.O. and Schaefer, J. (1975) J. Magn. Reson., 18, 560-563.
8. Harris, J.W. and Kellermeyer, R.W. (1970) The Red Blood Cell, Harvard University Press, Cambridge, Massachusetts, pp. 281-283.
9. Torchia, D.A. (1978) J. Magn. Reson., 3013-3016.
10. Minton, A.P. (1974) J. Mol. Biol., 82, 483-498.

Published 1980 by Elsevier North Holland, Inc.
Liu/Mamiya/Yasunobu, eds. Frontiers in Protein Chemistry

HIGH POWER PROTON DECOUPLED ^{13}C NMR STUDY OF MOLECULAR
MOTION IN SPECIFICALLY LABELLED COLLAGEN

LYNN W. JELINSKI AND D.A. TORCHIA
National Institutes of Health, National Institute of Dental Research, Laboratory
of Biochemistry, Bethesda, Maryland 20205

ABSTRACT

Collagen was labeled at specific backbone carbons ([1- and 2-^{13}C]gly) and at
certain sidechain carbons ([6-^{13}C]lys, [5-^{13}C]glu, [methyl-^{13}C]met, and [3-^{13}C]
ala) via chick calvaria culture. Lineshapes, nuclear Overhauser enhancements,
linewidths, and T_1 values, were measured for each collagen sample both as
fibrils and as native (helical) material in solution using high power proton
decoupled (dipolar decoupled) ^{13}C nuclear magnetic resonance (NMR). Analysis of
the relaxation data indicates that in solution, the collagen backbone undergoes
rotational diffusion ($R_1 \sim 10^7$ s^{-1}) about the long axis of the molecule, and
that in the fibrils, the rate of rotational diffusion is similar, although the
range of azimuthal reorientation is restricted to $\geq 30°$. In addition, the NMR
data indicate that the terminal carbons of lysine, glutamic acid, methionine,
and alanine are characterized by reorientation rates of $ca.$ 10^9-10^{10} s^{-1}. Taken
together, the NMR data provide strong evidence that the contact regions between
the helices in collagen fibrils are fluid and that there is not a unique set of
interactions between the amino acid sidechains. In this respect, these NMR
results support current concepts of globular protein structure which suggest
that a variety of conformations, in dynamic equilibrium, are responsible for the
structure and function of proteins.

INTRODUCTION

Collagen, a structural protein rich in proline and hydroxyproline, is com-
prised of Gly-X-Y triplet repeats except in the short nonhelical terminal regions.
Two identical polypeptide chains designated α1(I) and one highly homologous α2 chain comprise the triple helical collagen molecule of molecular weight
~285,000 and dimensions ~15 x 3000 Å.[1-3] In the triple helical structure,
the glycyl residues occupy positions within the core, or backbone, of the
molecule, whereas the amino acid sidechains are directed to the outside of the
molecule and occupy positions on the molecular surface (Fig. 1).[4,5] The molecules associate in $vivo$ and in $vitro$ to form fibrils 300-3000 Å in diameter
and many molecules in length. The association of collagen molecules into

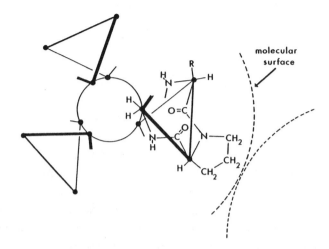

Fig. 1. Projection of the coiled-coil structure of the collagen molecule
derived from X-ray data (adapted from reference 4). The three triangular skele-
tons represent the α chains of collagen, and the solid circles correspond to
the α-carbons of the collagen backbone. The (Gly-Pro-Y) sequence overlaid on
one of the triangles shows that the glycine -CH_2- hydrogens reside within the
core of the molecule and that the nonglycine amino acid sidechains occupy positions on the molecular surface.

fibrils and the resultant highly regular banding pattern observed by electron
microscopy for the fibrils has prompted a number of investigations into collagen
fibrillogenesis and structure.[6,7] These studies have implicated interactions
between amino acid sidechains as being important for fibril formation, structure,
and stability.[8-13]

The purpose of this investigation was to study the mobility and interactions
of various amino acids in collagen, and from this information, to develop a
model for the dynamic structure of the collagen molecule, both in solution and
as fibrils. The technique of ^{13}C NMR of solids is particularly well-suited to
this investigation for a number of reasons. It is possible to produce ample
quantities of pure, specifically labeled collagen with a high degree of [13]C
incorporation by tissue culture procedures.[14] The ^{13}C label in the sample thus
provides a non-perturbing probe of the mobility of a specific amino acid residue.
In addition, the strong, angularly dependent [13]C-^{1}H static dipolar interaction
which greatly broadens the lines in normal (scalar decoupled, $\gamma H_2/2\pi \sim 3$ kHz)
spectra of collagen is averaged out using high power proton decoupling (dipolar
decoupling, $\gamma H_2/2\pi \sim 60$ kHz).[15-17] The linewidths observed in dipolar decoupled
spectra often allow one to determine residual chemical shift anisotropy.[16,17]

The observed shift anisotropies provide information about motions that occur on the millisecond time scale. T_1 and NOE values can also be obtained from dipolar decoupled spectra and provide information about motions on the time-scale of nanoseconds.[18] Thus, information about samples having a wide range of mobilities can be obtained using high power ^{13}C-^1H double resonance techniques.

The amino acids used for labeling collagen were chosen to probe both backbone motion ([1- and 2-^{13}C]glycine) and sidechain motion (terminal carbons of lysine, glutamic acid, alanine, and methionine)(Fig. 2). [6-^{13}C]Lysine is an example of a positively-charged sidechain, [5-^{13}C]glutamic acid is negatively charged; [3-^{13}C]alanine is an example of a small hydrophobic sidechain, whereas [methyl-^{13}C]methionine is an example of a large hydrophobic sidechain. The labeled

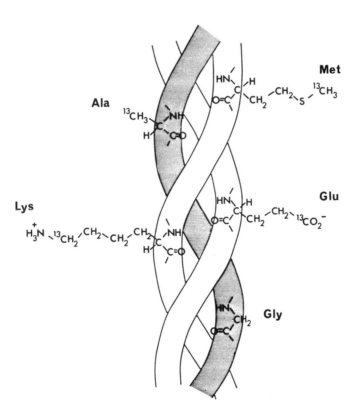

Fig. 2. Schematic representation of a collagen molecule showing the particular amino acids which were labeled for this ^{13}C NMR investigation. (Both the carbonyl carbon and the α-carbon of glycine were labeled in separate experiments.)

samples were studied both in solution and as fibrils. Lineshapes and linewidths, in addition to nuclear Overhauser enhancements and T_1 values, were measured for each labeled sample. Overall and internal rotational diffusion constants were calculated for collagen in solution from the measured NMR parameters, assuming a rigid ellipsoid model for collagen. Comparison of T_1 and NOE values for collagen in solution and as fibrils, in addition to the chemical shift aniso-tropy information, enabled the estimation of correlation times for various carbons in collagen fibrils.

MATERIALS AND METHODS

General

All ^{13}C labeled amino acids were purchased from Merck Isotopes, with the ex-ceptions of D,L-[6-^{13}C]lysine (KOR Isotopes) and D,L-[5-^{13}C]glutamic acid (Koch Isotopes). Each amino acid was checked by elemental analysis, mass spectroscopy (N-acetyl methyl ester derivative), ^{1}H NMR at 220 MHz, and ^{13}C NMR at 15.09 MHz prior to use in the labeling experiments. All radioactive isotopes employed as tracers were purchased from Schwarz/Mann, with the exceptions of D,L-[6-^{3}H]lysine and D,L-[5-^{14}C]glutamic acid, which were purchased from New England Nuclear.

Collagen sample preparation

Collagen samples were prepared by culturing 17-day-old embryonic chick cal-varia at 38-39°C in a 5% CO_2/air humidified atmosphere according to the method of Siegel.[19] Eagle's minimum essential medium, which was deficient in the amino acid to be labeled, was supplemented with the following (per 100 ml of medium): 10 mg L-ascorbic acid sodium salt, 5000 units penicillin-streptomycin, 12 mg glycine (except in the cases where glycine was the amino acid carrying the ^{13}C label), an appropriate ^{14}C labeled amino acid (see Table 1), and the ^{13}C labeled amino acid (see Table 1). After two 2-hr preincubation periods in this medium, the calvaria were incubated for two 48-hr periods in the above medium to which 5 mg/100 ml β-aminopropionitrile fumarate had been added. Collagen was extracted and purified as described previously.[20] Typically, ~20 mg of collagen were obtained from 25-dozen chick embryos.

The collagen samples were characterized by melting curves (Cary 60 spectro-polarimeter), amino acid analyses (Durrum and Beckman amino acid analyzers), and in some cases, by electron microscopy. Per cent incorporation of ^{13}C was determined from a ratio of the ^{14}C specific activity of the added amino acid (from a flow count of the amino acid analysis of the medium) to the specific activity of that amino acid in the protein hydrolyzate. In addition, the per cent incorporation of ^{13}C was determined by gas chromatography-mass spectroscopy

TABLE 1

AMOUNT OF ^{13}C LABELED AND ^{14}C LABELED AMINO ACID USED PER 100 ml MEDIUM

Amino acid	^{13}C Labeled amino acid		^{14}C Labeled amino acid		
	Optical isomers	amount[a]	Optical isomers and position of each	amount[a]	% Incorporation
[3-^{13}C]alanine	L	10.2 mg	L-uniform label	10 µCi	29%
[5-^{13}C]glutamic acid[b]	D,L	60.7	D,L-[5-^{14}C]	17	15
[1-^{13}C]glycine	-	16.0	uniform label	10	52
[2-^{13}C]glycine	-	16.0	uniform label	10	54
[6-^{13}C]lysine	D,L	25.8	L-uniform label	20	66
[methyl-^{13}C]methionine	L	5.3	L-[methyl-^{14}C]	20	85

[a] Amount per 100 ml Eagle's minimum essential medium.
[b] Glutamine withheld from the medium, 10.6 mg NH$_4$Cl added/100 ml medium.

analysis of the N-acetyl methyl ester derivatives of the protein hydrolyzates.
A Finnigan chemical ionization model 10150 gc-ms was used with a 3% OV-17 2 mm
x 5 ft glass column. Radiotracer analysis of the protein hydrolyzates and ^{13}C
NMR of heat-denatured collagen samples established that there was insignificant
radioactivity or ^{13}C associated with amino acids other than the desired labeled
amino acid, with two exceptions: 5-Hydroxy-[6-^{13}C]lysine, which is known to
occur as a postranslational modification during collagen biosynthesis,[21] was
present in the theoretically expected 1:3 ratio with [6-^{13}C]lysine in the lysine
labeled collagen sample. Approximately 30% of the ^{13}C label was associated with
[5-^{13}C]glutamine in the [5-^{13}C]glutamic acid labeled collagen sample.

Nuclear magnetic resonance sample preparation, instrumentation, and analysis

Collagen fibrils were packed into 5 or 8 mm NMR tubes by centrifugation at
2000 rpm. Excess 0.02 M Na_2HPO_4 was removed by pipet. Solution samples were
prepared in 8 mm NMR tubes by dissolving lyophilized collagen in 0.1 M acetic
acid so that the final collagen concentration was ~10 mg/ml. The temperature
was maintained at $16° \pm 2°C$ in the sample probe for all experiments involving
native collagen. Spectra of denatured collagen were obtained at 45°C.

^{13}C NMR spectra were obtained using a Nicolet Technology 15.09 MHz spectrometer which had been extensively modified for high power double resonance experiments. The spectrometer modifications and the experimental procedures employed
for dipolar decoupling and cross polarization have been described previously.[20]

Nuclear Overhauser enhancements (NOE's) were determined as described previously and are reported as $1 + \eta$.[22] Spectra for T_1 measurements were obtained
by inversion-recovery $(180°- t - 90°- T)_n$ or by progressive saturation $(90°- t -
90°- t)_n$.[23] Integrated intensities were used to measure the magnetization for
the T_1 determinations for the [1- and 2-^{13}C]glycine labeled samples. For the
collagen samples labeled at the amino acid sidechains, the peak height for each
value of t, corrected for natural abundance background, was used for the T_1
determinations. The T_1 values were determined from a least squares analysis of
the data points obtained in the initial part of the recovery curve in order to
avoid contributions from cross correlation effects.[24]

The chemical shift anisotropies for the polycrystalline amino acids were
determined by computer-simulation of the lineshapes. A Gaussian function
representing the calculated broadening due to ^{13}C-^{13}C and ^{14}N-^{13}C dipolar interactions[18] was superimposed on a theoretical lineshape for which σ_{11}, σ_{22}, and
σ_{33} were varied until a suitable match was obtained.

The per cent of the sample giving rise to the signal was determined by integration of the signal intensity of a known amount of collagen fibrils of known

^{13}C enrichment. This value was then compared with the integrated intensity arising from a known weight of ethylene glycol. For collagen fibrils labeled with [1- and 2-^{13}C]glycine and [6-^{13}C]lysine (the samples for which signal-to-noise limitations permitted accurate measurements), >95% of the labeled carbons contributed to the signal intensity.

RESULTS

Linewidths and lineshapes

Collagen backbone. The linewidths obtained for [1- and 2-^{13}C]glycine labeled collagen are listed in Table 2. All linewidths were taken as the width of the signal at half its maximum height, except for that of [1-^{13}C]glycine labeled collagen fibrils. An axially asymmetric chemical shift tensor pattern was observed for this sample, and the width was taken as $\sigma_{33} - \sigma_{11}$. Listed for comparison with the backbone labeled collagen samples are the full chemical shift anisotropies measured for polyglycine. These anisotropes were measured at high field (7 Tesla) in order to minimize the fractional contribution of the ^{14}N-^{13}C dipolar coupling to the ^{13}C linewidth.[25] Heteronuclear coupling contributes ~20 ppm to the linewidth at 73.96 MHz (7 Tesla), whereas at 15.9 MHz (1.4 Tesla), the contribution is ~100 ppm and obscures the chemical shift powder pattern. (Because of molecular motion in collagen, the ^{14}N T_1 was short enough to eliminate the heteronuclear interaction, and thus the ^{14}N-^{13}C interaction did not contribute to the linewidth at 15.09 MHz.)

Collagen sidechains. The linewidths obtained for collagen labeled with [3-^{13}C]alanine, [methyl-^{13}C]methionine, [5-^{13}C]glutamic acid, and [6-^{13}C]lysine are listed in Table 2. Included for comparison are the rigid lattice anisotropies for these amino acids. It was necessary to measure the rigid lattice anisotropy for [6-^{13}C]lysine at high field (7 Tesla, 73.96 MHz) in order to minimize contributions of ^{14}N-^{13}C dipolar coupling to the linewidth (see previous section).[25] The linewidth measured for the [5-^{13}C]glutamic acid labeled sample actually arises from an overlap of [5-^{13}C]glutamic acid and the resonance from biosynthetically produced [5-^{13}C]glutamine. A ^{13}C NMR spectrum of heat-denatured [5-^{13}C]glutamic acid labeled collagen indicates that the shifts for these resonances are separated by 2.4 ppm. Although biosynthetically produced 5-hydroxy-[6-^{13}C]lysine was present in the spectra for [6-^{13}C]lysine labeled collagen, it could be distinguished from the lysine peak and did not contribute to the measured [6-^{13}C]lysine linewidth (Fig. 5, a and b).

TABLE 2

COMPARISON OF THE LINEWIDTH OF THE LABELED CARBON OF VARIOUS COLLAGEN

SAMPLES WITH THE RIGID LATTICE ANISOTROPY FOR THAT CARBON

Collagen labeled with	Solution linewidth[a]	Fibril linewidth[b]	Rigid lattice anisotropy[c]	
			Model	Anisotropy
[3-^{13}C]alanine	4 ppm	14 ppm	[3-^{13}C]alanine	34 ppm
[5-^{13}C]glutamic acid	10[d]	21[d]	[5-^{13}C]glutamic acid	153
[1-^{13}C]glycine	8	103[e]	polyglycine	144[e]
[2-^{13}C]glycine	33	45	polyglycine	34
[6-^{13}C]lysine	3	7	[6-^{13}C]lysine	29
[methyl-^{13}C]methionine	2	5	[methyl-^{13}C]-methionine	31

[a] Linewidth taken as the width of the peak at half its maximal height; concentration ∼10 mg/ml in 0.1 M acetic acid; linewidth contains a 1-2 ppm contribution from digital filtering and magnet inhomogeneity; uncertainty ± 0.3 ppm (except as noted).

[b] Linewidth taken as the width of the peak at half its maximal height (except as noted); samples in equilibrium with 0.02 M Na$_2$HPO$_4$; linewidth contains a 2-3 ppm contribution from digital filtering and magnet inhomogeneity; uncertainty ± 0.3 ppm (except as noted).

[c] Spectra obtained at -40 to -99°C of 9 atom % ^{13}C polycrystalline enriched amino acids at 15.09 MHz, except for the spectra of unlabeled polyglycine and 9 atom % ^{13}C lysine which were recorded at ambient temperature at 73.96 MHz; anisotropy determined by computer-simulation of the lineshape as described in Materials and Methods; uncertainty ± 2 ppm.

[d] Contains a contribution from [5-^{13}C]glutamine; uncertainty ± 0.5 ppm.

[e] Value reported is the total chemical shift anisotropy measured as $\sigma_{33} - \sigma_{11}$; uncertainty ± 3%.

TABLE 3

T$_1$ VALUES[a] FOR THE LABELED CARBON OF VARIOUS COLLAGEN SAMPLES

Collagen labeled with	Solution[b] T$_1$	Fibril[c] T$_1$
[3-^{13}C]alanine	0.167 s	0.275 s
[5-^{13}C]glutamic acid[d]	--	< 1.5[e]
[1-^{13}C]glycine	0.535	2.01[f]
[2-^{13}C]glycine[g]	0.030	0.130
[6-^{13}C]lysine	0.226	0.146
[methyl-^{13}C]methionine	$\gtrsim$ 0.500[e]	$\gtrsim$ 0.500[e]

[a]Determined by inversion-recovery, except as noted

[b]~10 mg/ml in 0.1 M acetic acid; uncertainty $\pm$ 15%, except as noted.

[c]Reconstituted by precipitation from 0.02 M Na$_2$HPO$_4$; uncertainty $\pm$ 15%, except as noted.

[d]Contains an ~30% intensity contribution from [5-^{13}C]glutamine (see Results section).

[e]Estimated; uncertainty $\pm$ 25%.

[f]Determined by progressive saturation.

[g]These values have been corrected for natural abundance background (see text). The uncorrected values are 0.041 and 0.141 s for solution and fibrils, respectively.

T$_1$ values

Collagen backbone. The T$_1$ values for [1- and 2-^{13}C]glycine labeled collagen, both as fibrils and in solution, are listed in Table 3. Because the peak intensities were determined by integrating the signal from the labeled carbon (see Materials and Methods), the integrated intensity contains contributions from the natural abundance background. The carbonyl carbon of glycine and the naturally abundant carbonyl carbons in the molecule have approximately the same T$_1$ values. In the case of [2-^{13}C]glycine labeled collagen, however, the average T$_1$ for the naturally abundant carbons is much longer than the T$_1$ for the labeled carbons. Even though one-third of all the residues in collagen are glycine and the level of enrichment is 50-fold over natural abundance, about 15% of the carbons which contribute to the observed signal in the [2-^{13}C]glycine labeled collagen spectrum are aliphatic carbons in natural abundance. Because most of these carbons are on side chains at the helical surface, they can undergo rapid internal reorientation and thus have larger T_1 values than the backbone glycine -CH$_2$- carbons. The [2-^{13}C]glycine labeled collagen T_1 values in Table 3 have therefore been corrected for natural abundance background by fitting the experimental T_1 data to a two-component exponential function. The relative amplitudes of the two components were taken as 4:1 (rather than 0.85:0.15, since the inversion-recovery spectra are measured with an NOE, and the average NOE of the naturally abundant component is known).

$\underline{\text{Collagen sidechains}}$. The T_1 values for collagen labeled with [3-^{13}C]alanine and [6-^{13}C]lysine are listed in Table 3. Estimates for the T_1 values for collagen labeled with [methyl-^{13}C]methionine and [5-^{13}C]glutamic acid are also tabulated. Signal-to-noise limitations precluded accurate T_1 measurements for these latter two samples. Because the labeled carbon NMR signals were relatively sharp for the amino acid sidechains studied, the intensity contributed by the natural abundance background could be readily identified. Therefore, the labeled peak intensities were corrected for natural abundance background by visual inspection of the baseline, and it was unnecessary to fit the data to a multicomponent exponential function.

Nuclear Overhauser enhancements

Nuclear Overhauser enhancements for backbone and amino acid sidechain labeled collagen are listed in Table 4. The values for collagen labeled with [methyl-^{13}C]methionine and [5-^{13}C]glutamic acid are estimated because of signal-to-noise limitations.

Because the signal from the labeled carbon overlaps resonance of the naturally abundant sidechains, which have a larger average NOE value, it was necessary to correct the glycine α-carbon NOE for natural abundance background. Since the relative amounts of the labeled carbon and of the carbons from overlapping sidechains in natural abundance are known, and the average NOE of the natural abundance component is known, the corrected glycine α-carbon NOE could be calculated in a straightforward manner. The corrected NOE values for the [2-^{13}C]glycine labeled collagen sample are listed in Table 4.

Because of efficient proton spin diffusion in the collagen samples, it was often necessary to use long delay times (t >> 5 x ^{13}C T_1) between accumulations in order to obtain accurate Overhauser suppressed signal intensities. This was not the case for spectra obtained with an Overhauser enhancement, since the condition for this experiment is continuous saturation of the proton magnetization.[26] However, in the gated decoupling experiment (where the proton magnetization is saturated only during acquisition of the free induction decay), the

TABLE 4

NUCLEAR OVERHAUSER ENHANCEMENTS[a] FOR THE

LABELED CARBON OF VARIOUS COLLAGEN SAMPLES

Collagen labeled with	Solution[b] NOE	Fibril[c] NOE
[3-^{13}C]alanine	2.0	2.3
[5-^{13}C]glutamic acid	---	> 2 [d]
[1-^{13}C]glycine	1.52	1.60
[2-^{13}C]glycine[e]	1.44	1.74
[6-^{13}C]lysine	2.75	2.4
[methyl-^{13}C]methionine	---	> 2 [e]

[a]Values reported as 1+η; uncertainty ± 10%, except as noted.

[b] ~10 mg/ml in 0.1 M acetic acid.

[c]Reconstituted by precipitation from 0.02 M Na_2HPO_4.

[d]Estimated; uncertainty ± 25%.

[e]These values have been corrected for natural abundance background (see text). The uncorrected values are 1.52 and 1.79 for solution and fibrils, respectively.

delay time between acquisitions must be sufficiently long to allow both the ^{13}C and ^{1}H magnetizations to return to their equilibrium values. For samples where the carbon T_1 is short (0.03-0.2 s) compared to the proton T_1 (1-2 s), and where spin diffusion between water and protein protons is efficient, relaxation of the protons governs the return of the carbon magnetization to its equilibrium value. This effect of spin diffusion can be circumvented by obtaining the spectra in a D_2O, rather than in an H_2O solvent system.

DISCUSSION

Linewidths and lineshapes

   Collagen backbone. As fibrils, [1-^{13}C]glycine labeled collagen gives rise to an asymmetric chemical shift tensor pattern (Fig. 3a) with a total anisotropy of 103 ppm ($\sigma_{33} - \sigma_{11}$). Evidence that this lineshape represents a motionally collapsed asymmetric chemical shift tensor pattern comes from a comparison of this anisotropy with that of model compounds for the glycyl residue in collagen (Table 5). The carbonyl carbon of polyglycine has a rigid lattice anisotropy of 144 ppm (Fig. 4, Tables 2,5); the glycine carbonyl has a 143 ppm anisotropy;[27] and the

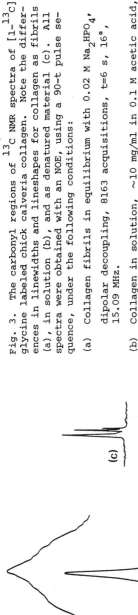

Fig. 3. The carbonyl regions of ^{13}C NMR spectra of [1-^{13}C] glycine labeled chick calveria collagen. Note the differences in linewidths and lineshapes for collagen as fibrils (a), in solution (b), and as denatured material (c). All spectra were obtained with an NOE, using a 90-t pulse sequence, under the following conditions:

(a) Collagen fibrils in equilibrium with 0.02 M Na$_2$HPO$_4$, dipolar decoupling, 8163 acquisitions, t=6 s, 16°, 15.09 MHz.

(b) Collagen in solution, ~10 mg/ml in 0.1 M acetic acid, dipolar decoupling, 4096 acquisitions, t=1 s, 16°, 15.09 MHz.

(c) Denatured collagen, ~10 mg/ml in 0.1 M acetic acid, dipolar decoupling, 10% D$_2$O, 2048 acquisitions, t=2 s, 50°, 25.01 MHz.

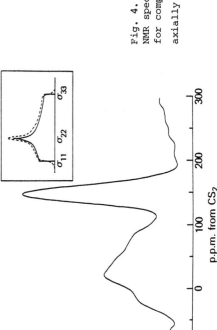

Fig. 4. A proton-enhanced natural abundance 73.96 MHz ^{13}C NMR spectrum of polyglycine powder at ambient temperature, for comparison with Fig. 3. The inset shows a theoretical axially asymmetric chemical shift tensor powder pattern.[16]

TABLE 5

PRINCIPAL VALUES OF CHEMICAL SHIFT TENSOR ELEMENTS[a]

FOR GLYCINE AND GLYCYL CARBONYL CARBONS

Compound	σ_{11}	σ_{22}	σ_{33}	$\sigma_{33} - \sigma_{11}$	σ_i
	parts per million				
[1-^{13}C]glycine labeled collagen fibrils	- 31	20	72	103	20.3
polyglycine	- 44	18	100	144	25
glycine[b,27]	- 57.6	11.0	85.0	143	12.8
glycylglycine[42]	- 51	16	105	156	23

[a]Relative to external CS_2, uncertainty $\pm$ 4 ppm for all compounds except glycine.

[b]The values from reference 27 have been converted to ppm from CS_2

peptide carbonyl of glycylglycine has a 156 ppm anisotropy (Table 5). Thus, the ca. 40 ppm narrowing of the carbonyl anisotropy in [1-^{13}C]glycine labeled collagen fibrils indicates that motion is fast in the fibril on a millisecond timescale. In addition, the lineshape shows no apparent distortion due to ^{13}C-^{14}N dipolar coupling. As a consequence of molecular motion, the ^{14}N T_1 is sufficiently short to decouple the ^{13}C-^{14}N interaction.[25]

In solution, where free rotation can take place about the long axis of the helix (in addition to head-over-head tumbling), the [1-^{13}C]glycine labeled collagen sample gives rise to a narrow line (Fig. 3b, Table 2) in the dipolar decoupled spectrum. A preliminary calculation shows that rotation about the long axis of the collagen molecule alone is sufficient to account for the rather narrow signal observed for [1-^{13}C]glycine labeled collagen in solution.[17,28] Distortion of this lineshape from the expected classical axially symmetric pattern may be due, in part, to the 3 ppm dispersion in isotropic chemical shifts for the glycine carbonyl carbons in random coil (denatured) collagen (Fig. 3c). This dispersion indicates that the principal values and orientations of the chemical shift tensors are not the same for all glycine carbonyl carbons in collagen. The three peaks observed in Fig. 3c are all attributed to [1-^{13}C]glycine in the sample, since radiotracer analysis indicates no significant contamination from other labeled amino acids. The broad lowest field peak is attributed to glycines which do not precede proline in the Gly-X-Y triplet. The variable nature in X produces the chemical shift dispersion which explains the greater width of this resonance. On the basis of chemical shifts of model peptides, the two upfield resonances are assigned to glycines which precede proline.[29] The relative integrated areas of the peaks are consistent with this assignment.

As fibrils and in solution, [2-^{13}C]glycine labeled collagen gives rise to unexpectedly broad dipolar decoupled ^{13}C NMR signals (Table 2). The labeled collagen linewidths are as broad, or broader, than the full powder linewidth of the methylene carbon in polycrystalline polyglycine. Random fluctuation (R_1 $\sim 10^7 s^{-1}$) of the strong dipolar field of the α-carbon arises as a consequence of collagen backbone motion. Unlike static dipolar interactions, which can be removed by dipolar decoupling (and therefore do not contribute to the linewidth), the random fluctuation of the dipolar field is a significant source of broadening of the α-carbon signal.[30,31] (This source of linebroadening is much less important in the case of the carbonyl carbon, because the magnitude of the broadening is inversely proportional to the sixth power of the [1]H-^{13}C internuclear distance.)

Collagen sidechains. In solution, the methyl signal in the alanine labeled collagen is a symmetric line with a 4 ppm width, indicating that the rigid body motion of the helix plus the methyl rotation are sufficient to completely average the methyl shift anisotropy. This result is not surprising since the calculated methyl anisotropy is less than 10 ppm (150 Hz) in the presence of the 3-fold motions of the methyl group in addition to the rotation about the long axis of the ellipsoid. Thus, head-over-head motion ($R_2 \sim 10^3$ s^{-1}) about the minor axis of the helix is required to further average this residual shift anisotropy to the experimentally observed value. However, the absence of a scalar decoupled spectrum shows that head-over-head motion does not average out the much stronger [13]C-^{1}H dipolar interactions.

The head-over-head motion of the helix is impossible in the fibrils, and furthermore, motion about the long axis is restricted (angular range $\gtrsim 30°$) (see analysis of T_1 and NOE data).[14,20] These restrictions on motion in the fibrils, as compared with the freedom in solution, are reflected in the observation of a fibril linewidth which is considerably broader than that found in solution. The width of the methyl signal in the fibrils is 14 ppm, a value which is about four-tenths as large as the 34 ppm anisotropy observed for the alanine methyl in a rigid lattice (Table 2). This partially averaged anisotropy is a result of rotation about the long axis of the molecule in the fibrils.

Since the alanine methyl moiety is bonded to the backbone alanine C2 carbon, rotation about the C2-C3 bond is the only local degree of freedom available to the alanine methyl group. Thus, the NMR parameters of the alanine methyl are determined as much by motion of the molecular backbone as by its local sidechain motion. Because the methyl group of methionine is at the end of a four atom sidechain (Fig. 2), this moiety can be expected to reflect primarily the local sidechain motions, if rotations about the sidechain bonds occur.

The methyl signal width (5 ppm) in [methyl-^{13}C]methionine labeled collagen fibrils is substantially smaller than the dipolar decoupled rigid lattice linewidth (31 ppm) and is also smaller than the linewidth observed for collagen fibrils labeled with [3-^{13}C]alanine (14 ppm), (Table 2). Absence of methionine methyl chemical shift anisotropy in collagen indicates that rotations occur about several methionine sidechain bonds in fibrils.

The 2 ppm width of the [methyl-^{13}C]methionine labeled collagen signal in solution is also small compared with the 31 ppm anisotropy observed for [methyl-^{13}C]methionine in a rigid lattice, and shows that molecular motion of collagen in solution (backbone motion plus local motion of the methionine sidechain) completely averages the methionine shift anisotropy. This result is expected since it has been shown above that for [3-^{13}C]alanine labeled collagen, the motion of collagen in solution plus methyl rotation will completely average a methyl shift tensor. This combination of motions, however, did not average out the strong static dipolar ^{13}C-^{1}H coupling in the alanine methyl group since a scalar decoupled spectrum of [3-^{13}C]alanine labeled collagen in solution was too broad to be observed. In contrast, the signal at 179 ppm remains in the scalar decoupled spectrum of [methyl-^{13}C]methionine labeled collagen in solution. Although the signal is 2-3-fold broader than in the dipolar decoupled spectrum, its width (5 ppm) is nonetheless small compared with the static ^{13}C-^{1}H dipolar interaction (~ 40 kHz rms). This result implies that collagen backbone motion plus internal rotation of the methionine sidechain produce an isotropic average of the strong dipolar fields at the methyl carbon.

The extensive motion of the methionine sidechain suggests that possible interactions involving this hydrophobic sidechain do not greatly hinder its motion either in solution or in fibrils. In contrast to methionine, the sidechain of lysine is charged at physiological pH, and therefore, NMR experiments of [6-^{13}C]lysine labeled collagen present an opportunity to study the local motions of a polar sidechain.

The dipolar decoupled ^{13}C NMR spectrum of hydroxylysine/lysine labeled collagen in solution exhibits a relatively sharp peak (3 ppm linewidth) centered at 153 ppm from CS_2 (Fig. 5a). The low field shoulder on the peak was assigned to hydroxylysine on the basis of its relative chemical shift[32] and on its relative intensity in the spectrum of denatured labeled collagen (Fig. 5c), in which [6-^{13}C]lysine and 5-hydroxy-[6-^{13}C]lysine exhibit well resolved resonances.

The narrow line exhibited by [6-^{13}C]lysine labeled native (helical) collagen (Fig. 5a) indicates that in solution there is essentially complete averaging of the [6-^{13}C]lysine chemical shift anisotropy (Table 2).

The lysine signal in the scalar decoupled spectrum of [6-^{13}C]lysine labeled collagen in solution contains only ~40% of the intensity of the corresponding dipolar decoupled spectrum (Fig. 5b). This indicates that although backbone and lysine sidechain motions in collagen are able to completely average the 29 ppm static chemical shift anisotropy, these motions are unable to completely average the much stronger (rms value in a rigid lattice ~35 kHz) ^{13}C-^{1}H dipolar interaction of the terminal lysine -CH$_2$- group. The observed reduction in signal intensity suggests that there may be heterogeneity in the motions of various lysine sidechains in the molecule. Possibly those residues close to negatively-charged amino acid sidechains suffer some loss in mobility because of electrostatic interactions. It is also possible that, for collagen in solution, there is segmental motion of the helix backbone which is variable along its length.

In [6-^{13}C]lysine labeled collagen fibrils, the lysine linewidth (7 ppm) is only twice that observed in solution (Table 2, Fig. 5, a and d). This result indicates that reorientation of the terminal lysine carbons in the fibrils is isotropic and fast on the timescale of milliseconds. Because backbone reorientation is limited in the fibrils as compared with solution, the isotropic reorientation of the terminal carbon must be due to rotations about the sidechain bonds. Because the lysine sidechains in collagen fibrils were found to be highly mobile, it was of interest to determine if the glutamic acid sidechains behaved similarly. To this end, [5-^{13}C]glutamic acid labeled collagen was prepared, and NMR spectra were recorded for the collagen both in solution and as fibrils.

In collagen fibrils labeled with [5-^{13}C]glutamic acid, the lineshape in the carbonyl region is difficult to determine because of the low degree of ^{13}C incorporation realized (therefore, the contribution from the natural abundance background is significant). However, when the background is removed by subtracting a spectrum of collagen fibrils obtained in natural abundance, the remaining signal in the carbonyl region has a width of 21 ppm (Fig. 6b, solid line; Table 2). This width is significantly smaller than the 153 ppm rigid lattice anisotropy of [5-^{13}C]glutamic acid, and indicates that extensive motions are present in the fibrils. Additional evidence that extensive motion of the glutamic acid sidechain occurs in the fibrils comes from a comparison of the linewidth of [5-^{13}C]glutamic acid labeled collagen with that of [1-^{13}C]glycine labeled collagen (Fig. 6b). Note that the full chemical shift anisotropy

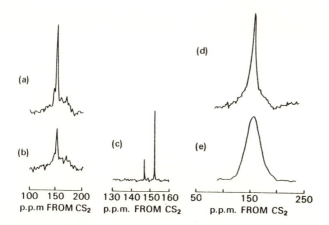

Fig. 5. ^{13}C NMR spectra of the methylene region of [6-^{13}C]lysine labeled collagen, in solution (a) and (b), as denatured material (c), and as fibrils (d),
for comparison with the lineshape for the polycrystalline amino acid, also
labeled in the 6-position (e). The spectra were obtained under the following
conditions:

(a) Native collagen in solution, ~5 mg/ml in 0.1 M acetic acid, dipolar decoup-
    ling, 16,384 acquisitions, 1 s repetition rate, 16°, with NOE, 15.09 MHz.
(b) Scalar decoupled spectrum of sample in (a) under identical conditions.
(c) Denatured collagen, ~10 mg/ml in 0.1 M acetic acid, scalar decoupling, 10%
    D_2O, 2048 acquisitions, 2 s repetition rate, 50°, with NOE, 25.01 MHz. The
    peak at 147.9 ppm is assigned to 5-hydroxy-[6-^{13}C]lysine, and the peak at
    153.3 ppm to [6-^{13}C]lysine.
(d) Collagen fibrils in equilibrium with Na_2HPO_4, dipolar decoupling, 8192
    acquisitions, 1 s repetition rate, with NOE, 16°, 15.09 MHz.
(e) Proton-enhanced spectrum of $\mathcal{D},L$-[6-^{13}C]lysine sodium phosphate salt, 4 ms
    contact time, 1 s repetition rate, ambient temperature, 73.96 MHz.

expected for the glycine carbonyl carbon is *ca.* 150 ppm, which is similar to
the 153 ppm value obtained for [5-^{13}C]glutamic acid. However, in [5-^{13}C]gluta-
mic acid labeled collagen fibrils, the observed linewidth is only one-fourth
that measured for [1-^{13}C]glycine labeled collagen fibrils. This nearly complete
collapse of the chemical shift anisotropy expected for [5-^{13}C]glutamic acid is a
consequence of the motion of the sidechain which is not possible for the backbone
glycine carbon.

As expected, the linewidth of [5-^{13}C]glutamic acid labeled collagen in solution is less than that of the fibrils, because collagen molecules have greater
freedom of motion in solution. The measured linewidth (10 ppm; Table 2) overestimates the true linewidth of [5-^{13}C]glutamic acid in collagen because of the
ca. 5 ppm chemical shift dispersion due to the presence of [5-^{13}C]glutamine in
this sample.

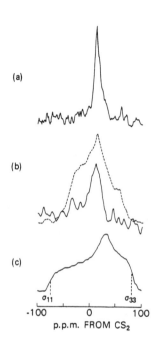

(a)

(b)

(c)

σ_{11} σ_{33}

-100   0   100

p.p.m. FROM CS$_2$

Fig. 6. ^{13}C NMR spectra of the carbonyl region of [5-^{13}C]glutamic acid labeled collagen in solution (a), and as fibrils (b) (solid line), for comparison with the carbonyl region of [1-^{13}C]glycine labeled collagen fibrils (b) (dashed line), and with the powder pattern for [5-^{13}C]glutamic acid (c). The spectra were all obtained at 15.09 MHz under the following conditions:

(a) Native collagen in solution, ~10 mg/ml in 0.1 M acetic acid, dipolar decoupling, 32,768 acquisitions, 2 s repetition rate, 16°, with NOE. The carbonyl peak arising from the solvent has been subtracted from this spectrum.

(b) Solid line: [5-^{13}C]glutamic acid labeled collagen fibrils in equilibrium with 0.02 M Na$_2$HPO$_4$ from which the natural abundance background was subtracted, dipolar decoupling, 11,491 acquisitions, 6 s repetition rate, with NOE, 16°. Dashed line: spectrum of [1-^{13}C]glycine labeled collagen fibrils, in equilibrium with 0.02 M Na$_2$HPO$_4$, dipolar decoupling, 4709 acquisitions, 6 s repetition rate, with NOE, 16°.

(c) Proton-enhanced spectrum of $\mathcal{D},\mathcal{L}$-[5-^{13}C]-glutamic acid, (90 atom % ^{13}C) powder, 1024 acquisitions, 1 ms contact time, 3 s repetition rate, -41°.

Analysis of T$_1$ and NOE data

Hydrodynamic model for collagen. Physico-chemical studies have shown that hydrated collagen is a rod-like triple helical molecule 3000 Å long by 15 Å in diameter.[33-36] Collagen can be treated as a rigid ellipsoid of revolution having a volume and length-to-diameter ratio equivalent to the physico-chemical cylinder. Rotational diffusion constants which describe the reorientation of the ellipsoid can be calculated according to the theory of Perrin[37] and Woessner.[31] R$_1$ and R$_2$, which are the respective diffusion constants for rotation about the long axis of the ellipsoid and about the axis perpendicular to the long axis, are given by:

$$R_1 = 3kT/16\pi\eta ab^2 \tag{3}$$

$$R_2 = (3kT/16\pi\eta ab^3)[2\ln(2/\rho)-1] \tag{4}$$

where k is Boltzmann's constant, T is the absolute temperature, and η is the solvent viscosity. Values of R_1 and R_2 calculated for the collagen ellipsoid model are listed in Table 7.

Collagen backbone in solution. One can derive experimental values of R_1 from measured T_1 and NOE values provided that the relaxation mechanism is known. In the present case several lines of evidence strongly indicate that the $^{13}C-^1H$ dipolar interaction is primarily responsible for the spin-lattice relaxation. First, NOE measurements have shown that in copolypeptides of proline and glycine, and in the collagen fragment α1-CB2, the dipolar mechanism is responsible for spin-lattice relaxation.[38] Second, the observation that the $[1-^{13}C]$glycine T_1 value is ca. 15 times larger than the $[2-^{13}C]$glycine T_1 value (Table 3) is consistent with the r^{-6} dependence of the dipolar relaxation mechanism. Third, on theoretical grounds, chemical shift anisotropy and $^{14}N-^{13}C$ dipolar coupling are the only mechanisms which might compete with the $^{13}C-^1H$ dipolar mechanism in producing relaxation.[18] However, at our field (1.4 Tesla), a simple calculation shows that $1/T_1$ due to these two mechanisms, even for the $[1-^{13}C]$glycine carbon, is less than 15% of the $1/T_1$ due to $^{13}C-^1H$ dipolar relaxation.[29]

Combining Solomon's dipole relaxation theory[30] with Woessner's expressions for spectral density appropriate to axially symmetric ellipsoids,[31] one obtains the following expression for the rate of relaxation of a carbon by a single proton.

$$\frac{1}{T_1} = (0.1) \, K \, r_{CH}^{-6} \, [J(\omega_H - \omega_C) + 3J(\omega_C) + 6J(\omega_H + \omega_C)] \tag{5}$$

In Eq. 5, K is $3.56 \times 10^{10} \, \text{Å}^6 \, \text{sec}^{-2}$, ω_H and ω_C are the respective proton and carbon-13 Larmor frequencies; r_{CH} is the length in Å of the C-H internuclear vector; and $J(\omega)$ is given by

$$J(\omega) = \frac{A\tau_A}{1 + \omega^2 \tau_A^2} + \frac{B\tau_B}{1 + \omega^2 \tau_B^2} + \frac{C\tau_C}{1 + \omega^2 \tau_C^2} \tag{6}$$

Here, $A = 1/4(3\cos^2\theta - 1)^2$, $\tau_A = (6R_2)^{-1}$; $B = 3\cos^2\theta\sin^2\theta$; $\tau_B = (5R_2 + R_1)^{-1}$; $C = 3/4 \sin^4\theta$; $\tau_C = (4R_1 + 2R_2)^{-1}$; and θ is the angle between the C-H internuclear vector and the unique rotation axis. An analytical expression for the NOE can also be obtained. However, the relationship between the NMR parameters, R_1, and θ, are more readily perceived when the results are presented graphically (Fig. 7). These plots are valid when motion is highly anisotropic, i.e., $R_1 \gg R_2$, as is the case for collagen. The R_1 values obtained from the measured NOE

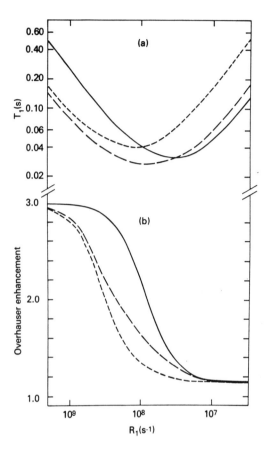

Fig. 7. NMR parameters calculated as a function of R_1 for an ellipsoid of revolution, showing the behavior of the T_1 (a) and of the nuclear Overhauser enhancement (b). The curves are calculated for $\theta = 30°$ (----), 60° (– –), and 90° (——), where θ is the angle between the C–H internuclear vector and the long axis of the ellipsoid. The curves apply to a carbon bonded to a single hydrogen with $r_{CH} = 1.09$ and apply for all values of R_2 less than 10^4 s^{-1}.

values and Fig. 7 are listed in Table 7. It should be noted that the NMR parameters are independent of R_2, because R_2 is five orders of magnitude smaller than the Larmor frequency. Hence, the magnetic field fluctuations produced by R_2 are too slow to affect the relaxation rate.

While the plots in Fig. 7 exhibit the dependence of T_1 and the NOE upon R_1, they cannot be used to determine R_1 from the T_1 when more than one proton contributes to the relaxation process. Equation 5 simplifies considerably when $\omega^2 \gg R_1^2$ and can be arranged to provide the following equation for R_1:

$$R_1 = \frac{\mathscr{C}}{T_1 r_{CH}^{-6} \sin^2\theta} \tag{7}$$

where $\mathscr{C} = 2.5 \times 10^5$ $\text{Å}^{-6}\text{s}^2$ and T_1 is in seconds. If more than one proton con-

TABLE 6

BOND LENGTHS AND INTERNUCLEAR C-H VECTOR ANGLES USED

FOR CALCULATION OF COLLAGEN ROTATIONAL DIFFUSION RATES

For collagen labeled with:[a]	r[b]	θ[c]	Source
[3-^{13}C]alanine	1.09 Å	75°[d]	Various models predict values of θ in the range 60°-90°.[1,4]
[1-^{13}C]glycine	2.1	103°, 135° (protons on Cα); 87° (N-H proton)	Calculated from the coordinates of poly-(gly-pro-pro)[1]
[2-^{13}C]glycine	1.09	60°, 120°	Calculated from the coordinates of poly-(gly-pro-pro)[1]

[a]Due to the r^{-6} dependence of the dipolar relaxation mechanism, only nearest neighbor protons were considered.

[b]Average length of the C-H internuclear vectors in Å.

[c]Angle made by the C-H internuclear vector and the long axis of the collagen molecule.

[d]Angle formed between the 3-fold axis of the methyl group (the C2-C3 bond axis) and the long axis of the collagen molecule.

tributes to the relaxation process, Eq. 7 can be immediately generalized to:[+]

$$R_1 = \frac{\mathscr{C}}{T_1 \sum_i r_{CH_i}^{-6} \sin^2\theta_i} \qquad (8)$$

Due to the sixth power distance dependence, only the nearby protons need be considered in the calculation of R_1. Further, in the present work relaxation times are the same when measured in D_2O and H_2O solution so that relaxation due to water protons is not significant.

[+]Cross-correlation contributions to the overall relaxation rate have not been included in this treatment. Cross-relaxation gives rise to a bi-exponential decay of the magnetization, where deviation from single exponential behavior becomes large at long delay times.[24] The experimental data used for these T_1 measurements were taken from the part of the magnetization decay curve which is characterized by a single exponential in order to minimize the effects of cross-correlation on the measured T_1 value.

TABLE 7

ROTATIONAL DIFFUSION CONSTANTS[a] FOR COLLAGEN

IN SOLUTION AND FOR COLLAGEN MODELS

Sample	Data Used[b]	R_1	R_2	R_{INT}
Collagen labeled with:				
[3-^{13}C]alanine	T_1, NOE	$4 \times 10^6 \text{ s}^{-1}$		$2 \times 10^{10} \text{ s}^{-1}$
[1-^{13}C]glycine	T_1	1.6×10^7		
[1-^{13}C]glycine	NOE	5.3×10^7		
[2-^{13}C]glycine	T_1	9.3×10^6		
[2-^{13}C]glycine	NOE	6.9×10^7		
Collagen model:				
hydrodynamic model[c]		1.7×10^6	4.6×10^2	
10-methylnonadecane[39]	10-methyl T_1			3.5×10^{10} [d]

[a]R_1 characterizes the rotational diffusion rate about the long axis of the ellipsoid; R_2 characterizes motion about the ellipsoid semi-minor axis (end-over-end motion); and R_{INT} refers to the internal methyl 3-fold jump diffusion.

[b]These experimental values were used in conjunction with Solomon-Woessner theory and with the bond lengths and internuclear C-H vector angles listed in Table 6.

[c]Calculated for an ellipsoid of revolution modeled to a 15 x 3000 Å cylinder, at 16°C.

[d]This value was calculated from the equation in Footnote 25, using the data in Table II (17°) of Lyerla and Horikawa.[39] It should be noted that due to an arithmetic error, the reorientation rates listed in Table IV of this reference are too large by a factor of 10^2.

The R_1 values obtained from the measured T_1 values are listed in Table 7. The values for r_{CH} and θ used for these calculations are listed in Table 6.

Agreement among the values of R_1 (Table 7) determined from the C^α and carbonyl carbon NMR parameters is reasonable considering the uncertainties involved. According to Eq. 8, R_1 depends upon r_{CH}^6 and $\sin^2 \theta_i$. In the case of the α-carbon, uncertainties in r_{CH} and θ_i produce errors of 25% in R_1, whereas in the case of the carbonyl carbon, the uncertainty in R_1 is twice as large because of larger uncertainties in the values of r_{CH} and θ_i. The values of R_1 determined from the carbonyl carbon T_1 data are overestimates because only nearest neighbor protons were included in Eq. 7. The uncertainty in the R_1 values derived from NOE measurements is particularly large since R_1 is a sensitive function of the NOE measured when the enhancement approaches its minimal value of 1.15 (see Fig. 7). Also contributing to the uncertainties in the determination of R_1 are the errors inherent in the T_1 and NOE measurements themselves. These are estimated to be $\pm$ 15% and $\pm$ 10%, respectively, from errors in integrated intensities. In view of the uncertainties discussed, it is fair to say that the analyses of the T_1 and NOE values indicate that R_1 is in the range 1-4 x 10^7 s^{-1} for the collagen helix in solution.

The maximum value of R_1 calculated from the Perrin theory[37] (Table 7) is four times smaller than the minimum value obtained from the NMR analysis. The observation that R_1 obtained from experiment is smaller than that calculated for rigid rod reorientation suggests that, in addition to rod-like reorientation, the collagen molecule undergoes torsional reorientation about its long axis.

Collagen sidechain motion in solution. The T_1 and NOE values obtained in solution for [3-^{13}C]alanine labeled collagen can be combined with the theory of Solomon[30] and Woessner[31] to determine the rotational diffusion constants which characterize the motion of a methyl group attached to an ellipsoid of revolution. Diffusion constants R_1 and R_2 characterize reorientations about the respective major and minor axes of the ellipsoid. The rate for three-fold jump

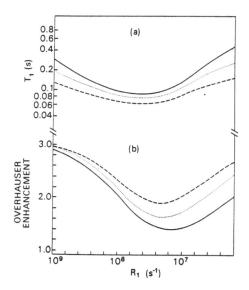

Fig. 8. NMR parameters calculated as a function of R_1 for an internally rotating methyl group. The 3-fold axis makes a 75° angle with the long axis of an ellipsoid of revolution for which R_1 = 10^6 R_2. T_1 (a) and nuclear Overhauser enhancement (b). The curves are calculated for R_{INT} = 5 x 10^{10} (——); R_{INT} = 2 x 10^{10} (....); and R_{INT}= 1 x 10^{10} s^{-1} (– – –).

diffusion of the methyl group is designated R_{INT}. The T_1 and NOE values depend on the angle between the 3-fold axis of the alanine methyl group (the C2-C3 bond axis) and the major axis of the ellipsoid. The bond angles and bond lengths used for this calculation are listed in Table 6, and the calculated results are shown in Fig. 8. As before, the calculated T_1 and NOE values do not depend on R_2, because this diffusion constant has a value ($\sim 10^3$ s^{-1}) which is five orders of magnitude less than the Larmor frequence ($\sim 10^8$ s^{-1}). The R_1 and R_{INT} values listed in Table 7 were obtained by combining the calculated results shown in Fig. 8 with the experimental T_1 and NOE values for [3-^{13}C]alanine labeled collagen in solution (Tables 3 and 4).

These results can be compared with the diffusion constants obtained for collagen labeled with [1- and 2-^{13}C]glycine and for the internal methyl of 10-methylnonadecane (a model for internal methyl rotation)[39] (Table 7). Agreement among these results is reasonable in consideration of the uncertainties in the assumed structural parameters, the errors in the relaxation data, and the simplicity of the motional models employed.

Collagen backbone motion in fibrils. The glycine α-carbon and carbonyl carbon T_1 values for the labeled fibrils are about four times larger than the values measured in solution (Table 3). This result implies that the dipolar relaxation mechanism is less effective in the fibers either because R_1 is smaller or because the angular range of reorientation is restricted. The NOE values measured in the fibrils are well in excess of the theoretical minimum value of 1.15 and are, in fact, slightly larger than the values measured in solution. These results indicate that the reorientation rate is the same in solution and in the fibrils, and that the larger T_1 value measured in the fibrils is a consequence of restrictions on the angular range of reorientation. Additional evidence for this conclusion comes from the observation of an axially asymmetric chemical shift tensor powder pattern for [1-^{13}C]glycine labeled collagen fibrils (see Discussion, above). Analysis of the T_1 data indicates that the angular range of azimuthal reorientation for the collagen backbone is restricted but must be larger than 30°.[14] Thus, although the intermolecular interactions in the fibrils do not "freeze" the azimuthal geometry, as a consequence of these interactions, certain orientations are more favored than others.

The alanine methyl T_1 and NOE values for the fibrils also provide evidence for motion of the molecular backbone. If there was no backbone motion (i.e., if the rapid 3-fold jumps of the methyl group ($R_{INT} = 2 \times 10^{10}$ s^{-1}) were the only motions permitted), the predicted NOE value would be 3, and the T_1 value would be 0.35 s. These values are inconsistent with the experimental results (Tables 3 and 4).

Collagen sidechain motion in fibrils. Additional evidence for reorientation of the amino acid sidechains in the fibrils comes from a comparison of the T_1 and NOE values obtained in solution with those obtained in the fibrils. The similarity of these parameters measured in these two environments (in spite of differences in backbone motions) suggests an overall similarity in motions for collagen in solution and as fibrils. The T_1 and NOE values correspond to correlation times in the range 10^{-9}-10^{-10} s, which implies that the terminal carbons of lysine, glutamic acid, methionine, and alanine in collagen fibrils have rotation rates of 10^9-10^{10} s^{-1}.

SUMMARY

Taken together, the NMR data provide strong evidence that rapid anisotropic reorientation of the helix backbone occurs in the fibrils, and that the terminal carbons in both charged and hydrophobic amino acid sidechains in the fibrils undergo rapid reorientation.

Collagen backbone. The spin-lattice relaxation times and nuclear Overhauser enhancements for [1- and 2-^{13}C]glycine labeled collagen in solution indicate that R_1 (diffusion constant for reorientation about the long axis of the molecule) is $\sim 2 \times 10^7$ s^{-1}. A substantially smaller value of R_1 (2.6×10^6 s^{-1}) was calculated for an axially symmetric ellipsoid of revolution having dimensions appropriate to the collagen helix. The discrepancy between the rigid ellipsoid model and experimental NMR values of R_1 suggests that the collagen molecule in torsional reorientation as well as rod-like reorientation about its long axis.

The T_1 and NOE values measured in the glycine labeled fibrils show that rapid axial motion ($R_1 \sim 10^7$ s^{-1}) persists in the fibrillar state, although the angular range of reorientation is restricted. In the collagen fibril, the full width of the glycyl carbonyl powder pattern is 103 ppm. This value is substantially smaller than the rigid lattice value, 144 ppm, which provides further evidence for motion in the fibril. The observed powder pattern is axially asymmetric, which shows that certain azimuthal orientations are energetically preferred in the fibril.

Rapid reorientation of the helix backbone in the fibrils indicates that the fibrillar structure does not require the existence of a unique set of intermolecular interactions at the helical surfaces and suggests that the collagen molecules in the fibrils have highly fluid surfaces. The NMR properties of the terminal carbons of several of the amino acid sidechains of collagen confirm the suggestion that the contact regions between the molecules in collagen fibrils are fluid.

114

Collagen sidechains. The bond rotations in collagen must be extensive enough to produce isotropic reorientation (on the timescale of milliseconds) of the terminal carbons of methionine, lysine, and glutamic acid. The large NOE values and efficient spin-lattice relaxation times show that some, and possibly all, bond rotations are fast on the nanosecond time scale. The rapid bond reorientations and the fluid nature of the contact regions between the molecules in collagen fibrils suggested by this NMR study are not in conflict with the known fact that collagen molecules associate specifically *in vivo* to produce well-organized fibrils. Rather, the NMR data suggest that organization and function of collagen fibers need not arise from a unique set of interactions between amino acid sidechains. In this respect, these results are in line with current views of globular protein structure, which suggest that a variety of conformations, in dynamic equilibrium, are responsible for the biological specificity and function of the protein.[40,41]

ACKNOWLEDGMENTS

We are grateful to Dr. R.G. Griffin for the 73.96 MHz ^{13}C spectrum of [6-^{13}C]-lysine; and to Dr. W.M. Egan for the 25.01 Mhz ^{13}C spectra of denatured collagen samples.

REFERENCES

1. Traub, W., and Piez, K.A. (1971) Adv. Protein Chem., 25, 243-352.
2. Piez, K.A. (1976) In Biochemistry of Collagen, (Ramachandran, G.N., and Reddi, A.H., eds., Plenum Publishing Co., New York, Chapter 1.
3. Fietzek, P.P., and Kühn, K. (1976) Int. Rev. Conn. Tissue Res., 7, 1-60.
4. Ramachandran, G.N. (1967) In Treatise on Collagen, Ramachandran, G.N., ed., Academic Press, New York, Vol. 1, Chapter 3.
5. Ramachandran, G.N., and Ramakrishnan, C. (1976) In Biochemistry of Collagen, Ramachandran, G.N., and Reddi, A.H., eds., Plenum Press, New York, Chapter 2.
6. Doyle, B.B., Hulmes, D.J.S., et al. (1974) Proc. R. Soc. B., 187, 37-46.
7. Doyle, B.B., Hukins, D.W.L., et al. (1974) Biochem. Biophys. Res. Commun., 60, 858-864.
8. Piez, K.A., and Miller, A. (1974) J. Supramol. Struc., 2, 121-137.
9. Piez, K.A., and Torchia, D.A. (1975) Nature, 258, 87.
10. Li, S.-T., Bolub, E., and Katz, E.P. (1975) J. Mol. Biol., 98, 835-839.
11. Trus, B.L., and Piez, K.A. (1976) J. Mol. Biol., 108, 705-732.
12. Piez, K.A., and Trus, B.L. (1977) J. Mol. Biol., 110, 701-704.
13. Piez, K.A., and Trus, B.L. (1978) J. Mol. Biol., 122, 419-432.
14. Torchia, D.A., and VanderHart, D.L. (1976) J. Mol. Biol., 104, 315-332.
15. Bloch, F. (1958) Phys. Rev., 111, 841-853.
16. Mehring, M. (1976) In NMR - Basic Principles and Progress, Diehl, P., Fluck, E., and Kosfeld, R., eds., Springer-Verlag, New York, Vol. II.
17. Pines, A., Gibby, M.G., and Waugh, J.S. (1973) J. Chem. Phys., 59, 569-590.
18. Abragam, A. (1961) The Principles of Nuclear Magnetism, Oxford University Press, London.
19. Siegel, R.C. (1974) Proc. Nat. Acad. Sci. USA, 71, 4826-4830.

20. Jelinski, L.W., and Torchia, D.A. (1979) J. Mol. Biol., 133, 45-65.
21. Prockop, D.J., Berg, R.A., et al. (1976) In Biochemistry of Collagen, Ramachandran, G.N., and Reddi, A.H., eds., Plenum Press, New York, Chapter 5.
22. Torchia, D.A., Hasson, M.A., and Hascall, V.C. (1977) J. Biol. Chem., 252, 3617-3625.
23. Farrar, T.C., and Becker, E.D. (1971) Pulse and Fourier Transform NMR, Academic Press, New York, Chapter 2.
24. Werbelow, L.G., and Grant, D.M. (1975) J. Chem. Phys., 63, 4742-4749.
25. Torchia, D.A., and VanderHart, D.L. (1979) In Topics in Carbon-13 NMR Spectroscopy, Levy, G.C., ed., John Wiley and Sons, Inc., New York, Vol. 3.
26. Noggle, J.H., and Schirmer, R.E. (1971) The Nuclear Overhauser Effect, Academic Press, New York.
27. Griffin, R.G., Pines, A., and Waugh, J.S. (1975) J. Chem. Phys., 63, 3676-3677.
28. Mehring, M., Griffin, R.G., and Waugh, J.S. (1971) J. Chem. Phys., 55, 746-755.
29. Torchia, D.A., and Lyerla, J.R., Jr. (1974) Biopolymers, 13, 97-114.
30. Solomon, I. (1955) Phys. Rev., 99, 559-565.
31. Woessner, D.E. (1962) J. Chem. Phys., 37, 647-654.
32. Horsley, W., Sternlicht, H., and Cohen, J.S. (1970) J. Am. Chem. Soc., 92, 680-686.
33. Thomas J.C., and Fletcher, G.C. (1979) Biopolymers, 18, 1333-1352.
34. von Hippel, P.H. (1967) In Treatise on Collagen, Ramachandran, G.N., ed., Academic Press, New York, Vol. 1, Chapter 6.
35. Tanford, C. (1961) Physical Chemistry of Macromolecules, John Wiley and Sons, Inc., New York, pp. 396-398.
36. Boedtker, H., and Doty, P. (1956) J. Am. Chem. Soc., 78, 4267-4280.
37. Perrin, F. (1936) J. Phys. Radium, 7, 1-11.
38. Torchia, D.A., et al. (1975) Biochemistry, 14, 887-900.
39. Lyerla, J.R., Jr., and Horikawa, T.T. (1976) J. Phys. Chem., 80, 1106-1112.
40. McCammon, J.A., Gelin, B.R., and Karplus, M. (1977) Nature, 267, 585-590.
41. McCammon, J.A., and Karplus, M. (1977) Nature, 268, 765-766.
42. Stark, R.E., Ruben, D.J., et al. (1980) J. Mag. Reson., submitted.

THE STRUCTURE OF THE CYTOCHROME a_3 - Cu_{a_3} SITE IN CYTOCHROME c OXIDASE

SUNNEY I. CHAN, TOM H. STEVENS, GARY W. BRUDVIG, AND DAVID F. BOCIAN
A.A. Noyes Laboratory of Chemical Physics, California Institute of Technology,
Pasadena, California 91125

ABSTRACT

Cytochrome a_3 and Cu_{a_3} comprise the oxygen binding site in cytochrome c oxidase. In the oxidized enzyme neither of these metal ions gives rise to an EPR signal due to strong antiferromagnetic coupling of the heme and copper centers. However, it is shown that both the heme and copper can be made EPR visible upon the binding of nitric oxide. Nitric oxide binds to Cu_{a_3} in the oxidized enzyme and results in the appearance of a high-spin heme EPR signal which can be assigned to cytochrome a_3. The competition of cyanide and fluoride binding with nitric oxide in the oxidized enzyme-NO complex indicates that interacting ligand binding sites are present at the cytochrome a_3 - Cu_{a_3} site. The addition of azide to the oxidized enzyme-NO-complex produces the one-electron reduction of cytochrome a_3 which results in a triplet EPR signal which can be interpreted in terms of a nitric oxide bridged ferrocytochrome a_3-NO-$Cu_{a_3}^{+2}$ species. The implications of these findings on the structure of the oxygen-binding site of the enzyme are discussed.

INTRODUCTION

Cytochrome c oxidase is a membrane-bound enzyme which mediates the transfer of electrons between reduced cytochrome c and molecular oxygen in mitochondria.[1] The overall reaction associated with this process is given by

4 Ferrocytochrome c + O_2 + $4H^+$ $\longrightarrow$ 4 Ferricytochrome c + $2H_2O$.

Intimately involved in this electron transfer are two heme iron centers and two copper ions. Both heme irons exist in the form of heme A. One of these, cytochrome a is low-spin and, from EPR data, it has been inferred that both axial ligands are histidine imidazoles.[2] The other heme, cytochrome a_3, is high-spin and is known to bind ligands, including N_3^-, CN^-, CO, NO, and SH^-.[3] The endogenous fifth ligand of cytochrome a_3 is thought to be a histidine.[4] This heme has been shown to be antiferromagnetically coupled to one of the copper centers, Cu_{a_3}.[5,6] The resulting S = 2 system does not give rise to an

EPR spectrum which can be monitored using conventional instrumentation. In any case, both of these metal centers appear to serve as the oxygen binding site.[7] The remaining copper center, Cu_a, is very unusual. EPR[8,9] and x-ray absorption[10,11] measurements of the oxidized protein indicate that this copper is either already reduced or is highly covalent, such that a high percentage of the electron spin density resides on an associated ligand. To account for these observations, it was recently proposed that this copper is ligated to two histidines and two cysteines arranged in a distorted tetrahedral geometry.[8] Evidence has also been obtained to suggest that this copper center is deeply buried in the membrane and that its inaccessibility may play a role in mediating the flow of electrons from cytochrome $\underline{a}$ to the cytochrome $\underline{a}_3$- Cu_{a_3} site.[9]

In an effort to delineate the structure of the cytochrome $\underline{a}_3$- Cu_{a_3} site, we recently undertook studies on the enzyme under conditions where the antiferromagnetic coupling between the two metal centers is "broken."[7,12] When the antiferromagnetic coupling between cytochrome $\underline{a}_3$ and Cu_{a_3} is broken, the two metal centers can be observed individually in conventional EPR experiments. From these experiments, much information has emerged about the structure of the two metal centers as well as insights into the nature of the interactions which render them coupled. We summarize some of these findings here.

EPR STUDIES ON THE INTERACTION OF NO WITH CYTOCHROME $\underline{c}$ OXIDASE

In this section, we describe the results of EPR experiments dealing with the interaction of nitric oxide (NO) with cytochrome $\underline{c}$ oxidase. NO is a paramagnetic molecule with an unpaired electron in an antibonding π_g molecular orbital. This paramagnetism may be taken advantage of in order to break the antiferromagnetic coupling between the cytochrome $\underline{a}_3$- Cu_{a_3} centers and to probe the electronic structures of the metal centers.

Interaction of NO with reduced cytochrome $\underline{c}$ oxidase. The binding of NO to ferrous hemoproteins is well known.[13] In the case of fully reduced cytochrome $\underline{c}$ oxidase, NO binds to cytochrome $\underline{a}_3$[4] and an NO-ferrocytochrome $\underline{c}$ oxidase complex is formed, which exhibits an EPR signal with $g_x = 2.091$, $g_z = 2.006$ and $g_y = 1.97$. In contrast, the fully reduced enzyme alone does not exhibit an EPR spectrum. When [14]NO is bound to reduced cytochrome $\underline{c}$ oxidase,[4,12] the EPR signal of the complex exhibits a nine-line superhyperfine pattern, which can be interpreted in terms of the superposition of *three* sets of three lines arising from two nonequivalent nitrogens (I = 1) interacting with the unpaired electron (Fig. 1A). The larger of the two superhyperfine coupling constants is

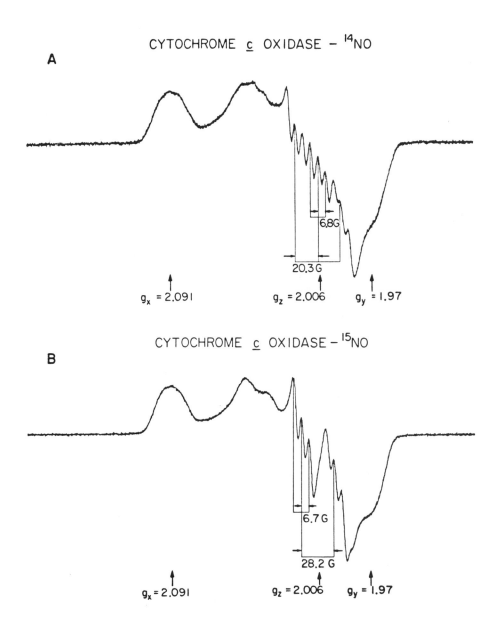

Fig. 1. The EPR spectra of (A) [14]NO-ferrocytochrome c oxidase and (B) [15]NO-ferrocytochrome c oxidase. Conditions: microwave power, 30 mW; microwave frequency, 9.15 GHz; modulation amplitude, 2 G; and temperature, 85 K (taken from reference 12).

20.3 G and the smaller 6.8 G. When ^{15}NO is used in this experiment, the ^{15}NO-bound protein exhibits an EPR spectrum with g values identical to those of the ^{14}NO-bound species, but with a superhyperfine pattern of *two* sets of three lines (Fig. 1B). This pattern is consistent with the presence of one ^{14}N and one ^{15}N nitrogen bound axially to cytochrome $\underline{a}_3$, with a 28.2 G splitting for the ^{15}N and a 6.8 G splitting for the ^{14}N ligand. The observed increase of the larger superhyperfine splitting from 20.3 G to 28.2 G upon substitution of ^{15}NO for ^{14}NO is expected on the basis of the relative magnetogyric ratios of the two nitrogen isotopes. These observations allow one to assign the larger of the superhyperfine coupling constants to the nitrogen of the bound NO and the smaller coupling constant to a nitrogen on an endogenous axial ligand of cytochrome $\underline{a}_3$.

The fifth nitrogen ligand on cytochrome $\underline{a}_3$ could be contributed by any of a number of amino acids, for example histidine, arginine, peptide nitrogen, and lysine. These potential fifth ligands vary greatly in their π-bonding capabilities, with histidine being a strong π-bonding ligand, and the remaining three ligands being predominantly σ-bonding. In this regard, Kon and Kataoka[14] have shown that the g values and nitrogen superhyperfine splittings of the EPR signals of NO-bound hemin are dependent on the π-bonding capability of the axially-bound nitrogen ligand opposite NO. With non-π-bonding ligands, such as amines, the EPR spectra exhibit axial symmetry, with $g_x = g_y = 2.07$ and $g_z = 2.008$. Also, the observed superhyperfine splitting attributed to the bound ^{14}NO nitrogen is typically 16 G, and no superhyperfine splitting is resolved from the nitrogen ligand opposite NO. In contrast, with strong-π-bonding ligands, such as pyridine, the EPR spectra exhibit rhombic symmetry, with g values and superhyperfine splittings similar to those of ^{14}NO-bound ferrocytochrome $\underline{c}$ oxidase. This result suggests that the endogenous axial ligand of cytochrome $\underline{a}_3$ is a strong π-bonding ligand, with histidine being the most likely candidate.[4]

Histidine is known to be an endogenous axial ligand to iron in the hemoproteins hemoglobin and cytochrome $\underline{c}$. The EPR spectra of ^{14}NO- and ^{15}NO-bound ferrocytochrome $\underline{c}$[14,15] and ferrohemoglobin[16] have been reported, and can be compared with the EPR spectra of NO-bound ferrocytochrome $\underline{c}$ oxidase. The similarities in the EPR spectra of NO-bound ferrocytochrome $\underline{c}$, NO-bound ferrohemoglobin and NO-bound ferrocytochrome $\underline{c}$ oxidase have led Van Gelder and Blokzijl-Homan[4] to suggest that the endogenous axial ligand to cytochrome $\underline{a}_3$ in cytochrome $\underline{c}$ oxidase is also a histidine. However, other nitrogen ligands cannot be unequivocally ruled out on the basis of the available EPR data. EPR studies of ^{15}N-histidine incorporated cytochrome $\underline{c}$ oxidase would, however, definitely confirm the presence of an endogenous axial histidine ligand to cytochrome $\underline{a}_3$. Efforts to this end are currently underway in our laboratory.

The above experimental observations do not permit us to draw conclusions regarding the position of the postulated histidine bound axially to cytochrome $\underline{a}_3$ vis-à-vis the copper center. It has been suggested, however, that the antiferromagnetic coupling between cytochrome $\underline{a}_3$ and Cu_{a_3} is provided by a bridging imidazole of a histidine residue.[17] This proposal has stimulated a flurry of activity among inorganic chemists towards the synthesis of binuclear Fe-Cu complexes linked by a bridging imidazole as models for the cytochrome $\underline{a}_3$ - Cu_{a_3} site of cytochrome $\underline{c}$ oxidase.

Interaction of NO with oxidized cytochrome $\underline{c}$ oxidase. The interaction of NO with oxidized cytochrome $\underline{c}$ oxidase has not been reported previously. However, NO is known to bind to certain ferric hemoproteins, the best known examples being ferricytochrome $\underline{c}$[18] and ferricytochrome $\underline{c}$ peroxidase.[13] NO also reacts with a number of oxidized copper proteins. The reaction of NO with oxidized ceruloplasmin, for example, leads to the formation of a reversible copper-NO complex.[19] In all the examples cited here, the resulting NO complex is either diamagnetic or EPR silent over a wide range of temperatures. These studies raise the possibility that NO might interact with at least one of the metal centers in the cytochrome $\underline{a}_3$ - Cu_{a_3} couple in fully oxidized cytochrome $\underline{c}$ oxidase. If indeed NO does interact with one of the metal centers of the site to form a *diamagnetic* NO complex, the antiferromagnetic coupling between cytochrome $\underline{a}_3$ - Cu_{a_3} will be broken and the "unaffected" metal center should reveal itself in the EPR spectrum.

The EPR spectrum of oxidized cytochrome $\underline{c}$ oxidase under anaerobic conditions in the presence of excess NO is shown in Fig. 2. There are no changes in the intensity or position of the low-spin cytochrome $\underline{a}$ or the Cu_a center EPR signals upon the addition of NO. However, a new rhombic high-spin heme EPR signal is observed. Since the cytochrome $\underline{a}$ EPR signal remains unchanged, the new high-spin heme EPR signal can be attributed only to cytochrome $\underline{a}_3$. This rhombic high-spin heme signal has g values of $g_x = 6.16$ and $g_y = 5.82$. The position of g_z is obscured by the Cu_a signal. The temperature dependence of the high-spin cytochrome $\underline{a}_3$ EPR signal indicates that the zero-field splitting parameter, D, is about 6 cm^{-1} (Fig. 3) which is characteristic of high-spin hemes.[20]

We have found that the optical spectrum of oxidized cytochrome $\underline{c}$ oxidase remains unchanged upon the addition of NO (Fig. 4). The lack of any effect by NO on the optical spectrum of oxidized cytochrome $\underline{c}$ oxidase indicates that no NO-heme interaction occurs. These optical results in conjunction with the EPR data suggest the formation of a cytochrome $\underline{a}_3^{+3}$, $Cu_{a_3}^{+2}$-NO complex.

122

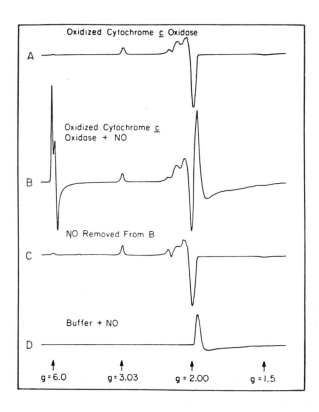

Fig. 2. The EPR spectra of (A) native oxidized cytochrome c oxidase, (B) NO added to A to a pressure of 723 mm Hg, (C) NO removed from B, and (D) NO added to the buffer only. The signals at g = 3.03, 2.21, and 1.5 are due to cyto-chrome a and the signals at g = 2.18, 2.03, and 1.99 are due to Cu_a. Conditions: microwave power, 0.2 mW; microwave frequency, 9.16 GHz; modulation amplitude, 10 G; and temperature, 7 K (taken from reference 7).

The binding of NO to oxidized cytochrome c oxidase is reversible (Fig. 2). In addition, the formation of the cytochrome a_3^{+3}, $Cu_{a_3}^{+2}$- NO complex and the nature of the complex are sensitive to exogenous ligands.[7] Thus, in the presence of F^-, the g = 6 signal is observed, but this signal is essentially axial. No high-spin heme EPR signal is observed at g ~ 6 when NO is added to the enzyme in the presence of CN^-. Instead, a low-spin heme signal at g ~ 3.5 character-istic of ferricytochrome a_3^{+3} - CN^- complex is observed. We have also noted that the enzyme *does not* exhibit a high-spin heme EPR signal when NO is added to the enzyme in the "oxygenated" state.[21]

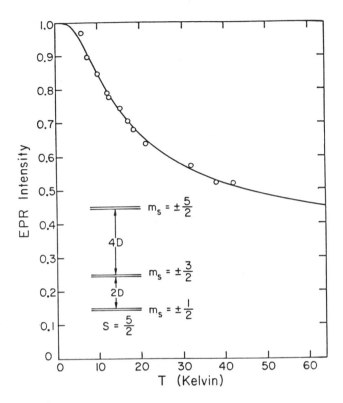

Fig. 3. The temperature dependence of the high-spin cytochrome $\underline{a}_3$ EPR signal observed upon the addition of NO to oxidized cytochrome $\underline{c}$ oxidase. The cyto-chrome $\underline{a}_3$ EPR intensity is expressed in terms of the cytochrome $\underline{a}$ EPR intensity. Thus, the points represent the fraction of cytochrome $\underline{a}_3$ giving rise to the g = 6 EPR signal. The temperature dependence of the high-spin cytochrome $\underline{a}_3$ EPR signal only reflects the zero-field splitting parameters. The solid line was calculated by assuming D = 6 cm^{-1} and E = 0.

Interaction of NO with oxidized cytochrome $\underline{c}$ oxidase in the presence of N_3^-. We have shown that the addition of N_3^- to the oxidized protein-NO complex results in the formation of a one-quarter reduced enzyme where only cytochrome $\underline{a}_3$ is reduced.[7] A possible scheme for the reduction of cytochrome $\underline{a}_3$ under these conditions is given by

$$\text{Cytochrome } \underline{a}_3^{+3} + N_3^- + NO \longrightarrow \text{Cytochrome } \underline{a}_3^{+2} + N_2O + N_2.$$

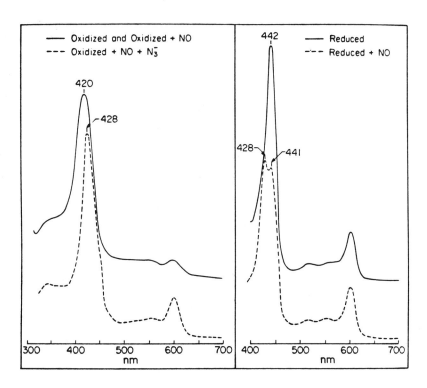

Fig. 4. The optical spectra of oxidized (left) and reduced (right) cytochrome $\underline{c}$ oxidase in the presence and absence of NO and N_3^- (taken from reference 7).

This scheme predicts the production of N_2O, which we have detected by mass spectroscopy. The substitution of ^{15}NO for ^{14}NO results in an N_2O parent peak located at M/e = 45, indicating that the nitrogen from NO appears in the N_2O molecule after the reaction. No N_2O is detected in the absence of cytochrome $\underline{c}$ oxidase, indicating that the redox reaction occurs at the ligand binding site.

The formation of the one-quarter reduced enzyme results in a dramatic change in the optical spectrum of the enzyme (Fig. 4). The Soret band shifts 8 nm to lower energy and narrows substantially while the α-band increases two-fold in intensity. Comparison of the optical spectra of the one-quarter reduced enzyme with that of the NO-bound reduced enzyme (Fig. 4) indicates that the effect of N_3^- together with NO is to reduce cytochrome $\underline{a}_3^{+3}$ followed by the binding of NO to cytochrome $\underline{a}_3^{+2}$.

We have obtained no evidence that the above process affects cytochrome $\underline{a}^{+3}$ or Cu_a, since the formation of this complex does not alter the low-spin

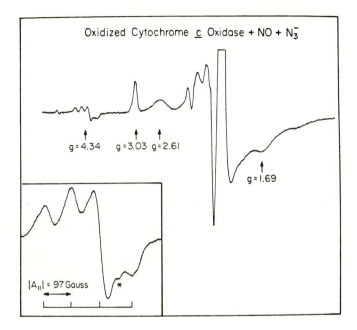

Fig. 5. The EPR spectrum of oxidized cytochrome $\underline{c}$ oxidase in the presence of N_3^- and NO. (Inset) Magnified view of the half-field transition region. The peak labeled (*) is not part of the triplet signal and is probably due to extraneous ferric iron. Conditions: microwave power, 2 mW (200 mW for the spectrum shown in the Inset); microwave frequency, 9.16 GHz; modulation amplitude, 10 G; and temperature, 7 K (taken from reference 7).

cytochrome $\underline{a}^{+3}$ or the Cu_a center EPR signals (Fig. 5). As expected, the high-spin heme EPR signal observed for the oxidized enzyme in the presence of NO is not observed when both NO and N_3^- are present in the solution. Also, no EPR signals typical of NO-$\textit{ferro}$hemoproteins are observed for our cytochrome $\underline{a}_3^{+2}$- NO complex. Instead, $\textit{new}$ EPR signals appear near g = 2 and at g = 4.34.

EPR transitions near g = 2 (ΔM_s = 1) and g = 4 (ΔM_s = 2) are characteristic of a triplet species with a small zero-field splitting. Accordingly, we have attributed the new EPR signals which we observe near g = 2 and at g = 4.34 to a triplet species. The ΔM_s = 2 transition which we observe exhibits a four-line hyperfine pattern with a splitting of 97 gauss due to a copper nucleus ($|A_{11}|$ = 0.020 cm^{-1}). This value of $|A_{11}|$ is indicative of a Type 2 copper ion.[22] We propose that the triplet signals originate from the magnetic coupling of the unpaired electron on the cytochrome $\underline{a}_3^{+2}$ - NO site with that of the $Cu_{a_3}^{+2}$. In this regard, the temperature dependence of the triplet EPR signal indicates that the two S = 1/2 sites are *anti*ferromagnetically coupled with an exchange interaction of about 5 cm^{-1}. The $\Delta M_s = 1$ transition near g = 2 indicates that $|D| \approx |3E| \approx 0.07$ cm^{-1}, where D and E are the axial and rhombic zero-field splitting parameters respectively.[23]

We have found that the process of NO binding in the presence of N_3^- cannot be reversed, in contrast to that observed in the absence of N_3^-. Furthermore, no changes are observed in the EPR or optical spectra upon the addition of O_2 to the sample. This result indicates that O_2 does not displace NO from this complex.

The above observations are consistent with the formation of a NO bridge between cytochrome $\underline{a}_3$ and Cu_{a_3} in the cytochrome $\underline{a}_3^{+2}$- NO-$Cu_{a_3}^{+2}$ complex. This conclusion is consistent with the known affinity of NO for *ferrous* cytochrome $\underline{a}_3$ as well as the affinity for *cupric* Cu_{a_3} established in these studies. The evidence (EPR and optical) that neither the removal of NO from the sample nor the subsequent addition of O_2 results in the displacement of NO in the cytochrome $\underline{a}_3^{+2}$- NO - $Cu_{a_3}^{+2}$ complex indicates that this bridge is a stable one. In contrast, the binding of NO in the case of the fully reduced and oxidized cytochrome $\underline{c}$ oxidase-NO complexes ($Fe_{a_3}^{+2}$- NO, $Cu_{a_3}^{+}$) and ($Fe_{a_3}^{+3}$, $Cu_{a_3}^{+2}$- NO), respectively, is reversible.

STRUCTURE OF THE CYTOCHROME $\underline{a}_3$- Cu_{a_3} SITE

The findings reported here permit us to make some definite conclusions regarding the structure of the cytochrome $\underline{a}_3$- Cu_{a_3} site. Two models which have been proposed for the cytochrome $\underline{a}_3$-copper site are shown in Fig. 6. In model A, a strongly-bound imidazole bridges the iron and the copper sites,[17] with the ligand binding site being the free axial position of the heme iron. In model B, the ligand binding site is between the two metal centers.[24] Our evidence that NO bridges the two metal centers in the cytochrome $\underline{a}_3^{+2}$- NO-$Cu_{a_3}^{+2}$ complex strongly argues in favor of model B.

Model B would allow formation of a bridged "peroxy" species as an intermediate in the reduction of molecular oxygen to water by cytochrome $\underline{c}$ oxidase. It also suggests the possibility of the formation of a μ-oxo bridge between these two metal centers after the complete reduction of O_2. It might be that the "oxygenated" enzyme produced by reaction of the reduced enzyme with O_2 consists of such a μ-oxo bridge at the cytochrome $\underline{a}_3$- Cu_{a_3} site. It is possible that the "oxygenated" enzyme differs from the oxidized "resting" enzyme only in

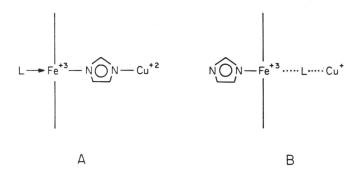

Fig. 6. Two proposed structures for the cytochrome $\underline{a}_3$- Cu_{a_3} couple in oxidized cytochrome $\underline{c}$ oxidase. L stands for a possible ligand (taken from reference 7).

terms of the nature of the bridging ligand. If this is the case, then the conformational difference normally associated with these two states of the enzyme might be understood. In any case, model B would seem to confer some degree of conformational flexibility on this part of the enzyme, inasmuch as cytochrome $\underline{a}_3$ apparently can accommodate a number of exogenous ligands of varying sizes as its sixth ligand.

Our finding that Cu_{a_3} is similar to a Type 2 copper in the cytochrome $\underline{a}_3^{+2}$ - $NO\text{-}Cu_{a_3}^{+2}$ complex suggests that Cu_{a_3} has a similar ligand environment in the native oxidized enzyme, i.e., square-planar coordination or octahedral coordination with a strong tetragonal distortion. For a d^9 copper involved in square-planar coordination, it is well known that the unpaired electron is in a $d_{x^2-y^2}$ orbital, with the orbital lobes directed towards the ligands in the square plane. The strong antiferromagnetic coupling between cytochrome $\underline{a}_3$ and Cu_{a_3} in the oxidized enzyme would be facilitated through a bridging ligand bound equatorially to Cu_{a_3} as is illustrated in Fig. 7a. This model raises the exciting possibility of modulating the interaction of Cu_{a_3} with the bridging ligand, and hence cytochrome $\underline{a}_3$, via the binding of strong exogenous ligands above or below the square plane of Cu_{a_3}. The attachment of a strong field ligand to the axial position of Cu_{a_3} should place the unpaired electron in a square plane containing the stronger field ligand (Fig. 7b). When this occurs, the interactions of the copper ion with cytochrome $\underline{a}_3$ will be modified. This model explains the results of our EPR studies of the NO-bound oxidized enzyme.

128

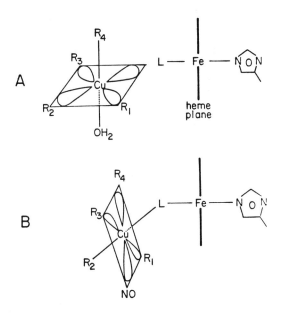

Fig. 7. Proposed structure of the cytochrome $\underline{a}_3$- Cu_{a_3} site in (A) native oxidized cytochrome $\underline{c}$ oxidase and (B) NO-bound oxidized cytochrome $\underline{c}$ oxidase. R_1, R_2, R_3, and R_4 denote endogenous ligands and L denotes the bridging ligand (which may or may not be endogenous).

If NO is a stronger field ligand than R_2 and L (Fig. 7), then binding of NO to Cu_{a_3} in the axial position would shift the axes of the ligand field. In this manner, the antiferromagnetic exchange interaction between Cu_{a_3} and cytochrome $\underline{a}_3$ would be greatly reduced; it would be eliminated if the Cu_{a_3} - NO interaction results in a diamagnetic center at the Cu_{a_3} site. In the latter case, the observation of an EPR signal from cytochrome $\underline{a}_3$ will be allowed. This model also predicts that the appearance of the cytochrome $\underline{a}_3$ high-spin EPR signal will depend on the nature of the bridging ligand, L, a prediction which is in accordance with our observations on the "oxygenated" enzyme and the effect of CN^- on the cytochrome $\underline{a}_3^{+3}$, $Cu_{a_3}^{+2}$- NO complex.

ACKNOWLEDGMENTS

This work, Contribution No. 6055, A.A. Noyes Laboratory of Chemical Physics, was partially supported by grant GM 22432 from the National Institute of General Medical Sciences, U.S. Public Health Service and by BRSG grant RR07003 awarded by the Biomedical Research Support Grant Program, Division of Research Sources, National Institutes of Health. Tom H. Stevens and Gary W. Brudvig are National Institutes of Health Predoctoral Trainees.

REFERENCES

1. Lemberg, M.R. (1969) Physiol. Rev., 49, 48-121.
2. Blumberg, W.E. and Peisach, J. (1979) in Cytochrome Oxidase, Vol. 5, King, T.E., Orii, Y., Chance, B., and Okunuki, K., eds., Elsevier-North Holland Biomedical Press, Amsterdam, pp. 153-159.
3. Nicholls, P., Petersen, L.C., Miller, M., and Hansen, F.B. (1976) Biochim. Biophys. Acta, 449, 188-196.
4. Blokzijl-Homan, M.F.J. and Van Gelder, B.F. (1971) Biochim. Biophys. Acta, 234, 493-498.
5. Falk, K., Vänngård, T., and Ångström, J. (1977) FEBS Lett., 75, 23-27.
6. Aasa, R., Albracht, S.P.J., et al. (1976) Biochim. Biophys. Acta, 422, 260-272.
7. Stevens, T.H., Brudvig, G.W., et al. (1979) Proc. Natl. Acad. Sci. USA, 76, 3320-3324.
8. Chan, S.I., Bocian, D.F., et al. (1978) in Frontiers of Biological Energetics, Vol. II, Dutton, P.L., Leigh, J., and Scarpa, A., eds., Academic Press, New York, pp. 883-888.
9. Chan, S.I., Bocian, D.F., et al. (1979) in Cytochrome Oxidase, Vol. 5, King, T.E., Orii, Y., Chance, B., and Okunuki, K., eds., Elsevier-North Holland Biomedical Press, Amsterdam, pp. 177-188.
10. Hu, V.W., Chan, S.I., and Brown, G.S. (1977) Proc. Natl. Acad. Sci. USA., 74, 3821-3825.
11. Powers, L., Blumberg, W.E., et al. (1979) in Cytochrome Oxidase, Vol. 5, King, T.E., Orii, Y., et al., eds., Elsevier-North Holland Biomedical Press, Amsterdam, pp. 189-195.
12. Stevens, T.H., Bocian, D.F., and Chan, S.I. (1979) FEBS Lett., 97, 314-316.
13. Yonetani, T., Yamamoto, H., Erman, J.E., et al. (1972) J. Biol. Chem., 247, 2447-2455.
14. Kon, H. and Kataoka, N. (1969) Biochemistry, 8, 4757-4762.
15. Kon, H. (1969) Biochem. Biophys. Res. Commun., 35, 423-427.
16. Kon, H. (1968) J. Biol. Chem., 243, 4350-4357.
17. Palmer, G., Babcock, G.T., and Vickery, L.E. (1973) Proc. Natl. Acad. Sci. USA, 73, 2206-2210.
18. Ehrenberg, A. and Szczepkowski, T.W. (1960) Acta Chem. Scand., 14, 1684-1692.
19. Van Lefuwen, F.X.R. and Van Gelder, B.F. (1978) Eur. J. Biochem., 87, 305-312.
20. Brackett, G.C., Richards, P.L., and Caughey, W.S. (1971) J. Chem. Phys., 54, 4383-4401.
21. Muijsers, A.O., Tiesjema, R.H., and Van Gelder, B.F. (1971) Biochim. Biophys. Acta, 234, 481-492.
22. Fee, J.A. (1975) Structure and Bonding, 23, 1-60.
23. Smith, T.D. and Pilbrow, J.R. (1974) Coord. Chem. Rev., 13, 173-278.
24, Winfield, M.W. (1965) in Oxidases and Related Redox Systems, Vol. 1, King, T.E., Mason, H.S., and Morrison, M., eds., John Wiley and Sons, New York, pp. 115-130.

ESR STUDIES OF SPIN LABELLED BOVINE LIVER MONOAMINE OXIDASE B

HUSSEIN ZEIDAN, KAZUHO WATANABE, LAWRENCE H. PIETTE, AND KERRY T. YASUNOBU
Department of Biochemistry-Biophysics, University of Hawaii School of Medicine,
Honolulu, Hawaii 96822

SUMMARY

Electron spin resonance (ESR) spinlabelling was used to probe the structure
of the active site of bovine liver monoamine oxidase. A series of spin labels
have been employed to explore the environment of the essential sulfhydryl groups
in the enzyme. The spin labels consisted of nitroxide coupled maleimide (or
iodoacetamide) derivatives in which the reporter nitroxide group is separated
by varying chain lengths. Both sets of spin labels preferentially bind to the
essential sulfhydryl group under appropriate conditions. From the change in the
ESR spectra, it appears that the essential sulfhydryl group is located in a
shallow crevice. The effect of a competitive inhibitor, pH, urea denaturation
on the active site conformation as measured by the spin labels was studied. The
use of double spin labelling in which a spin labelled competitive inhibitor and
spin labelled sulfhydryl reagent were used suggested that the essential sulfhydryl residues are a part of the active site.

INTRODUCTION

Monoamine oxidase is involved in the control of the levels of biogenic amines
in the brain and central nervous system.[1,2] This involvement has stimulated
considerable interest in the physicochemical and enzymatic properties of the
enzyme. The enzyme contains 4 sulfhydryl groups per subunit of enzyme only one
of which is thought to be essential.[3] Physicochemical studies have indicated
that the enzyme is made up of two identical subunits.[4] From the effect of pH on
the kinetic properties of the enzyme,[3] it was concluded that a cysteine residue
in the enzyme is catalytically important in the breakdown of the amines. Gomes
and Yasunobu[5] have shown that 5,5'dithiobis (2-nitrobenzoic avid) (DTNB) reacts
with several cysteine residues and causes inactivation of the enzyme, however,
the study concluded that only one sulfhydryl residue was essential per subunit
of the enzyme. It was suggested that this sulfhydryl residue may be a component
of the active site. To provide further evidence as to the exact function of the
residue we embarked upon a specific study of the environment of the sulfhydryl
residue using specific spin labels. Conformational changes that may take place
in the region of the sulfhydryl residue as a result of perturbations at the

active site could lend strength as to whether or not the essential sulfhydryl group is in fact at the active site.

The ESR spin-labelling technique is an extremely useful method in investigating the geometry of specific sites in a protein. The nitroxide radical, introduced through the use of a particular attaching group, acts as a reporter reflecting the motional freedom of the environment in which it resides.[6-8] In the present investigation, a series of nitroxide spin labels with a maleimide or iodoacetamide attaching group have been employed to monitor any conformational changes which take place in the environment of the essential sulfhydryl residues during the pH change, urea denaturation and binding of competitive inhibitor.[5] In addition, the "molecular dipstick" technique[8,9] was applied in which spin label probes of varying dimensions are used to probe the topology of the environment of the sulfhydryl residue.

MATERIALS AND METHODS

Materials. Enzyme was isolated as described previously.[10] The specific activity of the enzyme was 8600 μ/mg protein. The various nitroxide spin labelled sulfhydryl reagents were purchased from Syva Research Chemical Co. The other common reagents used were of reagent grade quality.

Modification of bovine liver monoamine oxidase using iodoacetic acid, N-ethylmaleimide reagents. About 2.5 mg of the enzyme in 1.0 ml was reacted with 93 mg of iodoacetic acid in 1.0 ml of potassium phosphate buffer, pH 7.4 and at 25°C. Aliquots (0.1 ml) were removed at the various time intervals and passed through a 1.3 cm column of Sephadex G-25. The enzyme was assayed for activity and also hydrolyzed with 5.7 N HCl for 24 hr at 110°. The hydrolyzates were then analyzed in the amino acid analyzer to determine the extent of labelling.

Modification of the enzyme using N-ethylmaleimide reagent was similar to iodoacetic acid reaction using 0.01 M NEM.

Methods. Electron paramagnetic resonance (EPR) measurements were recorded at room temperature on a Varian E-4 spectrometer. The field setting was 3415 gauss, the microwave frequency 9.5 GHz, and a modulation amplitude of 4.0 gauss was used throughout. Spin label concentrations were quantitated by double integration of the ESR spectra using an on-line V-72 16 K minicomputer interfaced with the ESR spectrometer. The spin label 2,2,6,6-tetramethyl-4-pyridinol-oxyl (Tempol) was used as the standard to calibrate the spin label concentrations. One important application of the computer in ESR is spectral subtraction which makes possible the resolution of a composite ESR spectrum into its component parts: strongly bound and partially immobilized spin labels.

Spectral subtraction requires that a spectrum of one or more of the components be obtained separately. Reference or standard spectra were obtained by taking the spectra of a standard maleimide spin label of known concentrations and dissolved in 95% glycerol at different temperatures. This method gives rise to ESR spectra of different immobilizations.

The digitalized spectra of the reference and the composite spectra were placed in different memory locations in the computer. If necessary, the computer shifts the spectra relative to each other so that the desired peaks are in register. Successive increments of the reference spectrum are subtracted until the component being subtracted from the composite spectrum disappears.

Modification of the enzyme with nitroxide spin labels was carried out by applying 5×10^{-4} moles of maleimide or iodoacetamide in 1 ml of 95% ethanol to a Whatman filter paper in a test tube which was then dried by a stream of nitrogen. Then 9.5×10^{-6} moles of modified enzyme dissolved in 1.0 ml of 0.05 M potassium phosphate buffer, pH 7.5 was added to the test tube. The reaction was allowed to proceed with gentle stirring at room temperature and under anaerobic conditions. It was then exhaustively dialyzed against 0.05 M potassium phosphate buffer, pH 7.5.

Amino acid analyses were performed on 24 hour acid hydrolyzates (110°), after the samples were passed through a 1.5 x 10 cm column of Sephadex G-25, on a Beckman Model 121 MB automatic amino acid analyzer as described by Spackman et al.[11]

RESULTS

Modification of enzyme. The enzyme was modified by reaction with either iodoacetic acid or N-ethylmaleimide as described in the experimental section. Three nonessential cysteine residues per subunit of enzyme reacted within 6 hours at pH 7.4 and 25°, with little loss in enzyme activity (Fig. 1a) As the remaining essential cysteine residues reacted, activity dropped progressively and after 24 hours, only 5% activity remained. Essentially similar results were obtained with iodoacetamide (results not shown). The N-ethylmaleimide, however, reacted differently (Fig. 1b). Both the essential and the nonessential sulfhydryl residues reacted at the same rates and complete inhibition occurred after 4 hours. From this study it was found that iodoacetate could be used as an effective blocking agent for the nonessential sulfhydryl residues and in all the experiments described, the $(Cys-Cm)_6$-enzyme derivative was first prepared prior to spin labelling the essential sulfhydryl groups.

Iodoacetamide spin-label and N-ethyl maleimide spin-label modification of the enzyme. A number of nitroxide containing spin labelled sulfhydryl reagents

134

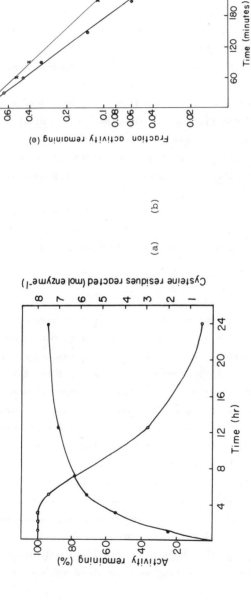

Fig. 1a. Kinetics of the inactivation and the S-β-carboxymethylation of bovine liver monoamine oxidase using iodoacetic acid. For the experiment, 2.0 mg of enzyme, specific activity of 10,000, was reacted with 9.3 mg of iodoacetic acid in 1.0 ml of potassium phosphate buffer, pH 7.4 and at 25°. Aliquots (0.1 ml) were removed at the indicated time intervals and passed through a 1.3 x 10 cm column of Sephadex G-25. The enzyme was assayed for activity and also hydrolyzed with 5.7 N HCl for 24 hours at 110°. The hydrolyzates were then analyzed in the amino acid analyzer. Enzyme activity (o-o-o-o), cysteine residues reacted (●-●-●).

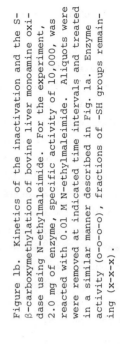

Figure 1b. Kinetics of the inactivation and the S-β-carboxymethylation of bovine liver monoamine oxidase using N-ethylmaleimide. For the experiment, 2.0 mg of enzyme, specific activity of 10,000, was reacted with 0.01 M N-ethylmaleimide. Aliquots were removed at indicated time intervals and treated in a similar manner described in Fig. la. Enzyme activity (o-o-o), fractions of -SH groups remaining (x-x-x).

have now been synthesized.[12] In the experiments to be described, the (Cys-Cm)$_6^1$- enzyme derivative was first synthesized in which 6 of the 8 sulfhydryl groups, one in each of the two subunits were blocked. This derivative was then reacted with either 3-(3-iodoacetamide)-2,2,5,5-tetramethyl-1-pyrrolidinyloxyl spin label (compound 120 in Table 1), or the 3-maleimido-2,2,5,5-tetramethyl-1-pyrrolidinyloxyl (compound 110 in Table 2) as described in section on materials and methods. This then put the spin label on the remaining two essential sulfhydryl groups. After dialysis to remove excess reagent, the ESR spectrum of the spin labelled monoamine oxidase was recorded as shown in Fig. 2 a,b. The spectrum of the bound labels exhibits two types of immobilization, one very strong represented by the 60 G splitting, and the partially immobilized type shown by the arrows. The contribution of the two forms was calculated as described in Materials and Methods. In the case of spin label 110, the strongly immobilized peak was 82%, and the partially immobilized peak was 18%. The results are presented in Fig. 3.

Reaction of enzyme with nitroxide containing iodoacetamide analogues and maleimide analogues of varying chain lengths. Several investigators have attempted to estimate the dimensions of the active sites of enzymes and immunoglobulins by means of spin labelling using a type of "molecular dipstick"[8,9,12,13] in which one can vary the dimension of the labelling probe.

The first requirement in this technique is that the particular residue of interest be in a restricted environment as indicated by the presence of a highly immobilized spectrum. Monoamine oxidase in which the nonessential sulfhydryl residues were blocked with iodoacetate and the essential cysteine residues were reacted with the maleimide derivative 110 and the iodoacetamide derivative 120 met this requirement. We have attempted to probe the environment of the essential sulfhydryls using maleimide derivatives (111, 112, 113, 114 and iodoacetamide derivatives 120, 122, 123, 124) (Tables 1 and 2) of varying lengths between the reactive portion and the nitroxide reporter.

The ESR spectrum of maleimide spin labelled 110 bound to bovine liver monoamine oxidase yields a strongly immobilized spectrum with a splitting of 64 gauss between the low and high field peaks (Fig. 4). When a methylene group is inserted between the nitroxide and the maleimide group as in 111 a spectrum with a maximum splitting of 58 gauss between the low and high field extrema (Fig. 4) is obtained. Increasing the separation between the reporter and maleimide merely changed the proportion of strongly immobilized to partially immobilized spins without affecting the extent of immobilization (Fig. 5b). The iodoacetamide derivatives show a similar effect of the change in the ratio of strong to partial immobilized spin as a function of increasing chain length (Fig. 5a).

136

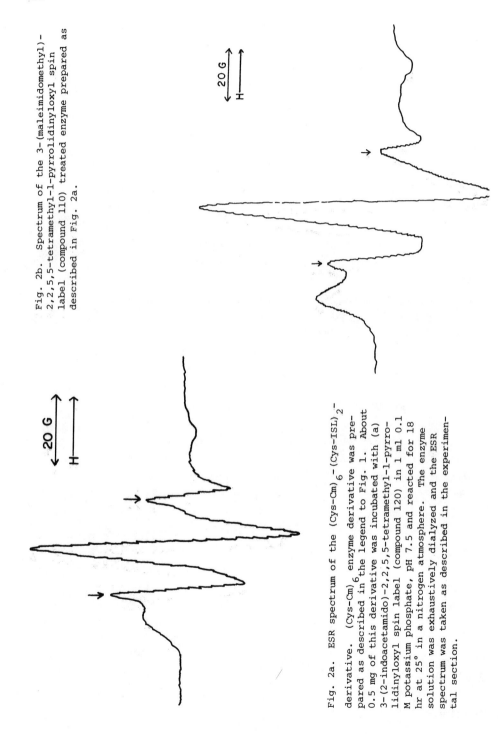

Fig. 2b. Spectrum of the 3-(maleimidomethyl)-2,2,5,5-tetramethyl-1-pyrrolidinyloxyl spin label (compound 110) treated enzyme prepared as described in Fig. 2a.

Fig. 2a. ESR spectrum of the $(Cys-Cm)_6-(Cys-ISL)_2$-derivative. $(Cys-Cm)_6$ enzyme derivative was prepared as described in the legend to Fig. 1. About 0.5 mg of this derivative was incubated with (a) 3-(2-indoacetamido)-2,2,5,5-tetramethyl-1-pyrrolidinyloxyl spin label (compound 120) in 1 ml 0.1 M potassium phosphate, pH 7.5 and reacted for 18 hr at 25° in a nitrogen atmosphere. The enzyme solution was exhaustively dialyzed and the ESR spectrum was taken as described in the experimental section.

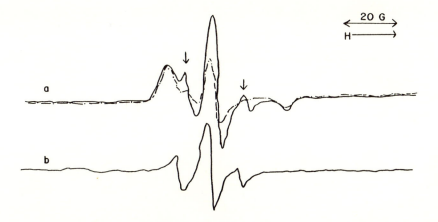

Fig. 3a. Resolution of the ESR spectrum of NEM SL 110 bound to bovine liver monoamine oxidase. The final concentration of spin label added was 3.0×10^{-4} M and the protein content was 0.5 mg/ml. The spectral subtraction was done on the Varian V-4 on-line computer. (a) Subtraction of the SI component (dotted line) from the composite, (b) the difference spectrum.

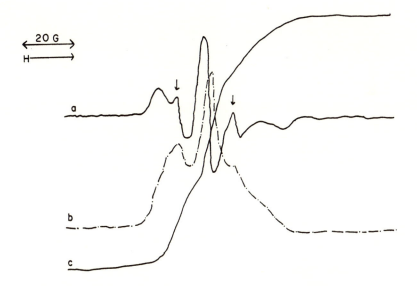

Fig. 3b. Integration of ESR spectra. (a) The strongly immobilized nitroxide is NEM SL 110 bound to $(Cys-Cm)_2$-bovine liver monoamine oxidase. The concentrations of spin label and enzyme were the same as in Fig. 2. (b) First integration of the strongly immobilized nitroxide NEM SL 110 in (a). (c) Second integration of spectrum (a)

138

TABLE 1

TYPES AND LENGTH OF NITROXIDE SPIN LABELLED SULFHYDRYL REAGENTS USED

Designation	Structure.	Systematic Name	Length ($\overset{o}{A}$)
120	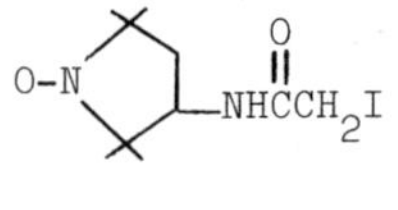 O–N⟨⟩—NHCCH$_2$I (O)	3-(2-Iodoacetamido)-2,2, 5,5-tetramethyl-1-pyrro-lidinyloxyl	5.1
122	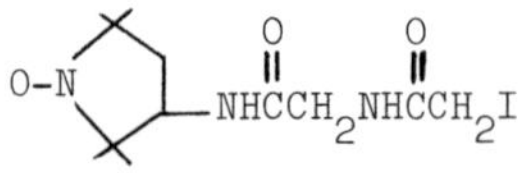 O–N⟨⟩—NHCCH$_2$NHCCH$_2$I (O,O)	3-[2-(2-Iodoacetamido)ace-tamido]-2,2,5,5-tetramethyl-1-pyrrolidinyloxyl	8.7
123	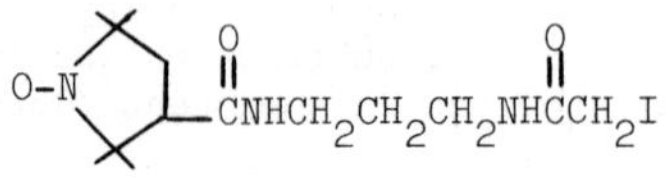 O–N⟨⟩—CNHCH$_2$CH$_2$CH$_2$NHCCH$_2$I (O,O)	3-{[3-(2-Iodoacetamido) propyl]carbamoyl}-tetra-methyl-1-pyrrolidinyloxyl	11.2
124	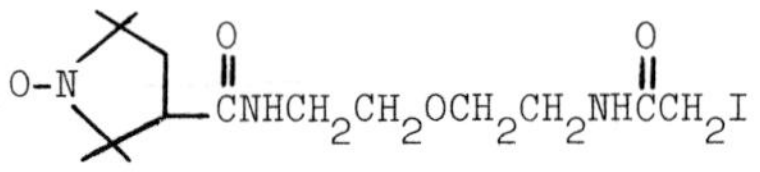 O–N⟨⟩—CNHCH$_2$CH$_2$OCH$_2$CH$_2$NHCCH$_2$I (O,O)	3-{{2-[2-(2-Iodoacetamido) ethoxyl]ethyl}carbamoyl}-2,2,5,5-tetramethyl-1-pyrrolidinyloxyl	13.5

Influence of pH on the ESR spectrum of spin-labelled bovine liver monoamine oxidase. The effect of varying the pH of solutions of spin labelled (Cys-Cm)$_6$ SPL 110 enzyme are presented in Fig. 6a. From pH 6.5 to 11, there is a gradual change in the spectrum of the 110 labelled enzyme.

The effect of raising the pH up to pH 11 resulted in the immediate appearance of spectrum Fig. 6d, characterized by an irreversible narrowing of all spectral lines. The effect of pH on enzyme activity has been examined, and it was observed that the enzyme maintains about 80% of its activity within the range of 6.5-9.5 and then loses about 95% of its activity at pH 11.5 (Fig. 6b). Thus, loss of activity with pH change parallels loss of immobilization.

Influence of urea on the ESR spectrum of (Cys-Cm)$_6$ SPL 110 derivative. An appropriate volume of a freshly prepared concentrated urea stock solution was added to a volume of spin-labelled enzyme to attain the desired final urea

TABLE 2

TYPES AND LENGTH OF NITROXIDE SPIN LABELLED SULFHYDRYL REAGENTS USED

Designation	Structure	Systematic Name	Length (Å)
110		3-Maleimido-3,3,5,5-tetra-methyl-1-pyrrolidinyloxyl	4.4
111		3-(Maleimidomethyl)-2,2,5,5-tetramethyl-1-pyrrolidinyloxyl	5.7
112		3-[2-(Maleimidoethyl)carbamoyl]-2,2,5,5-tetramethyl-1-pyrrolidinyloxyl	9.3
113		3-[2-(2-Maleimidopropyl)-carbamoyl]-2,2,5,5-tetra-methyl-1-pyrrolidinyloxyl	10.5
114		3-{2-[2-(Maleimidoethoxy)-ethyl]-carbamoyl}-2,2,5,5-tetramethyl-1-pyrrolidinyloxyl	12.9

concentration. Electron spin resonance spectra were taken repeatedly on each
sample to ensure that the unfolding was complete (usually 8-15 min); it was
essential to measure all spectra within 60-120 min after exposure to urea, as a

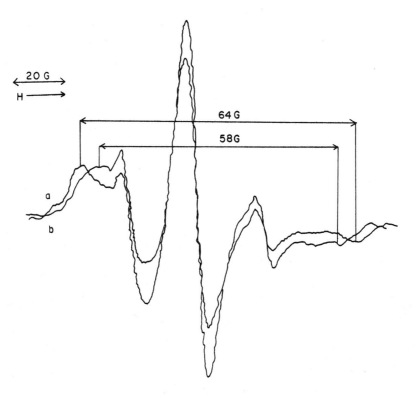

Fig. 4. ESR spectrum of the $(Cys-Cm)_6-(Cys-ISL)_2$ with spin label analogues of various side chain lengths. The exact conditions are described in the legend for Fig. 1 and were used to prepare the derivatives in which (a) spin label 110 and (b) spin label 111 were separately reacted with $(Cys-Cm)_6$-enzyme.

slow hydrolysis yields of free spin label may add to the spectral components superimposable with those measured for the urea unfolded enzyme. Electron spin resonance spectra of urea denatured enzyme is presented in Fig. 7. It is clear from the spectra that as the concentration of the urea increased, there was an increase in the ratio of partially to strongly immobilized spins, and that the sulfhydryl site opened up at 6 M urea concentration. The conformational change induced by urea was reversible up to 4 M urea, and at 6 M urea was irreversible. Our activity studies on the enzyme have demonstrated that the enzyme maintains about 70% of its activity at 4 M urea, and dropped to complete inactivation at 6 M urea (Fig. 7b). Thus, loss of activity by denaturation paralleled loss of immobilization.

Competitive inhibition. The spin label 110 which has a very short link be-tween reporter and sulfhydryl reagent is highly immobilized when reacted with

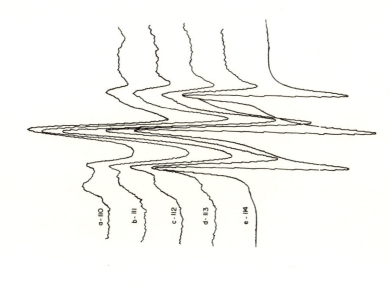

Fig. 5b. Spin-labelled maleimide analogues of varying chain lengths. The exact conditions described in Fig. 2 were used to prepare derivatives in which compounds 110-114 were separately reacted with the $(Cys-Cm)_6$-enzyme.

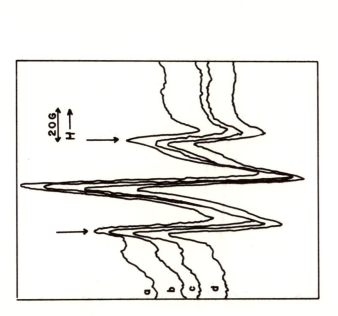

Fig. 5a. ESR spectrum of the $(Cys-Cm)_6$-enzyme reacted with (a) spin-labelled iodoacetic analogues of varying chain lengths. The exact conditions described in Fig. 2 were used to prepare derivatives in which compounds 120-124 (a-d)[22] was separately reacted with the $(Cys-Cm)_6$-enzyme.

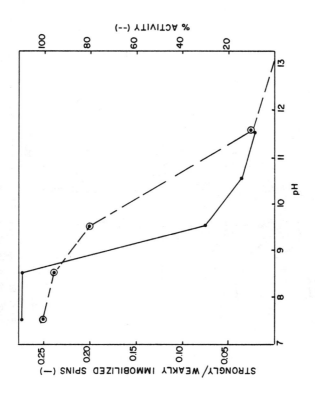

Fig. 6b. Influence of pH on bovine liver monoamine oxidase activity (--), the ratio of strongly to partially immobilized spins is plotted against the pH (---).

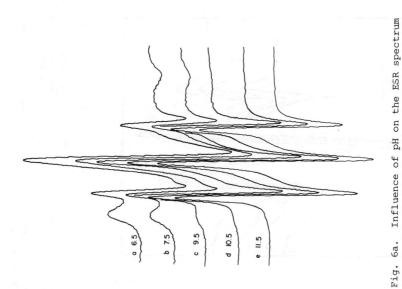

Fig. 6a. Influence of pH on the ESR spectrum of spin-labelled bovine liver monoamine oxidase. For these experiments, the pH of the (Cys-Cm)$_6$-enzyme (Cys-ISL)$_2$ solutions were adjusted to: (a) pH 6.5; (b) pH 7.5; (c) pH 9.5; (d) pH 11.5.

(Cys-Cm)$_6$ bovine liver monoamine oxidase due to the strong interaction of the reporter with the walls of the protein matrix. The combination of the ease of modifying the essential sulfhydryl residue of the enzyme and the strong immobilization of the label can be explained by moderate rigidity of the matrix, which allows small fluctuations in the size of the cleft of the essential sulfhydryl group to accommodate the radical. Due to the importance of the sulfhydryl residue, it was of interest to examine whether or not this essential sulfhydryl residue is at the active site. Previous investigations in our laboratory[14] have demonstrated that hydroxyamphetamine acts as a reversible competitive inhibitor with a dissociation constant of 2.5 x 10^{-4}. Huang et al. in 1976[15] were able to use the noncovalent spin labelled p-hydroxyamphetamine to demonstrate the presence of different forms of brain monoamine oxidase. We have studied the effect of the spin labelled p-hydroxyamphetamine, and nonlabelled p-hydroxyamphetamine on the environment of the essential sulfhydryl residue. Figure 7 shows a gradual conversion from the strongly immobilized state to the rapidly tumbling one with increasing concentration of added nonlabelled p-hydroxyamphetamine. The effect of the spin labelled p-hydroxyamphetamine on the local environment of the essential sulfhydryl group was the same as for the nonlabelled derivative, namely, to wipe out the strongly immobilized population leaving only the freely tumbling species (results are not shown). Upon dialysis of the p-hydroxyamphetamine treated enzyme, the spectrum of (Cys-Cm)$_6$ ISL protein returned to that shown in Fig. 2a.

DISCUSSION

The use of spin-labelling techniques has contributed significantly to investigations of structural, conformational, functional as well as kinetic characteristics of macromolecules, membranes, and tissues of biological systems (McConnell and McFarland, 1970).[6] It is well known that fast tumbling paramagnetic free radicals (molecules with unpaired electron spins) with anisotropic magnetic interactions in solution give rise to sharp, intense Lorenztian ESR absorption lines due to rotational averaging processes.[16-18] Radicals on the other hand, in which the tumbling motion is constrained, i.e., spin-labelled biological macromolecules in ordinary buffer, yield ESR absorption lines which are broad, and asymmetric due to the strong anisotropic interactions between the coupling of the nuclear spin and the electron spin in a nonspherical orbital. Thus environment changes that can perturb the motion of the free radical group are strongly reflected in the ESR spectrum.

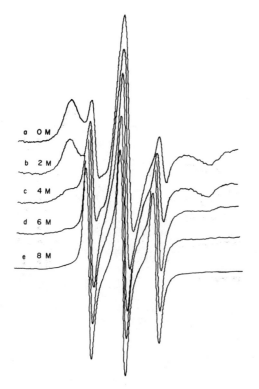

Fig. 7a.   Changes of ESR spectra of the Cys-6-(Cys-110)$_2$ enzyme with increasing urea concentrations.

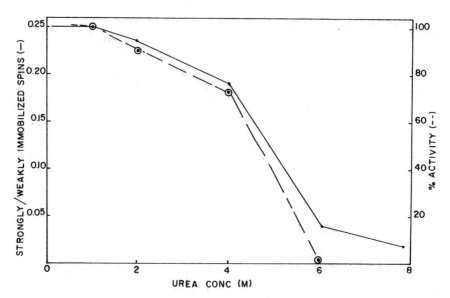

Fig. 7b.   Influence of urea on bovine liver monoamine oxidase activity (---) and the ratio of strongly-to-partially immobilized spins is plotted against urea concentrations (-).

The ESR spectrum of the bound labels in Fig. 2 exhibits two types of immobilization, one very strong represented by the 60 gauss in $(Cys-Cm)_6$-ISL-spin labelled, and 64 gauss for $(Cys-Cm)_6$-NEM-spin labelled enzyme and the partially immobilized type shown by the arrows. The two degrees of immobilization suggest two different environments. The contribution of the strongly immobilized peak is much greater than that of partially immobilized as it was calculated. In the case of spin label 110, the amount of strongly immobilized peak was 82% and the partially immobilized peak 18%.

Since the stoichiometry of the total spin label never exceeded 2-2.4 mole/mole of carboxymethylated protein, it was not possible to distinguish between two possibilities, namely, that the two types of spin label spectra, partially and strongly immobilized, respectively, resulted from: a) partial labelling of two amino acid residues in different environments, or b) from an equilibrium between two conformations that can be arrived at by the spin label being on a single amino acid residue, with the two different conformations leading to different motional freedom. Such an equilibrium between different orientations of the same spin label has been observed in spin-labelled hemoglobin,[18] and in spin-labelled muscle adenylate kinase.[19]

Amino acid analysis of the carboxymethylated enzyme indicates that carboxymethylation of only the cysteine residue occurs and none of the histidine nor lysine residues are carboxymethylated under these conditions.

The ESR spectrum of maleimide spin labelled 110 bound to bovine liver monoamine oxidase yields a strongly immobilized spectrum with a splitting of 64 gauss between the low and high field peaks (Fig. 4a). When a methylene group is inserted between the nitroxide and the maleimide group as in 111 a spectrum with a maximum splitting of 58 gauss between the low and high field extrema (Fig. 4b) is obtained. Thus, a single methylene group inserted between the nitroxide and the amino group of the maleimide forced the reporter group into a less restricted environment. Increasing the separation between the reporter and maleimide merely changed the proportion of strongly immobilized to partially immobilized spins without affecting the extent of immobilization (Fig. 5b). Thus, the change observed in going from 110 to 111 could be interpreted as the reporter group being forced out of the sulfhydryl binding site because of the increased distance, however, as the distance increases in going to 112, 113, 114, the extent of immobilization does not change but the ratio of strongly-to-partially immobilized spin changes (Fig. 5b) suggesting that either the equilibrium between these two environments is altered or the larger reagents cannot bind at the tightly immobilized site or finally that the larger reagents induce a greater conformational change in the protein in the vicinity of the sulfhydryl residue resulting in more partially immobilized spin over strongly immobilized. This conformational change need not involve the entire protein, but may merely be a conformational change at the sulfhydryl site.

The iodoacetamide derivative does not show the same change in immobilization (Fig. 5a) as a result of insertion of a single methylene group but merely shows the effect of the change in the ratio of strongly to partially immobilized spins as a function of increasing chain length. The change observed as a function of pH (Fig. 6a) is similar to that observed with the long chain sulfhydryl reagents namely, a gradual decrease in the ratio of the strongly to the partially-immobilized spins suggesting a conformational change at the sulfhydryl site. Raising the pH up to pH 11 resulted in the immediate appearance of spectrum (Fig. 6e) which shows an irreversible narrowing of all the spectral lines. As it has been previously reported[3] an essential sulfhydryl residue with a pka 10 is involved in breakdown of the amine. The ESR data reported here with the activity studies seems to support the suggestive evidence of the importance of the sulfhydryl residue and characterizes the environment where this residue is located. It is clear from the spectra (Fig. 7a) that as the concentration of the urea increased, there was an increase in the ratio of partially-to-strongly immobilized spins, and that the sulfhydryl site opened up at 6 M urea concentration. The conformational change induced by urea was reversible up to 4 M urea, and at 6 M urea was irreversible.

These data are consistent with a stepwise denaturation mechanism. The conformational change caused by the spin labelled p-hydroxyamphetamine and nonspin labelled (Fig. 8) p-hydroxyamphetamine can be explained in one of the following ways: a) both the spin-labelled p-hydroxyamphetamine and nonlabelled p-hydroxy-amphetamine bind at the same site as the sulfhydryl site, suggesting that the sulfhydryl group is at the active site, or b) the sulfhydryl group is located in another site close to the active site, and the addition of either labelled p-hydroxyamphetamine or nonlabelled p-hydroxyamphetamine causes a conformational change at this site. It was concluded by Oi and Yasunobu[3] that an essential sulfhydryl residue is involved in the breakdown of the amines. The ESR results strongly suggest that the essential sulfhydryl group is most likely located at the active site of the enzyme. In summary, our results conclude the following: 1) the loss of activity with pH change parallel loss of immobilization, 2) loss of activity by denaturation with urea-paralleled loss of immobilization, 3) use of long chain reagents induces loss of immobilization, and 4) competitive inhibition produces loss of immobilization. Therefore, we conclude that the change in immobilization must be reflecting a change in conformation at the active site.

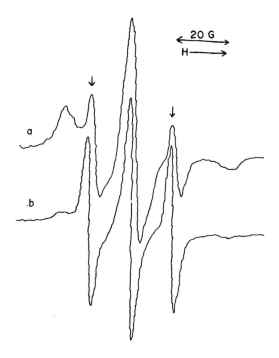

Fig. 8. Conformational changes during binding of the competitive inhibitor, p-hydroxyamphetamine were added to the $(Cys-Cm)_6-(Cys-120)_2$-enzyme in 1.0 ml of 0.05 M potassium phosphate buffer, pH 7.5. In the figure (a) represents zero; (b) 6×10^5 moles of inhibitor.

ACKNOWLEDGMENTS

This work was partially supported by grant MH 21539 and CA 15655 from the National Institutes of Health and by grant SER 77-06923 from the National Science Foundation.

REFERENCES

1. Garattini, S. and Shore, P.A., eds. (1968) Adv. in Pharmacol., Academic Press, New York, p. 1.
2. Usdin, E., ed. (1976) Neuropsychopharmacol. of Monoamines and their Regulatory Enzymes, Raven Press, New York.
3. Oi, S., Yasunobu, K.T. and Westley, J. (1971) Arch. Biochem. Biophys., 145, 557.
4. Igaue, I., Gomes, B. and Yasunobu, K.T. (1967) Biochem. Biophys. Commun., 29, 562-570.
5. Gomes, B., Igaue, I., et al. (1969) Arch. Biochem. Biophys., 132, 16.
6. McConnell, H.M. and McFarland, B.G. (1970) Q. Rev. Biophys., 3, 91.
7. Griffith, O.H. and Waggoner, A.S. (1969) Acc. Chem. Res., 2, 17.

8. Hsia, J.C. and Piette, L.H.   (1969) Arch. Biochem. Biophys., 129, 296.
9. Chignell, F.C., Starkweather, K.D. and Erlich, H.R.   (1972) Biochim. Biophys. Acta, 271, 6.
10. Minamiura, N. and Yasunobu, K.T.   (1978) Arch. Biochem. Biophys., 189, 481.
11. Spackman, D.H., Moore, S. and Stein, W.H.   (1958) Anal. Chem., 30, 1190.
12. Berliner, L.H., ed.   (1976) Spin Labelling, Academic Press, New York, p. 1.
13. Smith, I.C.P.   (1972) In Biological Applications of ESR Spectroscopy, Bothon, J.R.D., and Schwartz, H., eds., Wiley-Interscience, New York, p. 489.
14. Yasunobu, K.T. and Oi, S.   (1972) In Adv. Biochem. Psychopharmacol., Vol. 5, Costa, E. and Greengard, P., eds., Raven Press, New York, p. 91.
15. Huang, R.H., Eiduson, S. and Shih, J.N.   (1976) J. Neurochem., 26, 799.
16. Huang, R.H. and Kivelson, D.   (1974) J. Magn. Res., 14, 202.
17. Kivelson, D.   (1972) in Electron Spin Relaxation in Liquids, Plenum Press, New York, p. 213.
18. McConnell, H.M., Deal, W. and Ogata, R.T.   (1969) Biochemistry, 8, 2580.
19. Price, N.C. and Cohn, M.   (1975) J. Biol. Chem., 250, 644.

Published 1980 by Elsevier North Holland, Inc.
Liu/Mamiya/Yasunobu, eds. Frontiers in Protein Chemistry

[13]CARBON NUCLEAR MAGNETIC RESONANCE STUDY OF THE

POLYPEPTIDE CARDIAC STIMULANT ANTHOPLEURIN-A

RAYMOND S. NORTON[*] AND TED R. NORTON[**]
[*]Roche Research Institute of Marine Pharmacology, Dee Why, 2099, Australia;
[**]Department of Pharmacology, University of Hawaii, Honolulu, Hawaii 96816

ABSTRACT

Natural-abundance ^{13}C nuclear magnetic resonance (NMR) spectra of the polypeptide cardiac stimulant Anthopleurin-A are presented. The spectra contain a
number of resolved one- and two-carbon resonances, many of which have been
assigned to individual carbons in the protein. The effect of pH on the [13]C
spectrum has been investigated. In conjunction with the resonance assignments,
this yields estimates for the pKa values of the C-terminal and N-terminal residues, the side chain carboxylate of one of the two aspartic acid residues, and
the imidazolium groups of the two histidine residues. The results are discussed
in relation to a postulated model for the mode of action of Anthopleurin-A.

The sea anemone *Anthopleura xanthogrammica* contains a potent cardiac stimulant which is found to be a polypeptide consisting of 49 amino acids with a
molecular weight of about 5200.[1,2] The amino acid sequence of this polypeptide,
designated Anthopleurin-A (AP-A), has been determined.[2] Pharmacological
studies[3,4] indicate that AP-A exerts a selective, potent, positive inotropic
effect without affecting heart rate or blood pressure. Studies on its mechanism
of action on the heart suggest that it acts on the translocation of the intracellular Ca^{2+} which is available to the contractile elements.[5]

We have initiated a study by natural-abundance ^{13}C NMR spectroscopy of the
properties of AP-A in aqueous solution with the aims of characterizing its
structure in solution and its interaction, if any, with Ca^{2+} ions. In this
paper, we describe a number of resonance assignments and an analysis of the pH-
dependence of the spectrum.

RESULTS AND DISCUSSION

The AP-A was isolated as described previously.[1,2] Natural-abundance ^{13}C NMR
spectra were obtained at 15.04 MHz on a Jeol FX-60 spectrometer operating in
the pulse Fourier transform mode, with 10 mm o.d. spinning sample tubes and a
probe temperature of 30-32°C.

Figure 1 shows the ^{13}C NMR spectrum of AP-A in D_2O at pH 6.7. The overall
features of the spectrum are similar to those of other small native proteins at

150

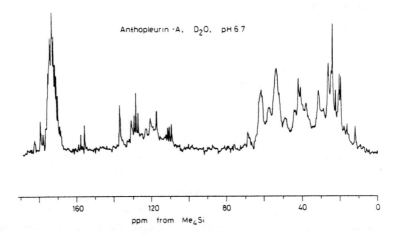

Anthopleurin ·A. D$_2$O. pH 6.7

160   120   80   40   0

ppm from Me$_4$Si

Fig. 1. Natural-abundance ^{13}C NMR spectrum of 17 mM AP-A in D$_2$O at pH 6.7.
Spectrum was recorded at 15.04 MHz with 52,000 accumulations and a recycle time
of 1.5 sec.

this magnetic field strength.[6-8] From measurements of the spin-lattice relaxation time (T$_1$) and nuclear Overhauser enhancement (NOE) of the α-carbon envelope
(45-60 ppm, see Fig. 1) we determined the rotational correlation time (τ$_R$) for
AP-A. The τ$_R$ value obtained, approximately 2 nsec, is favorable in terms of
sensitivity at our magnetic field strength of 14 kG, because it yields a substantial NOE in conjunction with T$_1$ values which are not significantly longer
than the minimum value attainable at 14 kG.[9]

The aliphatic region contains a few resolved resonances which can be assigned
to individual carbons in the molecule,[9] but spectra at higher magnetic field
strengths[10] will be required before individual aliphatic carbons can be studied
in detail.

By contrast, the aromatic carbon region contains many resolved resonances, as
shown in Fig. 2. Figure 2A was recorded under conditions of complete proton
decoupling. As expected, the spectrum consists of sharp resonances superimposed
on broader ones. Figure 2B was obtained with noise modulated off-resonance
proton decoupling, a technique which broadens resonances from methine carbons
but does not affect resonances from quaternary carbons.[7,8] Based on a comparison of Figs. 2A and 2B, we have labeled the resolved quaternary carbon resonances
Q1-Q13 and the well-resolved methine carbon resonances M1-M4. All 13 quaternary
carbon resonances (corresponding to 14 carbons) can be assigned unequivocally
to carbon types on the basis of their chemical shifts, integrated intensities, pH

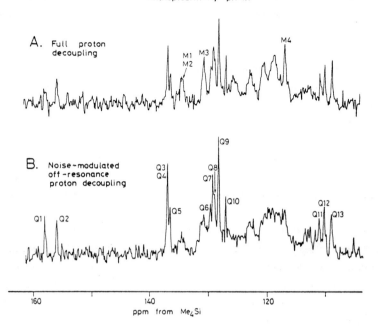

Fig. 2. Region of aromatic carbons and C of Arg-14 in natural-abundance ^{13}C NMR spectra of 17 mM AP-A in D_2O at pH 4.4. A, 30,000 accumulations, with recycle time 1.0 sec. B, 36,000 accumulations, with recycle time 2.0 sec.

dependence (see below) and their behavior in a partially-relaxed spectrum.[6-9] The four well-resolved methine aromatic carbon resonances are assigned on the basis of their chemical shifts and pH dependence.

The chemical shifts of the aromatic resonances from the single Tyr residue at position 25 in the sequence are almost identical with those of Tyr residues in small peptides and denatured proteins,[9] suggesting that the phenolic ring of Tyr-25 is not involved in strong intramolecular interactions. Furthermore, narrow two-carbon resonances are observed for the δ-carbons (peak M3) and ε-carbons (peak M4). This could arise from accidental coincidence of the chemical shifts of $C^{\delta 1}$ and $C^{\delta 2}$ at 130.9 ppm and of $C^{\epsilon 1}$ and $C^{\epsilon 2}$ at 117.3 ppm. However, it is more likely to be due to internal rotation about the $C^{\beta}- C^{\gamma}$ bond of Tyr-25 at a rate sufficiently fast to yield an exchange-averaged resonance. Indeed, it appears that the phenolic ring of Tyr-25 in AP-A may be flipping quickly enough to affect ^{13}C relaxation. Previous work[6] has shown that the linewidth of a methine aromatic carbon resonance is more sensitive to internal rotation than either the T_1 or NOE. The natural linewidths of the C^δ and C^ϵ resonances would be 10.6 Hz for τ_R = 2.2 nsec in the absence of rapid internal motion. There-fore, we expect experimental linewidths of about 13 Hz. This contrasts with the observed linewidths of about 9 Hz. Thus, it appears that the ring is undergoing internal motion at a rate comparable to the rate of overall molecular tumbling in solution. In order to determine the exact rate, it would be necessary to measure accurately the T_1, NOE and linewidth values for the C^δ and C^ϵ resonances.

We had intended to determine the pKa values for the phenolic ring of Tyr-25 and the imidazolium moieties of the histidine residues at positions 34 and 39 in the sequence. However, we found that although AP-A was stable at pH values as low as 1, it gradually lost activity at pH $\gtrsim$ 9.[9] Therefore, we have con-fined our examination of the pH-dependence of the ^{13}C spectrum to pH values below 7.5.

Peaks Q1-Q5, Q7, Q9-Q13, M3 and M4 undergo pH-dependent changes in chemical shift of 0.2 ppm or less in the pH range 2-7. At pH values > 5 peaks M1, M2, Q6 and Q8 experience pH-dependent downfield shifts which may be ascribed to imidazolium deprotonation. In a fresh sample of AP-A peak Q6 can be observed at pH values up to about 6.4 and peak Q8 at pH values up to 7. Above these respective pH values, peak broadening occurs which precludes accurate chemical shift measurements. This broadening could be due to the presence of traces of paramagnetic impurities[11] or to exchange between the imidazole and imidazolium forms at a rate comparable with the chemical shift difference between the exchanging species.[12] A significant contribution from paramagnetic impurities is indicated by the observation that broadening of peaks Q6 and Q8 occurred at lower pH values in previous samples of AP-A[9] and in the present sample after further handling. These effects are exacerbated by aggregation of AP-A which occurs above pH 9.[9]

In order to estimate the pKa values for the two His residues, we have assumed that the pH-dependence of their C^γ chemical shifts can be described by the Henderson-Hasselbalch treatment for a single noninteracting ionizable group, and that the titration shift for each C^γ resonance is 4.3 ppm, the value found in small peptides and denatured proteins.[7,13] This yields values of 6.7 $\pm$ 0.2 and 7.5 $\pm$ 0.3 for the pKa's of the His residues giving rise to peaks Q6 and O8, respectively. The occurrence of downfield shifts for both His C^γ resonances at high pH indicates that both imidazolium rings deprotonate to give the $N^{\epsilon 2}$- H imidazole tautomers.[13,14]

We have also examined the pH dependence of the carbonyl region of the spec-trum. This region contains a number of well-resolved resonances, as illustrated in Fig. 3. These consist of one two-carbon resonance, C2, and seven single-

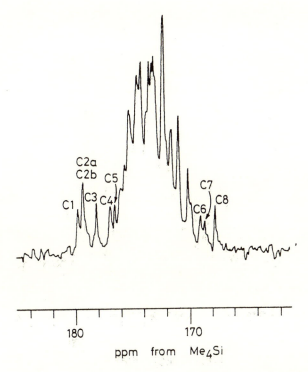

Fig. 3. Region of carbonyl carbons in natural-abundance ^{13}C NMR spectrum of 18 mM AP-A in D_2O at pH 4.6 (49,000 accumulations, recycle time 1.5 sec).

carbon resonances, C1 and C3-C8. One component of peak C2, designated peak C2b, is apparently extremely sensitive to paramagnetic impurities which broaden and/or shift the resonance, as it was not detected in our earlier work.[9] Because of these effects, we have not yet been able to examine the pH dependence of its chemical shift in detail, except to note that it is essentially pH-independent in the range 5-7. The effect of pH on the chemical shifts of the other component of peak C2 and of the resolved single carbonyl resonance is shown in Fig. 4. These data have been analyzed previously[9] and we may summarize the findings as follows: i) Peaks C1 and C2a arise from the δ- and α- carbonyl carbons, respectively, of the C-terminal Gln-49, which titrates with a pKa of 3.5; ii) Peak C3 arises from the γ-carbonyl of one of the two Asp residues (at positions 7 and 9 in the sequence). It has a pKa of about 1.7 and its titration is accomplished by significant changes in the chemical shifts of peaks C6 and C7; iii) Peak C8 corresponds to the peptide carbonyl of Gly-1, which has a pKa of approximately 8.

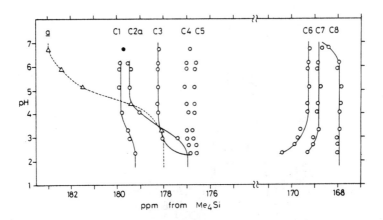

Fig. 4. Effect of pH on the chemical shifts of well-resolved carbonyl carbons of AP-A in D_2O. Protein concentrations were in the range 9-17 mM. Some samples contained an acetate impurity (not present in Fig. 3), represented by a triangle. Open circles indicate peaks that arise from one carbon in the protein, filled circles those from two carbons. The lines are titration curves.

Anthopleurin-A contains only three carboxylate groups, the side chain carboxylates of Asp-7 and -9, and the C-terminal carboxylate. We have identified carbonyl resonances from two of these moieties in our pH titration, but have not observed a second resolved resonance which titrates in the manner expected for an Asp side chain carboxylate carbon.[9,15] The chemical shift of C^{γ} of Asp in small peptides is 178.5 ppm at high pH.[15] Thus, it is possible that the second contribution to peak C2 comes from C^{γ} of the remaining Asp residue. In this case, a titrating resonance would not be observed because this peak is apparently affected by paramagnetic impurities. However, at this stage, we cannot eliminate the possibilities that the Asp C^{γ} resonance is shifted upfield and thus obscured by peak overlap, and/or that the Asp side chain carboxyl group does not titrate in the pH range examined. Further work is required to characterize the titration behavior of this residue. It should be noted that a suggestion made previously[9] that the second Asp might have a pKa of about 5 and that its titration may be indirectly responsible for broadening of one of the His C^{γ} resonances in the pH range 4.5-6 is shown to be incorrect by the data presented above.

We now consider the results of the pH titration in relation to the structure of AP-A in solution (to facilitate the discussion, the amino acid sequence of AP-A is shown in Fig. 5. The pKa of the C-terminal carboxylate and the chemical shifts of the α-COOH and δ-CONH$_2$ carbonyl resonances are close to values found in small peptides.[15] Thus, it appears that the C-terminal carboxylate is not

10
Gly - Val - Ser - Cys - Leu - Cys - Asp - Ser - Asp - Gly -

20
Pro - Ser - Val - Arg - Gly - Asn - Thr - Leu - Ser - Gly -

30
Thr - Leu - Trp - Leu - Tyr - Pro - Ser - Gly - Cys - Pro -

40
Ser - Gly - Trp - His - Asn - Cys - Lys - Ala - His - Gly -

49
Pro - Thr - Ile - Gly - Trp - Cys - Cys - Lys - Gln

Fig. 5. Amino acid sequence of AP-A.

involved in strong intramolecular interactions. The same appears to be true
for the N-terminal amino group, although our data here are incomplete. In contrast, one of the Asp side chain carboxylate groups has an unusually low pKa
value of less than 2, the value found in small peptides being about 3.9.[15] The
most likely explanation for this low pKa is that the carboxylate is close to a
positively charged group on the protein. In this case, protonation of the
carboxylate should affect ^{13}C resonances from the counterionic side chain. We
may therefore eliminate the N-terminus, Arg-14 and the two His residues from
consideration, as we can observe resonances from all of these groups at low pH
and none of them is affected significantly by pH in this range. The only other
residues with positively charged side chains at low pH are Lys-37 and Lys-48.
Thus, it appears that the side chain amino group of one of these lysine residues
may be in close proximity to the side chain carboxylate of either Asp-7 or
Asp-9. Recently, Wunderer[16] determined the disulfide bridges in the homologous
polypeptide Toxin-II from *Anemonia sulcata*. If AP-A were crosslinked in the
same manner, then Cys-6 would be attached to Cys-36. Yasunobu and co-workers
(unpublished results) have confirmed this experimentally. This suggests that
the ion pair may involve the side chains of Asp-7 and Lys-37.

Previously, Norton et al.[17] proposed that the side chain carboxylates of
Asp-7 and Asp-9 were in close proximity to the side chains of Lys-37 and His-39,
respectively. Although our data are consistent with the suggestion that a salt
bridge does exist between an Asp and a Lys residue, further work will be required

156

to demonstrate the existence of an interaction between the second Asp and His residue. Neither of the His C^γ resonances experiences any significant pH-dependent change in chemical shift below pH 5. Their imidazolium pKa values are not unusual, although the higher of the two may correspond to a His residue in the vicinity of a negatively charged group.

It has been postulated[17] that the side chain carboxylates of Asp-7 and Asp-9, together with the hydroxyl of Ser-8, may form a cation exchange site which might participate in calcium transfer. Our data on the pH-dependence of the spectrum suggest that only one of the Asp carboxylates is likely to bind cations, the other being involved in a salt bridge. Indeed, preliminary Ca^{2+} binding studies (R.W. Sleigh, unpublished results) indicate that Ca^{2+} binding to AP-A is very weak. We plan to examine the effect of Ca^{2+} on the ^{13}C NMR spectrum of AP-A.

ACKNOWLEDGMENTS

This work was supported in part by a National Institutes of Health Grant HL-15991 from the National Heart, Lung and Blood Institute to Ted R. Norton. Some of this work was carried out during the tenure of a Queen Elizabeth II Fellowship in Marine Science by Raymond S. Norton.

REFERENCES

1. Norton, T.R., Shibata, S., et al. (1976) J. Pharm. Sci., 65, 1368-1374.
2. Tanaka, M., Haniu, M., et al. (1977) Biochemistry, 16, 204-208.
3. Shibata, S., Norton, T.R., et al. (1976) J. Pharmacol. Exp. Ther., 199, 298-309.
4. Blair, R.W., Peterson, D.F., and Bishop, V.S. (1978) J. Pharmacol. Exp. Ther., 207, 271-276.
5. Shibata, S., Izumi, T., et al. (1978) J. Pharmacol. Exp. Ther., 205, 683-692.
6. Oldfield, E., Norton, R.S., and Allerhand, A. (1975) J. Biol. Chem., 250, 6368-6380.
7. Oldfield, E., Norton, R.S., and Allerhand, A. (1975) J. Biol. Chem., 250, 6381-6402.
8. Allerhand, A. (1978) Acc. Chem. Res., 11, 469-474.
9. Norton, R.S. and Norton, T.R. (1979) J. Biol. Chem., 254, in press.
10. Norton, R.S., Clouse, A.O., et al. (1977) J. Amer. Chem. Soc., 99, 79-84.
11. Wasylishen, R.E., and Cohen, J.S. (1974) Nature, 249, 847-850.
12. Dwek, R.A. (1973) Nuclear Magnetic Resonance in Biochemistry, Oxford University Press, London.
13. Reynolds, W.F., Peat, I.R., et al. (1973) J. Amer. Chem. Soc., 95, 328-331.
14. Ugurbil, K., Norton, R.S., et al. (1977) Biochemistry, 16, 886-894.
15. Keim, P., Vigna, R.A., et al. (1973) J. Biol. Chem., 248, 7811-7818.
16. Wunderer, G. (1978) Hoppe-Seyler's Z. Physiol. Chem., 359, 1193-1201.
17. Norton, T.R., Kashiwagi, M., and Shibata, S. (1978) in Drugs and Food From the Sea. Myth or Reality, Kaul, P.N., and Sindermann, C.J., eds., University of Oklahoma Press, Norman, Oklahoma, 37-50.

II
Chemical Synthesis and Chemical Studies

TOTAL SYNTHESIS OF BOVINE PANCREATIC RIBONUCLEASE A

HARUAKI YAJIMA AND NOBUTAKA FUJII
Faculty of Pharmaceutical Sciences, Kyoto University, Kyoto 606, Japan

ABSTRACT

Bovine pancreatic RNase A was synthesized by conventional techniques, using
amino acid derivatives bearing protecting groups removable by methanesulfonic
acid, i.e., Arg(MBS), Lys(Z), Cys(MBzl), Glu(OBzl) and Asp(OBzl). Relatively
small 30 peptide fragments of established purity served to construct the entire
amino acid sequence of RNase. Purity of each product was assessed by amino acid
analysis, in which Phe was selected as the diagnostic amino acid. After reduction of the sulfoxide of Cys(MBzl) partially formed during the synthesis and
Met(0) with thiophenol, all protecting groups were removed from the protected
RNase by methanesulfonic acid in the presence of m-cresol. After air oxidation
and affinity chromatographic purification, enzyme with 75% activity of RNase
was obtained.

The structure of bovine pancreatic RNase was determined in 1963 by Smith,
Stein and Moore.[1] In 1969, two groups of investigators reported the chemical
syntheses of materials with partial enzyme activity of RNase. Gutte and Merrifield[2] undertook the automated solid phase synthesis of RNase with protecting
groups removable by HF and reported that the supernatant solution of the ammonium sulfate fractionation of the trypsin treated product exhibited 78%
activity of natural RNase. The solution was estimated to contain 0.64 mg of
protein. However, this fraction claimed to be the most active component, has
not been fully chemically characterized. The Merck group[3] reported the synthesis of the S-protein (tetrahectapeptide, RNase 21-124) by conventional
techniques and obtained a solution containing ca. 2 γ of S'-activity when it was
combined with the S-peptide (RNase 1-20) derived from the natural source. This
minute amount made it impossible to carry out further purification and characterization of the end product. Unambiguous synthesis of RNase A or S-protein
remained to be accomplished.

Our synthesis of RNase was accomplished also by conventional chemical synthesis. Based on the strategy (Fig. 1) for applying our newly introduced
deprotecting reagent, methanesulfonic acid (MSA),[4] in the final step. Amino
acid derivatives bearing protecting groups removable by MSA were employed; i.e.,

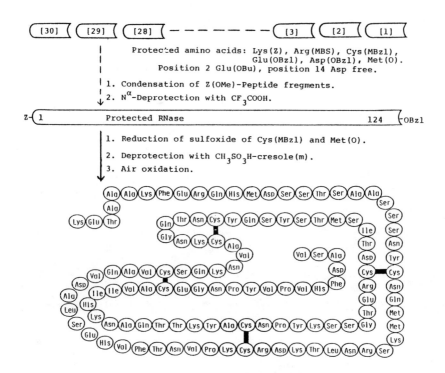

Fig. 1. Strategy for the total synthesis of RNase A.

Arg(MSA),[5] Lys(Z), Cys(MBzl), Glu(OBzl); and as the N^α-terminal protecting
group, the TFA labile Z(OMe) group[6] was employed. Relatively small 30 peptide
fragments of established purity served to construct the entire amino acid
sequence of RNase by the Rudinger's azide procedure[7] as the main tool.

Such small acyl components used in excess in each condensational step were
easily removed by precipitation from appropriate solvents or occasionally by gel-
filtration. Purity of each product was assessed by amino acid analysis, and
Phe was selected as a diagnostic amino acid, since this amino acid occurs only
three times at positions 8, 46 and 120. In the final step, the protected RNase
was treated with MSA in the presence of cresole (m) to remove all protecting
groups. After air oxidation and affinity chromatographic purification, the synthetic protein with 74% activity of native RNase was obtained. We wish to
report now, the outline of the synthesis of RNase, which we have been engaged
in for the past 3.5 years.

Synthesis of Z(OMe)-(RNase 110-124)-OBzl

 The C-terminal protected pentadecapeptide ester was synthesized by assembl-
ing 3 fragments as shown in Scheme 1. Among these fragments, construction of
the peptide bonds, Val-Pro-Val (116-118), offered some difficulty, because of
the combination of sterically hindered amino acids. Thus, the bond between
116-117 was established by the azide condensation of Z(OMe)-Pro-Tyr-Val-NHNH$_2$
and Pro and the bond between 117 and 118 was established by the fragment condensation of [2] and [1] with pentachlorophenyl trichloroacetate.[8] This reagent
was also employed for the subsequent condensation of [3].

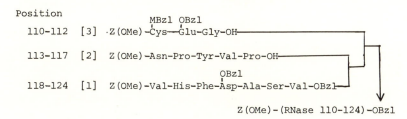

Position
		MBzl OBzl
110-112	[3]	·Z(OMe)-Cys—Glu-Gly-OH
113-117	[2]	Z(OMe)-Asn-Pro-Tyr-Val-Pro-OH
		OBzl
118-124	[1]	Z(OMe)-Val-His-Phe-Asp-Ala-Ser-Val-OBzl

Z(OMe)-(RNase 110-124)-OBzl

Scheme 1. Synthetic route to the protected pentadecapeptide ester, Z(OMe)-
(RNase 110-124)-OBzl.

Synthesis of Z(OMe)-(RNase 89-124)-OBzl

 Next, the chain elongation of Z(OMe)-(RNase 110-124)-OBzl was carried out for
peptide Z(OMe)-(890124)-OBzl by successive azide condensation of 6 peptide fragments ([4] to [9]) as shown in Scheme 2. Each condensation was performed with
ca. 1.2 to 2.5 equivalent of an acyl component, till the solution because ninhydrin negative. The progress of the reaction was examined by Shimadzu dual
wavelength tlc scanner CS-900 throughout the synthesis. A mixture of DMSO-DMF
or HMPA-DMF was employed as the solvent, depending on the solubility of amino
components. Incorporation of the fragment [5] was assessed after 72 hr acid
hydrolysis, because of the resistance of the Ile-Ile bond hydrolysis,[9] by the
usual 24 or 48 hr hydrolysis periods.

 Among the fragments tested, we first prepared Z(OMe)-Ile-Ile-Val-Ala-NHNH$_2$
and found it was not very soluble in DMF. Thus this peptide unit was subdivided into [4] and [5]. In the acid hydrolysate of Z(OMe)-(106-124)-OBzl, we
noticed the presence of D-allo-Ile[10] in 2.3%, which resulted from the racemization of Ile during the condensation of [5] with the Val-terminal amino component.
As pointed out by other authors,[11] we have to admit the possibility that racemization took place not only in the above coupling step, but also in every azide
condensation step of the peptide fragments employed. However, the above instance

Position

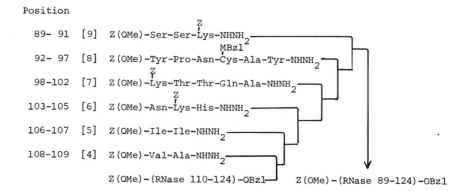

89- 91	[9]	Z(OMe)-Ser-Ser-Lys-NHNH$_2$ (Z)
92- 97	[8]	Z(OMe)-Tyr-Pro-Asn-Cys-Ala-Tyr-NHNH$_2$ (MBzl)
98-102	[7]	Z(OMe)-Lys-Thr-Thr-Gln-Ala-NHNH$_2$ (Z)
103-105	[6]	Z(OMe)-Asn-Lys-His-NHNH$_2$ (Z)
106-107	[5]	Z(OMe)-Ile-Ile-NHNH$_2$
108-109	[4]	Z(OMe)-Val-Ala-NHNH$_2$

Z(OMe)-(RNase 110-124)-OBzl   Z(OMe)-(RNase 89-124)-OBzl

Scheme 2. Synthetic route to the protected hexatriacontapeptide ester, Z(OMe)-(RNase 89-124)-OBzl.

seemed to be the worst case, since the condensation has occurred between sterically hindered amino acids, i.e., Ile and Val. In model experiments,[12] racemization was usually in less than 0.04%. We are still convinced that the azide coupling method compared with the other methods is the better method of choice for obtaining peptides of higher optical purity.

Synthesis of Z(OMe)-(RNase 69-124)-OBzl

Starting with Z(OMe)-(RNase 89-124)-OBzl, the peptide chain was elongated to protected peptide Z(OMe)-(RNase 60-124)-OBzl by successive azide condensations of 6 peptide fragments ([10] to [15]) as shown in Scheme 3. Among fragments, [10] was prepared using the substituted hydrazine, Troc-NHNH$_2$, introduced by us in 1971.[13] With the aid of this protecting group removable by Zn, the Asp(OBzl) could be incorporated into the hydrazide fragment, without exposing the ester group to hydrazine. The Met residue in the fragment [12] was oxidized to the corresponding sulfoxide[14] by tetrachloroauric (II) acid[15] to prevent partial oxidation during the synthesis and as well as during the partial alkylation step during the Z(OMe)-deprotection reaction.

In most instances, the azide condensation of these fragments were performed with 2.5 to 3.5 equivalents of an acyl component in the solvent system DMSO-DMF. Some impurity due to the incomplete removal of the acyl components was removed from Z(OMe)-(RNase 69-124)-OBzl by gel-filtration of Sephacryl S-200 with NMP-5% H$_2$O as eluent.

Synthesis of Z(OMe)-(RNase 41-124)-OBzl

Z(OMe)-(RNase 41-124)-OBzl was synthesized by further chain elongation of

Position

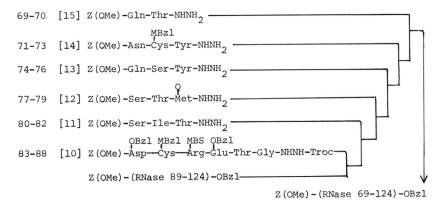

Scheme 3. Synthetic route to the protected hexapentacontapeptide ester Z(OMe)-(RNase 69-124)-OBzl.

Z(OMe)-RNase 69-124)-OBzl, for which 7 peptide fragments ([16] to [22]) were assembled successively by the azide procedure. Again, Troc-NHNH$_2$ was employed for the preparation of the fragments [19] and [20] containing Asp(OBzl) and Glu(OBzl), respectively. In addition, the fragment [16] containing the base-labile Asn-Gly bond[16] was prepared with an aid of this substituted hydrazine.

Position

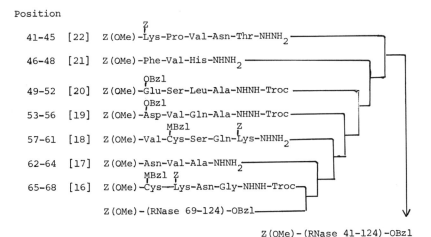

Scheme 4. Synthetic route to the protected tetraoctacontapeptide ester, Z(OMe)-(RNase 41-124)-OBzl.

Each condensation was performed with an increased amount of an acyl component
(3 to 5.5 equivalent) in a mixture of three solvents, such as DMP-DMSO-NMP or
DMF-DMSO-HMPA, because of the poor solubility of the amino peptide
components. At the step in which we introduced fragment [16], the protected
peptides and even the N^{α}-deprotected amino components remained at the origin
on the plate in any solvent systems so far examined during thin layer chromatography (tlc). Thus, in the acid hydrolysates, the ratios of newly incorporated
amino acids to that of Phe provided a method for assessing the homogeneity of
the peptide. One mole of Leu was incorporated after the first step of the
synthesis of the fragment [20] and thus this amino acid acted as an additional
diagnostic amino acid for assessment of homogeneity of products. Z(OMe)-(RNase
41-124)-OBzl was submitted to gel-filtration on Sephacryl S-200 for further
assessment of the homogeneity.

Synthesis of Z(OMe)-(RNase 21-124)-OBzl (Protected S-Protein)

Z(OMe)-(RNase 21-124)-OBzl was synthesized by further chain elongation of
Z(OMe)-(RNase 41-124)-OBzl, for which 6 peptide fragments ([23] to [28]) were
assembled as shown in Scheme 5. The fragment [23] containing Asp (OBzl) was
synthesized with an aid of Troc-NHNH$_2$. The Met residue in the fragment [25] or
[26] was also protected as its sulfoxide by oxidation with sodium metaperio-
date,[17] a mild oxidant now available. Fragment condensations of the acyl components (3 + 2 equivalents in 72 hr intervals), were performed respectively to
complete the reactions. However, for the condensation of the fragments [25],
[26] and [27], three additions of the acyl components (5 + 2 + 2 equivalents in
72 and 48 hr intervals) were employed and for fragment [28], four additions of
the acyl component (6 + 2 + 2 + 2 equivalents in 72, 48 and 48 hr intervals) were
necessary to drive the reaction to completion. The homogeneity of the protected
S-protein was ascertained by gel-filtration on Sephacryl S-200 and by amino acid
analysis.

Thus, after successive condensations of 28 peptide fragments during which
column chromatographic examinations were made at three points (positions 69, 41
and 21), we were able to obtain 4.9 g of the protected S-protein in a state
with a high degree of homogeneity. Hirschmann et al.[3] prepared the protected
S-protein by condensation of Boc-(21-64)-NHNH$_2$ and H-(65-124)-OH, but this
product, without purification and characterization, was submitted to final deprotection. No information about the coupling yield and the purity of the protected S-protein were reported.

Synthesis of Protected RNase

Synthesis of S-peptide and analogs were reported by Hofmann et al.,[18]

Position

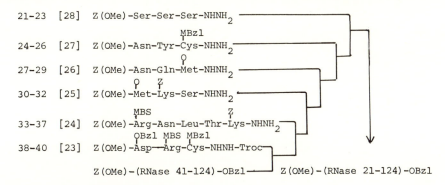

Scheme 5. Synthetic route to the protected tetrahectapeptide ester, Z(OMe)-(RNase 21-124)-OBzl (protected S-protein).

Position

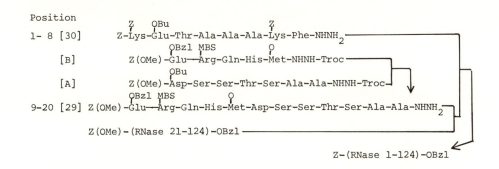

Scheme 6. Synthetic route to the protected tetracosahectapeptide ester, Z-(RNase 1-124)-OBzl (protected RNase).

Scoffone et al.[19] and others.[20] We prepared three peptide fragments necessary for the synthesis of the S-peptide and condensed it to Z(OMe)-(RNase 21-124)-OBzl as shown in Scheme 6.

The fragment [29] was prepared by condensation of fragments [A] (position 14-20) and [b] (position 9-13). The Asp(Bu^t) ester group in [A] was removed by TFA together with the Z(OMe) group prior to condensation to avoid the succinimide rearrangement[21] of this residue adjacent to Ser. After treatment with Zn, the resulting fragment [29] was submitted to condensation with the TFA treated sample of Z(OMe)-(RNase 21-124)-OBzl. Four additions of the azide peptide fragments (10 + 10 + 5 + 5 equivalents in 72, 48 and 24 hr intervals) were employed to obtain maximum condensation. After 7 days of reaction, the product

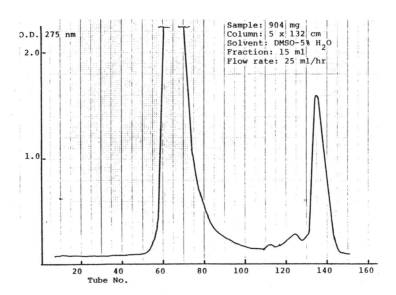

Fig. 2. Purification of the protected RNase by gel-filtration on Sepharcyl S-200.

was isolated and purified by gel-filtration on Sephacryl S-200 with DMSO-5% H_2O as eluent. The final condensation of the fragment [30] and subsequent purification of the product by gel-filtration (Fig. 2) were performed in essentially the same manner as described above. Starting with 1.66 g of the protected S-peptide, we were able to obtain 0.98 g of the protected RNase with a high degree of homogeneity. Amino acid ratios in acid hydrolysate of protected RNase and selected intermediates in the outline are listed in Table 1.

Synthesis of RNase A

Prior to deprotection, the Cys(MBzl) sulfoxide[22] formed partially during the synthesis and Met(O) were reduced by incubation with thiophenol. The protected RNase thus treated was exposed to MSA at 25-30° for 90 minutes in the presence of m-cresol. This treatment was repeated to remove all protecting groups completely. The deprotected peptide was then reduced with 2-mercapto-ethanol at pH 8.6 and subsequently subjected to air oxidation to form the 4 disulfide bridges, according to Anfinsen et al.[23] The crude product, isolated by gel-filtration on Sephadex G-75, exhibited 9% activity against yeast RNA. No sizable amount of the aggregated component was isolated in the above gel-filtration step. For further purification, affinity chromatography on agarose-5'-(4-aminophenylphosphoryl)-uridine-2'(3') phosphate was performed, according

TABLE 1

AMINO ACID RATIOS OF PROTECTED RNase AND INTERMEDIATES

Position Residue	110-124 (15)	89-124 (36)	69-124 (56)	41-124 (84)	21-124 (104)	1-124 (124)
Asp	2.03 (2)	4.20 (4)	6.14 (6)	10.20 (10)	14.20 (14)	15.34 (15)
Thr		1.60 (2)	5.76 (6)	6.88 (7)	7.69 (8)	9.39 (10)
Ser	0.90 (1)	2.47 (3)	4.51 (6)	6.46 (8)	9.59 (12)	11.93 (15)
Glu	1.07 (1)	2.08 (2)	5.13 (5)	8.23 (8)	9.39 (9)	12.64 (12)
Pro	2.19 (2)	2.62 (3)	2.76 (3)	3.79 (4)	3.81 (4)	4.11 (4)
Gly	0.99 (1)	1.08 (1)	2.17 (2)	3.18 (3)	3.02 (3)	3.09 (3)
Ala	1.02 (1)	3.89 (4)	3.98 (4)	6.89 (7)	6.85 (7)	12.11 (12)
Val	2.98 (3)	4.03 (4)	3.91 (4)	8.87 (9)	8.53 (9)	8.71 (9)
Met			0.67 (1)	0.73 (1)	2.41 (3)	3.31 (4)
Ile		1.84 (2)	2.44 (3)	2.47 (3)	2.01 (3)	1.99 (3)
Leu				1.03 (1)	1.97 (2)	2.00 (2)
Tyr	0.94 (1)	2.65 (3)	5.33 (5)	5.37 (5)	6.26 (6)	6.19 (6)
Phe	1.00 (1)	1.00 (1)	1.00 (1)	2.00 (2)	2.00 (2)	3.04 (3)
Lys		2.99 (3)	3.24 (3)	6.21 (6)	8.37 (8)	10.57 (10)
His	0.87 (1)	1.82 (2)	1.74 (2)	2.70 (3)	2.23 (3)	3.36 (4)
Arg			0.89 (1)	0.93 (1)	2.86 (3)	3.87 (4)
Cys	(1)	(2)	(4)	(6)	(8)	(8)
Recovery	89%	86%	88%	88%	89%	85%

Met (O) was not calculated.

to Wilchek and Gorecki.[24] Some fractions which passed through the column were
inactive. However, the product which emerged with 0.2 N AcOH (Fig. 3) exhibited
74% of the activity on the average of native RNase, when assayed against yeast
RNA at three different concentrations. The purified enzyme exhibited a single
band, with a trace of impurity, when examined by disc electrophoresis at pH 4.3
and its amino acid ratios in the 6 N HCl hydrolysate were in good agreement
with those of natural RNase (Sigma, 5 x cryst.) as shown in Table 2. Further
purification of synthetic RNase is in progress.

ACKNOWLEDGMENTS

The authors express their sincere appreciation to Dr. T.Y. Liu, Bureau of
Biologics, Food and Drug Administration, Bethesda, Maryland, U.S.A., for his

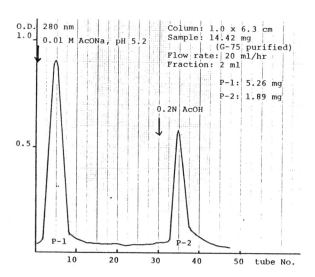

Fig. 3. Affinity chromatography of synthetic RNase A on agarose-5'-(4-amino-phenyl-phosphoryl)-uridine-2'(3')-phosphate.

valuable discussions throughout this investigation. The authors are also grateful to Dr. K. Koyama for preparation of necessary peptide fragments.

REFERENCES

1. Smith, D.G., Stein, W.H., and Moore, S. (1963) J. Biol. Chem., 238, 227-234.
2. Gutte, B. and Merrifield, R.B. (1969) J. Am. Chem. Soc., 91, 501-502; (1971) J. Biol. Chem., 246, 1922-1941.
3. Hirschmann, R., et al. (1969) J. Am. Chem. Soc., 91, 507-508.
4. Yajima, H., et al. (1975) Chem. Pharm. Bull. (Japan), 23, 1164-1166.
5. Nishimura, O. and Fujino, M. (1976) Chem. Pharm. Bull. (Japan), 24, 1568-1575.
6. Weygand, F. and Hunger, K. (1962) Chem. Ber., 95, 1-16.
7. Honzl, J. and Rudinger, J. (1961) Coll. Czech. Chem. Comm., 26, 2333-2344.
8. Fujino, M. and Hatanaka, C. (1968) Chem. Pharm. Bull. (Japan), 16, 929-932.
9. Jenkins, S.R., et al. (1969) J. Am. Chem. Soc., 91, 505-506.
10. Bodanszky, M. and Conklin, L. (1967) J.C.S. Chem. Comm., 773-774.
11. Sieber, P. and Riniker, B. (1973) In Peptides 1971, Nesvadba, H., ed., Proceeding of the 11th Peptide Symposium, North Holland Pub., p. 49-53, and other references cited therein.
12. Kemp, D.S., Bernstein, Z. and Rebek, Jr., J. (1970) J. Am. Chem. Soc., 92, 4756-4757.
13. Yajima, H. and Kiso, Y. (1971) Chem. Pharm. Bull. (Japan), 19, 420-423.
14. Iselin, B. (1961) Helv. Chim. Acta, 44, 61-78.
15. Bordignon, E., et al. (1973) J.C.S. Chem. Comm., 878-879.
16. Gráf, L., et al. (1971) Acta. Biochim. Biophys. Acad. Sci., Hung., 6, 415-148, and other references cited therein.
17. Fujii, N., et al. (1978) Chem. Pharm. Bull. (Japan), 26, 650-653.
18. Hofmann, K., et al. (1966) J. Am. Chem. Soc., 88, 4107-4109; Finn, F.M. and Hofmann, K. (1973) Accounts of Chem. Res., 6, 1969-1976 and other references cited therein.

TABLE 2

AMINO ACID RATIOS OF SYNTHETIC RNase A

	Sephadex G-75	Affinity passed F	Affinity eluted F	Natural RNase	theory
Asp	16.17	13.75	15.54	15.29	15
Thr	8.99	8.84	9.89	9.97	10
Ser	10.88	12.63	14.61	14.37	15
Glu	12.27	12.05	12.59	11.90	12
Pro	3.52	4.24	4.43	4.05	4
Gly	3.17	3.27	3.73	3.54	3
Ala	11.65	10.81	11.90	12.43	12
Cys	2.27	2.20	3.25	3.71	4
Val	8.71	8.00	8.81	8.55	9
Met	1.75	3.00	3.08	3.99	4
Ile	1.96	1.84	2.08	1.87*	3
Leu	2.00	2.00	2.00	2.00	2
Tyr	2.49	2.98	5.65	5.84	6
Phe	2.37	1.87	2.59	3.06	3
Lys	11.97	8.88	10.85	10.46	10
His	3.11	2.80	3.29	3.50	4
Arg	4.10	3.45	4.13	4.14	4
Recovery	70%	73%	77%	74%	

*Due to incomplete hydrolysis of Ile-Ile.

19. Filippi, B., Moroder, L., et al. (1975) Eur. J. Biochem., 52, 65-76 and other references cited therein.
20. Hotkamp, I.V., and Schattenkerk, C. (1979) Int. J. Peptide Protein Res., 13, 185-194 and other references cited therein.
21. Martinez, J. and Bodanszky, M. (1978) Int. J. Peptide Protein Res., 12, 277-283 and other references cited therein.
22. Yajima, H., Funakoshi, S., et al. (1979) Chem. Pharm. Bull. (Japan), 27, 1060-1061.
23. Anfinsen, C.B. and Haber, E. (1961) J. Biol. Chem., 236, 1361-1363.

STRUCTURE AND FUNCTION RELATIONSHIPS IN INSULIN: A SYNTHETIC APPROACH

P. G. KATSOYANNIS
Department of Biochemistry, Mount Sinai School of Medicine of the City University
of New York, New York, N.Y. 10029

The last two decades may well be characterized as the "Golden Age" of insulin.
Some 40 years had elapsed since Banting and Best discovered this hormone and,
in spite of the intensive efforts of many investigators all over the world, the
picture of insulin research did not change appreciably. Then in the last
twenty years, a rapid succession of breakthroughs in the chemistry and biology
of insulin spearheaded by the structure elucidation in the mid-50's, and cul-
minated by its chemical synthesis in the early 60's, have changed dramatically
our understanding of this protein. Indeed, insulin occupies now a unique
position among the other proteins. It is the first protein to be recognized as
a hormone and thus forced the acceptance of the fact that a protein can be a
hormone and vice versa; it is one of the first proteins to be crystallized; it
is the first protein whose structure has been elucidated; it is the first
protein-hormone whose three-dimensional structure has been determined; it is the
first protein-hormone where precursor processing was recognized as playing an
important role in its biogenesis; and, finally, it is the first protein to be
chemically synthesized.

This last aspect is of great importance in the clinical treatment of diabetes.
It has been emphasized from different quarters that, in the not too distant
future, the demand for insulin will most likely surpass the supply from natural
sources and hence the need for synthetic insulin or related substances will
become acute. It is true that chemically synthesized insulin is too expensive
to compete with insulin isolated from natural sources, but if the supply of the
natural hormone cannot meet the demand, then the cost of the synthetic hormone
becomes a less important factor. In addition, however, the chemical synthesis
of insulin has opened the way for the synthesis of insulin analogs and an effective approach to the problem of understanding the relationship between chemical
structure, biological activity and immunological behavior of this hormone.

Modification of the insulin molecule may alter its biological profile, either
by altering its binding ability to the receptor or by altering the "message
region" of the molecule so that the insulin analog-receptor complex cannot
initiate the biological response in the same fashion as the natural hormone-
receptor complex. Consequently, a broad spectrum of insulin analogs must be

synthesized and biochemically evaluated in terms of biological and immunological activity and receptor binding ability, in order to be able to draw meaningful conclusions regarding the functional areas of the insulin molecule. The availability of efficient methods, both for the synthesis of the insulin chains and their combination to produce the active hormone in highly purified form, make the synthesis of a variety of insulin analogs a readily attainable task. In our own laboratory, we have chemically synthesized thus far some 40 new insulins and the synthesis, at will, of any new insulin is only a matter of time. In other laboratories, also , a number of insulins were prepared from the natural hormone, either by modification of its functional groups with specific reagents or by enzymatic degradation of the insulin molecule often followed by synthetic manipulations of the degraded hormone. In this presentation, I will discuss only the biological behavior of some of the new insulins synthesized in our laboratory.

A brief outline of the procedures we currently employ for the synthesis of the insulin analogs is in order. The principle underlying these syntheses is that interation of the sulfhydryl form of the A chain with the S-sulfonated form of the B chain affords insulin in fairly good yields. This basic reaction is part of a procedure which we have worked out several years ago, during our work on the synthesis of insulin,[1-5] involving the cleavage of insulin to its individual chains and its resynthesis by the interaction of these chains.[5-8] The overall scheme of this process is shown in Fig. 1. Insulin is cleaved in its two chains by oxidative sulfitolysis, namely by treatment with sodium sulfite in the presence of a mild oxidizing agent such as sodium tetrathionate.[8] The resulting S-sulfonated chains are isolated and purified.[8] Conversion of the S-sulfonated A chain to its sulfhydryl form (on treatment with a thiol, i.e., mercaptoethanol) and interaction of the latter compound with the S-sulfonated B chain leads to the generation of insulin.[5-7] The yield of insulin produced is approximately 50 per cent of theory based on the amount of the B chain used. Isolation of insulin from the combination mixture of the A and B chains is accomplished by chromatography on a CM-cellulose column with acetate buffer (pH 3.3) and an exponential sodium chloride gradient.[5,7] The recovery of insulin by this procedure, based on the insulin activity present in the combination mixture, is approximately 50 per cent of theory.[5] Figure 2-I illustrates the chromatogram of natural bovine insulin in this chromatographic system, and Fig. 2-II shows the chromatographic pattern obtained from a combination mixture of natural bovine insulin A and B chains. As can be seen in Fig. 2-II, insulin is the exclusive product formed, among the many possible isomers, by combination of the A and B chains. Other products of the combination mixture of the

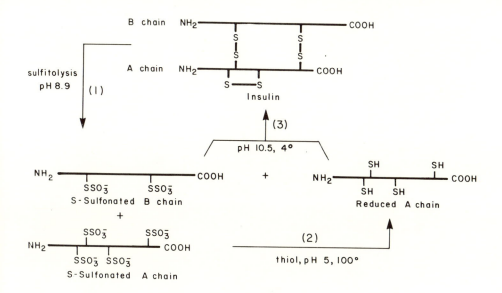

Fig. 1. Cleavage and resynthesis of insulin: (1) cleavage by oxidative sulfitolysis and separation of the resulting S-sulfonated A and B chains; (2) conversion of the S-sulfonated A chain to its sulfhydryl form; (3) interaction of
the reduced A chain with the S-sulfonated B chain to generate insulin.

A and B chains, besides the regenerated insulin (component c), are unreacted
derivative(s) of the A chain (component a) and unreacted derivative(s) of the
B chain (component b). The implication, therefore, arises that as far as
insulin is concerned, the information needed for the folding and orientation of
the A and B chains in a manner that permits their spontaneous combination to
form insulin is embodied into the primary structure of the chains. It must be
pointed out that, as in the case of the natural insulin chains (Fig. 2-II), the
chromatogram of the combination mixture of the A and B chains of every insulin
analog synthesized thus far in our laboratory does not indicate the presence of
other components, except the analog itself, that consist of A and B chains.
Even in cases when the modified insulin chains combined in low yields to produce
an insulin analog, no other component consisting of A and B chains was ever
detected.

It is interesting also to note that the unreacted derivative(s) of the A
chain (component a in Fig. 2-II) is readily isolated, resulfitolyzed and used
in combination experiments with B chain to produce insulin.[9] Theoretically,
then, all the available A chain can be converted to insulin. The unreacted B

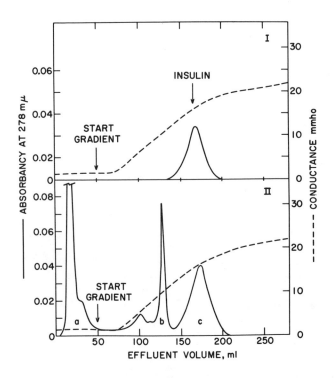

Fig. 2. Chromatography of natural bovine insulin (I) and of a combination mixture of the sulfhydryl form of the A chain with the S-sulfonated form of the B chain (II), on a 0.9 x 23 cm CM-cellulose column with acetate buffer (pH 3.3; Na⁺, 0.024 M) and an exponential NaCl gradient. The column effluent was monitored with a Gilford recording spectrophotometer and by a conductivity meter (reference 7, by permission).

chain derivative(s) (component b in Fig. 2-II), however, could not be converted to usable form for combination with A chain.[9] Obviously, it is polymerized or altered in an irreversible manner.

It is apparent from this discussion that the key intermediates for the synthesis of any insulin analog are the S-sulfonated derivatives of the corresponding A and B chains. For the synthesis of these derivatives the heneicosapeptide and triancontapeptide embodying the amino acid sequences found in the A and B chains, respectively, are synthesized with their functional groups, including the sulfhydryl functions, protected. Removal of the protecting groups and sulfitolysis of the resulting reduced chains leads to the respective S-sulfonated derivatives. The protected A and B chain derivatives are synthesized by the classical methods of peptide chemistry, namely, stepwise elongation and

fragment condensation (for a review, see reference 10). In order to improve
the yields during the various synthetic steps, we often vary the length of the
intermediate peptide fragments and employ a variety of blocking groups for the
protection of their functions. A typical example for the synthesis of an S-
sulfonated A chain analog is shown in Chart I[*] where the synthesis of the S-
sulfonated A chain of [arginine[21]-A] insulin is outlined.[11] This insulin analog
differs from the parent molecule in that the asparagine residue at position 21
of the A chain moiety has been replaced with arginine. Purification of the S-
sulfonated A chain analog is accomplished by chromatography on Ecteola-cellulose
columns with Tris·HCl buffer.[11] A typical example of the synthesis of an S-
sulfonated B chain analog is shown in Chart II, where the synthesis of the S-
sulfonated B chain of [Leucine[10]-B] insulin is outlined.[12] This insulin analog
differs from the parent molecule in that the histidine residue at position 10 of
the B chain moiety has been replaced with leucine. Purification of the S-
sulfonated B chain analog is accomplished by chromatography on CM-cellulose
columns with urea-acetate buffer at pH 4.0.[8] The purity of the synthetic modified chains is established by their chromatographic behavior in ion-exchange
columns, thin-layer electrophoresis and/or isoelectric focusing, and by amino
acid analysis after acid and enzymatic hydrolysis.

Combination of the chains to produce the insulin analog is carried out, as
was mentioned previously, by the interaction of the sulfhydryl form of the A
chain with the S-sulfonated form of the B chain.[5,6] Isolation of the insulin
analog from the combination mixture is carried out by the procedure illustrated
in Fig. 2, namely, chromatography on a CM-cellulose column with acetate buffer,
pH 3.3, and an exponential sodium chloride gradient.[5,7] A typical chromatogram
of an analog is shown in Fig. 3, and depicts chromatography of a combination
mixture of the sulfhydryl form of [Sarcosine[1], Isoasparagine[21]] A chain with the
S-sulfonated B chain of sheep insulin.[13] This analog differs from the parent
molecule in that the glycine at position 1 and asparagine at position 21 of the
A chain moeity have been replaced with sarcosine and isoasparagine, respectively.
It is apparent that, as in the case of the natural chains (Fig. 2-II), the
chromatogram of the combination mixture of the A and B chains of the analog does
not indicate the presence of other components, except the analog itself, that
consist of A and B chains. The synthetic analog is isolated from the column
effluent as the hydrochloride *via* picrate. Its purity is established by amino

---

*The following abbreviations are used: Boc, tert-butyloxycarbonyl; Bzl, benzyl;
Tos, tosyl; Bzh, diphenylmethyl; Z, benzyloxycarbonyl; Bu^t, tert-butyl.

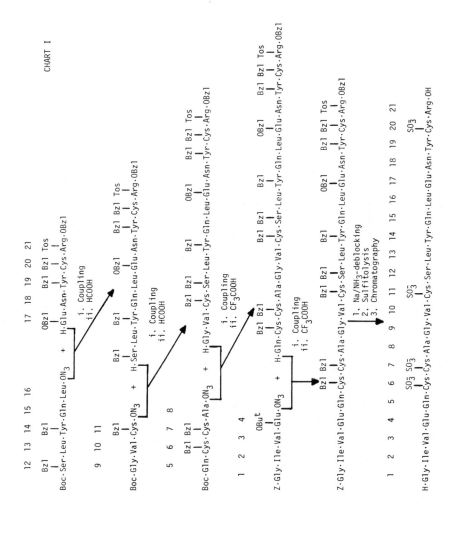

CHART I

177

CHART II

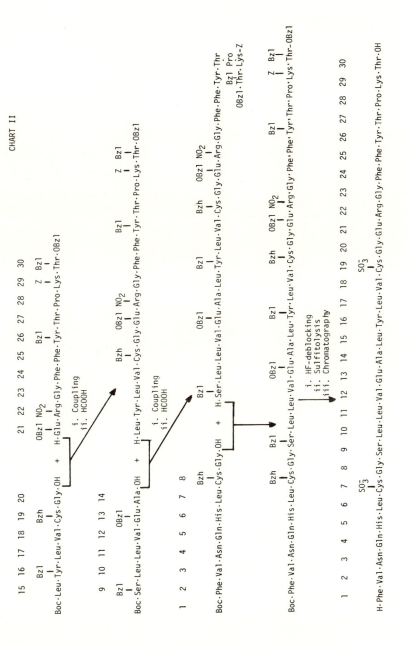

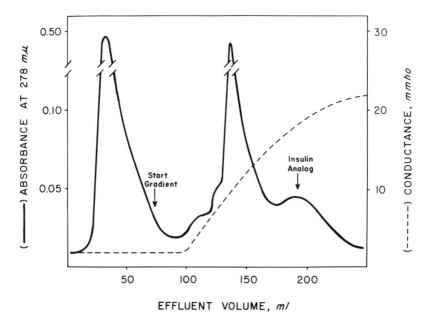

Fig. 3. Chromatography of a combination mixture of the sulfhydryl form of [Sarcosine[1], Isoasparagine[21]] A chain with the S-sulfonated form of B chain of sheep insulin on a 0.9 x 23 cm CM-cellulose column with acetate buffer (0.024 M, pH 3.3) and an exponential NaCl gradient. The column effluent was monitored by a Gilford recording spectrophotometer and a conductivity meter (reference 13, by permission).

acid analysis and by high-voltage, thin-layer electrophoresis and/or isoelectric focusing.

Our efforts thus far have been directed to the synthesis of insulin analogs (Fig. 4) involving modifications at: (a) the amino terminal regions of the A and B chains; (b) the carboxyl terminal regions of these chains; (c) the intra- and inter-chain disulfide bridges; and (d) amino acid residues, found in the interior of the insulin chains, which the X-ray structure predicts are involved in interactions of critical importance to the biological activity of the hormone.

Synthetic insulin analogs with modifications at the amino and carboxyl terminal regions of the A chain

Some of the analogs in this category are shown in Table 1. The des[Tetra-peptide A[1-4]] insulin,[14] an analog that lacks the N-terminal tetrapeptide sequence from the A chain, is inactive whereas the Deamino A[1] insulin,[15] which

TABLE 1

SYNTHETIC INSULIN ANALOGS WITH MODIFICATIONS AT THE AMINO AND CARBOXYL TERMINAL REGIONS OF THE A CHAIN*

Insulin Analog (S = Sheep) (P = Porcine)		Potency: IU/mg				Receptor binding % of insulin	References
		MC	GO	DGT	RIA		
Des[tetrapeptide A$^{1-4}$]	(S)	0	-	-	-	0.03 (LM)	14
Des[tetrapeptide A$^{1-4}$]	(P)	0	-	-	-	-	14
Deamino-A^1	(S)	7-10	-	-	-	-	15
[Sarcosine^1-A]	(S)	20	-	-	9	-	16
[L-Alanine^1-A]	(S)	7.5-9	2.3	3	-	10 (FC)	18
[D-Alanine^1-A]	(S)	10.5-12	23	24	-	100 (FC)	18
[Isoasparagine^{21}-A]	(S)	21	-	-	16	-	13
[D-Asparagine^{21}-A]	(S)	8	-	-	4	-	23
[Arginine^{21}-A]	(S)	10.5-12	9.5	-	8.6	40 (FC) ; 18 (LM)	11
[Asparaginamide^{21}-A]	(S)	17	3.5	-	2.6-4	50 (FC) ; 64 (LM)	9
[Sarcosine^1-,Isoasparagine^{21}-A]	(S)	15	-	-	7	-	13

*MC, mouse convulsion assay; GO, glucose oxidation in isolated fat cells; DGT, 2-deoxy-D-glucose transport assay; RIA, radioimmunoassay; LM, liver membranes; FC fat cells. The potency of natural insulin is 24-25 IU/mg.

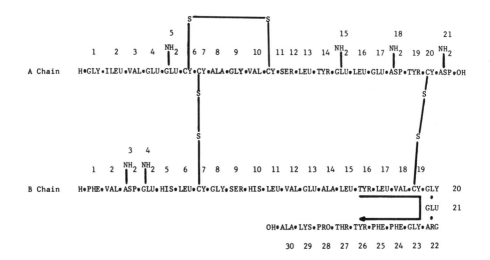

Fig. 4.   Structure of sheep insulin.

differs from the natural hormone in that the α-amino group of the A^1 glycine is replaced by a hydrogen, is ca. 35% as active as the natural protein.   The [Sarcosine^1-A] insulin,[16] in which the A^1 glycine of the parent molecule has been repaced with sarcosine, a modification that increases the basic character of that residue (pK_2 of glycine, 9.6, and of sarcosine, 10.0) is endowed with high biological activity but with a substantially decreased immunoreactivity. Investigations in other laboratories (for a review, see reference 17) also indicate that alterations at the A^1 position affect the biological activity of insulin and specifically a decrease in biological activity is concomitant with changes of the circular dichroism spectra of the analogs.   Obviously, the A^1 glycine is involved in the stabilization of the three-dimensional structure of insulin.   These findings are in agreement with the X-ray model which indicates that the A^1 glycine is involved in interactions with the carboxyl terminal region of the A chain that are important in the maintenance of the structure of this protein.   Replacement of the A^1 glycine with L- and D-alanine results in the analogs [L-Alanine^1-A] and [D-Alanine^1-A] insulins,[18] which in the in vitro assays, have relative potencies, as compared to insulin, of approximately 10% and 100%, respectively.   For both analogs the relative binding affinities to isolated fat cells are the same as the relative in vitro biological potencies. Our data are in agreement with data from other laboratories pointing to the suggestion that the A^1 glycine is one of the amino acid residues involved in

receptor binding.[19] A comparable situation regarding the binding affinity and
biological activity was encountered with several analogs which we have tested
and with analogs prepared in other laboratories.[20,21] This prompted the speculation that the insulin binding site and the site responsible for activating
cellular processes may reside in the same region of the molecule.[20,21] As we
will discuss later, however, some recent findings in our laboratory indicate
that this speculation might not be valid.

The X-ray structure of insulin indicates that the B chain folds itself
within the insulin molecule in a way that the A^{21} asparagine and the B^{22} arginine
are brought to such juxtaposition as to insure salt bridge formation between
the α-carboxyl group of the A^{21} asparagine and the guanidinium group of the B^{22}
arginine.[17,22] This structural arrangement appears to stabilize the tertiary
structure of insulin. Atomic models of the natural hormone for the area
involved in the salt bridge formation reveal that the α- and β-carboxyl groups
of A^{21} asparagine are nearly equivalent.[23] Atomic models for the same region
of [D-Asparagine^{21}-A] insulin,[23] however, reveal that the α-carboxyl group, as
compared to the natural hormone, is less favorably disposed for salt bridge
formation with the B^{22} arginine. In agreement with these structural features
are the findings that [Isoasparagine^{21}-A] insulin,[13] which has a free β-carboxyl
group and an amidated α-carboxyl group, has similar biological activity, while
[D-Asparagine^{21}-A] insulin has a considerably decreased biological activity
compared to the natural hormone.

The [Arginine^{21}-A] insulin,[11] an analog that in the A^{21} position has arginine
instead of asparagine which is present in the natural hormone, retains the A^{21}-
B^{22} salt-bridge capability, but it is only 50% as active as insulin. This
implies that other factors, perhaps steric hindrance, come into play which
involve the side chain of the A^{21} amino acid residue.

The biological behavior of [Asparaginamide^{21}-A] insulin,[9] an analog that has
the α-carboxyl group of the A^{21} asparagine amidated, is of considerable interest.
This analog possesses _in vitro_ and _in vivo_ activities approximately 15% and 70%,
respectively, of that of insulin. The low _in vitro_ activity demonstrates the
importance for biological activity of a free carboxyl group in the A^{21} position,
either for salt-bridge formation with the B^{22} arginine or for reasons not as
yet understood. The high _in vivo_ biological activity, on the other hand, may
indicate partial enzymatic deamidation of the α-carboxamide and generation of
the native hormone in the test system. Surprisingly enough, the binding
affinity of this analog to isolated fat cells or to liver membranes is considerably higher than its _in vitro_ biological activity. The disparity of
binding affinity and _in vitro_ biological activity found in this analog was

encountered with only one natural insulin, the hagfish insulin,[24] and is contrary to the observations with all other of our analogs and with all natural or
modified natural insulins. Apparently there is some degree of dissociation of
the structural features responsible for the insulin binding to the receptor and
for the expression of the biological activity of the hormone.

The [Sarcosine[1]-,Isoasparagine[21]-A] insulin,[13] an analog which differs from
the parent molecule in that the A^1 glycine and A^{21} asparagine have been
replaced with sarcosine and isoasparagine, respectively, exhibits a modestly
lower biological activity but a significantly lower immunoreactivity than the
natural hormone. A comparable situation exists with the aforementioned [Sar-
cosine[1]-A] insulin.[16] This may suggest that the change in the relative positive
charge at the amino terminus of the A chain is responsible for the considerable
decrease in the immunoreactivity of these analogs.

Synthetic insulin analogs with modifications in the intra- and inter-chain disulfide bridges

Analogs with modifications in the intra- and inter-chain cyclic systems of
insulin are shown in Table 2. Substitution of the A^6 and A^{11} cysteine residues,
which establish the 20-membered intrachain disulfide bridge in insulin, by
alanine residues results in [Alanine[6,11]-A] insulin.[25] This analog, which lacks
an intrachain disulfide bridge, has a biological activity of 8-10% of that of
insulin. The [Homocysteine[6,11]-A] insulin,[26] which because of the presence of
homocysteine instead of cysteine residues at positions A^6 and A^{11} has an
enlarged (22-membered) intrachain cyclic system, exhibits a biological activity
30% of that of the natural hormone. These findings imply that the intrachain
disulfide system does not participate functionally in the expression of the
biological activity of insulin but its presence may be necessary to generate
interactions commensurate with high biological activity. Our data are in
agreement with the X-ray structure which indicates that the intrachain bridge is
completely buried and is part of the hydrophobic core of the insulin molecule.[17]
Hence, its inaccessibility to the exterior of the insulin molecule makes it a
most unlikely candidate to be involved directly in any chemical interplay
between the hormone and its target cell. The [Homocysteine[7,20]-A] insulin,[27]
in which the cysteine residues at positions A^7 and A^{20}, which are involved in
the formation of the interchain disulfide bridges, have been replaced with
homocysteine residues and hence has an enlarged interchain cyclic system, is
biologically inactive. Apparently, the biological activity of insulin depends
critically on a particular geometry conferred on the molecule by the proper
disposition of its individual chains.

TABLE 2

SYNTHETIC INSULIN ANALOGS WITH MODIFICATIONS IN THE
INTRA- AND INTER-CHAIN DISULFIDE BRIDGES

| | Potency: IU/mg | | |
Insulin analog (sheep)	Mouse convulsion assay	Radio immuno- assay	References
[Alanine^{6,11}-A] (no intrachain -S-S-)	2-2.5	2-2.5	25
[Homocysteine^{6,11}-A] (enlarged intrachain -S-S-)	7-8	7-8	26
[Homocysteine^{7,20}-A] (enlarged interchain -S-S-)	0	2	27

The potency of natural insulin is 24-25 IU/mg.

Synthetic insulin analogs with modifications in the interior of the A chain

Our analogs, which differ from insulin in that amino acid residues found in
the interior of the A chain have been replaced by other residues, are recorded
in Table 3. The X-ray model of insulin indicates that in the insulin monomer
the A chain is folded upon itself so that the A^2 isoleucine, which is buried
and is part of the hydrophobic interior of the molecule, is in van der Waals'
contact with the A^{19} tyrosine which is on the surface of the molecule.[17] The
X-ray model further predicts that this arrangement generates interactions contributing to the maintenance of the structure and hence the biological activity
of insulin.[17] Our data are consistent with these predictions of the X-ray
model. Substitution of the A^2 isoleucine with norleucine causes a dramatic
decrease of the biological activity and the binding affinity in liver membranes,
as compared to insulin. The [Norleucine^2-A] insulin[9] shows, by circular dichroism measurements, a change in its secondary structure, and exists as a monomer
whereas the natural hormone under similar conditions exists as a dimer. Apparently the disruption of its secondary structure originates at the monomer-
monomer contact region. Modifications at the A^{19} position also generated
analogs with interesting biological behavior. Substitution of the A^{19} L-
tyrosine with the D-isomer ([D-Tyrosine^{19}-A] insulin[9]) leads to a drastic
decrease of the biological activity and the binding affinity whereas inversion
of the location of the A^{19} tyrosine and A^{18} asparagine, as is the case with
[Tyrosine^{18}-,Asparagine^{19}-,Arginine^{21}-A] insulin[9] results in total inactivation.
Similarly, replacement of the A^{19} tyrosine with phenylalanine leads to an

TABLE 3

SYNTHETIC INSULIN ANALOGS WITH MODIFICATIONS IN THE INTERIOR OF THE A CHAIN*

Insulin Analog (Sheep)	Potency: IU/mg			Receptor binding % of insulin	References
	MC	GO	RIA		
[Norleucine[2]-A]	1	0.22	3	0.6 (LM)	9
[D-Tyrosine[19]-A]	-	(0.38)	0.22	1.4 (LM)	9
[Tyrosine[18]-,Asparagine[19]-,Arginine[21]-A]	0	-	0.15-0.36	< 0.1 (LM)	9
[Phenylalanine[19]-A]	11-12	1.7	7-10	10 (LM)	9
[α,γ-Diaminobutyric acid[2]-,Glutamic acid[19]-A]	0	-	0	-	28
[Threonine[5]-A]	6.5-7.5	-	5.3	-	9
[Leucine[5]-A]	6-9	(4)	7-10	30 (LM)	9

*MC, mouse convulsion assay; GO, glucose oxidation in isolated fat cells; RIA, radioimmunoassay; LM, liver membranes; values in parentheses indicate preliminary data. The potency of natural insulin is 24-25 IU/mg.

analog, [Phenylalanine^{19}-A] insulin^9 with a considerably decreased *in vitro* biological activity and receptor binding affinity, as compared to insulin. The special steric relationship of the A^2 and A^{19} amino acid residues in the insulin molecule is further demonstrated by the fact that substitution of the weakly interacting hydrophobic residues A^2 isoleucine and A^{19} tyrosine with the potentially strongly interacting polar residues α,γ-diaminobutyric acid and glutamic acid, respectively ([α, γ-Diaminobutyric acid^2-Glutamic acid^{19}-A] insulin),[28] leads to total inactivation.

Another amino acid residue which attracted our attention is glutamine at position 5 of the A chain moiety of insulin. It is an invariant residue in insulin sequences from various species, located on the surface of the molecule[17] and, as far as is known, it has not been assigned any structural role. Substitution of this hydrophilic residue, however, either with the hydrophilic threonine or with the hydrophobic leucine, affords the [Threonine^5-A] and [Leucine^5-A] insulins,[9] respectively, which possess *in vivo* approximately 30% of the biological activity of insulin. The [Leucine^5-A] insulin, furthermore, exhibits a relative binding affinity to liver membranes and an *in vitro* biological activity of 30% and 17%, respectively, as compared to the natural hormone.[9] We are tempted to speculate that this surface residue is in the receptor-binding region of insulin.

Synthetic insulin analogs with modifications at the amino and carboxyl terminal regions of the B chain

Among the many structural features of the insulin molecule which are visualized in the X-ray model of the hormone, those of the B chain moiety are most interesting. It appears that in this chain the amino acid residues B$^{20-23}$ form a U-turn so that the carboxyl terminal segment of the chain is folded back and lied in antiparallel fashion against the central segment of that chain.[17] This arrangement generates interactions which appear to be important to the maintenance of the structure and hence the activity of insulin. The B$^{27-30}$ segment, however, does not appear to participate in interactions of any importance.[17] Similarly, the X-ray model shows that the amino terminal segment of the B chain is on the surface of the insulin molecule and from position B^7 is folded across lying in an antiparallel fashion to the central portion of the A chain. In this conformation the B$^{1-3}$ sequence does not participate in stabilizing interactions that might affect the structure and hence the activity of the hormone.[17] It appears, however, that the B^4 glutamine and particularly the B^5 histidine are involved in hydrogen bond interactions with A^{11} and A^7 cysteine residues, respectively, of importance to the structure and activity of the hormone.[17] The biological evaluation of our synthetic analogs shown in Table 4 provided data which are generally consistent with these predictions of the X-ray model. The Des[tripeptide B^{28-30}] insulin[29] is fully active, demonstrating that the carboxyl terminal tripeptide sequence of the B chain is not important to the biological activity of insulin. Subsequent elimination of amino acid residues from the carboxyl terminal region of the B chain, however, is accompanied by a gradual decline of the potency of the hormone. Des[tetrapeptide B^{27-30}] insulin[30] is approximately 54% as active as insulin and Des[pentapeptide B^{26-30}] insulin[31] is only approximately 36% as active as the natural hormone. These data demonstrate that the carboxyl terminal pentapeptide sequence of the B chain is not involved functionally in the biological activity of insulin. A modest increase of the biological activity of the latter two analogs can be effected by amidation of the carboxyl group of the newly exposed C-terminus of the truncated B chain. Thus [Des(tetrapeptide B^{27-30}),Tyr$(NH_2)^{26}$-B] and [Des (pentapeptide B^{26-30}),Phe$(NH_2)^{25}$-B] insulin[32] possess 70% and 44%, respectively, of the biological activity of the natural hormone, compared to 54% and 36%, respectively, for the nonamidated parent molecules. A possible explanation of this observation may be provided from certain predictions of the X-ray model. Indeed, as a result of the folding of the C-terminal region of the B chain, the nonpolar residues B^{26} tyrosine, B^{25} phenylalanine, B^{24} phenylalanine, B^{16} tyrosine and B^{12} valine form a nonpolar surface region which contributes to the stability of the insulin structure and appears to be the contact area between the insulin monomers in the formation of the insulin dimer.[17] It is speculated that this nonpolar region is involved in the binding of the insulin monomer to its receptor in target cells, in a fashion comparable to the insulin monomer-monomer interaction during the formation of the insulin dimer.[19,33] In view of these considerations, it is tempting to speculate that the gradual decline in the biological activity of insulin observed as amino acid residues are eliminated from the C-terminus of the B chain may be due to the proximity of a polar group (the carboxyl group of the newly formed C-terminal residue of the truncated B chain) to the hydrophobic core of the molecule, with consequent deleterious conformational changes.[32] One would then expect that conversion of the C-terminal carboxylate ion to the less polar carboxamide group will lessen its disruptive impact on the structure and hence on the activity of the molecule.

Selective Edman degradation of the B chain moiety of natural insulin led to the preparation of Des[tripeptide B^{1-3}] insulin which was found to possess approximately 70% of the biological activity of the natural hormone.[34] This indicates that the amino terminal tripeptide sequence of the B chain is not important to the biological activity of insulin. The synthetic Des[tetrapeptide

TABLE 4

SYNTHETIC INSULIN ANALOGS WITH MODIFICATIONS AT THE AMINO AND CARBOXYL TERMINAL REGIONS OF THE B CHAIN*

Insulin Analog	(H = Human) (B = Bovine)	Potency: IU/mg			Receptor binding % of insulin	References
		MC	GO	RIA		
Des[tripeptide B$^{28-30}$]	(B)	25	-	-	-	29
Des[tripeptide B$^{28-30}$]	(H)	17-20	-	-	80 (LM)	29
Des[tetrapeptide B$^{27-30}$]	(H)	11-15	-	7	-	30
Des[pentapeptide B$^{26-30}$]	(H)	8.5-9	-	11	-	31
[Des(tetrapeptide B$^{27-30}$), Tyrosinamide^{26}-B]	(H)	17	-	10.8	-	32
[Des(pentapeptide B$^{27-30}$), Phenylalaninamide^{25}-B]	(H)	10.5	-	14	-	32
Des[tetrapeptide B$^{1-4}$]	(H)	13	-	7.6	-	35
Des[pentapeptide B$^{1-5}$]	(H)	-	1.2	3.7	-	35

*MC, mouse convulsion assay; GO, glucose oxidation in isolated fat cells; RIA, radioimmunoassay; LM, liver membranes. The potency of natural insulin is 24-25 IU/mg.

B^{1-4}] insulin[35] (Table 4) was found to possess 54% of the activity of insulin, indicating that the B^4 glutamine, as is the case with the B^1 phenylalanine, B^2 valine and B^3 asparagine, is not important in the biological activity of the hormone. A drastic decrease of the biological activity of insulin occurs, however, when the B^5 histidine is also eliminated. Thus, the Des[pentapeptide B^{1-5}] insulin[35] is only 5% as active as the natural hormone. This, of course, demonstrates that the B^5 histidine, an invariant amino acid residue in all insulin species, plays a crucial role in the expression of the biological activity of insulin and supports the prediction of the X-ray model that the B^5 residue is involved in intramolecular interactions of structural importance.[17]

Synthetic insulin analogs with modifications in the interior of the B chain

Our synthetic analogs with modifications in the interior of the B chain are shown in Table 5. The [Lysine^{22}-B] insulin,[36] which differs from the parent molecule in that the arginine residue (pK of the guanidinium group, 12.48) at B^{22} has been replaced with a lysine residue (pK of the ε-amino group, 10.53) was found to possess approximately 56% of the biological activity of the natural hormone. On the other hand, guinea pig insulin, which has an aspartic residue (pK of the β-carboxyl group, 3.65) at B^{22} possesses only 9% of the activity of human insulin.[37,38] It thus appears likely that there is some relationship between biological activity and the pK of the functional group at the B^{22} position of insulin.[36] Whether this relationship involves a salt bridge formation between the A^{21} carboxylate ion and the B^{22} residue (B^{22}: Arg or Lys) as the X-ray model indicates,[17] or a hydrogen bond between the A^{21} carboxylate ion and the B^{22} Asp as might be the case with guinea pig insulin,[36] or some other not as yet understood interactions of the A^{21} and B^{22} residues, remains to be proven. Since, however, [Asparaginamide^{21}-A] insulin, which cannot form a salt bridge with the B^{22} residue, is still biologically active, the importance of this type of interaction is doubtful.

The serine residue at B^9 is the first member of the helical segment B^9-B^{19} of insulin and is on the surface of the hormone molecule.[17] It can be anticipated that substitution of the B^9 residue, since it is the first member of the helical segment and is on the outside of the protein molecule, would not be expected to significantly distort the helix and thus alter the biological profile of insulin. Yet substitution of the hydrophilic B^9 serine with the hydrophobic leucine results in the [Leucine^9-B] insulin[39] which has a potency 55% of that of insulin. The relative binding affinity of this analog, in liver membranes, is 30% of that of the natural hormone.[9] We attribute the lower biological activity of [Leucine^9-B] insulin, as compared to the natural protein, to

TABLE 5

INSULIN ANALOGS WITH ALTERATIONS IN THE INTERIOR OF THE B CHAIN*

Insulin analog (human)	Potency: IU/mg				Receptor binding % of insulin	References
	MC	GO	LIP	RIA		
[Lysine^{22}-B]	13-14	-	-	8.1	-	36
[Leucine^{9}-B]	13-14	-	-	11-12	30 (LM)	39
[Leucine^{10}-B]	10-11	-	-	9	-	12
[Lysine^{10}-B]	-	-	(5)	-	(19) (LM)	9
[Asparagine^{10}-B]	-	-	(9.5)	-	(60) (LM),(FC)	9
[Asparagine^{12}-B]	-	0.01-0.04	-	-	0.03-0.07 (LM)	9

*MC, mouse convulsion assay; GO, glucose oxidation in isolated fat cells; LIP, lipogenesis in isolated fat cells; RIA, radioimmunoassay; LM, liver membranes; FC, fat cells; values in parentheses indicate preliminary data. The potency of natural insulin is 24-25 IU/mg.

the perturbing influence that the bulky hydrophobic leucine may have on the interactions of that region with the aqueous environment or to steric effects that this bulky residue may generate during hormone-receptor interactions.

The histidine residue at position B^{10} is involved in the formation of zinc insulin hexamers.[17,22] Substitution of this residue with leucine leads to the formation of [Leucine^{10}-B] insulin[12] which was found to possess ca. 45% of the activity of the natural hormone. This demonstrates that the B^{10} histidine, in contrast to the B^5 histidine, is not involved directly in the expression of the biological profile of the hormone. The lower potency of the [Leucine^{10}-B] insulin, as compared to the natural protein, however, cannot be attributed to the substitution of the polar B^{10} histidine with the nonpolar leucine. This is evidenced by the fact that substitution of the polar B^{10} histidine with the polar lysine results in [Lysine^{10}-B] insulin, which, by preliminary assays, was found to possess only ca. 20% of the biological activity and ca. 19% of the binding affinity of the natural hormone.[9] On the other hand, replacement of the B^{10} histidine with asparagine results in [Asparagine^{10}-B] insulin,[9] which, on preliminary tests, was found to possess at least 38% of the biological activity of insulin and ca. 60% of its binding ability in liver membranes. These data indicate that the magnitude of the biological activity of insulin analogs with modifications at B^{10} position may depend primarily on the steric effects that the substituting amino acid residue at B^{10} may exert on the conformation of the protein molecule.

As was mentioned previously, the amino acid residues B^{12} valine, B^{16} tyrosine, B^{24} phenylalanine, B^{25} phenylalanine and B^{26} tyrosine, are all members of the hydrophobic core of the insulin molecule which, the X-ray model predicts,[17] is the contact region of the insulin monomers in the formation of the insulin dimer. Presumably, this hydrophobic core is part of the binding area of the insulin monomer to its receptor in the target cells.[19,33] Consistent with these predictions are the data obtained from the biochemical evaluation of the [Asparagine^{12}-B] insulin. This analog, in which the hydrophobic valine residue at B^{12} is substituted with the hydrophilic asparagine, exhibits only traces of *in vitro* biological activity and receptor binding affinity in liver membranes. These findings constitute the first evidence that supports the involvement of the hydrophobic core of insulin in receptor binding and, consequently, in the biological activity of the hormone.

Conclusion

The chemical synthesis of insulin opened the way for the synthesis of insulin analogs and the pursuit of structure-activity relationships. The synthesis and

biochemical evaluation of more than forty analogs in our laboratory already
have provided valuable information regarding the contributions of various
structural features of insulin to its biological activity and its receptor
binding affinity in target cells. The majority of our data and those obtained
in other laboratories indicate that changes in the biological activity of
insulin are related to changes in its binding affinity to the receptor of target
cells. This implies that the site involved in activating cellular processes and
the site involved in binding of insulin to its receptor may reside in the same
region of the hormone molecule. This region most likely includes the amino acid
residues A^1 glycine, A^5 glutamine, A^{19} tyrosine, A^{21} asparagine, B^{12} valine, B^{16}
tyrosine, B^{24} and B^{25} phenylalanines and B^{26} tyrosine. Recent findings, however,
indicate that the structural features of the insulin molecule involved in the
biological activity of the hormone and in its binding to the receptor, apparently
overlapping, may be dissociated. We believe that the synthesis and biochemical
evaluation of several more insulin analogs will facilitate the solution of this
problem. Finally, another interesting aspect of this work is the verification
to a remarkable degree of most of the predictions of the X-ray model regarding
intramolecular interactions and their possible effect on the biological behavior
of insulin.

ACKNOWLEDGMENTS

I am indebted to my co-workers, Drs. G. Schwartz, A. Cosmatos, N. Ferderigos,
Y. Okada, C. Zalut, J. Ginos, K. Cheng, and G. T. Burke, who have contributed
at various time intervals to this work. This research was supported by the
National Institute for Arthritis, Metabolism and Digestive Diseases, USPHS
(AM 12927).

REFERENCES

1. Katsoyannis, P.G., Tometsko, A., and Fukuda, K. (1963) J. Amer. Chem. Soc.,
   85, 2863-2865.
2. Katsoyannis, P.G., et al. (1964) J. Amer. Chem. Soc., 86, 930-932.
3. Katsoyannis, P.G., et al. (1966) J. Amer. Chem. Soc., 88, 164-166.
4. Katsoyannis, P.G., Tometsko, A., and Zalut, C. (1966) J. Amer. Chem. Soc.,
   88, 166-167; (a) Katsoyannis, P.G. (1966) Science, 154, 1509-1514.
5. Katsoyannis, P.G., et al. (1967) Biochemistry, 6, 2656-2658.
6. Katsoyannis, P.G., and Tometsko, A. (1966) Proc. Nat. Acad. Sci. USA, 55,
   1554-1561.
7. Katsoyannis, P.G., et al. (1967) Biochemistry, 6, 2642-2655.
8. Katsoyannis, P.G., et al. (1967) Biochemistry, 6, 2635-2642.
9. Unpublished data from this laboratory.
10. Katsoyannis, P.G., and Schwartz, G. (1977) in Methods in Enzymology,
    Hirs, C.H.W., and Timasheff, S.N., eds., Vol. 47, Part E, Academic Press,
    New York, pp. 501-578.
11. Ferderigos, N., et al. (1979) Int. J. Peptide Protein Res., 13, 43-53.

12. Schwartz, G., and Katsoyannis, P.G.   (1977) J. Chem. Res. (S) 220-221;
    J. Chem. Res., (M) 2453-2469.
13. Cosmatos, A., Okada, Y., and Katsoyannis, P.G.   (1976) Biochemistry, 15,
    4076-4082.
14. Katsoyannis, P.G., and Zalut, C.   (1972) Biochemistry, 11, 3065-3069.
15. Katsoyannis, P.G., and Zalut, C.   (1972) Biochemistry, 11, 1128-1132.
16. Okada, Y., and Katsoyannis, P.G. (1975) J. Amer. Chem. Soc., 97, 4366-4372.
17. Blundell, T., et al.   (1973) in Advances in Protein Chemistry, Anfrinsen,
    C.B., Edsall, J.T., and Richards, F.M., eds., Vol. 26, Academic Press, New
    York, pp. 279-402.
18. Cosmatos, A., et al.   (1978) J. Biol. Chem., 253, 6586-6590.
19. Pullen, R.A., et al.   (1976) Nature (London), 259, 369-373.
20. Freychet, P., Brandenburg, D., and Wollmer, A.   (1974) Diabetologia, 10, 1-5.
21. Gliemann, J., and Gammeltoft, S.   (1974) Diabetologia, 10, 105-113.
22. Blundell, T., et al.   (1971) in Recent Progress in Hormone Research,
    Astwood, E.B., ed., Vol. 27, Academic Press, New York, pp. 1-40.
23. Cosmatos, A., et al.   (1975) J. Chem. Soc., Perkin I, 2157-2163.
24. Emdin, S.O., Gammeltoft, S., and Gliemann, J.   (1977) J. Biol. Chem., 252,
    602-608.
25. Katsoyannis, P.G., Okada, Y., and Zalut, C.   (1973b) Biochemistry, 12,
    2516-2525.
26. Cosmatos, A., and Katsoyannis, P.G.   (1973) J. Biol. Chem., 248, 7304-7309.
27. Cosmatos, A., and Katsoyannis, P.G.   (1975) J. Biol. Chem., 250, 5315-5321.
28. Ferderigos, N., and Katsoyannis, P.G.   (1977) J. Chem. Soc., Perkin I,
    1299-1305.
29. Katsoyannis, P.G., et al.   (1971) Biochemistry, 10, 3884-3889.
30. Katsoyannis, P.G., et al.   (1973) J. Amer. Chem. Soc., 95, 6427-6434.
31. Katsoyannis, P.G., et al.   (1973) J. Chem. Soc., Perkin I, 1311-1317.
32. Cosmatos, A., Ferderigos, N., and Katsoyannis, P.G. (1979) Int. J. Peptide
    Protein Res., 14, 457-471.
33. Insulin Research Group, Academia Sinica (1974) Sci. Sinica, 17, 779-792.
34. Geiger, R., and Langner, D.   (1973) Hoppe-Seyler's Z. Physiol. Chem., 354,
    1285-1290.
35. Schwartz, G., and Katsoyannis, P.G.   (1978) Biochemistry, 17, 4550-4556.
36. Katsoyannis, P.G., Ginos, J., et al.   (1975) J. Chem. Soc., Perkin I,
    464-469.
37. Smith, L.F.   (1966) Amer. J. Med., 40, 662-666.
38. Zimmerman, A.E., Kells, D.I.C., and Yip, C.C. (1972) Biochem. Biophys. Res.
    Commun., 46, 2127-2133.
39. Schwartz, G., and Katsoyannis, P.G. (1976) Biochemistry, 15, 4071-4076.

MASS SPECTROMETRIC METHODS FOR THE STRUCTURE DETERMINATION
OF NEUROPEPTIDES, PROTEINS AND GLYCOPROTEINS

HOWARD R. MORRIS, ANNE DELL, TONY ETIENNE AND GRAHAM W. TAYLOR
Department of Biochemistry, Imperial College of Science and Technology,
London SW7, U.K.

INTRODUCTION

The solution of some modern structural problems on the frontiers of protein
chemistry, e.g., in neuro- or cell surface-chemistry, requires the analysis of
complex, "blocked" or novel structures, and/or the determination of picomolar
quantities of peptides.

Over the years, classical methodology, based in the main upon Edman degradation, has developed in two directions; (i) towards greater automation using
solid-phase or spinning cup sequencers, and (ii) towards greater sensitivity
using radiolabelling techniques. There has been a period of consolidation and
refinement of chemical methods, but both the frustration and the fascination
of life for the protein chemist is that no two peptides/proteins will behave
identically under a given set of chemical conditions, and the choice of a
sequencing procedure is, in general, an arbitrary decision. Of course certain
factors, solubility, quantity, presence of methionine, lysine, etc., may point
in one direction or another but no single classically-based method will give a
facile solution to the range of problems facing the modern protein chemist.
Thus, even for general structural studies, the development of a non-Edman
approach rouses interest, and more obviously, the efficient solution of some
difficult problems, e.g., blocked peptides, structures containing new or modified amino acids, demands a new strategy.

During the past few years, we have developed and perfected such a strategy,
a mass spectrometric one; since it is not based on Edman chemistry, problems
of repetitive yield, coupling efficiencies (either chemically or to resins)
blocked structures, quantitation of PTH's etc. do not present themselves. The
purpose of this account is to describe some of the salient features of this
new methodology and look at the results obtained by its application to genuine
problems in the fields of protein chemistry and neurochemistry.

PROTEIN SEQUENCE ANALYSIS

A current prerequisite for studying any polar molecule, e.g., peptide or
carbohydrate, in the mass spectrometer is the preparation of a volatile deriva-
time. (This is a slight over-generalization since we can obtain some information

i.e., molecular weight, but usually not structure, on underivatized substances.)
The chemical problems associated with converting a water soluble peptide (or
carbohydrate) into a chloroform soluble product in high yield took many years
to overcome, due mainly to the wide variety of functional groups found in
proteins. However, the procedures we presently use are applicable routinely
between the 5-100 nmole levels and are based upon hydrazine treatment--only for
arginine containing peptides[1]--a "short acetylation" procedure[2] (Morris, H.R.,
et al., unpublished work), followed by a "short permethylation" reaction[1,3]
(see Fig. 1 for reagent details).

It should be noted here that the methodology to be described, in general
does not require complete, or even 95% conversion of any one amino acid residue
to its derivative, because the principle of "repetitive yield", so important
in, for example, sequencer studies, does not apply to the strategy described
later.

FRAGMENTATION

The one feature of the early work which suggested that mass spectrometry may
be useful for peptide sequencing was the observation of fragmentation at the
peptide bond. This "sequence ion" formation first observed for acyl peptide
esters is an even more pronounced fragmentation pathway when the amide bond is
methylated as described above. Recognition of this important fragmentation
pathway is not, however, the only thing needed for the successful interpretation of the mass spectrum. Often, the true molecular ion is not present, and
we must be aware of other fragmentation pathways including radical or neutral
losses from the molecular ion or other fragment ions, and also formation of
"N-C cleavage fragments". The exact nuances of interpretation are too detailed
for discussion here (see the reference list for details) but suffice to say
that current ideas on peptide sequence assignment go well beyond simple recognition of "sequence ions", to an understanding and recognition usually of every
signal in the spectrum. This degree of sophistication has been generated by
the study of hundreds of peptides in protein sequence analyses, and means that
valuable sequence information can now be obtained from spectra, the interpretation of which may well have been abandoned ten years ago because of "ambigu-
ous" assignments.

STRATEGY

A most important aspect of this research is which strategy should be developed to make the method successful. Should we use "classical" enzymes, e.g.,
trypsin, for initial digestion of the protein, purifying all the peptides and

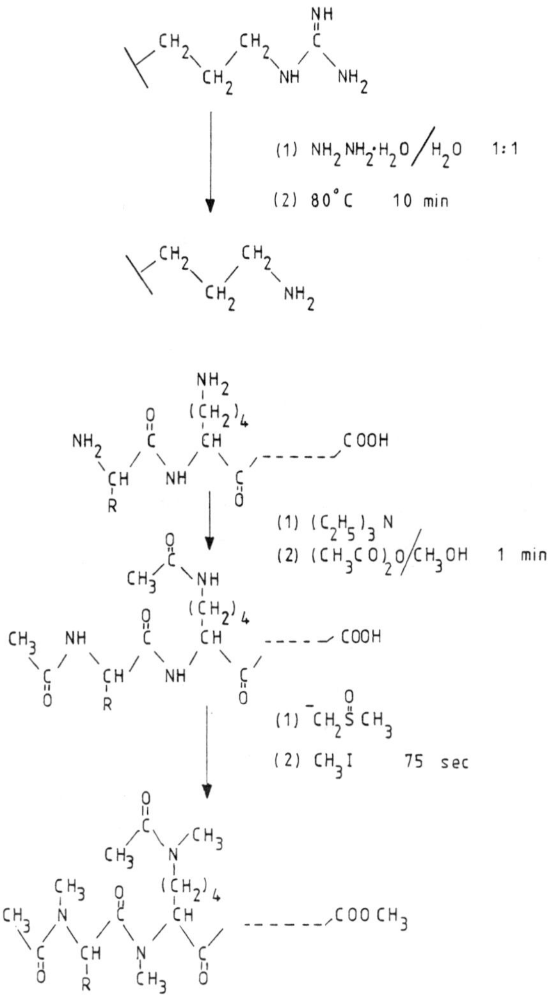

Fig. 1. General derivative-forming reactions applicable to peptides. Arginine-containing peptides require an initial hydrazinolysis step.

redigesting the larger ones? What purification methods will be compatible with mass spectrometry? Should we purify the peptides or the peptide derivatives, for example, ion exchange in nonsalt buffers or gas chromatography (G.C.) or high pressure liquid chromatography (HPLC) of derivatives? What type of deriva- tive? How can we best obtain information from the mass spectrometer, e.g., high resolution, low resolution, electron impact (E.I.), chemical ionization (C.I.), etc.? Some of these questions are interrelated; for example the initial choice of say a dipeptidyl aminopeptidase digest would lead to a certain choice of derivatives (not permethylation) and probably G.C.-M.S. separation and identification of dipeptides. Such procedures have been used on modest sized problems but as yet have not been demonstrated on unknown proteins, normally defines as a polypeptide chain greater than 100 amino acids in length.

Faced with this question of strategy some years ago, we decided that low resolution direct probe analysis would be the approach to develop. A proper justification for this choice would involve many detailed comparisons, outside the scope of the present discussion, but some advantages of this approach can be stated briefly. A key word is "mixtures." The ability to unambiguously sequence a mixture of peptides[4,5,6] is unique to mass spectrometry, but the significance of this statement is only grasped when one realizes that the isolation and preparation of individual purified peptides is the rate determining step in any classical sequencing method; even when using an automated spinning cup sequencer, rarely more than 50-100 amino acids can be placed in sequence by studying the intact protein, and then the laborious digestion, isolation and purification of individual peptides must be resorted to in order to finish the sequence. Mixture analysis provides us with a much needed way of bypassing this rate limiting step in the overall strategy. The principle is one of fractional distillation of the individual peptides from the mixture on the probe tip within the mass pectrometer using either the probe or ion source heater; the signals associated with any one peptide in the mass spectrum will rise and fall together, and depending upon the separation achieved, at different rates to other components in the mixture. The result is that each component can normally be sequenced unambiguously, the method being best suited to mixtures of two to five peptides.

To complete the strategy we need a method of producing peptides within the effective mass range of mass spectrometry, i.e., less than ~1200 MW. Cathepsins can be used but by generating such small fragments (dipeptides) we would be forfeiting the valuable "overlap" information present in larger peptides. In contrast, trypsin or chymotrypsin may produce fragments too large for the mass spectrometer. The successful solution is to choose a nonspecific protease[7,8] such as thermolysin, elastase or subtilisin.[9]

The overall strategy developed is well illustrated by our study of the sequence of dihydrofolate reductase from MTXR-resistant *L. casei*[10,11] (Morris, H.R., et al., unpublished work). Here several digests were employed, including elastase and subtilisin, and the large cyanogen bromide fragments were isolated in order to overlap and place the many peptides produced into a final sequence, Fig. 2. Peptide digests were, in general, separated by only one cation exchange

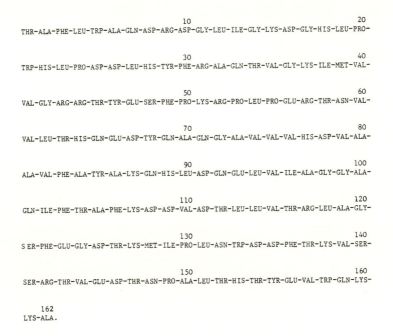

```
                                    10                                    20
THR-ALA-PHE-LEU-TRP-ALA-GLN-ASP-ARG-ASP-GLY-LEU-ILE-GLY-LYS-ASP-GLY-HIS-LEU-PRO-

                                    30                                    40
TRP-HIS-LEU-PRO-ASP-ASP-LEU-HIS-TYR-PHE-ARG-ALA-GLN-THR-VAL-GLY-LYS-ILE-MET-VAL-

                                    50                                    60
VAL-GLY-ARG-ARG-THR-TYR-GLU-SER-PHE-PRO-LYS-ARG-PRO-LEU-PRO-GLU-ARG-THR-ASN-VAL-

                                    70                                    80
VAL-LEU-THR-HIS-GLN-GLU-ASP-TYR-GLN-ALA-GLN-GLY-ALA-VAL-VAL-VAL-HIS-ASP-VAL-ALA-

                                    90                                    100
ALA-VAL-PHE-ALA-TYR-ALA-LYS-GLN-HIS-LEU-ASP-GLN-GLU-LEU-VAL-ILE-ALA-GLY-GLY-ALA-

                                    110                                    120
GLN-ILE-PHE-THR-ALA-PHE-LYS-ASP-ASP-VAL-ASP-THR-LEU-LEU-VAL-THR-ARG-LEU-ALA-GLY-

                                    130                                    140
S ER-PHE-GLU-GLY-ASP-THR-LYS-MET-ILE-PRO-LEU-ASN-TRP-ASP-ASP-PHE-THR-LYS-VAL-SER-

                                    150                                    160
SER-ARG-THR-VAL-GLU-ASP-THR-ASN-PRO-ALA-LEU-THR-HIS-THR-TYR-GLU-VAL-TRP-GLN-LYS-

162
LYS-ALA.
```

Fig. 2. Mass spectrometrically determined amino acid sequence of dihydrofolate reductase (*L. casei* MTX-R). Heterogeneity was discovered at some positions in this sequence.

column step, the effluent being screened analytically by high voltage paper electrophoresis prior to pooling of peptide mixture fractions, and derivatization. A typical spectrum from the elastase digest is seen in Fig. 3. We see two sequences emanating from normal sequence ions at m/e 114 (Gly) and m/e 128 (Ala) but also other signals at m/e 242, 440 and 585 which do not belong to any obvious N-terminus. Working backwards from m/e 585, we can see that m/e 440 corresponds to a methionine mass difference (145 m.u.) and similarly m/e 242 corresponds to a lysine mass difference from m/e 440 (198 m.u.). Both m/e 242 and m/e 440 can be seen to lose methanol indicating that they must be preceded by a hydroxy amino acid, either serine or threonine. M/e 242 minus serine (115 m.u.) gives m/e 127- a meaningless number for an "N-terminal" mass. However, m/e 242 minus threonine (129 m.u.) gives m/e 113 which we assign to N-C cleavage at aspartic acid; from experience we know that m/e 113 is not a well stabilized fragment and is normally of very low abundance. In addition to the two obvious sequences, we can in this way assign a valuable sequence Asp-Thr-Lys-Met. At higher source temperatures, the sequence shown could be extended and gave the sequence later assigned as residues 119-129 in Fig. 2.

198

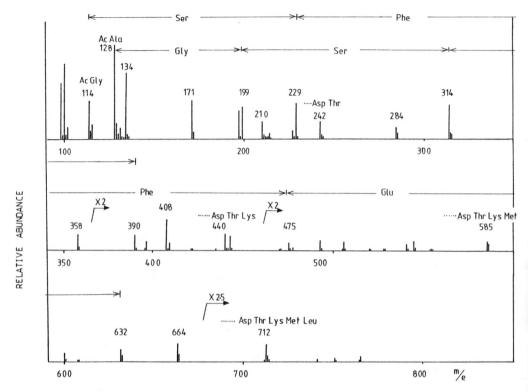

Fig. 3. Electron impact mass spectrum of an N-acetyl N,O-permethyl derivative
of a peptide fraction obtained by ion exchange chromatography of an elastase
digest of DHFR.

Dihydrofolate reductase is the first protein of unknown sequence to be fully
sequenced by mass spectrometry without the aid of classical methods. Interestingly, with a molecular weight of 18,000, this must be the largest molecule
ever to have had its structure fully determined by mass spectrometry!

A number of important proteins carry naturally blocked N-termini; an ingen-
ious modification of the above methods can be used to give (i) the nature of
the blocking group, e.g., N-acetyl, and (ii) the N-terminal sequence of up to
ten residues, and the scheme used is summarized in Fig. 4. This shows a refined version of an earlier method applied to glutamate dehydrogenase[12] and is
applicable, given 50-100 nmoles of protein, without isolating the blocked N-
terminal peptide.

One important improvement to this general sequencing method is the automatic
interpretation of mass spectra by computer, thus removing one of the skills of

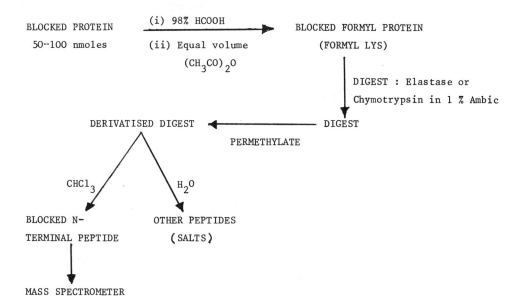

Fig. 4. Procedure for the identification of the blocked N-terminus of a protein.

the procedure and making it a more viable alternative for the protein chemist. Good mass spectra can be produced from 50-100 nmoles of peptide, and in more experienced hands 10-30 nmoles is all that is required for sequence assignment. It is still normal to purify several micromoles of protein prior to a sequence study, but studies now in progress on derivative forming technique and new ionization methods could reduce this even further. The dihydrofolate reductase work described above took place over several years, mainly because of difficulties in protein availability and purification; the total time spent during this interval amounted to some 24 man months. This compares favorably with the time needed for classical study. With our current developments in computer automation and automatic sample loading we are now working to dramatically reduce the time needed for the sequence analysis of an enzyme of this size to approximately 2 man months, using the strategy and techniques described in this section.

GLYCOPEPTIDE ANALYSIS

An increasingly important group of substances to the biochemist are glycoproteins which provide cell surface recognition sites and other important functions. Structure/function studies pose an especially difficult problem at the structural level, so much so that few glycoproteins have been fully characterized.

The problem is basically five-fold. We need information on:

1. Protein sequence
2. Carbohydrate sequence
3. Nature of protein-carbohydrate linkage; e.g., O-glycosidic
   to serine
4. Nature of saccharide-saccharide linkage; e.g., 1 ⟶ 4, 1 ⟶ 3
5. Configuration of linkage(s).

We have been studying ways of using mass spectrometric methods in a minimum number of experiments to obtain information in the above five areas. Our work can be illustrated by reference to "antifreeze glycopeptide 8" (AF8) a glyco-peptide isolated from Antarctic fish blood plasma, and one of a family of gly-coproteins some members of which are able to depress the freezing point of water some 500 times more effectively than NaCl on a molar basis. In this work[13] we found it possible to identify each of points 1 to 4 above by the study of just two derivatives.

The peptide sequence was established by studying the electron impact (N-terminal) and chemical ionization (C-terminal) spectra of an acetyl permethyl derivative of glycopeptide AF8. The spectrum showed evidence of heterogeneity in several positions of the 14 residue peptide sequence, and the points of attachment of carbohydrate were clearly visible by the assignment of dehydro-amino acids, formed by β-elimination of carbohydrate during addition of the base in the permethylation reaction. Without any necessity for pre-separation the eliminated permethyl carbohydrate was observed at a different source temperature (fractional vapourization) and the sequence assigned as hexose-aminohexose. Interpretation of linkage between saccharides is normally based upon relative abundances of fragment ions which are more pronounced in petri-methylsilyl rather than permethyl derivatives. In a separate experiment, the carbohydrate was eliminated from the peptide using sodium hydroxide, and again without any purification or separation, a trimethylsilyl derivative was pre-pared. Comparison of the spectra produced with those of standard 1 ⟶ 4 and 1 ⟶ 3 linked carbohydrates allowed an assignment of a 1 ⟶ 3 linkage in the amino disaccharide. Thus, one can deduce a remarkable amount of structural information from the two suggested experiments. While the overall structure assigned to glycopeptide AF8 in Fig. 5 is quite complex, it is worth noting that had the carbohydrate portions been composed of, say, just hexoses or larger branched saccharides, then, of course, the problem would have been much greater.

One of our present studies is the determination of the carbohydrate struc-tures in the N-terminal region of the glycoprotein prothrombin. This major task requires a number of different chemical and enzymic approaches, but

```
ALA-ALA-THR-ALA-ALA-THR-PRO-ALA-THR-ALA-ALA-THR-PRO-ALA
     |              |              |              |
     X              X              X              X


A LA-ALA-THR-ALA-ALA-THR-ALA-ALA-THR-PRO-ALA-THR-PRO-ALA
     |              |              |              |
     X              X              X              X


ALA-ALA-THR-ALA-ALA-THR-ALA-ALA-THR-ALA-ALA-THR-PRO-ALA
     |              |              |              |
     X              X              X              X
```

X = β-D-GALACTOSYL(1→3)α-N-ACETYLGALACTOSAMINE

Fig. 5. Structure of glycopeptide "AF8" as determined by mass spectrometry.

Figs. 6 and 7 show how the mass spectrometer can be used to characterize large fragments or mixtures thereof. As the figures illustrate, useful data can be obtained from quite large carbohydrate chains, and we are developing a mixture-analysis approach similar to that described earlier for peptides, in order to minimize purification after enzymic or chemical degradation. Carbohydrate structure can, of course, be quite complex and these results are preliminary, but nevertheless there seems to be good reason for optimism that at least points 1 to 4 above may be determined in present studies on complex glycopeptides using the mass spectrometric approach outlined here.

NEW AMINO ACIDS

Another important area of biochemistry in which mass spectrometry is playing a significant role is in enzymic regulation and control in the blood coagulation system.

Perhaps the most obvious strength of the mass spectrometric method is in the area of new structures where no reference compounds exist with which to make a classical interpretation based upon comparative t.l.c. mobility etc. Just such a situation arose quite surprisingly in 1974 in a sequence study of one of the blood coagulating enzymes, prothrombin. The outcome of this work was the discovery of a new amino acid γ-carboxyglutamic acid (Gla) (Fig. 8) and its sequence location in ten positions in the N-terminal region of prothrombin.[14-16] This new amino acid was missed in classical sequencing procedures (in fact, it was assigned as ordinary glutamic acid) and we can now rationalize this as due to a facile decarboxylation during Edman degradation or hydrolysis in the dansyl procedure. Since this discovery, there has been considerable

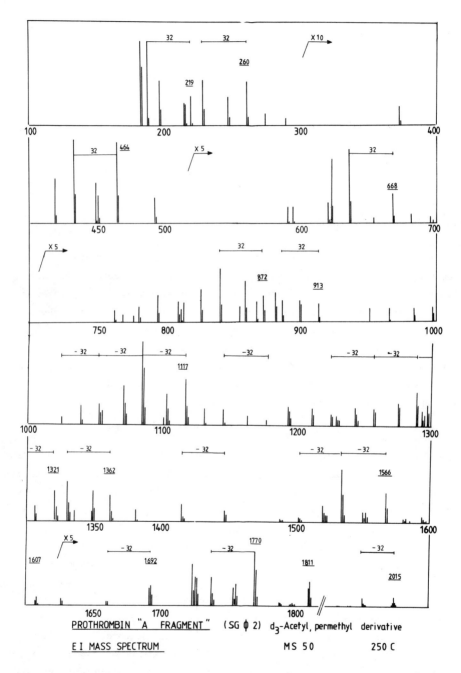

Fig. 6. Electron impact mass spectrum of the N-acetyl N,O-permethyl derivative of a neutral hydrolytic carbohydrate-containing fraction from bovine prothrombin A fragment.

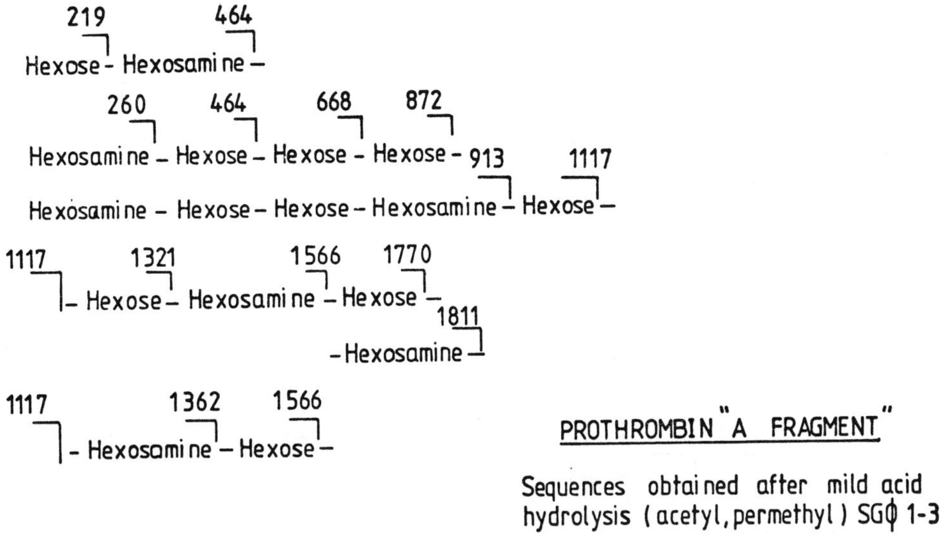

```
 219         464
   ┐           ┐
Hexose- Hexosamine —
         260       464       668       872
           ┐         ┐         ┐         ┐
     Hexosamine — Hexose — Hexose — Hexose - 913        1117
                                                 ┐          ┐
     Hexosamine — Hexose — Hexose — Hexosamine — Hexose —

1117        1321          1566     1770
  ┐           ┐             ┐        ┐
  |_ Hexose— Hexosamine — Hexose —
                                    1811
                                      ┐
                          -Hexosamine —

1117         1362      1566
  ┐            ┐         ┐
  |- Hexosamine — Hexose —
```

PROTHROMBIN "A FRAGMENT"

Sequences obtained after mild acid
hydrolysis (acetyl,permethyl) SG∅ 1-3

Fig. 7. Mass spectrometrically determined carbohydrate sequences of bovine prothrombin A fragment.

interest in the function of Gla and, in relation to this, finding whether other proteins contain thé amino acid.

Factor X is another of the important zymogens in blood, and this protein has been shown to exist in two chromatographically distinct forms, Factors X_1 and X_2.

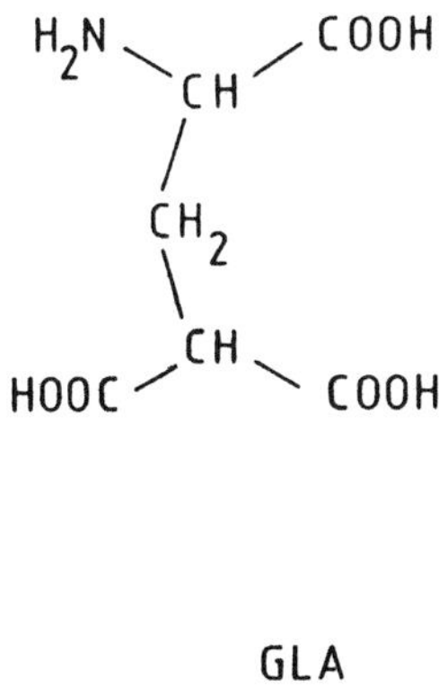

```
 H₂N        COOH
    \      /
     C H
    /
   CH₂
     \
      CH
HOOC /    \ COOH
```

GLA

Fig. 8. Structure of γ-carboxyglutamic acid as determined by mass spectrometry.

REFERENCE	PROTEIN	GLA	BASIS OF ASSIGNMENT
Enfield et al. (PNAS '75, 72, 16-19)	X_1	14?	Automated Sequencer: Poor yields of PTH's
Howard & Nelsestuen (PNAS '75, 72, 1281-5)	Unseparated X	14	Diborane Reduction and Quantitation of Dihydroxyleucine
Bucher et al (FEBS LETT. '76, 68, 293-6)	Mixed X_1 and X_2	11	Sequencer : MS of Me-PTH's
Neal et al (FEBS LETT. '76, 66, 257-60)	X_1 X_2	7-8 13	Amino Acid Analysis of Alkaline Hydrolysates

Fig. 9. Attempts to determine the Gla-content of Factors X_1 and X_2.

Both forms have biological activity, and an explanation for this phenomenon has been sought for a number of years. The discovery of Gla in prothrombin led to several groups screening for the new amino acid in Factor X (Fig. 9). As we can see from the figure, different groups have obtained conflicting results, and attempts to explain the difference between Factors X_1 and X_2 as being due to different Gla content are unconvincing.

In our original work on the identification of Gla[14,16] we established two unique mass spectrometric signals, and two unique mass differences which are diagnostic for the new amino acid, given trivial names X, X_c, Y, Y_c. The observation of these signals in a mass spectrum allows a definitive assignment not just of the presence of Gla in a peptide but also its exact location in the sequence. In a recent study of Factor X_1 and X_2[17] we have isolated peptides from the N-terminal 42 residues of each protein and sequenced them by mass spectrometry using the methods outlined earlier. Our results, shown in Fig. 10, demonstrate that the data in Fig. 9, derived mainly by automated sequencer studies, are incorrect and that each of Factors X_1 and X_2 contain twelve residues of the new amino acid Gla, and their sequences are identical.[17] The difference between these two proteins must therefore reside elsewhere in the

BF.X : ALA-ASN- -SER-PHE-LEU-<u>GLA</u>-<u>GLA</u>-VAL-LYS-GLN-GLY-ASN-LEU-<u>GLA</u>-ARG-<u>GLA</u>-CYS-LEU-<u>GLA</u>-<u>GLA</u>-ALA-
            1   2   3   4   5   6   7   8   9 10 11 12 13 14 15 16 17 18 19 20 21 22

BPT : ALA-ASN-LYS-GLY-PHE-LEU-<u>GLA</u>-<u>GLA</u>-VAL-ARG-LYS-GLY-ASN-LEU-<u>GLA</u>-ARG-<u>GLA</u>-CYS-LEU-<u>GLA</u>-<u>GLA</u>-PRO-

BF.X : CYS-SER-LEU-<u>GLA</u>-<u>GLA</u>-ALA-ARG-<u>GLA</u>-VAL-PHE-<u>GLA</u>-ASP-ALA-<u>GLA</u>-GLN-THR-ASP-<u>GLA</u>-PHE-TRP-
      23 24 25 26 27 28 29 30 31 32 33 34 35 36 37 38 39 40 41 42

BPT : CYS-SER-ARG-<u>GLA</u>-<u>GLA</u>-ALA-PHE-<u>GLA</u>-ALA-LEU-<u>GLA</u>-SER-LEU-SER-ALA-THR-ASP-ALA-PHE-TRP-

Fig. 10. Mass spectrometric sequences of calcium-binding regions of bovine prothrombin (BPT) and factors X_1 and X_2 (BF.X).

molecule, possibly in the carbohydrate attachment, and this is under investigation.

    It is interesting to note that Factor X, for some reason as yet unknown,
has two γ-carboxyglutamic acid residues more than prothrombin but that we find
ten of the twelve residues in identical positions in both sequences. This basic
structural information is now being used to formulate hypotheses for the role
of Gla in blood coagulation control mechanisms. The formation of Gla is vitamin
K-dependent and post ribosomal, and it is reasonable to assume that one of its
functions is to bind calcium. However, a large number of Ca^{++}-binding proteins
exist which do not require Gla in their sequences, making the necessity for this
new amino acid more intriguing.

NEUROCHEMISTRY

    Another area where mass spectrometry has played and is playing an important
role is in studies on neuropeptides isolated from the central nervous system.
Imagine that you have just isolated a substance from a brain homogenate which
has morphine-like activity. This activity is lost on incubation of the substance with certain proteases, suggesting that it is a peptide. After several
purification steps, the amino acid analysis suggests that you have a few tens
of nanomoles of material, but also that the substance still is not necessarily
pure and ultraviolet fluorescence data suggests that the molecule may contain
a "modified" tryptophan residue which would not have been detected by conventional amino acid analysis. This is the situation in which Hughes and co-workers
found themselves after isolating the endogenous ligand for the opiate receptor.
Classical dansyl-Edman studies gave only partial data, not correlating well with
the amino acid analysis. This situation is common in neurochemical studies,
where the confidence of sequence assignment normally associated with protein
sequence studies is not available when the sample is not necessarily of protein
origin, and therefore may have unusual amino acids, unusual linkages, or may
even not be a peptide at all.

The mass spectrometer gives a further degree of confidence in such structural assignments because it can handle "impure" materials using the mixture analysis principle, and it operates on more exact physico-chemical properties of the compound in question than just a relative mobility on a t.l.c. plate.

The assignment of the definitive structure of the above substance, Enkephalin, was made by studying the mass spectrum of an acetyl permethyl derivative, and Enkephalin was found to be a mixture of two pentapeptides, Tyr-Gly-Gly-Phe-Met and Tyr-Gly-Gly-Phe-Leu.[18] The amino acid analysis data had indicated a longer peptide, but once this mass spectrometric assignment had been made and the presence of any tryptophan analogue discounted, then the classical data was found not to be in disagreement with this interpretation.

A number of biologically active peptides have been shown to possess blocked structures, and these pose considerable problems for classical methodology, but fewer problems for mass spectrometry since free peptides would be "blocked" prior to study by making derivatives. A recent example of this type of problem was the study of Adiopokinetic Hormone, isolated from locust and responsible for lipid mobilization during flight.[19] Amino- and carboxy-peptidases failed to destroy the biological activity, indicating a blocked peptide, and it was not possible to determine the full structure by classical methods for neither N- or C-terminal degradation was possible.

The mass spectrum of a permethyl derivative of adiopokinetic hormone was complex but nevertheless interpretable, showing a pyrrolidone carboxylate (m/e 98 and 126) blocking the N-terminus, and a signal at m/e 428 interpreted as arising from N–C cleavage at tryptophan and containing a C-terminal amide, ...Trp-Gly-Thr-NH$_2$. A further small sample of adipokinetic hormone was then derivatised using CD$_3$I in order to confirm assignments by comparing mass shifts. The spectra could be interpreted as arising from a fully blocked decapeptide,

PCA-Leu-Asn-Phe-Thr-Pro-Asn-Trp-Gly-Thr-NH$_2$.

In contrast to the Enkephalin story, tryptophan had not been expected in this structure, but was, of course, observed as a clear (well stabilized) signal and placed in sequence without difficulty!

Our present work in the neuropeptide area is following two lines. Firstly, the development of an efficient purification method, applicable to the picomolar quantities of peptide encountered in central nervous system studies. This we have now done, and some details and features of this important new method are described below (Fig. 11). Secondly, we are searching for new mass spectrometric methods which will enable us to detect and even deduce structures on picomolar quantities of peptide. Preliminary studies in this area are encouraging and an example is given below.

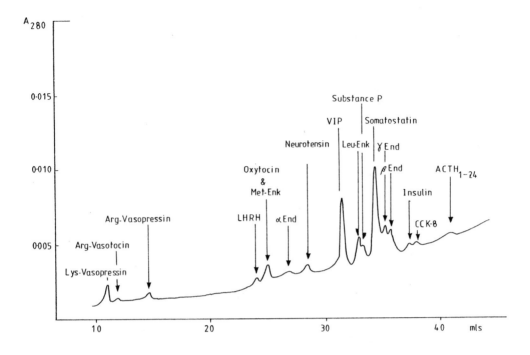

Fig. 11. HPLC purification of neuropeptides on microbondapak C-18. Conditions: 30 min. concave gradient (7 of WATERS programmer) from 10:90 (v/v) to 40:60 (v/v) propan-1-ol/5% acetic acid. Flow rate 1 ml/min.

The development of a new purification procedure arose from work in testing our hypothesis that incorrect levels of neuropeptides may be responsible for the genesis or expression of mental illness, for example, schizophrenia. The most sensitive quantitation method for low levels of certain peptides in, for example, cerebral spinal fluid is radioimmunoassay (RIA), but the specificity of RIA methods is questionable. We therefore combine RIA with our high-resolution, high pressure liquid chromatography (HPLC) purification system to produce a virtual chemical characterization of the peptide being quantitated. The HPLC system we have developed uses non-salt buffers (propanol/acetic) and so is compatible with both bioassay and RIA; it gives high yields ex-column (e.g. 90% for β-endorphin) even at the low picomolar level, and gives repro-ducible high-resolution chromatographs from either cerebral spinal fluid or tissue extracts. We are currently using this method together with the mass spectrometric methods described previously for the identification of new peptides, having opioid activity, derived from brain tissue.

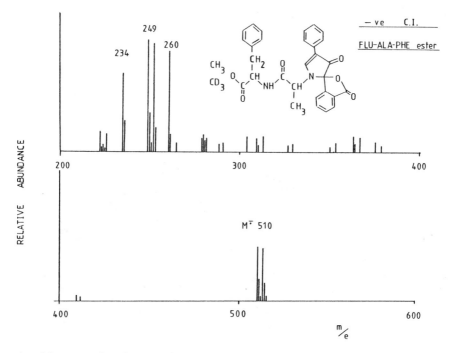

Fig. 12. Negative ion chemical ionization spectrum of less than one nanomole of a fluorescent derivative of Ala-Phe.

Radioimmunoassay can be used to determine only known substances, and there is now a clear need in neurochemistry for a procedure for chemical characterization of new peptides at the picomole level. An interesting recent development in this connection is the study of negative chemical ionization. When acyl permethyl peptides are ionized in a conventional chemical ionization (CI) source, the resulting fragmentation pathways differ from those induced by electron impact (EI), yielding C-terminal fragments of value to the structure elucidation. The methane gas normally used for CI studies produces thermal electrons when ionized and these can be used to generate low internal energy *negative ions*. The fragmentation of such acyl permethyl peptide negative ions differs from that observed in the positive ion mode, but can nevertheless be used for structure elucidation (Buko,A., Hunt, D.F. and Morris, H.R., unpublished work). The attractive feature of negative ion CI is the possibility for much enhanced sensitivity over the positive ion spectrum for components with positive electron affinities and large capture cross sections. The capture of thermal electrons can be promoted by "tagging" the molecule with an efficient electron capturing group, e.g., a fluoroacyl group, and detailed studies of the effects of various blocking groups on peptide negative ion spectra are now under way. In some recent work, we have attempted to combine a useful feature of protein chemistry, that is tagging the peptide with a fluorescent label, with negative ion CI to produce a spectrum of the 1:1 CH_3OH/CD_3OD labelled peptide ester, Fig. 12. The spectrum of approximately one nanomole of this sample shows a remarkably abundant M^- signal at m/e 510, whereas no molecular ion is visible in the positive CI spectrum. The applicability of this technique to real problems in protein chemistry is now under investigation.

CONCLUSION

The mass spectrometer is now a useful complement and a viable alternative to classical protein sequencing methods, and the first total protein sequence analysis by mass spectrometry alone has been accomplished. In other areas, e.g., neuropeptides and coagulation, important problems have already been solved, and we have presented evidence that mass spectrometry is the method of choice here. Current work on HPLC, new ionization/derivative combinations and computer automation promises to give us even greater sensitivity and to allow the solution of further important chemical and biochemical problems beyond the scope of present technology.

REFERENCES

1. Morris, H.R., Dickinson, R.J. and Williams, D.H. (1973) Biochem. Biophys. Res. Commun., 51, 247.
2. Morris, H.R. (1974) Biochem. Soc. Trans., 2, 806.
3. Morris, H.R. (1972) FEBS Lett., 22, 257.
4. Morris, H.R., Geddes, A.J. and Graham, G.N. (1969) Biochem. J., 111, 38p.
5. Geddes, A.J., Graham, G.N., et al. (1969) Biochem. J., 114, 695.
6. Morris, H.R., Williams, D.H. and Ambler, R.P. (1971) Biochem. J., 125, 189.
7. Morris, H.R., Williams, D.H., et al. (1974) Biochem. J., 141, 701.
8. Morris, H.R., Batley, K.E., et al. (1974) Biochem. J., 137, 409.
9. Dell, A. and Morris, H.R. (1977) Biochem. Biophys. Res. Commun., 78, 874.
10. Batley, K.E. and Morris, H.R. (1977) Biochem. Biophys. Res. Commun., 4, 1010.
11. Batley, K.E. and Morris, H.R. (1977) Biochem. Soc. Trans., 5, 1097.
12. Morris, H.R. and Dell, A. (1975) Biochem. J., 149, 754.
13. Morris, H.R., Thompson, M.R., et al. (1978) J. Biol. Chem., 253, 5155.
14. Magnusson S., Sottrup-Jensen, L., et al. (1974) FEBS Lett., 44, 189.
15. Stenflo, J., Fernlund, P., et al. (1974) Proc. Nat. Acad. Sci. U.S.A., 71, 2730.
16. Morris, H.R., Dell, A., et al. (1976) Biochem. J., 153, 663.
17. Thorgersen, H.C., Petersen, T.E., et al. (1978) Biochem. J., 175, 613.
18. Hughes J., Smith, T.W., et al. (1975) Nature, 258, 577.
19. Stone, J.V., Mordue, W., et al. (1976) Nature, 263, 207.

Published 1980 by Elsevier North Holland, Inc.
Liu/Mamiya/Yasunobu, eds. Frontiers in Protein Chemistry

THE DESIGN AND UTILIZATION OF PHOTOAFFINITY PROBES

IN THE INVESTIGATION OF NUCLEOTIDE BINDING REGIONS

RICHARD JOHN GUILLORY
Department of Biochemistry and Biophysics, University of Hawaii,
Honolulu, Hawaii 96822

ABSTRACT

Advantages and disadvantages of the photoaffinity approach are examined using
as a background studies with nucleotide probes developed in this laboratory.
Questions of the rational design and characterization of photoprobes of adenine
and guanosine nucleotides of the pyridine nucleotides and of Coenzyme A are
considered.

Application of these probes to investigations of the catalytic site for
soluble and membrane bound ATPase, the regulatory sites of glutamic dehydrogenase
and the pyridine nucleotide binding sites present in NADH–CoQ$_1$ reductase (complex I) of mitochondria are outlined. In addition, photoprobes of Coenzyme A
and acetyl–Coenzyme A have been synthesized and partially characterized with
respect to their structures and interaction with the phosphotransacetylase from
Clostridium kluyveri. The potential use of photoaffinity techniques in the
labeling of different protein allosteric states is advanced.

INTRODUCTION

An understanding of the mechanism of action of enzyme catalyzed reactions
requires the identification of amino acids at active sites. Such amino acids
may be distinguished as being those involved in the binding of ligands as well
as amino acids directly involved in catalysis. The classical chemical approach
to acquiring such information has been the utilization of reagents which bring
about chemical modification of specific residues. Inactivation correlated with
the modification of a single residue is considered to be associated with interaction at the "active site." There is more often, however, destruction of a
number of residues and the relationship of inhibition to interaction at the
catalytic site is often tenuous.[1]

One requires for affinity labels of active site regions compounds capable of
utilizing the binding characteristics (affinity) of the active site for the
transport of a reactive group to the vicinity of the active site. In theory,
binding of such an analogue would be followed by a secondary reaction allowing
adjacent nucleophilic amino acid residues to carry out displacement reactions

resulting in covalent binding of the reagent. Most chemical reagents used to modify proteins in this way are weakly electrophilic in order to avoid rapid reactivity with water; they consequently react preferentially with nucleophiles at active sites.[2,3] Subsequent degradation of the inactive protein can result in the identification of the modified residue and provide information as to the mechanism of the enzymatic reaction or characteristics of the binding region.

In their analysis of the basic theory for general affinity labeling, Wolsey et al.[4] stress the requirement for the initial formation of a noncovalent reversible complex similar to the Michaelis complex of the natural ligand. The insertion reaction consequently depends upon the reactivity of the transported electrophile as well as the reactivity of nucleophiles present at the binding site. A major factor limiting the classical labeling procedure is the fact that there is no reason to expect reactive nucleophilic centers at generalized ligand binding sites.[5,6] In addition, depending upon the reactivity of the electrophile one might expect varying degrees of interaction with protein nucleophiles not situated at the binding site under study.

Photoaffinity labeling represents a special category of the general affinity technique in which the reactive chemical grouping is replaced by an inert but photoactive reagent. The application of such photoaffinity probes to active site labeling has recently received a great deal of attention due primarily to the pioneering work of Westheimer and his group who, in 1962,[7] initiated experiments utilizing diazoacetyl precursors of active carbene and to Knowles et al.[8] in the application of arylnitrene precursors to the investigation of the interacting site for antibody recognition. These two initiating studies were instrumental in pointing out the advantages and weaknesses of the photoaffinity approach. Carbene and nitrene precursors are the principal reagents utilized for photoaffinity studies in biological systems, they are much less selective towards covalent bond formation than electrophiles, are very reactive, and capable of carbon-hydrogen bond insertion.[9]

The potential for generating a reactive reagent *in situ* by photoexcitation and for labeling of sites not necessarily rich in reactive nucleophilic residues provides photoaffinity labeling with a special advantage in the study of ligand binding sites. Covalent insertion as a result of photogeneration of a reactive reagent at a specific binding site would be expected to be a first order process assuring specificity of attachment. Similar reactivity at secondary (i.e., non-specific) sites would not be dependent upon the binding potential between the ligand and reactive site and consequently would be a second order reaction. In addition, the reactivity of the reagent is masked and under the control of the experimentalist. Control of activation, however, does not necessarily result in time control over covalent insertion.

Requirements of a photoaffinity labeling reagent

Theoretical considerations indicate certain requirements which must be satisfied in order to assure that the design of an active site directed irreversible inhibitor will function in an expected manner.[10] If precise information about molecular interactions is desired, it is imperative to establish that (i) *the probe not only binds to the natural ligand binding site, but also that the noncovalent binding is in a mode identical to that of the true ligand.* The reagent must consequently contain the structural elements required for noncovalent binding. In order to assure this, the reagent should be designed with some knowledge of the enzyme's active site. Since knowledge of the composition of the binding site for a particular ligand is not always available, the structure of the substrate, reactivity of modified substrates or the action of competitive inhibitors may provide useful information. Active site interaction is usually tested for kinetically by examining whether the natural ligand blocks noncovalent binding or covalent incorporation of the probe.

Even with modification of the structure of a natural ligand without reducing selectivity or binding potential, one must still be concerned with the attitude of the photolytic portion relative to the topology of the binding site. If the probe is directed away from the binding site and into the solvent phase, one would not expect covalent insertion to take place.

(ii) *Covalent bond formation should be equal to the loss of reversible binding sites for the native ligand.* This requires that there be a reliable quantitative assay for natural ligand binding unperturbed by analogue interaction.

(iii) *The reaction which labels the receptor covalently following formation of the reversible complex should be indiscriminatory.* The photoaffinity reagents do have low selectivity towards covalent bond formation. In addition to direct insertion into C-H bonds carbene and nitrene reactions of abstraction, cycloaddition, attack by nucleophiles and rearrangement are possible. While direct insertion reactions are considered rare, if a nitrene or carbene is generated *in situ* at a binding region which in effect restricts movement, intramolecular insertion could represent the major reaction pathway. Although covalent insertion may occur with a wide array of amino acid side chains, the stability of the reaction products could limit their usefulness. In this regard, the stability of nitrene reaction products might be considered to be less than that for carbenes.

The best security at present against nonspecific interaction due to these variable reaction pathways is a rapid insertion reaction, assured with analogues having firm binding with the ligand site. There is always the possibility that the label covalently binds to a site different from the "true" (i.e., tightest)

receptor site, and blocks natural ligand binding by steric interference or by
allosteric effects.

(iv) *The active photogenic speci must have a short lifetime.* This is necessary in order to allow that covalent insertion reaction at the site(s) of generation take place at rates comparable to or less than the rate of equilibrium
among the sites. Knowles[5] has pointed out the necessity of having the reagent,
upon irradiation, give rise to reactive species when situated within the binding
site, noncovalently bound to the receptor.

Arylnitrene half-lives vary with substituents and increase by nearly two
orders of magnitude when going from p-acetyl (short lived) to a p-N-morpholine
(long lived) derivative.[11] A range of reagents of different reactivities can
therefore, in principle, be synthesized. In comparison to nitrene probes which
decay in the millisecond range, the upper limit of carbene lifetimes is of the
order of 1 μsec. The increased selectivity of alkyl- and arylazides, isocyan-
ates and acylazide nitrenes relative to carbene reagents is a reflection of the
relatively long lifetime for nitrenes. Singlet nitrenes react preferentially
by insertion into oxygen-hydrogen or nitrogen-hydrogen bonds and have electrophilic character. Electron withdrawing groups, particularly para to the nitrene,
increase the electrophilic character and hence, reactivity of the singlet nitrene. Electron releasing substituents decrease the electrophilicity and enhance
the yield of triplet state reactants.[12] In the less reactive states, insertion
into carbon-hydrogen bonds becomes more probable.

Such lifetime considerations are particularly important in problems inherent
in photoaffinity labeling of "loose" receptor ligand complexes. In principle,
if the dissociation constant of an analogue is low (10^{-5} to 10^{-6}), the binding
competitive with that of the natural ligand and the "lifetime" of the reactive
species generated *in situ* short (10^{-5} sec or less) specific labeling of a
receptor can be expected. Nonspecific labeling arising from relatively long
"lifetimes" of the reactive species generated upon photolysis of aromatic azides
has been demonstrated by Singer and his group and has been termed pseudophoto-
affinity labeling.[13] Reactions of the free radical in an active site may have
what Singer[14] has described as a "shotgun" effect. A number of different residues within a peptide region may appear to be simultaneously modified.

(v) *The reagent should not react too rapidly with water.* With general
affinity labeling, stereochemical restrictions favor the use of highly reactive
labeling reagents. The practical upper limit of reactivity of a photoaffinity
reagent is determined by the rate of reaction with a residue within a
binding site region relative to its reaction rate with solvent.

(vi) *Activation of the photoprobe should be accomplished at wavelengths of irradiation remote from protein damaging regions.* Alkylnitrenes, which undergo major rearrangement reactions, have absorption maxima about 290 nm, and are thus activated at wavelengths in the ultraviolet region. In comparison, suitably substituted arylazides have strong absorption bands above 400 nm and can be photolyzed to nitrenes at wavelengths where photoinduced protein changes should be minimal.[15,16]

Limitations of the photoaffinity labeling method

(i) Specific vs nonspecific reactivity. As mentioned above, one major limitation of photoaffinity labeling is the inability to distinguish absolutely between nonspecific labeling and that due to binding at specified loci. This problem can be attacked experimentally by what has been termed differential labeling. The enzyme is pretreated with a nonlabeled analogue in the presence of a large excess of substrate or competitive inhibitor. Complete modification of all reacting residues other than at the active site is thus achieved. The protective agent is then removed and a radioactively tagged form of the reagent added. Stoichiometric inactivation can then be observed, if exhaustive modification of nonspecific groups has been initially achieved. In addition to this uncertainty there are three other limiting considerations which may arise.

(ii) Low yields of photodependent insertion. In comparison with electrophilic affinity labeling, photoaffinity labels have in the past been associated with rather low yields of insertion products. This appears to be less of a problem with those probes having a close relationship structurally and kinetically to the natural ligand.

(iii) Photosensitized destruction. Destruction of a particular receptor site may take place due to photosensitized destruction caused by the photolabile ligand even when irradiated at wavelengths where the receptor itself does not absorb. In this situation, the ligand's binding site would be modified and not representative of the normal state of the protein.

(iv) Limited knowledge of the photochemical mechanism giving rise to covalent incorporation. In order to examine the photochemical mechanism which gives rise to insertion, a knowledge of the structure of the labeling reagent is an absolute necessity. However, coupled to the complexity of the biological receptor site is that of the photogenic reaction products formed upon irradiation. This is especially true of arylazido analogues which appear to undergo multiple forms of degradation dependent upon the time of irradiation and the type of irradiating source (Fig. 1). Thus, diversified means of activation of photoaffinity probes could result in significant differences in labeling due to potentially different intermediates formed in the activation process.

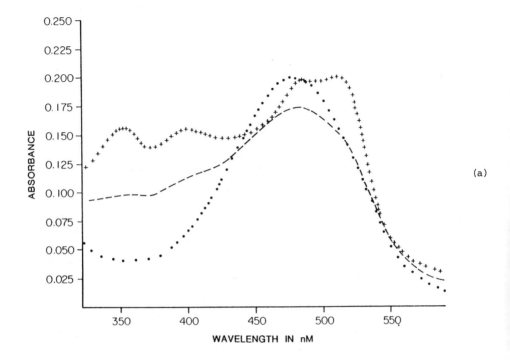

(a)

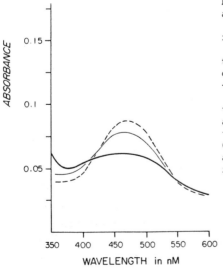

(b)

Fig. 1. (a) Photoirradiation of an aqueous solution of arylazido-β-alanine (15.0 μM) in a pyrex glass vessel using flashes generated by an xenon flash lamp (FL 1-BALCAR, 1200 watts, U-shaped flash tube lamp) with flashes of less than 1 ms duration. ooo, no flash; ---, 9 flashes; +++, 18 flashes.
(b) Photoirradiation of an aqueous solution of arylazido-β-alanine (14.2 μM) in a pyrex glass vessel equidistant between two tungsten halogen projector lamps, 650 watts, DVY 3400°K, spaced 20 cm apart. ---, no irradiation; ——, 6 min irradiation; ▬, 10 min irradiation.

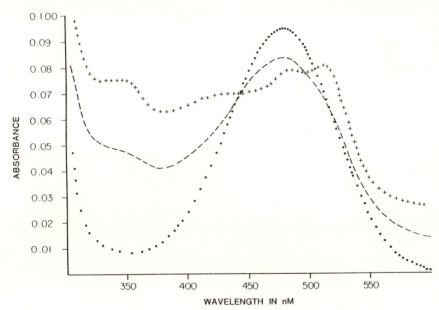

Fig. 1c. Photoirradiation of an aqueous solution of arylazido-β-alanine (22.9 μM) in a pyrex glass vessel positioned 12.5 cm from a 250 wall medium pressure mercury lamp (Applied Photophysics, Ltd., London, England) with a 340 nm and 3 cm path water filter. ●●●, no irradiation; ---, 60 sec irradiation; +++, 270 sec irradiation.

Considerations in the design of an ATP photoaffinity probe

The three criteria of chemical inertness of the precursor, apparent lack of major rearrangement of the reactive reagent, and suitable absorptive character- istics suggest that arylnitrene generating reagents are the most ideal photo- affinity reagents presently available.[8] A word of caution is, however, neces- sary; while arylnitrenes are less susceptible to photochemical rearrangements, these are known to occur. Arylnitrenes are in addition quite selective with respect to insertion reactions; insertion into tertiary C-H bonds occurs, for example, 100 times faster than into primary C-H bonds. One might expect as well directed labeling into hydrophobic pockets of proteins and membranes due to the aromatic and partial electrophilic character of the arylazido grouping.

The above considerations were used in our initial search for a general re- agent which might be utilized in a comparative study of the nucleotide binding region of a variety of energy transducing systems.[17] A valuable approach to the study of the mechanism of myosin ATPase was the utilization of a large number of structural analogues of ATP.[18-20] We were able to make on the basis of these

studies, an assessment of those components of the nucleotide required for binding and to evaluate the portions of the molecule which could be modified without reducing binding efficiency. A most important aspect of the
actomyosin-nucleotide analogue studies was the finding that the ribose ring can
be drastically modified without influencing markedly the rate of myosin-
catalyzed hydrolysis.[21] In view of these facts, it was decided to develop
methods by which an active photosensitive adjunct could be attached to the
ribose portion of the nucleoside triphosphate. An additional restraint was that
the photogenerated species should be formed at wavelengths of light remote from
regions of protein damaging radiation, a restriction which also directed us
to arylazido photoaffinity reagents.

The utilization of carbodiimidazole to facilitate formation of the activated
carboxylic acids[22] of azidonitrophenyl-β-alanine was used in synthetic schemes
resulting in the esterification of adenosine nucleotides at the ribose hydroxyl.[17]
Analysis based upon visual, ultraviolet and infrared spectra coupled with chemical degradation and mass spectroscopic data resulted in the structural assignment 3'-O-{3'-[N-(4-azido-2-nitrophenyl)-amino]propionyl}adenosine-5'-triphos-
phate for the product of this reaction (Fig. 2.)

The esterification step can provide for 2' or 3' isomers of the ATP analogue.
From a combination of kinetic and thermodynamic reasoning, it was considered

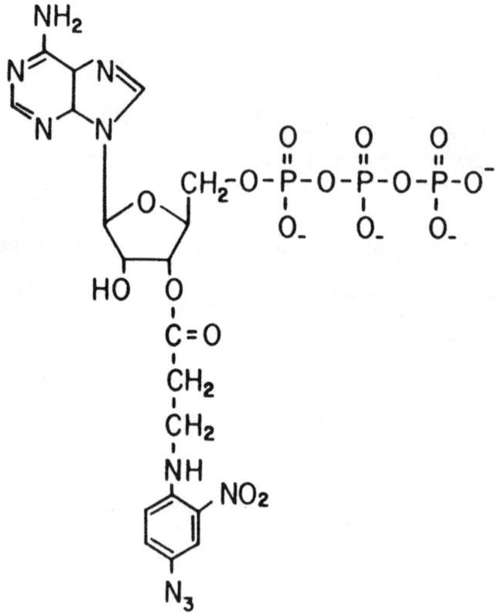

3'-O-{3-[N-(4-azido-2-nitrophenyl) amino]propionyl}adenosine-5'-triphosphate

Fig. 2. Arylazido-β-alanyl ATP.

that the 3' isomer was the specific component isolated under the synthetic conditions used.[23] Recently, the examination of the nuclear magnetic resonance spectra of arylazido-β-alanyl derivatives of a number of nucleotide analogues has been shown to be consistent with this assignment.[24]

Myosin ATPase activity is substantially inhibited when the protein is photoirradiated in the presence of the nucleotide analogue.[17] The photodependent chemical interaction between myosin and arylazido-β-alanyl ATP was shown to be covalent in nature. Jeng[25] reported that in addition to the Ca^{++} dependent ATPase activity of myosin, the heavy meromyosin (HMM), the subfragments prepared with papain (i.e., $S_1(P)$) or with chymotrypsin (i.e., $S_1(CT)$) as well as the EDTA stimulated ATPase of $S_1(P)$ were all inhibited in a photodependent manner by arylazido-β-alanyl ATP. In all cases, the presence of ATP during irradiation protected against the analogue-dependent inhibition. The analogue is able to function as a substrate for the ATPase active site as indicated by the catalytic hydrolysis of arylazido-β-alanyl [γ^{32}P] ATP exhibiting a K_m of 6.15 μM for $S_1(P)$.

Arylazido-β-alanyl [^{3}H] ATP is bound photolytically to HMM and $S_1(P)$ at maximum molar ratios of protein-to-analogue of 1:2 and 1:1, respectively. These ratios were evaluated following a stepwise procedure of irradiation assaying for activity loss, filtration through sephadex gel, and measuring radioactivity as a function of protein concentration. It would appear that the arylazido-β-alanyl ATP does not distinguish between the two heads of HMM as being different with respect to nucleotide binding. The sodium dodecylsulfate gel pattern of $S_1(P)$ photoirradiated in the presence of tritium containing arylazido-β-alanyl ATP demonstrates that the light chains of myosin are not labeled by the nucleotide analogue.[25] While only the subfragment 1 heavy chains are photolytically labeled, this does not exclude the possible involvement of the light chains in the enzymatic activity of the myosin but would appear to eliminate them as direct participants in nucleotide binding.

Thus, the design of a labeling reagent with regards to the three-dimensional aspects of a specific binding site, in order to allow for steric complementarity to the active site, has been successful. Presently, examination of the cyanogen bromide cleavage products of the labeled protein is under way. The initial finding that more than one peptide is labeled, in light of the exact stoichiometry of binding, may be an experimental example of an excited nitrene situated at an active site inserting simultaneously into a number of amino acid residues.[14]

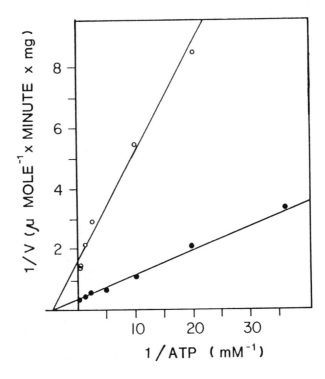

Fig. 3. The photodependent inhibitory profile of submitochondrial ATPase following arylazido-β-alanyl ATP treatment. Submitochondrial particles at 1.07 mg protein per ml in 20 mM Tris SO_4 pH 7.5 were irradiated with a total of 18 flashed in the presence of 8 μM arylazido-β-alanyl ATP using a successive addition procedure (i.e., six flash periods with 1.33 μM increments of analogue). The ATPase activity was measured in the presence of varying concentrations of ATP. The reciprocal plot of activity versus the ATP concentration gives a K_m(ATP) of 190 μM for both control and analogue treated preparation and V_{max} values of 3.06 μmol Pi liberated min^{-1}. mg $protein^{-1}$ for the control (●-●) and 0.66 μmol Pi liberated min^{-1} mg $protein^{-1}$ for the analogue treated sample (o-o).[27]

Photoinhibition and labeling of integrated forms of mitochondrial ATPase by arylazido-β-alanyl ATP

Arylazido-β-alanyl ATP has been shown to bring about a photodependent inhibition of the mitochondrial F_1ATPase associated with a specific covalent labeling of the soluble enzyme.[26] Protection against the photodependent inhibition by ATP was readily demonstrated as was the analogue's effective action as a substrate for this ATPase activity. Both effects are consistent and required criteria for an effective active site label.

In Fig. 3, the ATPase activity of submitochondrial particles is shown to be inhibited following photoirradiation in the presence of arylazido-β-alanyl ATp and assayed at varying levels of the natural substrate. No change in K_m(ATP) is observed for the inhibited preparation; however, there is a decrease of 21% in the maximum velocity. Thus, following photodependent inhibition, the natural substrate is not capable of protecting against inhibition nor of reversing the inhibition even at a large excess of the ATP.

A comparison of the pH profile of the ATPase activity with that of the photo-dependent inhibition by arylazido-β-alanyl ATP showed close similarity. The similar pH profiles indicate that the same ionizable group(s) appear to be impli-cated in both the binding step for the nucleotide analogue and for ATP hydrolysis.[27]

Experiments with submitochondrial particles represent an examination of the effectiveness of arylazido-β-alanyl ATP as a photoaffinity probe on an inte-grated form of the mitochondrial ATPase. In this case, the complex ATPase enzyme is coupled with electron transport and can produce ATP from ADP and Pi as well as exchange the phosphate of ATP with inorganic phosphate. In order to investigate possible secondary influences of photoproducts on the incorporation of the arylazido-β-alanyl ATP into this membrane-bound system, photoirradiation was carried out using the flash generated by an xenon lamp (FT 1-BALCAR, U-shaped flash tube) with flashes of less than one msec duration. Filtration of the light prevented the passage of more than 99% of radiation below 300 nm. Under these conditions, the ATPase and ^{32}[P]Pi-ATP exchange activities (both of which are photoinhibited in the presence of the analogue) were inhibited maxi-mally 5% with 20 flashes in the absence of the adenine analogue. When arylazido-β-alanyl ATP was added in several aliquots such addition being followed by photo-irradiation, the efficiency of the photodependent inhibitor was increased over that system in which the adenine nucleotide analogue was added in a single addi-tion and the mixture then subjected to the same number of irradiation flashes (Fig. 4). Presumably, following a few flashes at high analogue concentrations, the concentration of "free" photoreacted nucleotide analogue is high compared to the successive addition system in which there is a lower analogue concentration of which a significant proportion is protein bound. In the former case, the accumulated photoreacted free analogue can act as a competitor against arylazido-β-alanyl ATP binding. Thus, with respect to the number of binding sites for arylazido-β-alanyl ATP on the ATPase protein, there may exist but one type but this becomes more difficult to saturate with photoactive analogue. This in-creasing competitive concentration of the light product of arylazido-β-alanyl ATP is indicated by the inability to inhibit the ATPase activity 100%. In this case, the criteria for complete inactivation cannot be fulfilled due to the

222

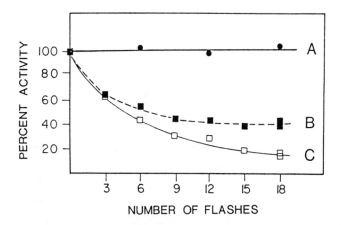

Fig. 4. Photodependent inhibitory effect of arylazido-β-alanyl ATP accomplished by a successive addition procedure. During photoirradiation, submitochondrial particles were at 1.55 mg protein per ml in 25 mM Tris SO_4 pH 7.5. Following irradiation of a specific interval as indicated by the number of flashes on the abscissae, an aliquot of the irradiated mixture was tested for ATPase activity and a comparison made with the activity of a dark control taken to represent 100% activity. Under condition A, no arylazido-β-alanyl ATP was present during irradiation; for condition B, the nucleotide analogue was added to 15 μM and the mixture subjected to a three-flash irradiation. An aliquot was then taken and kept in the dark for subsequent ATPase determinations while the remaining mixture was irradiated again with sets of three flashes up to a total of 18 flashes. Under condition C, arylazido-β-alanyl ATP was added at 2.5 μM, the mixture subjected to an additional three flashes and so on up to 18 flashes and a final arylazido-β-alanyl ATP concentration of 15 μM. Control ATPase was assayed at 2.55 μmol Pi min^{-1} mg $protein^{-1}$.[27]

photoproduction of modified nonincorporated reagent. In many systems, inactivation kinetics have been described as the sumation of two different rates representing multiple site interaction. Such multiple reactivities may, in some cases be explained on the basis of a single site reactivity with the non-bounded photoaffinity product competing for the binding site with the still active photoaffinity probe.

The ATPase complex is another integrated form of the mitochondrial ATPase, in which the enzyme extracted from the mitochondrial membrane is still associated with membranous subunits and phospholipid vesicles.[28] In this case, the enzyme is capable of ATP induced proton translocation as well as [32][P]Pi-ATP exchange and ATPase but is not able to couple phosphorylation to electron transport. In this preparation, the ATPase activity is enhanced by the presence of phospholipid vesicles; similar enhancement is observed for the [32][P]Pi-ATP exchange.

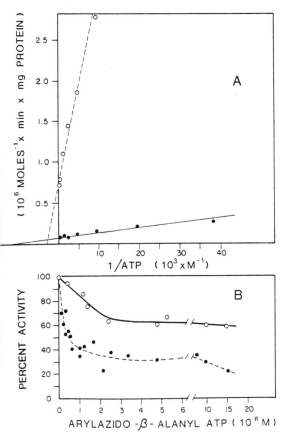

Fig. 5. The effect of phospholipids on the ATPase activity and on the arylazido-β-alanyl ATP photodependent inhibition in the ATPase complex. The ATPase activity of the partially purified ATPase complex was determined in the absence and presence of phospholipids. When used, phospholipids (Asolectin) were at 0.32 mg per ml. This concentration, emulsified by sonication, gave optimal activity when tested at 4 mM ATP. (5a) The reciprocal plot of activity as a function of the ATP concentration gives values of $Vm = 1.4$ μmol phosphate min^{-1} mg protein^{-1}; K_m (ATP) = 330 μM in the absence of phospholipids (o-o-o) and $Vm = 13.8$ μmol phosphate min^{-1} mg protein^{-1}; K_m (ATP) = 77 μM in presence of phospholipids (●-●-●) (5b) The ATPase complex at 0.53 mg protein per ml in 50 mM Tris SO_4 pH 7.5 was irradiated with 18 flashes in the presence of varying concentrations of arylazido-β-alanyl ATP using the successive addition procedure described in Fig. 1. Irradiation was either in the presence (●-●-●) or the absence (o-o-o) of phospholipids. The ATPase activity was determined in all cases in the presence of phospholipids (0.32 mg/ml) and was compared with controls treated in the same way except the arylazido-β-alanyl ATP was omitted. Control ATPase activity was assayed at 3.02 μmol phosphate liberated min^{-1} mg protein^{-1}.[27]

The activating effect of phospholipids consists of an increase in the V_{max} and a decrease in the K_m for ATP (Fig. 5a). It was consequently of interest to determine if this activation effect on enzymatic activity would have a comparable influence on the interaction with the ATP photoaffinity analogue. As can be seen from Fig. 5b, a parallel increase in the affinity and in the inhibitory effect of arylazido-β-alanyl ATP was observed when the ATPase complex was photoirradiated in the presence of phospholipids. In the presence of phospholipids arylazido-β-alanyl ATP concentrations 3- to 6-fold in excess of the ATPase complex are able to inhibit up to 70% of the ATPase activity.

This influence of phospholipids on the increased sensitivity of the membrane bound ATPase to the ATP analogue is consistent with the ten-fold higher titer of the analogue required to inhibit the soluble enzyme. It is proposed that this differential sensitivity is due to two different conformational attitudes exhibited by the enzyme; a membrane dependent or natural conformation (one in which the enzyme carries out its energy conservation function) and an "unnatural" conformation exhibited in the soluble state. Further work in this laboratory is directed towards an evaluation of possible different labeling patterns by arylazido-β-alanyl ATP for the soluble and membrane bound forms as a measure of possible different conformational attitudes.

An attempt was made[27] to measure the stoichiometry of arylazido-β-alanyl ATP binding to the ATPase complex by the parallel measurement of the degree of inhibition of the ATPase activity and the quantity of arylazido-β-alanyl ATP bound following photoirradiation in the presence of varying concentrations of tritium-labeled analogue. Figure 6 shows that 80% of the ATPase activity can be inhibited by the binding of 1.6 to 2.0 nmol of arylazido-β-alanyl ATP per mg protein of the ATPase complex. The range of possible values assuming reasonable values for the molecular weight of the complex indicates that the arylazido-β-alanyl ATP is bound to a restricted number of sites, approximating one site per molecule of ATPase protein. This figure illustrates as well an attempt to analyze the binding data on the basis of the method of Scatchard.[29] Following photoinsertion, the theoretical basis for the use of the Scatchard equations, i.e., the reversible binding of ligands to an acceptor allowing measurements to be made at equilibrium, would not be expected to hold. However, it is reasoned that under limiting conditions such an analysis may be applicable.[*]

For binding at a given concentration of free ligand (where S is arylazido-β-alanyl ATP and E is the F_1ATPase)

$$E + S \underset{k_2}{\overset{k_1}{\rightleftharpoons}} ES \qquad\qquad (1)$$

photolytic insertion would then be represented by

$$ES \xrightarrow[h\upsilon]{k_3} E \cdot S' \qquad\qquad (2)$$

In this case, $E \cdot S'$ may or may not represent a measure of the equilibrium state of the ES complex depending upon the ratio of k_1/k_2 relative to the value

---

[*] In collaboration with Dr. J. Cosson.

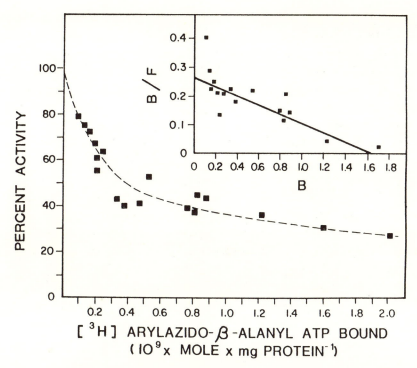

Fig. 6. Determination of the "maximal" quantity of arylazido-β-alanyl ATP bound to the ATPase complex. The ATPase complex at 1.33 mg protein per ml in 50 mM Tris SO$_4$ pH 7.5 was irradiated with 18 flashes in the presence of phospholipids (Asolectin at 0.32 mg per ml) and arylazido-β-alanyl ATP at concentrations varying from 0.3 to 0.8 µM, using the successive addition procedure described in Fig. 1. For each concentration of arylazido-β-alanyl ATP, the ATPase activity was determined and referred to the control irradiated under the same conditions but without arylazido-β-alanyl ATP. This control ATPase activity tested in the presence of phospholipids was assayed at 11.9 µmol phosphate liberated min^{-1} mg protein^{-1}. Following irradiation each sample was treated with SDS and dialyzed. In the insert, the experimental points are plotted according to the method of Scatchard.[29] The abscissae representing the quantity of nucleotide analogue bound (nmol per mg protein) and the ordinate representing the ratio of arylazido-β-alanyl ATP bound (nmol per mg protein) and the free analogue (µM). An analysis with a linear regression program showed that by taking all experimental points into account, an intercept with the B axis is reached at 1.69 with a correlation coefficient of 0.81 (J. Cosson and R. Guillory, unpublished observations).

of k_3. The value of k_3 is a function of the conditions of irradiation and the specific characteristics of the insertion reaction. If the insertion reaction is very rapid compared to the rate at which ES is equilibrating with E and S, then one can, so to speak, freeze the ES complex into the E·S' complex without disturbing the conditions of Eq. 1, i.e., a measure of E·S' is a measure of ES at equilibrium. In order for this to be the case, the photoeffect must convert completely all of the bound S to S' and the insertion reaction must be near 100% effective, i.e., all of ES must be converted to E·S'. In addition, there should be minimal noncompeting labeling sites having different kinetic constants of interaction.

The ATPase complex is formed by the association of 9 to 10 species of protein subunits and some confidence in the specificity of interaction of the ATP analogue with the membrane protein complex could be realized if a restricted number of the subunits were found to be covalently labeled by the analogue. The partially purified ATPase complex was irradiated in the presence of arylazido-β-[3-^{3}H]-alanyl ATP and the labeled protein subjected to electrophoretic treatment as described in Fig. 7. Essentially, only the α and β subunits of the complex contain significant label, a labeling which is associated with an 80% inhibition of ATPase activity.[27] The binding of 0.8 to 0.95 mol of arylazido-β-alanyl ATP per mol of soluble F_1 results as well in 80% inhibition of enzymatic activity.[26] These results indicate that the occupancy of the summation of a single nucleotide binding site by the covalently bound derivative is sufficient to inhibit a major portion of the ATPase activity. If the α + β subunit pair carries the active site, how can one explain the fact that more than one α + β subunit pair appears to be present per molecule of F_1ATPase? It may be that the catalytic site for ATPase activity is composed of a combination of the α and β subunits and that the point of insertion of the photoactivated nitrene has an equal possibility for interaction with either an α or β subunit region of the catalytic site. Alternatively, occupancy of one ATPase site may provoke inhibition of the catalytic activity of other sites.[27] A quantitative assessment of the amount of nucleotide analogue bound to both α and β subunit as a function of ATPase inhibition and the possible presence of post-photolytic insertion reactions (see below) may help to clarify this problem.

Applications of the photoaffinity technique to the study of the enzyme reactions of complex I

Controlled incubation of N-4-azido-2-nitrophenyl-β-alanine and carbodiimidazole in the presence of nicotinamide adenine dinucleotide resulted in the formation of a NAD^+ analogue which was found to be capable of undergoing reduction in the presence of ethanol and yeast alcohol dehydrogenase.[30] Evidence was presented for the structural assignment, A3'-O-{3-[N-(4-azido-2-nitrophenyl)-amino]propionyl} NAD^+ (Fig. 8). The presence of a single arylazido-β-alanyl

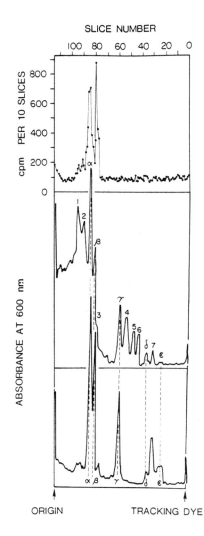

Fig. 7. The electrophoretic pattern of the arylazido-β-alanyl ATP treated ATPase complex. The partially purified ATPase complex at 1.57 mg protein per ml in 50 mM Tris SO_4 pH 7.5 was irradiated with 18 flashes in the presence of phospholipids (Asolectin at 0.35 mg pr ml) and ^3H aryl-azido-β-alanyl ATP (5.8×10^6 cpm μmol^{-1}) at a total concentration of 6.4 μM aryl-azido-β-alanyl ATP with 3 flashes follow-ing each addition). The inhibition of the ATPase activity amounted to 61% of that of the control irradiated in the absence of arylazido-β-alanyl ATP. The control activity was 13.2 μmol Pi min^{-1} mg pro-tein^{-1}. When the treated protein was dialyzed, a total of 6,900 cpm was found associated with 1.49 mg protein. Follow-ing lyophilization, the sample was resus-pended in 1 ml of 1% SDS, 50 mM dithiothre-itol and 50 mM Tris Cl pH 7.5 and prepared for gel electrophoresis. Following the 3 min treatment at 100°C glycerol was added to 25% final concentration and bromophenol blue to 0.08 mg ml^{-1}. Aliquots (0.1 ml) containing 149 μg of protein were placed on the top of ten different acrylamide gels for electrophoretic separation. Following staining the gels were scanned (center figure) sliced and counted (top figure). A control gel illustrated in the bottom figure was run in parallel containing the F_1ATPase extracted from the ATPase com-plex permitting the experimental identifi-cation of F_1 subunits of the ATPase com-plex.[27]

group per pyridine nucleotide unit was evidenced by hydrolysis of the arylazido-β-alanine from the nucleotide analogue, followed by isolation and quantitative assessment of the hydrolyzed components. The site of esterification of the arylazido-β-alanyl group to the adenine ribose of NAD$^+$ was revealed by studying the products formed upon treatment of the analogue with nucleotide pyrophospha-tase.

In addition to arylazido-β-alanyl NAD$^+$, an NADP$^+$ analogue has been prepared[*] utilizing the same methods developed for the adenine nucleotides and for arylazido-β-alanyl NAD$^+$. Structural evaluation was carried out in much the

---

[*] Chen, S. and Guillory, R.J. (1980) J. Biol. Chem., 255, 2445-2453. See as well, reference 24.

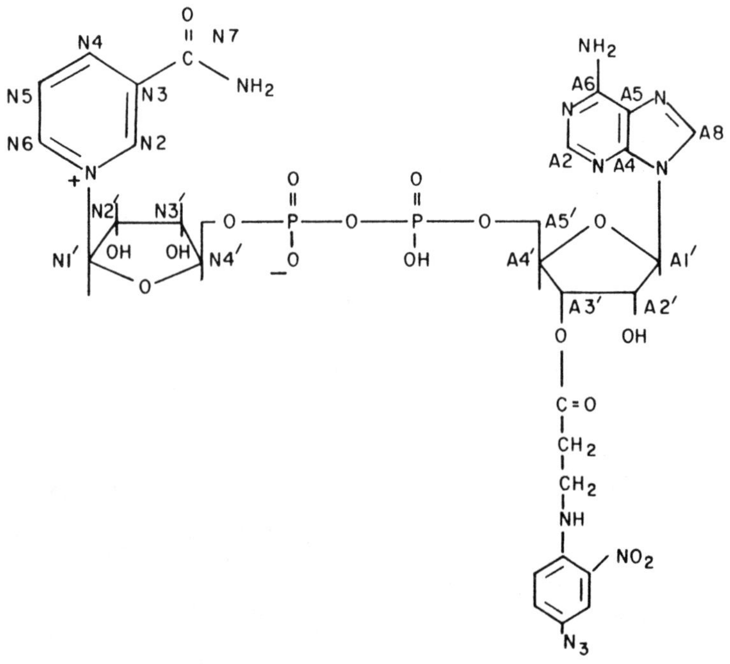

Fig. 8. Arylazido-β-alanyl NAD$^+$,
A3'-O-{3-[N-(4-azido-2-nitrophenyl)amino]propionyl} NAD$^+$

same manner as outlined for the NAD$^+$ analogue. In this case, the nucleotide
analogue has a single arylazido-β-alanyl group associated with the NMN portion
of NADP$^+$; the structural assignment is N3'-O-{3-[N-(4-azido-2-nitrophenyl)-
amino]propionyl} NADP$^+$ (Fig. 9).

Alcohol dehydrogenase was used to assess the effectiveness of the NAD$^+$
analogue as a photoaffinity probe for pyridine nucleotide binding sites. The
analogue acts as a substrate for the enzyme in the dark (K_m 0.052 mM, V_{max}
4.4 μmol mg^{-1} min^{-1} at pH 7) and inhibits the reduction of the natural substrate.
Photoirradiation in the presence of arylazido-[3-^{3}H]-β-alanyl NAD$^+$ resulted in
the covalent labeling of 1.9 mol of arylazido-β-alanyl NAD$^+$ to 1 mol of dehydrogenase (MW 141,000) associated with 91% inhibition of enzymatic activity.[30]

The reduced arylazido-β-alanyl NAD$^+$, formed by reaction with alcohol and
yeast alcohol dehydrogenase in the dark, was tested for its substrate and photodependent effects on the membrane bound NADH dehydrogenase system present in
complex I. The NADH-CoQ reductase (complex I) of ox heart mitochondria represents one of the major enzyme complexes of the mitochondrial electron transport
system.[31] The rate of CoQ$_1$ reduction catalyzed by arylazido-β-alanyl NADH

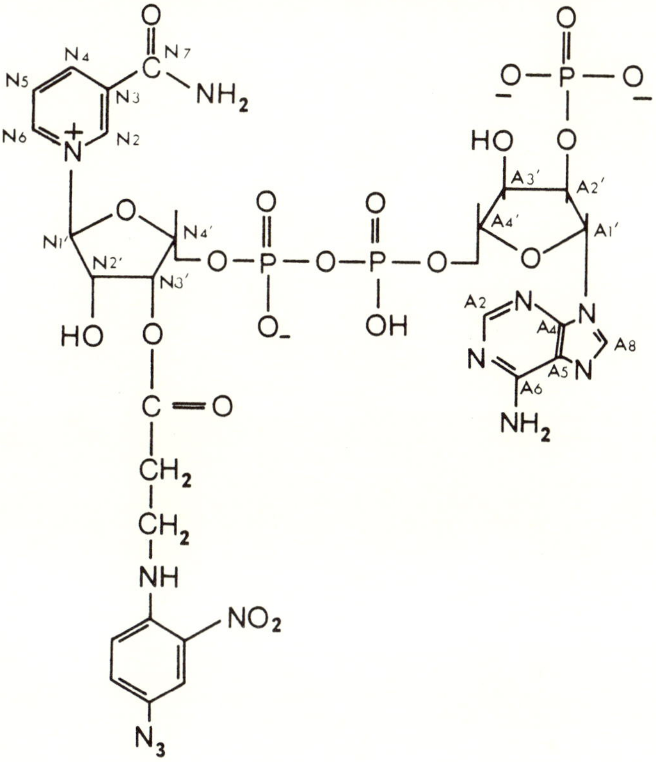

Fig. 9. Arylazido-β-alanyl NADP$^+$, N3'-O-{3-[N-(4-azido-2-nitrophenyl)amino]propionyl} NADP$^+$

(70% of the NADH rate) was shown to be rotenone sensitive, supporting the specificity of action of the nucleotide analogue as a substrate for the NADH-CoQ$_1$ reductase (K$_m$ 33 μM).

A combined analysis of the kinetic and photodependent inhibition effects of the two photoprobes has been carried out at three levels of reactivity for the enzyme systems present in complex I. At the membrane level (submitochondrial particles) at the level of resolved particles (complex I) and at the level of purified enzymes (NADH dehydrogenase and NADPH-NAD$^+$ transhydrogenase).

Table 1 shows that of the four major activities measurable in complex I, all but the NADPH-AcPyAD$^+$ transhydrogenase activities are inhibited sizably by the NAD$^+$ analogue. The competitive inhibitory effect with respect to CoQ$_1$ is considered to be indirect since CoQ$_1$ at concentrations two-fold greater than that of NADH protect only 10%. The added hydrophobicity of arylazido-β-alanyl NAD$^+$ may influence the reactivity of the analogue at the CoQ$_1$ site.

TABLE 1

THE EFFECTS OF ARYLAZIDO-β-ALANYL NAD$^+$ ON PYRIDINE NUCLEOTIDE
DEPENDENT ACTIVITIES OF COMPLEX I FROM OX HEART MITOCHONDRIA

Reaction	Concentration of fixed substrate (μM)	Concentration of Arylazido-β-alanyl NAD$^+$ (μM)	Inhibitory kinetic effect	$K_{m,app}$ (μM)	$K_{i,app}$ (μM)
NADH-CoQ$_1$ reductase	30 (CoQ$_1$)	147	Competitive	7.5 (NADH)	68
	80 (NADH)	140	Competitive	11.2 (CoQ$_1$)	171
NADH-AcPyAD$^+$ transhydrogenase	200 (AcPyAD$^+$)	19	Noncompetitive	16.7 (NADH)	---
	120 (NADH)	21	Competitive	248 (AcPyAD$^+$)	9.5
NADPH-CoQ$_1$ reductase	30 (CoQ$_1$)	14	Mixed	385 (NADPH)	---
NADPH-AcPyAD$^+$ transhydrogenase	300 (AcPyAD$^+$)	157	Competitive	71 (NADPH)	384

Chen, S and Guillory, R.J., (1979) J. Biol. Chem., 254, 7220-7227.

The noncompetitive inhibition exhibited by the analogue with respect to NADH for NADH-AcPyAD$^+$ transhydrogenation is interesting in view of the competitive inhibition observed with NADH-CoQ reduction. Since arylazido-β-alanine is on the ribose of the AMP portion in this analogue, the transhydrogenase protein may "recognize" the photoaffinity reagent as an NADP$^+$ analogue. This is consistent with the competitive inhibition observed with respect to AcPyAD$^+$ since in this assay, the latter substitutes for NADP$^+$.[31]

Upon photolysis the NAD$^+$ analogue is a potent, noncompetitive, irreversible inhibitor of the NADH-CoQ reductase, NADPH-CoQ$_1$ reductase and NADH-AcPyAD$^+$ transhydrogenase. Under identical conditions, there was no influence on the NADPH-AcPyAD$^+$ transhydrogenase activity.[*] It is clear that the photodependent inhibition due to arylazido-β-alanyl NAD$^+$ is not simply a result of nonspecific effects of arylazido-β-alanine or to a photosensitizing effect of NAD$^+$.

The low inhibitory titer following photolysis combined with the specificity of interaction are strong indications of a covalent nature for the photolytic dependent inhibitions. This point is further supported by an examination of the radioactive labeling pattern of complex I photolyzed in the presence of arylazido-[3-^{3}H]-β-alanyl NAD$^+$ and then subjected to gel electrophoresis (Fig. 10). Of the 12 to 14 peptides of complex I resolved by sodium dodecyl gel electrophoresis, a limited number of three peptides, those of 71, 50 and 30.5 thousand molecular weights, were labeled. When complex I was photoirradiated in the presence of a 50-fold excess of arylazido-[3-^{3}H]-β-alanine it was clear that no major specificity existed for labeling by arylazido-β-alanine (Fig. 11). At this concentration, arylazido-β-alanine does not inhibit NADH-CoQ reductase, NADH-AcPyAD$^+$ transhydrogenase or NADPH-AcPyAD$^+$ transhydrogenase. It is consequently reasonable to assume that the labeling pattern observed for arylazido-β-alanyl NAD$^+$ is a primary result of the active site binding specificity for the pyridine nucleotide portion of arylazido-β-alanyl NAD$^+$.

Interaction of arylazido-β-alanyl NADP$^+$ with complex I

A kinetic analysis of the effects of arylazido-β-alanyl NADP$^+$ on the pyridine nucleotide dependent activities present in complex I revealed that, in contrast to its ineffectiveness on the NADH-AcPyAD$^+$ transhydrogenation, the analogue is a very potent competitive inhibitor (with respect to NADPH) for NADPH-AcPyAD$^+$ transhydrogenation. The low $K_{i,app}$ value for the NADP$^+$ analogue (6 μM) can be compared to the $K_{m,app}$ for NADPH of 161 μM. The analogue is a noncompetitive inhibitor with respect to AcPyAD$^+$. The NADPH-AcPyAD$^+$ transhydrogenase is the

---

[*] Chen, S. and Guillory, R.J. (1980) J. Biol. Chem., 255, 2445-2453.

232

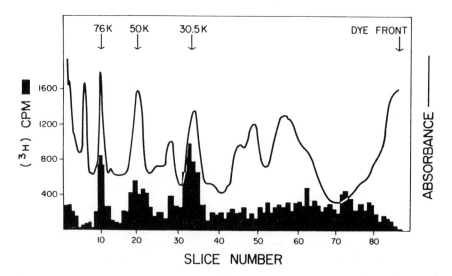

Fig. 10. Sodium dodecylsulfate polyacrylamide gel electrophoresis of complex I following irradiation with arylazido-[3-^{3}H]-β-alanyl NAD$^+$. 0.6 mg of complex I in 0.18 ml 0.25 M phosphate buffer pH 7.0 was photolyzed (2 min) in the presence of 0.14 μmol arylazido-[3-^{3}H]-β-alanyl NAD$^+$, specific activity 7.3 x 10^7 cpm/μmol. The photoirradiated sample was dialyzed against 1 liter of a solution containing 1% (w/v) sodium dodecylsulfate and 1% (v/v) β-mercaptoethanol. Following overnight dialysis in the dark at 25°C the solution was applied to 10% polyacrylamide sodium dodecylsulfate gels for electrophoresis according to the procedure described by K. Weber and M. Osborn (J. Biol. Chem., 244, 4406-4412, 1969). The gels stained with coomassie blue were scanned at 550 nm and sliced with a multi-blade slicer. The combined slices (1 mm) from twelve gels were digested overnight with 0.3 ml of 30% H$_2$O$_2$ at 65°C, and the radioactivity of each sample measured by liquid scintillation in 5 ml of Aquasol (New England Nuclear). Molecular weights were evaluated on sodium dodecylsulfate gels using glucose-6-phosphate dehydrogenase (104,000), E.C.1.1.1.49; pyruvate kinase (57,000), E.C.2.7.1.40; liver alcohol dehydrogenase (41,000), E.C.1.1.1.1; lactate dehydrogenase (37,000), E.C.1.1.1.27; and cytochrome C (12,300), as standards. The evaluated molecular weights for the major labelled peptides are indicated by a postscript above each band.[24]

only pyridine nucleotide dependent activity of complex I influenced by this analogue.

Photodependent effects of arylazido-β-alanyl NADP$^+$ on the pyridine nucleotide dependent activities of complex I

Following a one min photoirradiation of the enzyme complex (0.45 mg) with 44 nmol arylazido-β-alanyl NADP$^+$, the NADPH-AcPyAD$^+$ transhydrogenase activity was inhibited 54% compared to the nontreated preparation. (None of the other pyridine

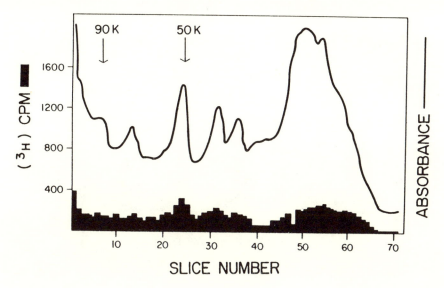

Fig. 11. Sodium dodecylsulfate polyacrylamide gel electrophoresis of complex I following irradiation with arylazido-[3-^{3}H]-β-alanine. The conditions were identical to those described for Fig. 10 except that arylazido-[3-^{3}H]-β-alanyl NAD$^+$ was replaced by 0.211 μmol arylazido-[3-^{3}H]-β-alanine of specific activity 6.9 x 10^7 cpm μmol^{-1}.[24]

nucleotide dependent reactions was influenced.) However, the enzyme complex mixed with an identical concentration of arylazido-β-alanyl NADP$^+$ but not sub-jected to light irradiation was also inhibited (41%). In view of the potent competitive inhibition demonstrated for this activity ($K_{i,app}$ 6.0 μM with a K_m(NADPH) of 161 μM), the inhibition is not unexpected. With submitochondrial particles a clear photodependent arylazido-β-alanyl NADP$^+$ inhibition of trans-hydrogenation is observed following centrifugation and resuspension of the photoirradiated particles.[32] Centrifugation effectively removes the excess unbound nucleotide analogue. Instability as a result of centrifugation prevents this approach in the case of complex I.

That there is covalent labeling of complex I by arylazido-β-alanyl NADP$^+$ which may be primarily on a limited number of peptides and on the one(s) associ-ated with the NADPH-AcPyAD$^+$ transhydrogenase is indicated by the experiment illustrated in Fig. 12. Under photoirradiation conditions only those peptides of molecular weight 90,000, 76,000, 37,000 and 26,500 are significantly labeled. The mitochondrial pyridine nucleotide transhydrogenase has been purified by

234

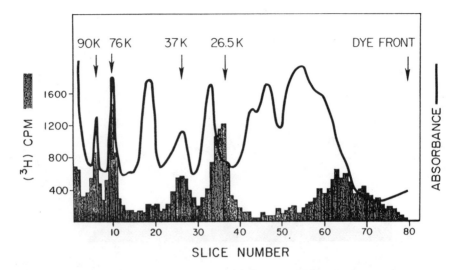

Fig. 12. Sodium dodecylsulfate polyacrylamide gel electrophoresis of complex I
following irradiation with arylazido-[3-^{3}H]-β-alanyl NADP$^+$. The conditions
were identical to those described for Fig. 10 except that arylazido-[3-^{3}H]-β-
alanyl NAD$^+$ was replaced by arylazido-[3-^{3}H]-β-alanyl NADP$^+$. (Chen, S. and
Guillory, R.J., (1980) J. Biol. Chem., 255, 2445-2453.

Anderson and Fisher[33] and Hojekerg and Rydstron[34] and both reports indicate
that the transhydrogenase has a molecular weight of from 90,000 to 110,000.
Since the NADPH-AcPyAD$^+$ transhydrogenase activity is the only enzymatic activity inhibited by this analogue, it is suggested that the 90,000 peptide is the
transhydrogenase protein. The NAD$^+$ analogue, arylazido-β-alanyl NAD$^+$, does
not label the 90,000 molecular weight peptide consistent with the lack of an
inhibitory effect of this analogue on NADPH-AcPyAD$^+$ transhydrogenase activity.

In collaboration with R. Fisher and W. Marshall Anderson of the University of
South Carolina, the pyridine nucleotide analogues were tested by Dr. S. Chen on the
purified pyridine nucleotide transhydrogenase isolated in Fisher's laboratory.[33]
The analogues influence the enzymatic activity of the transhydrogenase protein
in a manner identical to that described for the NADPH-AcPyAD$^+$ transhydrogenase
activity present in complex I. When arylazido-β-anyl NADP$^+$ was added to
transhydrogenase irradiation resulted in 75% inhibition of enzymatic activity
(Table 2); arylazido-β-alanyl NAD$^+$ or arylazido-β-alanine provided only 24% and
29% inhibition, respectively. The protein labeled with arylazido-[3-^{3}H]-β-alanyl
NADP$^+$ comigrated on sodium dodecylsulfate gel electrophoresis with the 90,000
molecular weight peptide of complex I.

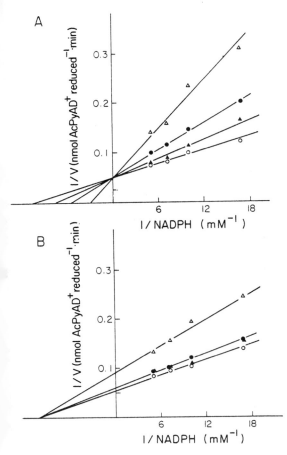

Fig. 13. Kinetic analysis of the arylazido-β-alanyl NADP$^+$ inhibition of the NADPH-AcPyAD$^+$ transhydrogenase activity associated with NADH-CoQ reductase (complex I). (A) In addition to a control sample (o) containing 0.5 mg of complex I in a final volume of 0.2 ml, three other samples were prepared containing in addition to complex I 56.4 nmol of arylazido-β-alanyl NADP$^+$. One of the latter samples was subjected to light irradiation for 3 min (Δ), a second was covered with aluminum foil, and maintained in the dark (●) and with the third sample (▲) the arylazido-β-alanyl NADP$^+$ was first photolyzed for 3 min prior to mixing with complex I.

For analysis of enzymatic activity a 10 μl aliquot of each mixture was assayed for enzymatic activity at different NADPH concentrations in the presence of 300 μM AcPyAD$^+$.

(B) The samples from A were incubated at 0-4° for 5 hours and then assayed as outlined above. (Chen, S. and Guillory, R.J., (1980) J. Biol. Chem., 255, 2445-2453.

The interaction of arylazido-β-alanyl NADP$^+$ with the site responsible for NADP-AcPyAD$^+$ transhydrogenase is not a simple ligand interaction as indicated by an interesting time dependent change in the kinetic order of the inhibitory effect of arylazido-β-alanyl NADP$^+$. A competitive inhibition with respect to NADPH is observed following irradiation of the analogue with complex I (Fig. 13). The $K_{i,app}$ values for arylazido-β-alanyl NADP with and without light irradiation were 1.04 μM and 1.94 μM, respectively, and 7.33 μM when the analogue was pre-photolyzed. While a major portion of the inhibitory profile is light dependent, there is a large nonlight dependent inhibitory effect.

When the assays for transhydrogenation are carried out following an extended incubation of the complex I-analogue mixture at 0°C (following photoirradiation), the kinetic profile indicated is that of *noncompetitive* inhibition (Fig. 13b). The presence of arylazido-β-alanine or NADP$^+$ alone or in combination did not

TABLE 2

THE PHOTODEPENDENT EFFECT OF ARYLAZIDO-β-ALANYL PYRIDINE

NUCLEOTIDES ON PURIFIED ENZYME PREPARATIONS FROM COMPLEX I[32]

Enzyme Preparation	Addition			
	Arylazido-β-alanyl NAD$^+$		Arylazido-β-alanyl NADP$^+$	
	μmoles	% inhibition	μmoles	% inhibition
NADH dehydrogenase[a]	8.8×10^{-3}	90	44.0×10^{-3}	12
Pyridine nucleotide[b] transhydrogenase	8.9×10^{-3}	24	11.0×10^{-3}	75

[a]The photolysis mixture, 0.12 ml, contained in addition to the analogue 1 mg of NADH dehydrogenase. Following an irradiation of one minute at 5°C, the preparation had a specific activity of 43.5 μmol NADH oxidized·min^{-1} mg protein^{-1}.

[b]The photolysis mixture, 0.3 ml, contained in addition to the analogue 6 mg of transhydrogenase. NADPH-AcPyAD$^+$ reductase activity was measured as described.[32] This preparation had a specific activity of 2.7 μmol AcPyAD$^+$ reduced·min^{-1} mg protein^{-1}.

induce the time dependent change in the kinetic profile. This interesting result is tentatively ascribed to a conformational change occurring at the active site of the NADPH-AcPyAD$^+$ transhydrogenase induced by the nucleotide analogue.

Studies of the influence of the pyridine nucleotide photoaffinity probes on the purified NADH dehydrogenase preparation are seen in Table 2. Arylazido-β-alanyl NAD$^+$ inhibits 90% of the NADH-CoQ$_1$ reductase activity in a photodependent manner. At a 5-fold higher concentration arylazido-β-alanyl NADP$^+$ inhibits activity but 12%. The purified NADH dehydrogenase when subjected to sodium dodecylsulfate gel electrophoresis has three subunits with molecular weights of 50,000, 30,500 and 14,500. Preliminary experiments utilizing arylazido-[3-^{3}H]-β-alanyl NAD$^+$ indicate that the peptide with a molecular weight of 50,000 was labeled.

Interaction of arylazido-β-alanyl pyridine nucleotides with submitochondrial particles

Submitochondrial particles in addition to pyridine nucleotide oxidoreduction (NADH oxidase) have three types of transhydrogenase activities, NADPH-NADH$^+$ and NADH-NADP$^+$ transhydrogenation and the energy dependent reduction of NADP$^+$ by NADH. Arylazido-β-alanyl NADP$^+$ is a potent competitive inhibitor with respect to NADP(H) for all three transhydrogenase activities (Table 3) while it is a

TABLE 3

THE EFFECT OF ARYLAZIDO-β-ALANYL NADP$^+$ ON THE PYRIDINE NUCLEOTIDE

TRANSHYDROGENASE ACTIVITIES PRESENT IN THE SUBMITOCHONDRIAL PARTICLES[a][32]

Reaction	Concentration of fixed substrate (μM)	Concentration of NADP$^+$ Analogue (μM)	Inhibitory kinetic effect	$K_{m,app}$ (μM)	$K_{i,app}$ (μM)
NADPH-NAD$^+$[b]	100 (NADPH)	5.64	Noncompetitive	100 (NAD$^+$)	--
	200 (NAD$^+$)	5.64	Competitive	28 (NADPH)	0.81
NADH-NADP$^+$	80 (NADH)	22.56	Competitive	66.7 (NADP$^+$)	6.36
	400 (NADP$^+$)	22.56	Noncompetitive	6.8 (NADH)	--
NADH-NADP$^+$	80 (NADH)	22.56	Competitive	33.7 (NADP$^+$)	4.15
Energy-linked	400 (NADP$^+$)	22.56	Uncompetitive	17.2 (NADH)	--

[a]The submitochondrial particles (EDTA particles) are prepared as described by C.P. Lee and L. Ernster (1967), Meth. Enzymol. X, 543-548.

[b]Transhydrogenase activities were assayed as described by A. Teixeira Da Cruz, J. Rydstrom and L. Ernster (1971), Eur. J. Biochem., 23, 203-211; and J. Rydstron, A. Teixeira Da Cruz and L. Ernster (1971) Eur. J. Biochem., 23, 212-219.

noncompetitive inhibitor with respect to $NAD^+(H)$ for the $NADPH-NAD^+$ and $NADH-$
$NADP^+$ transhydrogenase, identical to the inhibitory profile for the $NADP^+$
analogue on the $NADPH-AcPyAD^+$ transhydrogenase of complex I. The uncompetitive
inhibitory pattern for the energy linked reaction (with respect to NADH) is
similar to that reported for adenylate.[35]

Photoirradiation of submitochondrial particles in the presence of arylazido-
β-alanine did not significantly influence NADH oxidase, $NADP-NAD^+$ transhydrogenase nor succinate oxidase activity.[32] Under the same conditions, arylazido-
β-alanyl NAD^+ inhibited the NADH oxidase in a photodependent manner, while succinate oxidase and $NADPH-NAD^+$ transhydrogenase were not significantly inhibited.
Significantly, arylazido-β-alanyl $NADP^+$ inhibited all three activities in a
photodependent manner. In the case of complex I, the NADPH dependent transhydrogenation is the only activity effectively influenced by the $NADP^+$ analogue.
The possibility of energy coupling between transhydrogenation and the electron
transport chain in submitochondrial particles being responsible for the sensitivity to arylazido-β-alanyl $NADP^+$ is made unlikely by the fact that the inhibition of the oxidase activity is not influenced by uncoupling agents. One
might consider the possibility that different structural restraints within submitochondrial particles as compared to complex I may be responsible for the
increased sensitivity of the former preparation to the $NADP^+$ analogue.

Photoaffinity binding at enzyme activity modifier sites

It is in the study of allosteric or general ligand binding sites at which
the primary event of catalysis does not take place that photoaffinity probes
may prove to be most useful.

Two well known modifiers of glutamic dehydrogenase are ADP and GTP, the
former raising the apparent dissociation constant of NADPH with the enzyme and
the latter hindering such dissociation.[36] In this manner ADP stimulates and
GTP inhibits the reduction of NADP in the overall enzymatic reaction. Arylazido-
β-alanyl ADP was found to effectively mimic the stimulatory profile of ADP with
glutamic dehydrogenase.[*] Arylazido-β-alanyl GTP, synthesized utilizing the
carbodiimidazole catalyzed esterification of the ribose hydroxyl group with
arylazido-β-alanine, was found to be an effective inhibitor of glutamic dehydrogenase. The nucleotide analogue displayed an inhibitory profile similar to
that of GTP.

The proposed structure for "arylazido-β-alanyl GTP" is based on an analogy
with that of the other synthesized arylazido-β-alanyl nucleotides, together with

[*] Jeng, S., Guillory, R.J. and Sund, H., unpublished observations.

a lack of an interaction of arylazido-β-alanine with the nucleotide base and a 1:1 stoichiometry between GTP and arylazido-β-alanine.

An interesting aspect of the interaction of the arylazido-β-alanyl GTP analogue and the arylazido-β-alanyl ADP analogue with glutamic dehydrogenase is the finding that photoinsertion of both into what is presumably their allosteric binding sites appears to freeze the enzyme into a specific configuration. Following photoirradiation with the nucleotide analogues the enzyme was cleared of residual nucleotide by gel filtration. In the case of the arylazido-β-alanyl GTP treated preparation the enzyme had undergone irreversible inhibition consistent with a GTP effect while the arylazido-β-alanyl ADP treated preparation retained a specific activity characteristic of the ADP activated state.

Preliminary data on the synthesis of arylazido-β-alanyl coenzyme A derivatives

Incubation of N-4-azido-2-nitrophenyl-β-alanine and carbodiimidazole in the presence of acetyl-Coenzyme A or Coenzyme A resulted in the formation of reaction products containing the characteristic arylazido absorption and distinct from starting material.

Both Coenzyme A and acetyl-Coenzyme A derivatives were tested for possible photodependent effects on the catalytic activity of phosphotransacetylase. As can be seen from Table 4, the acetyl-CoA derivative was able to bring about a substantial photodependent inhibition of enzymatic activity. The CoA derivative on the other hand was ineffective. Phosphotransacetylase is sensitive to ultraviolet irradiation and when it is subjected to photoflashing without a glass shield there results a sizable destruction of enzymatic activity. The Coenzyme A derivative while ineffective as a photoinhibitory probe is able to provide a major degree of protection against photodestruction (Fig. 14).

An examination of the nuclear (proton) magnetic resonance spectra of the two derivatives indicates that the CoA analogue contains, in comparison to the acetyl-Coenzyme A derivative, two moles of arylazido-β-alanine per mole of Coenzyme A. This stoichiometry is consistent with absorptive spectroscopy. The Coenzyme A molecule itself has two potential hydroxyl groups available for esterification, that at the 2' hydroxyl of the ribose unit and the secondary hydroxyl present in the pantothenic acid portion of the molecule. In addition, the thiol of Coenzyme A is a potential site for esterification. Work with the NADP[+] derivative indicates that a ribose hydroxyl when adjacent to phosphate is resistant to esterification. Consequently, it is assumed that in the acetyl-Coenzyme A derivative the site of esterification is at the secondary hydroxyl of the pantothenic acid. While this is reasonable, an examination of the nuclear magnetic resonance spectra of the arylazido-β-alanyl analogue does not

TABLE 4

THE EFFECT OF "ARYLAZIDO-β-ALANYL-COENZYME A"

DERIVATIVES ON PHOSPHOTRANSACETYLASE

Condition	ΔAbsorbance 232 mμ·min^{-1}	% Control
Control	0.0164	100
Irradiated	0.0140	85
Irradiated with "Arylazido-β-alanyl Acetyl CoA"	0.0034	21
Control with "Arylazido-β-alanyl Acetyl CoA"	0.0166	101
Irradiated with "Arylazido-β-alanyl CoA"	0.0156	95

Phosphotransacetylase activity was measured by the change in absorbance at 232 mμ with the phosphotransacetylase (Sigma Chemical Co.) (0.5 μg) in a total volume of 2 ml containing 0.09 M Tris buffer pH 7.4, 0.001 M ammonium sulfate, 1 mg glutathione, 0.3 mg coenzyme A and 2.5 mg acetylphosphate. Irradiation was carried out in a total volume of 0.5 ml containing 0.025 M Tris pH 8.0 and 2.5 μg enzyme according to the method described by J. Cosson and R.J. Guillory[27] using 8 flashes. When present, the coenzyme derivatives were at 100 μM concentration during irradiation.

indicate a chemical shift expected for resonance (g), corresponding to the proton on the carbon containing this hydroxyl. A sizable upfield shift of the acetate signal in arylazido-β-alanyl acetyl-Coenzyme A[37] indicates interaction between the acetate and the arylazido-β-alanyl structure.

It is tentatively assumed that in the case of the Coenzyme A derivative one of the arylazido-β-alanyl adjuncts is esterified to the free sulfhydryl group. This large bulky group situated at a very sensitive region for enzyme substrate interaction prevents the formation of an ES complex. In the case of the acetyl-Coenzyme A derivative, the presence of the acetyl group prevents esterification at the tiol and allows arylazido-β-alanyl acetyl-Coenzyme A to be an effective photoaffinity probe. Further work is continuing on illucidating in a more satisfactory manner the exact structure of these two analogues.

CONCLUSION

Some of the advantages, disadvantages and problems arising in the use of the photoaffinity approach for receptor site labeling have been outlined using as a background studies with arylazido-β-alanyl nucleotide probes developed in this laboratory.

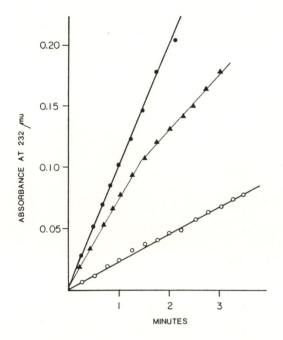

Fig. 14. Protection afforded
by "arylazido-β-alanyl Coenzyme A" against photoirradiation inhibition of phosphotransacetylase. The conditions for the assay of
enzymatic activity and photoirradiation are those
described for Table 4 except
that the protecting glass
shield was removed during
photoirradiation. ooo,
control; ●●●, irradiated
with three flashes;
ΔΔΔ, irradiated with three
flashes in the presence of
100 μM "arylazido-β-alanyl
Coenzyme A."

Knowledge of the active site structure and possible mechanism of action of
myosin ATPase has been used in the design of a photoaffinity probe for ATPase
enzymes. Substrate reactivity is considered a strong indication of proper
orientation within the binding region. The stoichiometric labeling pattern
exhibited for arylazido-β-alanyl ATP with the myosin and mitochondrial ATPase
enzymes supports the idea that the analogue undergoes an initial specific noncovalent interaction which so orients the arylazido adjunct that direct insertion is the only course open for the bound analogue upon photolysis.

As demonstrated here, if a photoprobe exhibits strong binding for the ligand
site, the degree of solvent reactivity may be unimportant and may in fact be
helpful in reducing nonspecific photolabeling.

The photodependent labeling pattern observed for heavy meromyosin and subfragment 1 with arylazido-β-alanyl ATP indicates that there is no major difference between the two active sites on myosin with respect to ATP binding. The
fact that more than a single cyanogen bromide peptide derived from photoirradiated subfragment 1 in the presence of radiolabeled analogue contains radioactivity is considered to represent an aspect of the flexibility of the active

site or of the analogue within the binding site. Design of photoprobes with shorter activation half-lives may make it possible to investigate time dependent changes (i.e., conformational) occurring at the active site of enzymes and within ligand binding sites of complex membrane systems. Most investigative methods used to gather structural information are limited by the static picture provided of actual conformationally dynamic systems, photoaffinity reagents have the potential of acting as probes of such dynamic states in solution in which the rapid reaction of a photogenerated reagent could in principle be used to "trap" short lived conformational forms.

Studies with integrated forms of the mitochondrial ATPase illustrate that the photoaffinity technique is able to bring about the labeling of specific reactive sites within membrane systems composed of complex multipeptide components. It would appear that the membrane environment controls both the rate of binding and the magnitude of interaction of the nucleotide photoprobe to the F_1ATPase. Experiments designed to investigate the labeling pattern of this probe as a function of different physiological states of the mitochondria membrane may aid in an understanding of the mechanism by which the membrane controls the conformational attitude of proteins.

The study of the kinetic and photodependent inhibitory influence of two pyridine nucleotide photoaffinity probes on the integrated and resolved components of mitochondrial complex I complement themselves in showing clearly the specificity of interaction of the two probes. The NADH dehydrogenase interacts only with arylazido-β-alanyl NAD$^+$ and the transhydrogenase only with arylazido-β-alanyl NADP$^+$. On the basis of the similarity of response towards the two analogues, it is clear that the NADH-CoQ$_1$ reductase and NADPH-AcPyAD$^+$ transhydrogenase activities of complex I are measures of mitochondrial oxidoreductase and transhydrogenation, respectively. Our results also suggest that the active site of the resolved NADH dehydrogenase and transhydrogenase enzymes must be of a similar configuration to that present within complex I. It is anticipated that the further resolution of complex I with the aid of the pyridine nucleotide photoprobes will assist in characterizing the relationship of its enzymatic activities to the different protein subunits observed in complex I.

The overlapping labeling pattern exhibited by complex I for the analogues is being further investigated with the view that such labeling represents juxtapositioning of the specific peptides to either the NADH dehydrogenase or the NADPH transhydrogenase of this multicomponent system.

Photoirradiation of arylazides in aqueous solution results in the formation of dark reaction products as evidenced by spectral changes occurring as a function of time following photoirradiation.[*] Such changes which can continue for

---

[*] Newcomb, R., unpublished observations

minutes are completely quenched by scavengers such as dithiothreitol. Comple-
menting these dark reactions one finds time dependent covalent labeling of proteins which are added following photoirradiation. In preliminary work, Dr. F.
Shore has shown that addition of F_1ATPase to an aqueous solution of arylazido-β-
alanyl ATP ($α^{32}$Pi) following irradiation of the nucleotide solution results in
covalent binding of the photoprobe to the enzyme. Such binding which is prevented by scavengers is presumably not involved with direct active site insertion since there is no added postphotolytic inhibition of enzymatic activity.
Electrophilic reactive species produced upon photolysis of the aromatic azide
may label nonspecifically nucleophilic centers on the protein. Consistent with
these observations are the facts that acid and copper ion catalyzed decomposi-
tion of diazo groups results in covalent labeling of enzymes[38] and that a
generalized labeling of a detergent solubilized acetylcholine receptor has been
ascribed to the diffusion of a long lived aromatic nitrene from specific sites
into neighboring subunits.[39]

Since postphotolytic quenching of these reactions is possible, such controls
should be a routine aspect of the use of photoaffinity reagents. On the other
hand, arylazides are known to be rapidly reduced to their corresponding amines
by dithiothreitol at room temperature and physiological pH[40,41] and such reagents
must be used with caution in order not to negate completely the effectiveness
of the photoprobe.

A comparison of the effectiveness of the pyridine nucleotide photoprobes on
the enzymatic reactivities of a membrane bound system (submitochondrial particles) with more resolved system (complex I) indicated that while in the
partially resolved system arylazido-β-alanyl $NADP^+$ was only effective against
transhydrogenation, in the intact system, the $NADP^+$ analogue inhibited NADH
oxidase as well as the $NADPH-NAD^+$ transhydrogenase. This inconsistency may have
a number of explanations, one of which could be the change in conformational
attitude of the protein in its firmly bound state as compared to its more
soluble form. The use of photoaffinity reagents when coupled with structural
analysis may consequently be helpful in evaluating those forces responsible for
establishing different conformational states.

Photoaffinity labeling is thus likely to emerge as one of the major techniques
for elucidating the relationship between structure and function in biological
membranes as well as for enzymatic and allosteric reactive proteins. Its
potential in viewing dynamic aspects of biological phenomena at the molecular
level has not yet been tested and is perhaps the most exciting frontier.

244

ACKNOWLEDGMENTS

My appreciation is extended to my present and past associates, Drs. S.J. Jeng,
S. Chen, J. Cosson and F. Shore, for their contributions both experimental and
intellectual to the work described in this paper. This work was supported by
grants from the Hawaii Heart Association, the American Heart Association, the
United States Publich Health Service (National Institutes of Health) and the
National Science Foundation (NSF).

REFERENCES

1. Means, G.E. and Feeney, R.E.   (1971) Chemical Modification of Proteins,
   Holden-Day, San Francisco.
2. Cohen, L.A.   (1968) Ann. Rev. Biochem., 87, 695-726.
3. Vallee, B. and Riodean, F.   (1969) Ann. Rev. Biochem., 38, 733-794.
4. Wolsey, L., Metzger, H. and Singer, S.J.   (1962) Biochem., 1, 1031-1039.
5. Knowles, J.R.   (1972) Acct. Chem. Res., 5, 155-160.
6. Bayley, H. and Knowles, J.R.   (1977) In Methods in Enzymology, Jakoby, W.B.
   and Wilchek, M., eds., Vol. XLVI, pp. 69-114.
7. Singh, A., Thornton, E.R. and Westheimer, F.H.   (1962) J. Biol. Chem., 237,
   3006-3008.
8. Fleet, G.W.J., Porter, R.R. and Knowles, J.R.   (1969) Nature, 224, 511-512.
9. Doering, W.V.E., Buttery, R.C., et al.   (1956) J. Am. Chem. Soc., 78, 3224.
10. Gilchrist, T.L. and Pres, C.W.   (1969) In Carbenes, Nitrenes and Arynes,
    Gilchrist, T.L. and Rees, C.W., eds., Pitman Press, London, pp. 10-27.
11. Reiser, A. and Leyshorn, L.   (1970) J. Am. Chem. Soc., 92, 7487.
12. McRobbie, I.M., Meth-Cohn, O. and Suschitzky, N.   (1976) Tetrahedron Letters,
    12, 926-928.
13. Ruoho, A.F., Kiefer, H., et al.   (1973) Proc. Nat. Acad. Sci. USA, 70,
    2567-2571.
14. Singer, S.J.   (1967) Adv. Protein Chem., 22, 1-54.
15. Doering, W.V.E. and Odmn, R.A. (1966) Tetradedron, 22, 81-93.
16. Wentrup, C. and Crow, W.D.   (1970) Tetrahedron, 26, 3965-3981.
17. Jeng, S.J. and Guillory, R.J.   (1975) Supramolecular Structure, 3, 448-468.
18. Tonomura, Y.   (1973) Muscle Proteins, Muscle Contraction and Cation Trans-
    port, University Park Press, Baltimore.
19. Azuma, N., Ikegara, M., et al.   (1962) Biochim. Biophys. Acta, 60, 104-111.
20. Yamashita, T., Soma, Y., et al.   (1964) J. Biochem., 55, 576-577.
21. Hiratsuka, T. and Uchida, K.   (1973) Biochim. Biophys. Acta, 320, 635-647.
22. Gottikh, B.P., Krayevsky, A.A., et al.   (1970) Tetrahedron, 26, 4419.
23. Zamecnik, P.C.   (1962) Biochem. J., 85, 257.
24. Guillory, R.J., Jeng, S.J. and Chen, S.   (1979) Ann. N.Y. Acad. Sci.,
    Conference on Applications of Photochemistry in Probing Biological Targets,
    in press.
25. Jeng, S.J. and Guillory, R.J.   (1976) Abstract, Am. Chem. Soc., 172d Meeting,
    Biol. 102, San Francisco, California.
26. Russell, J., Jeng, S.J. and Guillory, R.J.   (1976) Biochem. Biophys. Res.
    Commun., 70, 1225-1233.
27. Cosson, J.J. and Guillory, R.J.   (1978) J. Biol. Chem., 254, 2946-2955.
28. Serrano, R., Baruch, I.K. and Racker, E.   (1976) J. Biol. Chem., 251,
    2543-2461.
29. Scatchard, G.   (1949) Ann. N.Y. Acad. Sci., 51, 660.
30. Chen, S. and Guillory, R.J.   (1977) J. Biol. Chem., 252, 8990-9001.
31. Hatefi, Y. and Hanstein, W.G.   (1973) Biochemistry, 12, 3515-3522.
32. Chen, S. and Guillory, R.J.   (1979) in Membrane Bioenergetics, Lee, C.P.,
    Schatz, G. and Ernster, L., eds., Addison-Wesley Publishing Co., Inc., 61-80.

33. Anderson, W.M. and Fisher, R.R. (1978) Arch. Biochem. Biophys., 187, 180-190.
34. Hojeberg, B. and Rydstrom, J. (1977) Biochem. Biophys. Res. Commun., 78, 1183-1190.
35. Rydstron, J. (1972) Eur. J. Biochem., 31, 496-504.
36. Fisher, H.F. (1973) Adv. in Enzymol., Meister, A., ed., 39, 369-417.
37. Mieyal, J.J., Webster, L.T. and Siddiqui, U.A. (1974) J. Biol. Chem., 249, 2633-2640.
38. Rajagopalau, T.G., Stein, W.H. and Moore, S. (1966) J. Biol. Chem., 241, 4295-4297.
39. Witzemann, V. and Raftery, M.A. (1977) Biochemistry, 16, 5862-5868.
40. Staros, J.V., Bayley, H., et al. (1978) Biochem. Biophys. Res. Commun., 80, 568-572.
41. Cartwright, I.L., Hutchinson, D.W. and Armstrong, V.W. (1976) Nucleic Acid Res., 3, 2331-2339.

MICROENVIRONMENT AROUND HISTIDINE RESIDUES IN RIBONUCLEASES
ANALYZED BY HYDROGEN-TRITIUM EXCHANGE TITRATION

K. NARITA, K. KANGAWA[*], S. KIMURA, K. MIYAMOTO[+], N. MINAMINO[*] AND H. MATSUO[*]
Institute for Protein Research, Osaka University, Yamada-kami, Suita, Osaka 565,
Japan

ABSTRACT

A novel method for analysis of microenvironments around individual histidine
residues in proteins is described. The principle of the method is based upon
measurement of the pH-dependent pseudo-first-order rate constant for exchange of
the hydrogen atom attached to the $C^{\varepsilon 1}$ position of the imidazole ring of a histidine residue with tritium, which occurs during incubation of a protein in buffered tritiated water at 37°C. A sigmoid curve is obtained when the rate constants measured at various pH's are plotted against the pH of the reaction and
its midpoint corresponds to pKa of the histidine. A neighboring anionic group
raises pKa of the histidine, while a neighboring cationic group lowers it. The
magnitude of the second-order (with respect to the concentrations of histidinium
cation and hydroxide anion in the medium) rate constant is shown to correspond
to the solvent accessibility of the histidine residue in the three-dimensional
structure of the protein, which has been deduced from the application of the
present method to bovine pancreatic ribonuclease of known conformation. The
microenvironments around the three histidine residues in ribonuclease T_1 and the
two histidine residues in ribonuclease St were analyzed and histidine residues
involving in the active sites in the two enzymes were assigned.

The hydrogen atoms attached to the $C^{\varepsilon 1}$ positions of the imidazole rings of
histidine residues in proteins undergo exchange with tritium, when they are
incubated in tritiated water at 37°C.[1] The overall reaction rate in the exchange
reaction follows pseudo-first-order (rate constant, $\underline{k}_\psi$) kinetics with respect to
the total concentration of histidine and histidinium cation; both are in equilibrium (constant, Ka) in a certain pH range. According to Vaughan et al.,[2] the
hydrogen atom attached to the $C^{\varepsilon 1}$ position of histidinium cation is abstracted
by a hydroxide anion to yield an intermediate ylide, which quickly takes up
tritium from the medium, as shown in Fig. 1. The formation of the ylide is the

*Present address: Miyazaki Medical College, Kiyotake, Miyazaki 889-16.
+Present address: School of Medicine, Gunma University, Maebashi, Gunma 371.

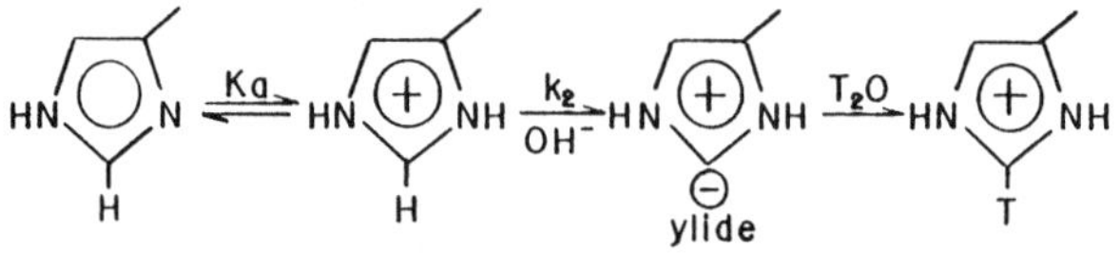

Fig. 1. Reaction mechanism for hydrogen-tritium exchange at $C^{\varepsilon 1}$ position of the imidazole ring of histidine residue (according to Vaughan et al.[2])

rate-determining step and is second-order (constant, $\underline{k}_2$) with respect to the concentrations of histidinium cation and hydroxide anion. In the equation

$$\underline{k}_\psi = \underline{k}_2 Kw/(Ka + [H^+]) \tag{1}$$

Kw is the ion product of water. Although the $\underline{k}_\psi$ value is pH-dependent, no exchange takes place at acidic pH but the value increases alkaline pH values until it finally becomes constant regardless of the pH of the medium.

$$\underline{k}_\psi^{max} = \underline{k}_2 Kw/Ka \tag{2}$$

Between the two extreme pH's, $\underline{k}_\psi$ changes as a function of the pH of the reaction according to Eq. 2. The correlation of $\underline{k}_\psi$ with the pH of the reaction is expressed as a sigmoid curve and its midpoint corresponds to pKa of the histidine residue.[1] The values of pKa of histidine residues in several proteins measured by the present hydrogen-tritium exchange titration range between 5 and 8, but their values converge to about 6.5 upon denaturation of these proteins, indicating that pKa values of histidine residues in native proteins reflect differences in their microenvironments.[3] One of the major factors which affect on pKa value is concluded to be neighboring charged groups, a positively charged group lowers pKa of the histidine residue, while a negatively charged group raises it.[4]

When values of pKa and $\underline{k}_\psi^{max}$ are measured experimentally, $\underline{k}_2$ can be calculated from Eq. 2. The Bronsted relationship (shown in Eq. 3, α and β are constants) which correlates log $\underline{k}_2$ with pKa for base-catalyzed reactions can be applied to the hydrogen-tritium exchange reaction for several imidazole derivations and oligopeptides containing a histidine residue, but not for histidine residues in native proteins.

$$log\ \underline{k}_2 = \alpha pKa + \beta \tag{3}$$

The magnitude of the deviation of the histidine residue in question from the Bronsted law corresponds to the solvent inaccessibility of the imidazole ring of the histidine in a native protein.[4] Upon denaturation of proteins, k_2 values for all histidine residues converge to definite values. We can, therefore, analyze the microenvironment around individual histidine residues in the three-dimensional structures of native proteins in solution by measuring values of pKa and k_2, since the present exchange reaction can be carried out without introduction of any chemical reagent except tritiated water provided that the proteins were stable enough over entire pH range of exchange titration.

Procedure of the tritium exchange titration of proteins. Protein samples (0.3 μmole are incubated in sealed tubes with buffered tritiated water (0.2 ml, 4 mCi) at various pH's and 37°C in solutions of definite ionic strength for 24-48 hours. After the incubation for a scheduled period, the reaction is terminated by addition of 0.1 ml of formic or acetic acid. The excess tritiated water is removed *in vacuo*. The removal of the tritium incorporated into amide nitrogen and other positions is repeated several times by dissolving the dried samples in 0.1 ml of the 10-30% acid and subsequent evaporation. The radioactive protein preparations are subjected to gel filtration on a column of Sephadex G-25 to remove salts and then are lyophilized. The tritiated protein preparations at various pH's are oxidized with performic acid and are digested with proteolytic enzymes for 4 hours at 37°C. The resulting histidine-peptides are separated by the two-dimensional peptide mapping method and are extracted from the map with 30% acetic acid. The extracted peptides are hydrolyzed as usual and a part of the hydrolyzate is subjected to amino acid analysis and another part for radioactivity determination.

The pseudo-first-order rate constant, k_ψ, in the tritium exchange reaction is calculated at various pH's by using the following equation:

$$\ln[(a - x)/a] = -k_\psi t \tag{4}$$

where a and x are the specific radioactivities of the tritiated histidine at time ∞ and t at 37°C. The former value is obtained by the use of acetylhistidine and the same lot of the tritiated water used for the protein at several reaction times at pH 8-9. The procedure is described in detail by Kimura et al.[5]

Ribonuclease A.[3,6] Bovine pancreatic ribonuclease (RNase) A is one of the most extensively studied proteins and the three-dimensional structure has been elucidated by X-ray crystallographic analysis.[7] Four histidines in RNase A are located at positions 12, 48, 105, and 119 from the N-terminus and His-12 and His-119 are known to be active site residues. The protein was subjected to the tritium exchange reaction at 37°C for 38 hours in 0.1 ionic strength at various

250

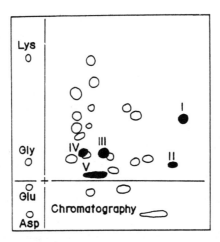

Fig. 2. Peptide map of the tryptic-chymotryptic digest of performic acid-oxidized RNase A.[3] Dark spots represent peptides containing histidine residues (Pauly reagent positive). Paper chromatography was carried out first with l-butanol-pyridine-acetic acid-water (15:10:3:12, v/v) followed by electrophoresis at pH 3.5 (pyridine-acetic acid-water = 1:10:289, v/v) and 50 volts/cm.

pH's. The peptide map of the tryptic-chymotryptic digest of the performic acid oxidized, tritiated enzyme is shown in Fig. 2. The peptides I-V were radio-active and Pauly reagent positive and only peptide V was ninhydrin negative. The results of amino acid and N-terminal analyses of each peptide established the following sequences for the five peptides, respectively.

Peptide I : Val-Pro-Val-His-Phe (position 119)

II : His-Ile-Ile-Val-Ala-Cys(O_3H)-Glu-Gly-Asn-Pro-Tyr (position 105)

III : Val-His-Glu-Ser-Leu-Ala-Asp-Val-Gln-Ala-Val-Cys(O_3H)-Ser-Gln-Lys (position 48)

IV : Gln-His-Met(O_2)-Asp-Ser-Ser-Thr-Ser-Ala-Ala-Ser-Ser-Ser-Asn-Tyr (position 12)

V : < Glu-His-Met(O_2)-Asp-Ser-Ser-Thr-Ser-Ala-Ala-Ser-Ser-Ser-Asn-Tyr (position 12)

Peptides IV and V contained His-12 and thus the two peptides were combined for

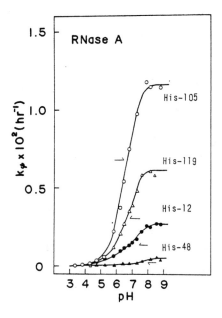

Fig. 3. The hydrogen-tritium exchange titration curves for the four histidines in RNase A.[3]

subsequent analyses. Figure 3 shows k_{ψ}-pH correlations for the four histidines and their respective pKa values obtained from the midpoints of the respective sigmoid plots and are listed in Table 1, together with their k_{ψ}^{max} and $log\ k_2$ values. The sigmoid plots for His-12, His-48 and His-119 did not fit theoretical curves, suggesting that they interact with neighboring charged groups, respectively. However, the interactions between the histidine residues and the charged groups did not seem to be strong, since their pKa values are not shifted greatly from the value for unperturbed histidine (pKa, about 6.5), which was measured in the presence of 6 M guanidine hydrochloride (Table 1).

The Bronsted plots for the four histidines are shown in Fig. 4, together with those for some imidazole derivatives. The plots for the model compounds, which possess different pKa values, are on a straight line with a slope of about -0.7, suggesting that they follow the Bronsted law. However, the plots for the four histidines in native RNase A molecule deviated from a straight line. The perpendicular distances between the line and the values for the four histidines correspond to the solvent inaccessibilities of the histidine residues in the enzyme's three-dimensional structure which was deduced by X-ray crystallographic analysis.[7] A schematic drawing of the ribonuclease is shown in Fig. 5.

Ribonuclease T_1.[5] We can predict the microenvironments around individual histidine residues in proteins, whose three-dimensional structures have not been elucidated, by using the information described above for RNase A. We applied

TABLE 1

pKa's AND THE RATE CONSTANTS FOR THE FOUR HISTIDINE RESIDUES
IN RNase A DETERMINED BY THE HYDROGEN–TRITIUM EXCHANGE TITRATION
IN THE ABSENCE AND PRESENCE OF 6 M GUANIDINE HYDROCHLORIDE AT 37°C[3]

	0.1 Ionic Strength			6 M Guanidine HCl		
	pKa	$k_{-\psi}^{max}$ x 10^2	log k_2 $(M^{-1}h^{-1})$	pKa	$k_{-\psi}^{max}$ x 10^2 (h^{-1})	log k_2 $(M^{-1}h^{-1})$
His-12	6.6	0.27	4.43	6.8	0.70	4.66
His-48	6.7	0.04	3.50	6.5	0.71	4.97
His 105	6.5	1.15	5.16	6.5	0.68	4.95
His-119	6.6	0.61	4.79	6.6	0.49	4.71

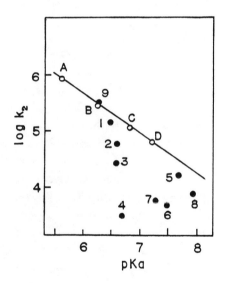

Fig. 4.   The Bronsted plot for imidazole derivatives and histidine residues in
proteins.
Model compounds: A, *trans*-urocanic acid; B, Ac-His-NHMe; C, Ac-His; D, imidazole
propionic acid
RNase A: 1, His-105; 2, HIS-119; 3, His-12; 4, His-48
RNase T_1: 5, His-40; 6, His-92; 7, His-27
RNase St: 8, His-91; 9, His-60

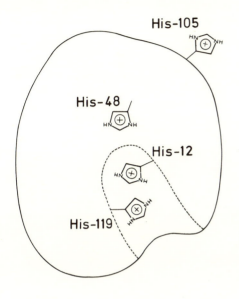

His-105

His-48

His-12

His-119

Fig. 5. Schematic drawing of the
three-dimensional structure of
RNase A.[4]

the present hydrogen-tritium exchange titration procedure to RNase T_1 (produced
by *Aspergillus oryzae*) which consists of 104 amino acid residues in a single
polypeptide chain cross-linked by two disulfide bridges.[8] In this case, RNase
T_1 was incubated in tritiated water at 37°C for 30 hours in solution of 0.2
ionic strength.

The peptide map of the tryptic-thermolytic digest of performic acid-oxidized
RNase T_1 is shown in Fig. 6. The amino acid sequences of the histidine-containing peptides extracted from the map were deduced from amino acid and N-terminal
analyses.

```
                        92
Peptide I-A : Val-Ile-Thr-His-Thr

                         92
        I-B : Val-Ile-Thr-His-Thr-Gly

                                 92
        II  : Val-Gly-Ser-Asn-Ser-Tyr-Pro-His-Lys

                   27
        III : Leu-His-Glu-Asp-Gly-Glu-Thr
```

The tritium exchange titration curves for the three histidine residues are
shown in Fig. 7. The apparent exchange rate for His-40 was fast compared with
the other two histidine residues. However, all the sigmoid curves for the three

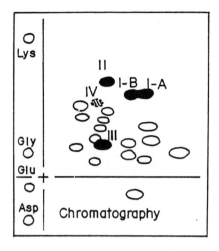

Fig. 6. Peptide map of the tryptic-thermolytic digest of performic acid-oxidized RNase T_1. Peptide IV is a deamidated form of peptide II.[5] For explanations, see legend to Fig. 2.

histidines changed their slopes simultaneously above pH 9 and their $k_{-\psi}^{max}$ values converged to about 0.01 h^{-1}, which is the intrinsic value for histidine residues in denatured proteins (Table 1) and suggested that the microenvironments around the all histidines were becoming denatured. The pKa values for the three histidines shown in Table 2 were very high compared with intrinsic pKa value of histidine residues in proteins (about 6.5, Table 1), suggesting that neighboring carboxylate groups interact strongly with the three histidine residues in native RNase T_1 molecule. Such interactions can be clearly seen in the ^1H NMR titration curves of the three histidines[9,10] as inflection points in the acidic

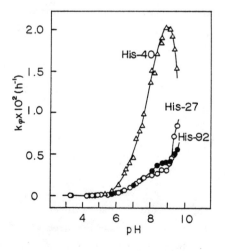

Fig. 7. The hydrogen-tritium exchange titration curves for the three histidine residues in RNase T_1.[5] Above pH 9, the enzyme was irreversibly denatured.

TABLE 2

pKa's AND THE TRITIUM EXCHANGE RATE CONSTANTS FOR THE THREE HISTIDINES IN

RNase T_1 IN THE ABSENCE AND PRESENCE OF 3'-GMP at 37°C AND AT 0.2 IONIC STRENGTH[5]

	3'-GMP	pKa	$\underline{k}_{-\psi}^{max}$ x 10^2 (h^{-1})	log $\underline{k}_2$ $(M^{-1}h^{-1})$	pKa determined by NMR titration $(32°C)^{9,10}$
His-27	−	7.3	0.31	3.75	7.2
	+	7.5			7.2
His-40	−	7.7	2.13	4.22	7.9
	+	8.5			8.6
His-92	−	7.6	0.42	3.70	∼8
	+	8.3			8.5

region, which is beyond the tritium exchange titration range as shown in Fig. 8.

The Bronsted plots for the three histidines are also shown in Fig. 4. His-tidine-40 appears to be exposed on the molecular surface, while His-27 and His-92 seem to be located in solvent inaccessible regions as noted for His-12 in RNase A.

The hydrogen-tritium exchange titration curves for His-40 and His-92 in the presence of four-fold molar excess relative to RNase T_1 of the competitive inhibitor, 3'-GMP, shifted the curve to the alkaline side (Fig. 9). This

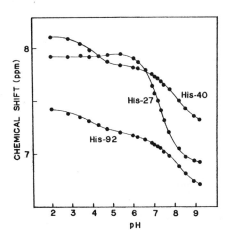

Fig. 8. Proton NMR titration curves for the three histidines in RNase T_1 at 32°C.[9]

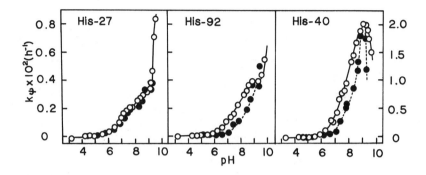

Fig. 9. The hydrogen-tritium exchange titration curves for the three histidines in RNase T$_1$ in the absence (open circles) and the presence (closed circles) of 3'-GMP.[5]

finding suggested that the two histidines are present in the active site of RNase T$_1$. For further details, see the paper by Kimura et al.[5]

Ribonuclease St.[11] Next, the tritium exchange titration was applied to RNase St produced by *Streptomyces erythreus*. This RNase consists of 101 amino acid residues arranged in a single polypeptide chain and is cross-linked by one disulfide bond. The revised amino acid sequence of the published sequence[12] is shown in Fig. 10. Histidine residues are present at positions 60 and 91 from the N-terminus of the protein. RNase St has the same substrate specificity as RNase T$_1$ and it splits the 3'-phosphodiester linkages of guanylic acid residues in RNA. The three-dimensional structure of the enzyme has not been elucidated. RNase St was incubated in tritiated water at 37°C for 40 hours at various pH values and the ionic strength was 0.2. The tritiated RNase St preparations at various pH values were oxidized with performic acid first and then digested at pH 8 successively with trypsin and thermolysin in each case for 2 hour periods. The resulting peptides were separated by the peptide mapping technique and were extracted from the map. The amino acid sequences of the two histidine-containing peptides were deduced from amino acid and N-terminal analyses to be:

$$\begin{array}{l}
\qquad\qquad\qquad 60 \\
\text{Peptide} \quad \text{I : Tyr-Tyr-His-Glu-Tyr-Thr} \\
\\
\qquad\qquad\qquad\qquad\qquad 91 \\
\qquad\quad \text{II : Phe-Tyr-Thr-Glu-Asp-His-Tyr-Glu-Ser}
\end{array}$$

The hydrogen-tritium exchange titration curves obtained in the presence and

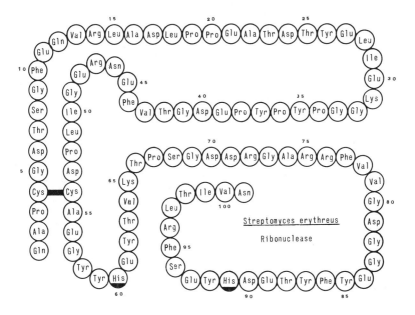

Fig. 10. The primary structure of RNase St.

absence of 3'-GMP for the two histidines are shown in Fig. 11. The alkaline shift of the sigmoid curve for His-91 in the presence of 3'-GMP clearly shows that it is an active site residue of the enzyme.

The pKa values and the exchange rate constants for the two histidines are listed in Table 3. The pKa value for His-91 is slightly lower than the intrinsic value, which suggests that a weak interaction with a neighboring cationic group exists, e.g., with the guanidyl group of an arginine residue or the ammonium group of a lysine residue. This interaction seems to accelerate the exchange reaction as can be seen in the Bronsted plot shown in Fig. 4. The pKa value for His-60 is higher than the intrinsic value which suggests the presence of a strong interaction with a negatively charged carboxylate group. The imidazole ring of His-60 also seems to be embedded in RNase St to somewhere between the extent that His-119 and His-12 are buried in native RNase A.

Concluding remarks. The hydrogen-tritium exchange titration of histidine residues reported here provides a novel method for the analyses of the microenvironments around the individual histidine residues in proteins provided they are stable over the entire pH range used in the exchange titration experiments. It is possible to determine whether or not there exists charged groups which interact with the histidine residue and the solvent accessibility of the

TABLE 3

pKa's AND THE TRITIUM EXCHANGE RATE CONSTANTS FOR THE TWO HISTIDINES IN RNase St
IN THE ABSENCE AND PRESENCE OF 3'-GMP at 37°C AND AT 0.2 IONIC STRENGTH[11]

	3'-GMP	pKa	k_ψ^{max} x 10^2	$log\ k_2$ $(M^{-1}h^{-1})$	pKa determined by NMR titration
His-60	-	7.95	1.78	3.90	8.12,8.26*
	+	8.03	1.84	3.85	8.46*
His-91	-	6.30	1.60	5.50	6.40, 6.50*
	+	7.54	1.44	4.96	8.01*

*Measured at 32°C.

histidine residue in question by measuring the hydrogen-tritium exchange rate
constants at various pH's. The technique is simple but tedious since many peptide maps of the proteolytic digests of the tritiated protein at various pH
values were necessary. The information thus obtained is of particular importance in ellucidating the molecular mechanism of catalysis when the histidine
residues are present in active site of an enzyme of unknown three-dimensional
structure. Furthermore, if the present exchange titrations are carried out
simultaneously with ^{1}H NMR titration experiments under the same conditions, the
histidine NMR signals can be assigned easily to the correct histidine residues
in the primary structure of the enzyme.

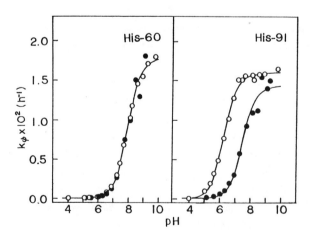

Fig. 11. The hydrogen-
tritium exchange titration
curves for the two histi-
dine residues in RNase St
in the absence (open
circles) and the presence
(closed circles) of 3'-
GMP.[11]

ACKNOWLEDGMENTS

This work was supported in part by grants from the Ministry of Education, Science and Culture of Japan αNo. 138013), the Tanabe Amino Acid Foundation and the Naito Foundation αNo. 75-115).

REFERENCES

1. Matsuo, H., Ohe, N., et al. (1972) J. Biochem., 72, 1057-1060.
2. Vaughan, J.D., Mughrabi, Z., and Wu, E.C. (1970) J. Org. Chem., 35, 1141-1145.
3. Kangawa, K., Matsuo, H. and Narita, K. (1974) Abstract of the 26th Symposium on Protein Structure (in Japanese), pp. 89-92.
4. Minamino, N., Matsuo, H. and Narita, K. (1977) in Peptide Chemistry, Shiba, T., ed., Protein Research Foundation, Osaka, pp. 85-90.
5. Kimura, S., Matsuo, H. and Narita, K. (1979) J. Biochem., 86, 301-310.
6. Ohe, M., Matsuo, H., et al. (1974) J. Biochem., 75, 1197-1200.
7. Carlisle, C.H., Palmer, R.A., et al. (1974) J. Mol. Biol., 85, 1-58.
8. Takahashi, K., (1965) J. Biol. Chem., 240 PC, 4117-4119.
9. Arata, Y., Kimura, S., et al. (1976) Biochem. Biophys. Res. Commun., 73, 133-140.
10. Arata, Y., Kimura, S., et al. (1979) Biochemistry, 18, 18-24.
11. Miyamoto, K., Arata, Y., et al. (1977) Abstract of the 28th Symposium on Protein Structure (in Japanese), pp. 73-76.
12. Yoshida, N., Sasaki, A. and Inoue, K. (1971) FEBS Lett., 15, 129-132.

III
Proteases and Other Hydrolytic Enzymes

Published 1980 by Elsevier North Holland, Inc.
Liu/Mamiya/Yasunobu, eds. Frontiers in Protein Chemistry

SUBUNIT STRUCTURE OF RAT KIDNEY γ-GLUTAMYL TRANSPEPTIDASE
AND ITS TOPOLOGY ON BRUSH BORDER MEMBRANE

SEIKOH HORIUCHI, MASAYASU INOUE AND YOSHIMASA MORINO
Department of Biochemistry, Kumamoto University Medical School, 2-2-1, Honjo,
Kumamoto 860, Japan

ABSTRACT

γ-Glutamyl transpeptidase (EC. 2.3.2.2) of rat kidney is an integral membrane
glycoprotein, composed of two nonidentical (large and small) subunits. The
enzyme was affinity labeled by 6-diazo-5-oxo-L-norleucine which was exclusively
incorporated into the small subunit. This enzyme in brush border membrane
vesicles was effectively inhibited by impermeable glutathione analog, and was
released from the membrane by treating vesicles with immobilized papain, indicating that the active site on the small subunit faced towards the outside of the
membrane. Molecular weights of these nonidentical subunits both from the
detergent-solubilized and the papain-solubilized enzyme were compared using
sodium dodecylsulfate polyacrylamide gel electrophoresis. The difference in the
molecular size between the detergent-solubilized and the papain-solubilized
enzyme was largely accounted for by the size difference between the large subunits. This view is consistent with the notion that the enzyme was anchored
into the brush border membrane mainly via the large subunit of the enzyme. The
large subunit separated from the papain-solubilized enzyme gave rise to an enzymically active preparation. Thus, the large subunit seems to have an active site
which remains latent in the native state of the enzyme. The small subunit alone
failed to restore the enzymic activity. However, the active site of the small
subunit was formed only in the presence of the large subunit. Thus, a cooperative interaction between the two nonidentical subunits during the refolding
process should be important for the formation of this oligomeric enzyme.

INTRODUCTION

γ-Glutamyl transpeptidase catalyzes the transfer reaction of γ-glutamyl moiety
of glutathione or its S-substituted derivatives to certain L-amino acids and
peptides.[1,2] This enzyme, a glycoprotein in nature, is believed to be an
integral component of brush border membrane of transmural tissues such as kidney and small intestine.[3-5] Several proposals have been made on the possible
participation of this membrane-bound enzyme in biological processes such as
secretion in general,[6,7] ammoniagenesis in metabolic acidosis,[8,9] amino acid and
peptide transport[10-12] and mercapturate biosynthesis.[13,14]

The general catalytic properties of this enzyme seems to have been extensively studied. However, the topological and molecular organization of this enzyme in the membrane is not sufficiently understood.

We describe herein the topology of this membranous enzyme and the use of an affinity label for the study of its subunit structure and its subunit-subunit interaction during the refolding process.

EXPERIMENTAL PROCEDURES

Assay for γ-glutamyl transpeptidase

Activity was measured with L-γ-glutamyl-p-nitroanilide as substrate according to the method of Orlowski et al.[15] Briefly, the reaction mixture contained, in a total volume of 1.0 ml, 2.5 mM L-γ-glutamyl-p-nitro-anilide, 50 mM glycylglycine and 0.1 M Tris-HCl buffer (pH 9.0). The reaction was started by adding the enzyme sample and the increase in absorbance at 410 nm was recorded in a Hitachi spectrophotometer model 124 or in a high-sensitivity spectrophotometer model SM-401 (Union Giken Co. Ltd., Osaka). One unit enzyme activity was defined as the amount of the enzyme required for the formation of 1 μmol p-nitro-aniline/min at 25°C. Specific activity was expressed as units/mg protein.

Purification of the papain-solubilized and the detergent-solubilized enzymes

The papain-solubilized enzyme was purified from Wistar rat kidney according to a modification of the method described by Hughey and Curthys.[16] The specific acitivity of the purified enzyme was approximately 2,000 units/mg at 37°C. Upon SDS-polyacrylamide gel electrophoresis, only two polypeptide chains were detected (refer to the text).

Detergent (Lubrol WX)-solubilized enzyme was purified by a modification[17] of the procedure described by Tate and Meister.[18] Details for the purification of the detergent-solubilized enzyme have been reported previously.[19]

Purification of brush border membrane from rat kidney

The brush border membrane was purified from rat kidney according to the method of Booth and Kenny.[20] Brush border membrane thus obtained showed a 15-fold increase in specific activity for the marker enzymes (alkaline phosphatase and γ-glutamyl transpeptidase) when compared with those in the homogenate. Brush border membrane was demonstrated to be composed of closed vesicles upon electron microscopic observation.[20]

Other experimental procedures for this study are described in each legend to figure and table.

RESULTS

Affinity labeling of rat kidney γ-glutamyl transpeptidase

In an oligomeric enzyme composed of nonidentical polypeptide chains, an affinity label is very useful for determining the subunit on which the active site resides. For this purpose, 6-diazo-5-oxo-L-norleucine (DON) was applied to rat kidney γ-glutamyl transpeptidase which is composed of two nonidentical (large and small) subunits. Incubation with DON resulted in a progressive decrease in the activity of γ-glutamyl transpeptidase (Fig. 1A). The inactivation followed pseudo-first-order kinetics (Fig. 1B). The rate constant for

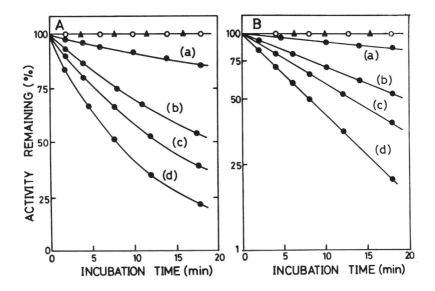

Fig. 1. Inactivation of γ-glutamyl transpeptidase with 6-diazo-5-oxo-L-nor-leucine (●) and 5-diazo-4-oxo-L-norvaline (▲). (A): The reaction mixture con-tained, in 0.2 ml, γ-glutamyl transpeptidase (2-10 units), 0.1 M Tris-HCl buffer (pH 7.4) and the norleucine or norvaline derivatives. The reaction was initi-ated by adding the enzyme sample. At various intervals, aliquots (5-20 µl) were removed for the determination of remaining activities of the enzyme and residual activities were plotted against the incubation time. Incubation was performed at 25°C. These reagents were synthesized by the method reported previously.[17] (a) 10 mM norleucine derivative (DON) and 10 mM reduced glutathione, (b) 1 mM DON, (b) 2 mM DON, and (d) 5 mM DON. The concentration of the norvaline de-rivative used was 50 mM. Open circles denote the data for the control experi-ments in the absence of these reagents.
(B): A semilogarithmic plot of (A). (Taken from reference 17.)

the inactivation was determined from the semilogarithmic plots as shown in
Fig. 1B by the use of the equation k = 0.693/t, where k is the first-order rate
constant for inactivation and t the time required for inactivating one-half of
the enzyme initially present. Double-reciprocal plots of the inactivation rate
constants versus varying concentrations of DON yielded a value of 1.8 mM for
the concentration of DON giving a half-maximum velocity of inactivation. The
inactivation of the enzyme with DON was effectively protected by reduced
glutathione, a natural substrate for this enzyme, in a competitive fashion
(Fig. 2). These findings indicate that DON binds to the active site to form a
dissociable complex with the enzyme prior to inactivation. 5-diazo-4-oxo-L-
norvaline, an asparagine analog, did not affect the enzyme activity when incubated with the enzyme at a concentration of 50 mM, reflecting the high specificity of DON (Fig. 1).

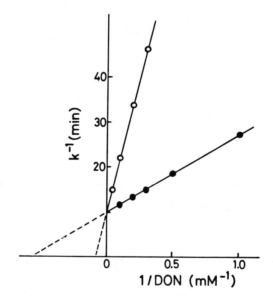

Fig. 2. Effect of reduced glutathione on the inactivation by DON. Double-
reciprocal plots of the first order rate constant (k) for the inactivation of
the enzyme versus varying concentration of DON in the presence (O——O) or
absence (●——●) of 2 mM reduced glutathione. The inactivation condition was as
described in the legend to Fig. 1. The rate constant for the inactivation (k)
was determined from the semilogarithmic plots (Fig. 1B) by the use of equation,
k = 0.693/t, where k is the first order rate constant for inactivation and t,
time required for the inactivating one-half the enzyme initially present.
(Taken from reference 17.)

To study whether the inactivation resulted from a covalent labeling of the enzyme active site by DON or not, an enzyme preparation inactivated by incubating with radioactive DON was exhaustively dialyzed, followed by incubating in 10 mM Tris-HCl buffer (pH 7.2) containing 0.1% sodium dodecylsulfate (SDS) and 2-mercaptoethanol for 2 hours at 37°C. After further dialysis, enzyme sample thus treated was determined for its radioactivity (Table 1). The result indicated a stoichiometric incorporation of the inactivator into the enzyme protein via a covalent linkage.

Subunit structure of rat kidney γ-glutamyl transpeptidase

Two enzyme forms of rat renal γ-glutamyl transpeptidase (the detergent-solubilized and the papain-solubilized enzyme) purified to a homogeneity were shown to be composed of two nonidentical subunits (the large and small subunit) on SDS-polyacrylamide gel electrophoresis (Fig. 3). Figure 3 shows that with either of these two enzyme forms, the small subunit was exclusively labeled upon the inactivation with DON. This indicated that the active site of this enzyme resides on the small subunit.[17,19,21]

TABLE 1

COVALENT INCORPORATION OF RADIOACTIVE 6-DIAZO-5-OXO-L-NORLEUCINE INTO γ-GLUTAMYL TRANSPEPTIDASE

The reaction mixture contained, in a total volume of 2.0 ml, 2 M sodium maleate, 0.1 M Tris-HCl buffer (pH 7.4), 2 μmol 6-diazo-5-oxo-L-[6-^{14}C]norleucine (1.0 x 10^6 counts/min/μmol), and 2.0 mg of the papain-solubilized enzymes from three different batches. After incubation at 27°C, the remaining enzyme activity was less than 0.1%. Then the reaction mixtures were dialyzed extensively against 10 mM Tris-HCl buffer (pH 7.2). The enzyme solutions thus dialyzed were incubated in 10 mM Tris-HCl buffer (pH 7.2) containing 0.1% SDS and 2-mercapto-ethanol for 2 hours at 37°C and dialyzed against the same denaturing solution for 20 hours at 4°C. (From reference 19, which contains experimental details.)

	Specific activity of enzymes used (units/mg)	Protein/DON bound (g/mol DON)	Stoichiometry* (mol DON/mol enzyme)
Experiment 1	1,920	69,000	1.05
Experiment 2	1,970	67,000	1.02
Experiment 3	2,000	63,000	0.95

* The molecular weight of the enzyme is assumed to be 66,000 (see Fig. 4).

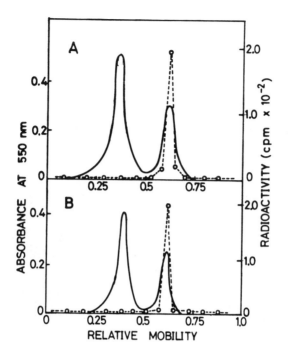

Fig. 3. SDS-polyacrylamide gel electrophoresis of two enzyme forms of γ-glutamyl transpeptidase inactivated with radioactive DON. Two enzyme forms of γ-glutamyl transpeptidase purified by the method described in Materials and Methods were subjected to SDS-polyacrylamide gel electrophoresis in 7.5% gel concentration. The procedures for staining and densitometry (————) were those reported previously.[17] The determination of the radioactivity derived from ^{14}C-DON was performed by cutting the gel into slices, and then solubilizing with 30% hydrogen peroxide at 40°C for 12-24 hours. The radioactivity was shown by (O----O). (A): the detergent-solubilized enzyme, (B): the papain-solubilized enzyme. (Taken from reference 17).

In order to assess the anchoring portion of the membrane-bound enzyme, we attempted to estimate the molecular weight of the nonidentical subunits from two enzyme forms of γ-glutamyl transpeptidase using the Ferguson procedure, which has recently been shown to permit a reliable estimate for molecular sizes of glycoproteins.[22] The molecular weights for the large and small subunits of the detergent-solubilized enzyme were 66,000 ± 8,000 and 25,000 ± 4,000, respectively, while the molecular weights for the corresponding subunits from the papain-solubilized enzyme were 45,000 ± 5,000 and 23,000 ± 3,000, respectively (Fig. 4). These data clearly indicate that the size difference between the detergent-solubilized and the papain-solubilized enzymes results largely from the size difference between the large subunits in these two enzyme forms. Thus,

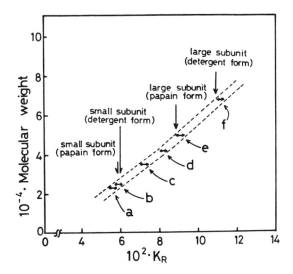

Fig. 4. Molecular weights of the small and large subunits of two enzyme forms of γ-glutamyl transpeptidase by Ferguson plots on SDS-polyacrylamide gel electro-phoresis. A Ferguson plot of standard marker proteins [(a) γ-globulin (light chain), (b) chymotrypsinogen, (c) pepsin, (d) ovalbumin, (e) γ-globulin (heavy chain), and (f) bovine serum albumin] and the nonidentical subunits of γ-glutamyl transpeptidase was performed as previously reported.[19] The retardation coefficient (K_R) of each protein was obtained by Ferguson plots. The dashed lines define the limit of deviation for the K_R values for standard proteins. The K_R values for the large and small subunit from both the detergent-solubilized and the papain-solubilized enzymes are indicated by vertical arrows. (Taken from reference 19.)

it is likely that the oligomeric γ-glutamyl transpeptidase was anchored into the brush border membrane mainly via the large subunit of the enzyme.

Sidedness of the active site on renal brush border membrane

Determination of the sidedness of the active site of the enzyme in the membrane seems to be imperative for understanding the physiological function of this membranous enzyme. For this purpose, the first approach was to test whether or not γ-glutamyl transpeptidase activity is inhibited by an S-acetyl polymer (molecular weight: 215,000) derivative of glutathione, which is believed to be impermeable through plasma membranes. As shown in Fig. 5, the rate of p-nitroaniline release was depressed in the presence of this glutathione deriva-tive of high molecular weight. The extent of inhibition was dependent on the concentration of the inhibitor, exhibiting a typical hyperbolic saturation curve.

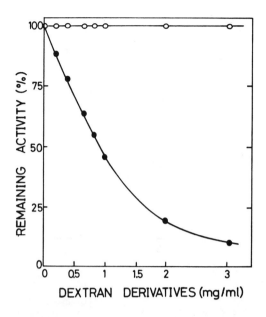

Fig. 5. Effect of unpermeant dextran derivative of glutathione on the γ-glutamyl transpeptidase activity of brush border membrane. Assay mixtures contained, in a total volume of 1.0 ml, 2.5 mM L-γ-glutamyl-p-nitroanilide, 50 mM glycylglycine and various concentrations of the S-acetyl dextran polymer derivative of glutathione or underivatized dextran polymer in Hanks' balanced solution (pH 7.4). The reaction was started by adding 0.02 ml brush border membrane (5 mg/ml) in Hanks' balanced solution and the increase in absorbance st 410 nm was determined. S-acetyl dextran polymer derivative of glutathione (●—●—●); dextran polymer only (O—O—O). (Taken from reference 19.)

The underivatized dextran polymer itself had no effect on the enzyme activity bound to the renal brush border membrane. This finding indicates that the synthetic substrate of a low molecular weight as well as glutathione linked to the fibrous polymer of a high molecular weight are equally accessible to the catalytic site of this membrane-bound enzyme.

The second approach was to determine whether this enzyme on the brush border membrane was proteolytically solubilized or not by papain immobilized to Sephadex G-10 beads. Incubation of rat renal brush border membrane vesicles with a very low concentration of papain resulted in an effective release of γ-glutamyl transpeptidase. After incubation for 90 min at 25°C with 40 µg papain, approximately 80% of the original, membrane-bound enzyme was released from the vesicles.[19] This proteolytic release of the membrane-bound enzyme was observed also with papain-immobilized Sephadex G-10 beads (Fig. 6). The increase in the absorbance at 280 nm appeared to reach a plateau within 30 min whereas the

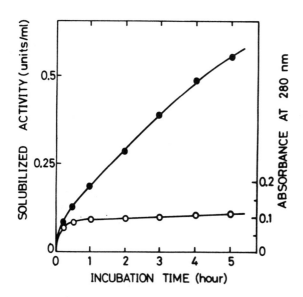

Fig. 6. Effect of immobilized papain on γ-glutamyl transpeptidase of brush border membrane. A preparation of papain conjugated to Sephadex G-10 beads (0.9 ml) (2.5 mg papain/ml Sephadex gel) was suspended in 10 mM mannitol, 1 mM dithiothreitol and Tris-HCl buffer (pH 7.4) in a total volume of 1.0 ml. The reaction was initiated by adding 0.1 ml of the brush border membrane fraction (1.8 mg protein containing 4.2 units of γ-glutamyl transpeptidase activity). Incubation was performed at 37°C. At the indicated time intervals, 4 ml ice-cold incubation medium was added and the mixture was centrifugated at 160,000 x g for 20 min at 2°C. The supernatant solution was determined for enzyme activity at 25°C (●——●——●) and proteins (○——○——○). (Taken from reference 19.)

amount of the enzyme solubilized into the supernatant solution continued to increase steadily with the incubation time, indicating a selective release of this membrane-bound enzyme by papain. γ-Glutamyl transpeptidase thus solubilized by the immobilized papain preparation was effectively adsorbed on to a concanavalin A-Sepharose column. And the adsorbed enzyme was specifically eluted by α-methyl D-glucoside, a haptenic inhibitor of this lectin.[19]

These two observations showed unequivocally that both the catalytic site and the oligosaccharide moiety of the membrane-bound enzyme molecule were topologically located in the portion facing outside of brush border membrane.

An active site in the large subunit

The separation of the large and small subunits was successfully achieved by chromatography of a denatured preparation of the papain-solubilized enzyme on a Sephadex G-150 column equilibrated with 6 M urea containing 1 M propionic acid

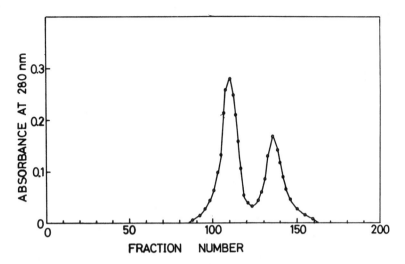

Fig. 7. Subunit separation of the papain-solubilized γ-glutamyl transpeptidase. The solubilized purified enzyme of papain (7.5 mg) was denatured in 6 M urea, 1 M propionic acid and 0.1 mM dithiothreitol at 37°C for 2 hours and then chromatographed on a column of Sephadex G-150 (2.5 x 80 cm) previously equilibrated with the same denaturing solution as that used for the enzyme denaturation. Fractions of 1.2 ml were collected and analyzed for absorbance at 280 nM of proteins. (Taken from reference 23.)

and 0.1 mM dithiothreitol (Fig. 7). SDS-Polyacrylamide gel electrophoresis of aliquots from the two protein peaks in Fig. 7 showed that the first and second protein peaks corresponded to the large and small subunit, respectively.

The large subunit but not the small subunit was found to exhibit the enzymic activity upon renaturation under the present condition (Table 2). This was an unexpected observation, since the active site of the native oligomeric enzyme was unequivocally demonstrated to reside on the small subunit (see Fig. 3). Therefore it was suspected that the renatured enzyme activity might be attributable to the small subunit contaminating the large subunit preparation used for the renaturation experiment. To rule out this possibility, the following two sets of experiments were performed. First, the renatured enzyme preparation from the large subunit was irreversibly inactivated by radioactive DON. Upon a gel filtration on a Sephadex G-150 column as shown in Fig. 7, the radioactivity derived from the norleucine derivative was found to comigrate with the protein peak corresponding to the large subunit.[23] Secondly, a pure preparation of γ-glutamyl transpeptidase which had been completely inactivated by affinity labeling with DON was dissociated into the small subunit and the large subunit under the denaturating conditions described in Fig. 7. Upon renaturation,

TABLE 2

RENATURATION OF THE LARGE AND SMALL SUBUNIT FROM THE

PAPAIN-SOLUBILIZED γ-GLUTAMYL TRANSPEPTIDASE

After the large and small subunits were denatured in 6 M urea, 50 mM Tris-HCl
buffer (pH 7.5) and 0.1 mM dithiothreitol for 2 hr at 37°C, they were separately
transferred to dialysis tubes and then submitted to successive dialysis against
the solutions containing urea at stepwise decreasing concentrations (3 M, 1.5 M,
0.5 M and, then 0 M in this order). Each dialysis step required about 6 hours.
All dialysis solutions contained 50 mM Tris-HCl buffer (pH 7.5) and the reagents
(dithiothreitol and reduced glutathione) as indicated. Concentrations of these
reagents were as follows; dithiothreitol (0.1 mM) and reduced glutathione (1 mM).
Dialysis was performed at 4°C. Each value for the renatured specific activity
was the mean of triplicate runs in which the final protein concentrations (post-
dialysis) of the subunits were within the range of 0.23-0.35 mg/ml (from reference 23).

| Renaturation conditions | Renatured specific activity[*] | |
	Large subunit	Small subunit
Without reagents	1.7	0.001
Dithiothreitol	3.9	0.001
Dithiothreitol and Gutathione	13.1	0.001

[*]Specific activity of the renatured enzyme (units/mg protein).

only the large subunit restored the enzymic activity (submitted for publication).
Furthermore, the renatured preparation of the large subunit was covalently
labeled with DON. The large subunit preparation thus inactivated was denatured
and then subjected to renaturation under the same conditions as described in
the legend to Table 2. However, no restoration of enzyme activity was observed
with DON-treated large subunit under the conditions where the large subunit
isolated from the untreated native enzyme could restore its enzymic activity
(submitted for publication). Thus, all these observations demonstrate unequivocally that the renatured active site is indeed derived from the large subunit.

Subunit-subunit interaction

   To inquire into the molecular size of the enzymically active molecules of
the renatured large subunit polypeptide chain, the renatured large subunit
preparation was passed over a column of Sephadex G-200 in 50 mM Tris-HCl buffer
(pH 7.5) and 0.1 mM dithiothreitol. As shown in Fig. 8A, almost all the
applied protein emerged in a void volume fraction whereas the enzymically
active species appeared as a small peak in an inner volume fraction. Molecular

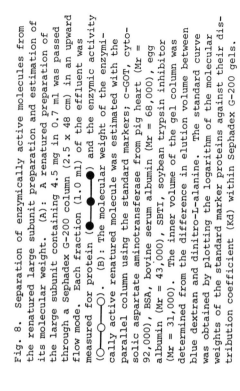

Fig. 8. Separation of enzymically active molecules from the renatured large subunit preparation and estimation of its molecular weight. (A): A renatured preparation of the large subunit containing 4.5 mg in 0.7 ml was passed through a Sephadex G-200 column (2.5 x 48 cm) in an upward flow mode. Each fraction (1.0 ml) of the effluent was measured for protein (○—○) and the enzymic activity (●—●). (B): The molecular weight of the enzymically active renatured molecule was estimated with the parallel column using the standard markers; c-GOT, cytosolic aspartate aminotransferase from pig heart (Mr = 92,000), BSA, bovine serum albumin (Mr = 68,000), egg albumin (Mr = 43,000), SBTI, soybean trypsin inhibitor (Mr = 31,000). The inner volume of the gel column was determined from the difference in elution volume between blue dextran and dinitro-L-alanine. The standard curve was obtained by plotting the logarithm of the molecular weights of the standard marker proteins against their distribution coefficient (Kd) within Sephadex G-200 gels.

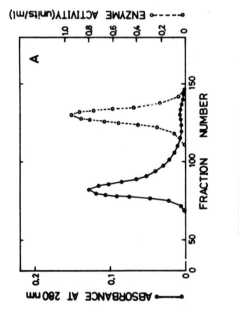

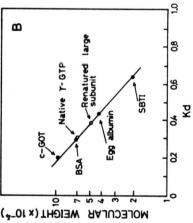

weight of this enzymically active large subunit was estimated to be 48,000 ±
2,000, which was practically indistinguishable from the molecular weight (45,000
± 5,000) estimated by Ferguson plots for the large subunit polypeptide chain
(see Fig. 4).   Thus, the enzymically active molecules of the renatured large
subunit preparation seems to be of a monomeric structure.

As described above, the small subunit per se did not exhibit any enzymic
activity under the renaturation conditions examined so far.   However, when a
denatured preparation of small subunit was added to the denatured large subunit
and the mixture was subjected to the renaturation as described under Materials
and Methods, the recovery of the enzymic activity was found to be significantly
higher than that obtained with the large subunit alone (Fig. 9).   Although the
data were not shown here, this accelerating effect of the small subunit showed
a concentration dependency, and the enzymically active molecule isolated from

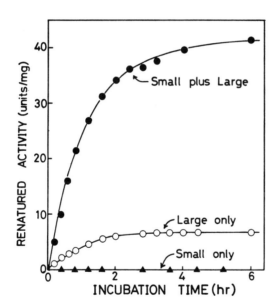

Fig. 9.   Subunit-subunit interaction between the nonidentical subunits of
γ-glutamyl transpeptidase.   The denatured samples contained, in a total volume
of 0.03 ml, the large subunit (40 µg) or the small subunit (15 µg), or both, in
7 M urea, 1 M propionic acid and 0.1 mM dithiothreitol.   These denatured samples
were added to 1 ml of the renaturation medium containing 0.1 mM Tris-HCl buffer
(pH 8.5), 0.1 mM reduced glutathione and 0.05 mM dithiothreitol and incubated
at 23°C.   At the indicated times, an aliquot of 5-10 µl was withdrawn from the
renaturation mixture and determined for the renatured enzyme activity at 25°C.
The renaturation of each denatured sample was as follows: small subunit (▲——▲),
large subunit (○——○) and small subunit plus large subunit (●——●).

TABLE 3

EFFECT OF DON-LABELED SMALL SUBUNIT ON THE RESTORATION OF ENZYMIC

ACTIVITY OBSERVED UPON INTERACTION WITH THE LARGE SUBUNIT

Experiment 1: A mixture containing, in a total volume of 0.03 ml, the large subunit (54 µg) and small subunit (27 µg) in 6 M urea, 1 M propionic acid and 0.1 mM dithiothreitol added to 1 ml of the renaturation medium (0.1 M Tris-HCl buffer, pH 8.5 and 0.1 mM reduced glutathione). The renaturation procedure was performed at 23°C. At various time intervals, an aliquot of 5-10 µl was withdrawn from the renaturation mixture and determined for the enzyme activity at 25°C. Experiment 2: performed by the same way as described above for Experiment 1 except that a DON-labeled small subunit sample replaced the nontreated small subunit preparation. Experiment 3: the small subunit sample was omitted from the reaction mixture in Experiment 1. Denatured preparations of the large and small subunits were obtained by the method described in Fig. 7. DON-labeled small subunit preparation was obtained as described previously.[23]

Experiment		Renatured activity (units/ml)
Experiment 1	(large plus DON-labeled small subunit)	0.7
Experiment 2	(large plus small subunit)	15.0
Experiment 3	(large subunit only)	0.65

the renatured enzyme preparation in Fig. 9 was indistinguishable from the native γ-glutamyl transpeptidase in terms of its molecular weight, subunit structure and chemical stabilities (manuscript in preparation). All these findings indicate that a cooperative interaction between the small and large subunit polypeptide chains during the folding process is important for the acquisition of a proper structural integrity of this oligomeric enzyme.

Then, a question arises; which subunit, the large or the small, manifests its active site during this renaturation process? As described in Fig. 3, the active site of the small subunit can be selectively labeled by inactivating the native oligomeric enzyme with DON. A denatured preparation of the small subunit thus labeled with DON was mixed with the denatured large subunit preparation and the mixture was subjected to renaturation as described above. As can be seen in Table 3 (Experiment 1), the extent of the restoration of enzymic activity was much less than that observed when the unlabeled small subunit preparation was employed and did not differ much from that observed when the large subunit alone (Table 3, Experiments 2 and 3) was subjected to renaturation. This result is consistent with the view that the active site of the small subunit is responsible

TABLE 4

CORRELATION OF THE EXTENT OF REFOLDING OF EACH SUBUNIT WITH THE

RESTORABLE ACTIVITY OBSERVED UPON COMBINATION OF BOTH SUBUNITS

15 µl of the denatured small subunit (30 µg) in 6 M urea containing 1 M propionic acid and 0.1 mM dithiothreitol was added to 0.5 ml of the renaturation
medium containing 0.1 mM Tris-HCl buffer (pH 8.5), 0.1 mM reduced glutathione
and 0.05 mM dithiothreitol and incubated at 23°C. At the indicated times, the
solution which contains the renaturing small subunit was mixed quickly with
15 µl of the denatured large subunit and 0.5 ml of the renaturation medium. The
renatured enzyme activity was measured as described in the legend to Fig. 9.
The plateau level of the renatured activity was reached within 6 hrs of incubation. The preincubation of the large subunit in the renaturation medium was
carried out as described above except that the order of addition of each subunit was reversed.

Subunit preincubated	Preincubation time (min)	Renatured activity (units/ml)	Capacity of renaturation* (%)
Small subunit	0	0.90	100
	10	0.54	60
	30	0.45	50
	60	0.38	42
Large subunit	0	0.90	100
	10	0.90	100
	30	0.87	97
	60	0.88	98

*
The renatured activity obtained in the absence of preincubation (Time 0) was
taken as 100%.

for the striking increase in the restorable enzyme activity observed in the
presence of both the small and large subunit.

To find a possible correlation of the restorable level of the enzymic
activity with the extent of refolding of each subunit, the following experiments
were performed. The denatured small subunit samples exposed to the renaturation
medium for various periods were mixed with a denatured preparation of the large
subunit. At one hour intervals, aliquots were withdrawn from the mixtures and
the enzymic activity determined. With all runs, the enzyme activity reached
plateau levels within 5-6 hours and the values obtained at this time were
assumed to represent the restorable activity for the individual runs. As can
be seen in Table 4, the recoverable activity decreased with the increase in the
time of preincubation of the small subunit preparation in the renaturation

medium. Thus, the exposure of the small subunit alone to the renaturation medium seems to lead to the formation of a species which no longer interacts with the denatured large subunit to give rise to a functional integral oligomeric structure. In other words, the small subunit alone tends to refold erroneously and the large subunit assists in the proper folding of the small subunit poly-peptide chain. A similar experiment was performed to examine the effect of the extent of refolding of the large subunit on the formation of the active oli-gomeric enzyme. The denatured large subunit exposed to the renaturation medium for varying periods was mixed with a denatured small subunit preparation and the recoverable enzyme activity was measured on these mixtures as described above. In contrast to the previous experiment, the difference in the stage of refolding of the large subunit did not affect the level of the restorable enzymic activity. Thus, it seems that the large subunit polypeptide chain is capable of refolding more or less independently without being assisted by the small subunit.

DISCUSSION

Since membranes represent interfaces between the cell and its environment, all reactions of the cell to extrinsic or intrinsic stimuli are mediated by the functional membrane molecules, which must exert their functions in the semi-solid environment of the membranes. Thus, in order to elucidate the physio-logical role of membrane-bound enzymes, comprehensive understanding of their topological features on the membranes seems to be essential in addition to the conventional enzymological approach.

In the present investigation, we demonstrated the topological characteristics of rat kidney γ-glutamyl transpeptidase on the brush border membrane in terms of its subunit structure. Results are schematically illustrated in Fig. 10. This membrane-bound enzyme was composed of two nonidentical (large and small) sub-units of a glycoprotein nature. In the native state of the enzyme, the small subunit has an active site reactive toward 6-diazo-5-oxo-L-norleucine (DON) which has been demonstrated to satisfy several criteria as an affinity label for this enzyme. This active site is topologically localized on the outer (luminal) surface of the brush border membrane. The large subunit, in contrast to the small subunit, was suggested to be involved mainly in the anchorage of this membrane-bound enzyme into the lipid bilayer. However, in addition to this structural aspect of the large subunit, the renaturation of the separated sub-units from the papain-solubilized enzyme indicates that the large subunit also has another active site distinct from the active site on the small subunit. Latent nature of the active site on the large subunit was shown by its

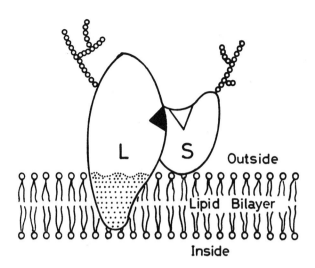

Fig. 10. A schematic representation of the topological organization of γ-glutamyl transpeptidase on renal brush border membrane. The large subunit and small subunit were expressed by L and S, respectively. Dotted area in the large subunit denotes the anchoring portion of enzyme. Open triangle denotes the active site on the small subunit, the darkened triangle denotes the latent active site on the large subunit. Chains of open circles denote the oligosaccharide moiety of the enzyme.

inaccessibility to DON in the native state of the enzyme.

Although the data were not shown, the renaturation of the large subunit was markedly accelerated by the presence of either reduced glutathione or S-methyl glutathione and was accompanied by conformational change as judged both by circular dichroism and immunochemical reactivity (submitted for publication).

From the reconstitution experiment, a subunit-subunit interaction was demonstrated to occur between these nonidentical subunits of the enzyme. The data strongly suggest that the generation of the active site on the small subunit must be assisted by the large subunit polypeptide chain. These results were obtained with the nonidentical subunits separated from the papain-solubilized enzyme which is thought to lack the anchoring domain of this enzyme. Therefore, a systemic study using the nonidentical subunits obtained from the detergent-solubilized γ-glutamyl transpeptidase is needed to elucidate the molecular assembly of this enzyme and its interaction with the renal brush border membrane.

The asymmetric orientation of the enzyme active site on brush border membrane made us[19,24] and other investigators[16,25,26] postulate that the enzyme may be constrained to interact only with γ-glutamyl compounds outside the membrane. In fact, intravenous administration of radioactive γ-glutamyl compound such as S-carbamidomethyl glutathione resulted in predominant incorporation of the radioactivity into the kidney, where γ-glutamyl transpeptidase is shown to exert its action in concert with renal peptidase(s) for sequential hydrodysis of γ-glutamyl compounds to its constituent amino acids and mercapturate biosynthesis (M. Inoue et al., submitted for publication).

ACKNOWLEDGMENTS

This work was supported in part by a grant in aid for scientific research from the Ministry of Education, Science and Culture of Japan and by the Yamada Science Foundation.

REFERENCES

1. Binkley, F. (1961) J. Biol. Chem., 236, 1075-1082.
2. Orlowski, M. and Meister, A. (1965) J. Biol. Chem., 240, 338-247.
3. Curthoys, N.P. and Lowry, O.H. (1973) J. Biol. Chem., 248, 162-168.
4. Curthoys, N.P. and Shapiro, R. (1975) FEBS Lett., 58, 230-233.
5. Kuhlenschmidt, T. and Curthoys, N.P. (1975) Arch. Biochem. Biophys., 167, 519-524.
6. Binkley, F., et al. (1975) FEBS Lett., 51, 168-170.
7. Binkley, F. and Wiesemann, M.L. (1975) Life Sci., 17, 1359-1362.
8. Curthoys, N.P. and Kuhlenschmidt, T. (1975) J. Biol. Chem., 250, 2099-2105.
9. Anderson, N.M. and Alleyne, G.A.O. (1977) FEBS Lett., 79, 51-53.
10. Orlowski, M. and Meister, A. (1970) Proc. Natl. Acad. Sci. USA, 67, 1248-1255.
11. Meister, A. (1973) Science (Washington, D.C.), 180, 33-39.
12. Tate, S.S. and Meister, A. (1974) Proc. Natl. Acad. Sci. USA, 71, 3329-3333.
13. Bray, H.G., Franklin, T.J. and James, S.P. (1959) Biochem. J., 71, 690-696.
14. Boyland, E., Ramsay, G.S. and Sims, P. (1961) Biochem. J., 78, 376-384.
15. Orlowski, M. and Meister, A. (1963) Biochem. Biophys. Acta., 73, 679-681.
16. Hughery, R.P. and Curthoys, N.P. (1976) J. Biol. Chem., 251, 7863-7870.
17. Inoue, M., Horiuchi, S. and Morino, Y. (1977) Eur. J. Biochem., 73, 335-343.
18. Tate, S.S. and Meister, A. (1975) J. Biol. Chem., 250, 4619-4627.
19. Horiuchi, S., Inoue, M. and Morino, Y. (1978) Eur. J. Biochem., 87, 429-437.
20. Booth, A.G. and Kenny, A.T. (1974) Biochem. J., 142, 575-581.
21. Tate, S.S. and Meister, A. (1977) Proc. Natl. Acad. Sci. USA, 74, 931-935.
22. Rodbard, D. and Chrambach, A. (1971) Anal. Biochem., 40, 95-134.
23. Horiuchi, S., Inoue, M. and Morino Y. (1978) Biochem. Biophys. Res. Commun., 80, 873-878.
24. Inoue, M., Horiuchi, S. and Morino, Y. (1977) Eur. J. Biochem., 78, 609-615.
25. Moldeus, P., et al. (1978) Biochem. Biophys. Res. Commun., 83, 195-200.
26. Orlowski, M. and Wilk, S. (1976) Eur. J. Biochem., 71, 549-555.

INHIBITION OF A TISSUE PEPTIDASE BY BRAIN
ENDO-OLIGOPEPTIDASE A ANTIENZYME ANTIBODY

A.C.M. CAMARGO[+], I.F. CARVALHO[+], K.M. CARVALHO[*],
M.A. CICILINI[*], AND H.L. COELHO[*]
+Protein Chemistry Laboratory, Department of Pharmacology, Faculty of Medicine
of Ribeirão Preto, USP, 14.100 Ribeirão Preto, S.P., Brazil; *Postgraguate
students in the Department of Pharmacology, Faculty of Medicine of Ribeirão Preto,
Preto, USP, 14.100 Ribeirão Preto, S.P., Brazil.

ABSTRACT

A goat was immunized against functionally pure rabbit brain endopeptidase A.
The antibodies to endopeptidase A inhibit this enzyme (total inhibition) but do
not affect other kininases present in nervous tissue. Titration of endopeptidase A in the supernatant fraction of rabbit brain homogenate indicates that
at least 60% of the kininase activity present in the cytosol of nervous tissue
is due to the activity of brain endopeptidase A. The presence of this enzyme
in other rabbit tissues was confirmed for heart, skeletal and smooth muscle,
liver and gonads. The results also confirmed the absence of this enzyme in
lung, spleen and kidney.

Brain endopeptidase A is a thiol-activated neutral endopeptidase isolated
from rabbit brain by Camargo et al.[1] The enzyme hydrolyzes bradykinin (Arg^1-
Pro^2-Pro^3-Gly^4-Phe^5-Ser^6-Pro^7-Phe^8-Arg^9) at the Phe^5-Ser^6 bond and several
synthetic peptides having 6 to 17 amino acid residues which contain part or the
whole sequence of bradykinin.[2] However, this endopeptidase does not hydrolyze
large substrates such as denatured hemoglobin, the bradykinin moiety in
denatured kininogen,[1,2] or a peptide related to bradykinin (Gly-Gly-Gly-Arg-
bradykinin) when this polypeptide is covalently bound to a high molecular weight
carrier through its terminal amino group.[3] Brain endopeptidase A and endopeptidase B, a second endopeptidase also isolated from brain,[2] hydrolyze
several peptide hormones such as angiotensin I, angiotensin II, substance P,
neurotensin, and LH-RH.[4] The existence of enzymes in other tissues exhibiting
the same specificity for bradykinin as brain endopeptidase A has been suggested.[5]

In this paper we demonstrate the presence of endopeptidase A-like enzymes in
other tissues of rabbit by using monospecific antibody against purified rabbit
brain endopeptidase A.

MATERIALS

Bradykinin, Arg-Pro-Pro-Gly-Phe, Ser-Pro-Phe-Arg, Arg-Pro-Pro-Gly-Phe-Ser-Pro and Phe-Arg were supplied by Dr. A.C.M. Paiva (Escola Paulista de Medicina, São Paulo, Brazil). The peptides had integral molar ratios of constituent amino acids after acid hydrolysis and were homogeneous by high voltage electrophoresis at pH 3.5.

METHODS

Enzyme preparations. Brain endopeptidase A was purified from the 25,000xg supernatant fraction of rabbit brain homogenate after precipitation at pH 5 as described by Oliveira et al.[2] An analytical system based on the amino acid analyzer was used to show that endopeptidase A hydrolyzes bradykinin at a single peptide bond.[2] Purified brain endopeptidase A released stoichiometric amounts of Arg-Pro-Pro-Gly-Phe and Ser-Pro-Phe-Arg from bradykinin. The specific activity of the purified enzyme was 130 nmol bradykinin/min/mg protein. Enzyme preparations from tissue homogenates were obtained as described by Cicilini et al.[5]

Antibody preparation. Antiserum was produced with 125 µg of purified brain endopeptidase A emulsified with complete Freund's adjuvant injected intradermally at multiple sites of a goat abdomen every 2 weeks. Sera (heat-treated at 56° for 30 min) were tested for anticatalytic activity and by immunodiffusion at regular time intervals. Anticatalytic activity of antiserum was obtained 60 days after immunization began. Monospecific immunoglobulin G antibodies (IgG) for brain endopeptidase A were obtained by immunoadsorption against protein contaminants and by chromatography on DEAE-cellulose.

Bioassay for bradykinin. Bradykinin activity was determined by measuring the isotonic contraction of the isolated guinea pig ileum suspended in aerated Tyrode solution, containing 3.5×10^{-7} M atropine and 1.7×10^{-6} M diphenylhydramine, at 37° C. Bradykinin was measured by a matching technique.[6]

Enzyme assay. Bradykinin was incubated with enzyme in 0.05 M Tris-HCl buffer, pH 7.5, containing 0.1 M NaCl for 20 min at 37° C. The enzyme concentration was selected so as to give 40 to 60% inactivation in 15 min. The bioassay on the isolated guinea pig ileum was used to measure the rate of bradykinin inactivation and an analytical system based on quantitative determination of fragments from bradykinin hydrolysis was used to detect the size of cleavage. Both methods and conditions are given in references 2 and 5.

RESULTS AND DISCUSSION

The antibody had no anticatalytic activity toward brain endopeptidase B, whereas 0.4 mg of protein from the immunoglobulin fraction (IgG) of antisera inhibits completely the activity of brain endopeptidase A. These results permitted us to use the antibody to titrate brain endopeptidase A in the 25,000 xg supernatant fraction of rabbit brain homogenate. The amino-terminal portion of bradykinin (Arg-Pro-Pro-Gly-Phe) released by brain endopeptidase A, or Arg-Pro-Pro-Gly-Phe-Ser-Pro released by brain endopeptidase B is very slowly degraded by enzymes present in the soluble fraction of rabbit brain.[1,2] Thus, the release of these peptides is directly proportional to the activity of brain endopeptidases A and B on bradykinin. The results presented in Fig. 1 indicate that 0.4 mg of antibody reduces the activity of brain endopeptidase A by 100% and has no effect on the activity of endopeptidase B.

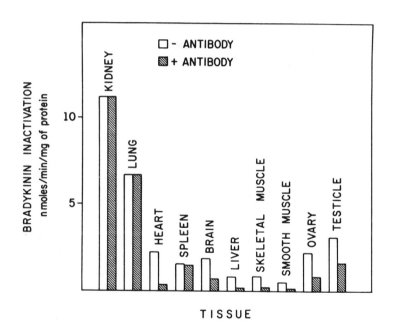

Fig. 1. Effect of antibrain endopeptidase A antibody on the 25,000 xg super-natant fraction of rabbit brain homogenate. Aliquots of 50 μl of the 0.25 M sucrose supernatant fraction of rabbit brain homogenate were added to 1 ml of 0.05 M Tris-HCl buffer, pH 7.5, containing 5×10^{-4} M dithiothreitol and 0.1 M NaCl. This mixture was preincubated for 15 min at 37° C with 0.1, 0.2, 0.4 and 0.84 mg of immunoglobulin G (IgG) from anti-endopeptidase A antibody, and 9.15 nmol of bradykinin was added. The enzymatic reaction was carried out at 37° C and bradykinin inactivation was measured by assay on the isolated guinea pig ileum. When bradykinin has been inactivated by about 50% the reaction was stopped by addition of 1 ml of 15% polyethylene glycol in 0.32 M citrate buffer, pH 2.2 One ml of this incubation mixture was submitted to amino acid and peptide analysis as described in Oliveira et al.[2]

   Since at least 80% of the kininase activity present in the soluble fraction
of rabbit brain is due to the activity of brain endopeptidases A and B (Martins
et al., submitted to J. Neurochem.), we can conclude that brain endopeptidase A
accounts for more than 60% of the kininase activity of the cytosol of rabbit
brain.

   Bradykinin inactivating enzymes (kininases) are present in several preparations of rabbit tissues. Properties of tissue kininases were compared in terms
of inhibition or activation by several compounds and by-product identification.[5]
As a result of this work and on the basis of the recovery of the complementary
peptide Arg-Pro-Pro-Gly-Phe and Ser-Pro-Phe-Arg we suggested the presence of
brain endopeptidase A-like enzymes in heart and smooth muscle. No evidence for
the presence of this enzyme was found in lung, spleen, kidney or plasma. In
this paper we use the anticatalytic activity of antiserum against brain endopeptidase A to screen for the presence of this enzyme in other rabbit tissues.

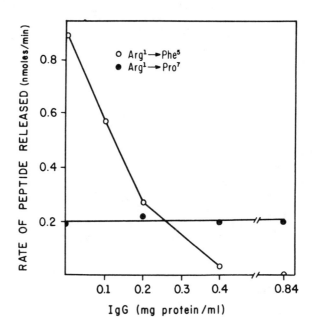

Fig. 2. Effect of anti-endopeptidase A antibody on kininase activity in the
25,000 xg supernatant fraction of rabbit tissue homogenates. Aliquots containing 0.23 mU of kininase activity[2] were preincubated at 37° C for 15 min
with 2.38 mg of IgG from anti-endopeptidase A antibody in 1 ml of 0.05 M
TrisHCl, pH 7.5, 0.1 M NaCl. Incubation was carried out at 37° C after
addition of 40 nmols of bradykinin. Control tubes were prepared in the same
manner, except for the use of 2.38 mg of IgG from pre-immune serum in place of
IgG from anti-endopeptidase A antibody. The reaction rate was followed by
measuring bradykinin inactivation with the isolated guinea pig ileum.

Figure 2 shows that monospecific antibody antirabbit brain endopeptidase A has no inhibitory effect on kininases from the soluble fraction of kidney, lung or spleen, but it inhibits the kininase activity of heart, brain, liver, smooth and skeletal muscle, ovary and testicle by more than 50%.

The same dependence on substrate size that we described for purified brain endopeptidases A and B[3] has also been suggested for tissue kininases[4] and brain arylamidases.[7] This view of the specificity of intracellular peptidases depending on both amino acid sequence and size of substrate reconciles the early concepts of Fisher-Bergman and Waldschmidt-Leitz[8] and strongly suggests the existence of two distinct proteolytic systems, i.e., proteolysis of high[9] and low molecular weight polypeptides. In these circumstances the peptidases which we have described may be part of the intracellular chain of low molecular weight proteolysis that modulates the conversion and inactivation of biologically active peptides.

ACKNOWLEDGMENTS

Research supported by Conselho Nacional de Pesquisa (CNPq), Grant No. 9532/78, and by Fundacão de Amparo à Pesquisa do Estado de São Paulo (FAPESP), Grant No. 75/163.

REFERENCES

1. Camargo, A.C.M., Shapanka, R., and Greene, L.J. (1973) Biochemistry, 12, 1838-1844.
2. Oliveira, E.B., Martins, A.R., and Camargo, A.C.M. (1976) Biochemistry, 15, 1967-1974.
3. Camargo, A.C.M., Caldo, H., and Reis, M.L., J. Biol. Chem., in press.
4. Camargo, A.C.M., Martins, A.R., and Greene, L.J., in Limited Proteolysis in Microorganisms, Cohen, G.N. and Holzer, H., eds., Department of Health, Education and Welfare Publication No. (NIH)78-1591, Washington, D.C.
5. Cicilini, M.A., Caldo, H., Berti, J.D., and Camargo, A.C.M. (1977) Biochemical Journal, 163, 433-439.
6. Camargo, A.C.M., Ramalho-Pinto, F.J., and Greene, L.J. (1972) J. Neurochem. 19, 37-49.
7. Suszkiew, J.B. and Bucher, A.S. (1970) Biochemistry 9, 4008-4017.
8. Bergman, M. (1972) Adv. Enzymol., 2, 49-68.
9. Schimke, R.T. (1975) in Intracellular Protein Turnover, Schimke, R.T. and Katunuma, N., eds., Academic Press, New York, 173-186.

Liu/Mamiya/Yasunobu, eds. Frontiers in Protein Chemistry

287

PHOSPHOLIPASES A$_2$ OF SEA SNAKE *Laticauda Semifasciata*:
ROLE OF TRYPTOPHAN IN THEIR REACTION KINETICS

N. TAMIYA, S. NISHIDA, H.S. KIM AND H. YOSHIDA
Department of Chemistry, Tohoku University, Aobayama, Sendai, Japan 980

ABSTRACT

Three phospholipases A$_2$ (PLases I, III and IV) have been isolated from the venom of a sea anake *Laticauda semifasciata* and their amino acid sequences determined. PLases I, III and IV consist of 118, 119 and 119 amino acid residues, respectively. These enzymes share 29 invariant and 8 type-conserved residues with other vertebrate phospholipases A$_2$. PLases III and IV are different from PLase I at 20 and 24 positions, respectively, and from each other at 6 positions. PLase I has one tryptophan residue at position 64, whereas PLases III and IV have none.

PLase I has the highest specific activity and shows normal reaction kinetics, whereas PLases III and IV are less active and show lag period in their action. The latter enzymes are activated by the reaction products and by oleate, with the lag times shortened.

On modification of the single tryptophan residue by N-bromosuccinimide, PLase I is converted into an enzyme similar to PLase III or IV, with smaller specific activity and with a lag period in the action. The tryptophan residue lies just outside of the crevice of the enzyme molecule, where catalytically essential histidine is found.

INTRODUCTION

Snake venom phospholipases are of special interest, because their physiological activities are very different from each other. β-Bungarotoxin,[1,2] notexin,[3] taipoxin[4] and sea snake myotoxin[5] are phospholipases with marked toxic activities, whereas some other phospholipases in the venoms are hemolytic but without special toxic effects.

The sea snake *Laticauda semifasciata* venom also contains phospholipases A$_2$,[6] which block neuromuscular transmission at rather high concentrations.[7] Present paper describes the structure-function relationships of the phospholipases.

EXPERIMENTAL RESULTS AND DISCUSSION

The structure of phospholipases A$_2$ from the venom of sea snake *Laticauda semifasciata*.

The sea snake *L. semifasciata* venom contains at least four phospholipases A$_2$,

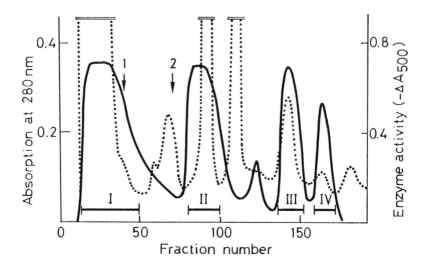

Fig. 1. Fractionation of the venom gland extracts of *L. semifasciata* by CM-cellulose column chromatography. The extracts prepared from 28.9g of the venom glands, containing 1,550 A_{280} units of protein in 168 ml, were loaded on a column (2.6 x 40 cm) of CM-cellulose (Brown) equilibrated with 0.01 M phosphate buffer (pH 6.4). After the nonadsorbed material had been washed out with the same buffer, elution was carried out with 0.01 M NaCl in the buffer (arrow 1), then with a linear gradient of NaCl from 0.01 to 0.20 M over 2.0 liters (arrow 2). Fractions (14.4 ml) were collected and 0.1 ml aliquots were assayed for enzyme activity. ...., A_{280}; — ● —, phospholipase A activity. Ea and Eb stand for erabutoxins a and b, respectively.[19] The peak heights in A_{280} of the flow-through fraction, Ea and Eb were 4.91, 1.97, and 2.57, respectively. Fractions indicated by the bars were pooled and used for subsequent purification.

which are named phospholipases A_2 I, II, III and IV (abbreviated as PLases I, II, III and IV) (Fig. 1). PLases I, III and IV can be isolated in pure state by repeated CM-cellulose column chromatography (Figs. 1 and 2) and desalted by gel filtration through Sephadex G-25 in 0.1 M acetic acid. The enzymic activity was followed by the hydroxamate method[8] during the course of purification. The content ratio of PLases I:III:IV was 100:56:12 on a weight basis. The purified enzyme preparations gave single bands on disc gel electrophoresis at pH 4.3. The molecular weights of PLases I, III and IV are the same being 14000 as measured by sodium dodecyl sulfate disc gel electrophoresis. The amino acid compositions of PLases I, III and IV were established. The absence of tryptophan in PLases III and IV was confirmed by UV absorption[9] and by N-bromosuccinimide method.[10]

The amino acid sequences of PLases I, III and IV have been determined (Fig. 3).

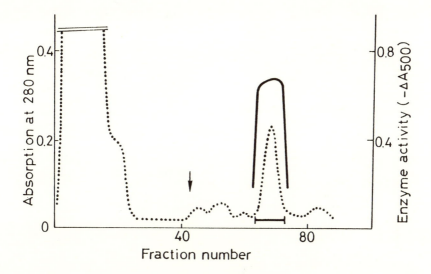

Fig. 2. Rechromatography of the PLase I fraction on a column of CM-cellulose. The fraction indicated by bar I in Fig. 1 was concentrated to about 200 ml with a Diaflo apparatus (Amicon Corp.) using a UM2 membrane, then dialyzed against 0.01 M phosphate buffer (pH 6.0). The solution, containing 244 A_{280} units of protein in 230 ml, was applied to a column (1.5 x 28.5 cm) of CM-cellulose (CM-52, Whatman) equilibrated with the buffer. Elution was carried out with a linear gradient of NaCl concentration from 0 to 012 M over 1.0 liter. The arrow indicates the start of the gradient. Fractions (17.2 ml) were collected and 0.1 ml aliquots were assayed for enzyme activity. ...., A_{280}; — ● —, phospholipase A activity. The peak height in A_{280} of the flow-through fraction was 0.886.

The details of the sequence determination will be described elsewhere. PLase I consists of 118 amino acid residues and PLases III and IV consist of 119 amino acid residues. PLases III and IV are different from PLase I at 20 and 24 positions, respectively, while they are different from each other at 6 positions. PLases I, III and IV share 29 invariant residues including 10 half-cystine residues with other vertebrate phospholipases A_2, namely with those of *Bitis gabonica*,[11] *Notechis scutatus scutatus* (Notexin[12] and Notechis 5[13]), *Hemachatus haemachatus* (DE-I[14]), *Naja melanoleuca* (DE-I, DE-II[15] and DE-III[16]), *Naja mossambica mossambica* (CM-I, CM-II and CM-III[17]), horse,[18] porcine,[19] bovine,[20] *Crotarus adamanteus*[21] and *Bungarus multicinctus* (A chain of β-bungarotoxin[2]). There are eight more positions among these phospholipases A_2, where residue-type conserved amino acids are found (Figs. 3 and 4).

Recently, the three-dimensional structure of a phospholipase A_2 from bovine pancreas was worked out.[22] The above mentioned invariant and type-conserved

```
        1      2      3
    5   0  5   0  5   0
I   NLVQFSNVIQCNLKGSRASYHYADYGCYCG
III       TYL    ANS K
IV        SYL    ANT K
```

```
        4      5      6
    5   0  5   0  5   0
    AGGSGTPVDELDRCCKIHDNCYGEAEKMGC
                     Q
```

```
        7      8      9
    5   0  5   0  5   0
    YPKWTLYTYESCTDTSP-CDE-KTGCQGFV
      L M N -Y GTZ  Y  BT    ZRY
      L M N -Y GTZG Y  BT KT ZRY
```

```
        1      1      1
        0      1      2
    5   0  5   0  5   0
    CACDLEAAKCFARSPYNNKNYNIDTSKRCK
```

Fig. 3. Amino acid sequences of PLases I, II, and III. For PLases III (III) and IV (IV), only the residues different from PLase I (I) are shown. Solid and dotted underlines indicate invariant and residue-type conserved positions, respectively. Tryptophan -64 of PLase I is shown by double underlines

residues are shown in Fig. 4 on the bovine enzyme. It is of interest that most of these residues are found around the crevice, where histidine-48[*], essential for the enzyme activity,[23] is located. The position corresponding to tryptophan-

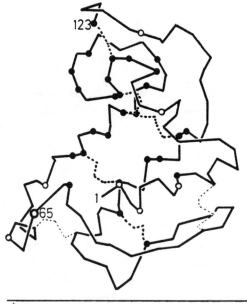

Fig. 4. Invariant and residue-type conserved positions of bovine pancreatic phospholipase A_2. Invariant (o) and residue-type conserved (●) positions are indicated on the stereo-drawing obtained by Dijkstra et al.[22] The position-65 corresponding to tryptophan-64 of PLase I is also shown.

[*] Residue numbers are given by those in Fig. 3.

TABLE 1

FATTY ACID COMPOSITIONS OF THE HYDROLYSIS PRODUCTS OF PHOSPHOLIPASES A I AND III

Fatty acid	Original lecithin	Phospholipase A I		Phospholipase A III	
		Free fatty acid	Lysolecithin	Free fatty acid	Lysolecithin
16:0	30.8	5.0	56.2	6.0	59.8
16:1	3.9	2.8	3.1	2.6	3.9
18:0	14.3	1.5	27.5	1.0	25.9
18:1	30.9	50.9	11.3	52.8	8.4
18:2	16.2	27.9	1.9	29.0	2.0
20:4	3.8	11.9	0	8.6	0

The results are expressed in percent of peak area on the gas chromatogram, thus in weight percent.

-64* of PLase I is also shown in Figs. 3 and 4. This position is occupied by a leucine residue in PLases III or IV and by one of isoleucine, valine, methionine and threonine residues in other phospholipases. The residue lies just outside of the crevice near two invariant tyrosine residues at positions 52 and 67.

Enzymic properties of PLases I, III and IV

Specificity. Reaction mixture (5 ml) containing lecithin (20 µmol), Triton X-100 (40 µmol), Tris HCl buffer (pH 7.5, 250 µmol), $CaCl_2$ (10 µmol), EDTA (35 µmol) and PLase (I, 0.030 A_{280} unit or III, 0.12 A_{280} unit) was incubated at 37°C for 45 min. The reaction products were extracted with chloroform and fatty acids and lysolecithin separated by silica gel column chromatography. Free fatty acids and fatty acids in lysolecithin were separately converted into their methyl esters and analyzed by gas chromatography.[24] The results are given in Table 1. With both enzymes, most of the unsaturated fatty acids were found liberated and saturated fatty acids in lysolecithin. As the positional distribution of saturated and unsaturated fatty acids in lecithin is known, it is concluded that both PLases I and III are phospholipases A_2.

Time course. The enzymic reaction was followed by titrimetric method in the following experiments. To a solution (1.8 ml) containing lecithin (12 µmol), Triton X-100 (24 µmol), $CaCl_2$ (60 µmol) and EDTA (2.1 µmol), 5 mM NaOH was added until a desired pH value (7.5 unless otherwise stated) was reached. The volume of the mixture was adjusted to 2.9 ml with water and the mixture warmed up to 37°C. The reaction was started by the addition of enzyme solution (0.1 ml).

* Residue numbers are given by those in Fig. 3.

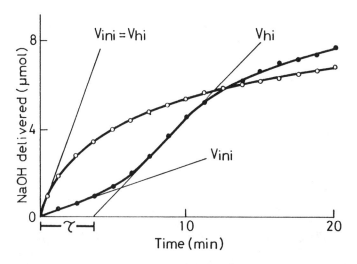

Fig. 5. Time course of the reaction catalyzed by PLases I and III. The reaction was carried out with 0.004 or 0.030 A_{280} unit of PLase I (o) or III (●), respectively.

The reaction was carried out in nitrogen atmosphere and followed with a pH stat using 5 mM NaOH as a titrant. Below pH 8.0, corrections for titration efficiency of fatty acid were made. The results on the initial phase of the reaction are shown in Fig. 5.

The reaction with PLase I followed a normal time course, whereas that with PLase III showed a lag time (τ). Essentially the same results were obtained with PLase IV as those with PLase III.

The initial and highest velocity for PLase I reaction and the highest velocity reached after a lag period (τ) for PLase III reaction were proportional to the enzyme concentrations (Fig. 6). With PLase III, lag time decreased rapidly with increasing enzyme concentration.

Activation of PLase III by reaction products. PLase III was activated by reaction products, namely, by lysolecithin and by fatty acids (Fig. 7). These reaction products showed also lag time reducing effect. In the presence of reaction products at 20 mol % substrate, PLase III became twice as active and showed a normal reaction time course without a lag period. These results explain the biphasic reaction progress curve of PLase III. PLase I, on the other hand, was hardly affected by the reaction products.

Palmitate and oleate, which are the main saturated and unsaturated fatty acid components, respectively, of the egg yolk lecithin, were tested for their activating capacity. Oleate was found to be more effective than palmitate (Table 2).

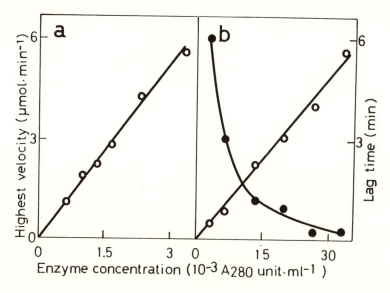

Fig. 6. Dependence of the highest velocity and lag period on enzyme concentration. Assays were performed with varying amounts of enzymes. Figures a and b are for PLases I and III, respectively. (o) Highest velocity; (•) lag period. No lag period was observed with PLase I.

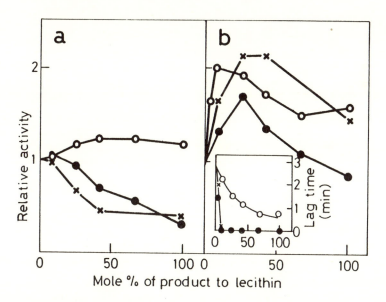

Fig. 7. Effects of reaction products on the activities of PLases I and III. Assays were carried out with 0.004 and 0.030 A_{280} unit of PLases I (a) and III (b), respectively. In the inset in Fig. b, the effect on the lag period is also shown. Additions: (o) Lysolecithin; (•) fatty acid; (x) equimolar mixture of lysolecithin and fatty acid.

294

TABLE 2

EFFECTS OF PALMITATE AND OLEATE ON PHOSPHOLIPASE A III.

Addition	Initial velocity (μmol/min)	Highest velocity (μmol/min)	Lag period (s)
None	1.05	3.78	36
Palmitate (1 mM)	1.72	4.62	27
Oleate (1 mM)	3.58	4.42	5

Assays were carried out with the additions described, under otherwise standard conditions using 0.062 A_{280} unit of the enzyme.

The unsaturated fatty acid may be the main factor to exhibit the biphasic reaction progress curve in Fig. 5.

Kinetic parameters. The initial velocity of the reaction catalyzed by PLase I, III or IV was studied as a function of lecithin concentration in the range of 2-20 mM, while the mol ratio of Triton X-100 to lecithin was kept at 2.0. For each enzyme, the pH and calcium ion concentration were set at the optimal values. Each enzyme gave a linear double-reciprocal plot and the kinetic parameters were determined as shown in Table 3. With PLases III and IV, the lag period was relatively independent of the substrate concentration within the range studied.

Conversion of PLase I to an enzyme like PLase III by modification with N-bromosuccimide

PL PLase I was treated with N-bromosuccinimide essentially according to Spande and Witkop,[10] but in 0.05 M acetate buffer, pH 4.1 without the addition of urea. When maximum decrease of A_{280} (20.1%) was attained, the modified enzyme was separated from the reagent by Sephadex G-25 gel filtration in 0.1 M acetic acid and freeze-dried. The results of amino acid analysis showed that 100% of the single tryptophan residue and 19% of 10 tyrosine residues were lost, whereas other amino acids remained intact.

The enzymic properties of the NBS-modified PLase I are shown in Table 3. It can be seen that the NBS-modified PLase I has very similar properties to PLases III and IV. Because all the ten tyrosine residues of PLase I are present at corresponding positions of PLases III and IV, the modification of tryptophan-64 of PLase I seems to be responsible for the change of enzymic properties. Its artificial modification as well as its natural substitution results in a phospholipase A with different properties, namely, higher Michaelis constant for

TABLE 3

ENZYMIC PROPERTIES OF MULTIPLE FORMS OF PHOSPHOLIPASE A
FROM *L. semifasciata* VENOM

	Phospholipase A			
	I	III	IV	I modified with NBS
Progress curve	Normal	Biphasic	Biphasic	Biphasic
Michaelis constant for lecithin (mM)	8.3	65	40	50
Maximum initial velocity (μmol/min A_{280} unit)	2,000	66	51	110^a
Optimum pH	7.6	7.8	8.2	8.3
Optimum calcium ion concentration (mM)	20	3	2	2

[a]For NBS-modified I, one A_{280} unit refers to the amount derived from one A_{280} unit of native I.

lecithin, lower enzymatic activity, higher optimum pH, lower optimum calcium ion concentration and activation by reaction products. It may be better to say that phospholipase A_2 of *L. semifasciata* is 20 times more active because of the introduction of a tryptophan residue at position 64, since this is the only tryptophan found at this position among the phospholipases whose sequences are known.

REFERENCES

1. Strong, P.N., Goerke, J., et al. (1976) Proc. Natl. Acad. Sci. USA, 73, 178-183.
2. Kondo, K., Narita, K. and Lee, C.Y. (1978) J. Biochem., 83, 101-115.
3. Karlsson, E., Eaker, D. and Ryden, L. (1972) Toxicon, 10, 405-413.
4. Fohlman, J., Eaker, D., et al. (1979) Eur. J. Biochem., 94, 531-540.
5. Fohlman, J. and Eaker, D. (1977) Toxicon, 15, 385-393.
6. Yoshida, H., Kudo, T., et al. (1978) J. Biochem., 85, 379-388.
7. Harvey, A.L., Rodger, I.W. and Tamiya, N. (1978) Toxicon, 16, 45-50.
8. Augustyn, J.M. and Elliot, W.B. (1969) Anal. Biochem., 31, 246-250.
9. Goodwin, T.W. and Morton, R.A. (1946) Biochem. J., 40, 628-632.
10. Spande, T.F. and Witkop, B. (1967) In Methods in Enzymology, Hirs, C.H.W., ed., Vol. 11, Academic Press, New York, pp. 498-506.
11. Botes, D.F. and Viljoen, C.C. (1974) J. Biol. Chem., 249, 3827-3835.
12. Halpert, J. and Eaker, D. (1975) J. Biol. Chem., 250, 6990-6997.
13. Halpert, J. and Eaker, D. (1976) J. Biol. Chem., 251, 7343-7347.
14. Jourbert, F.J. (1975) Eur. J. Biochem., 52, 539-554.

15. Jourbert, F.J.   (1975) Biochim. Biophys. Acta, 379, 345-359.
16. Jourbert, F.J.   (1975) Biochim. Biophys. Acta, 379, 239-244.
17. Jourbert, F.J.   (1977) Biochim. Biophys. Acta, 493, 216-227.
18. Evenberg, A., Meyer, H., et al.   (1977) J. Biol. Chem., 252, 1189-1196.
19. Puijk, W.C., Verheij, H.M. and deHaas, G.H.   (1977) Biochim. Biophys. Acta.
    492, 254-259.
20. Fleer, E.A.M., Verhey, H.M. and deHaas, G.H.   (1978) Eur. J. Biochem., 82,
    261-270.
21. Heinrikson, R.L., Krueger, E.T. and Keim, P.S.   (1977) J. Biol. Chem., 252,
    4913-4921.
22. Dijkstra, B.W., Drenth, J., Kalk, K.H., and Vandermaelen, P.J.   (1978)
    J. Mol. Biol., 124, 53-60.
23. Volwerk, J.J., Pieterson, W.A. and deHaas, G.H.   (1974) Biochem., 13, 1446-
    1454.
24. Yamakawa, T.   (1974) in Seikagaku Jikken Koza (in Japanese) Yamakawa, T.
    and Nojima, S., eds., Vol. 3, Tokyo Kagaku Dojin, Tokyo, pp. 198-200.

PHOSPHOLIPHASES A$_2$: STRUCTURE, FUNCTION AND EVOLUTION

ANNE RANDOLPH*, THOMAS P. SAKMAR, AND ROBERT L. HEINRIKSON
Department of Biochemistry, The University of Chicago, 920 East 58th Street,
Chicago, Illinois 60637

ABSTRACT

As part of an ongoing study of crotalid venom phospholipases A$_2$, the amino
acid sequence has been determined for the enzyme from western diamondback rattlesnake, *Crotalus atrox*. The *C. atrox* phospholipase sequence shows strong homology to that of the enzyme from the venom of the eastern diamondback rattlesnake
(*Crotalus adamanteus*) determined earlier in our laboratory [Heinrikson, R.L.,
Krueger, E.T. and Keim, P.S., J. Biol. Chem. (1977) 252, 4913]. In fact, the
C. atrox phospholipase A$_2$ differs in only six amino acid substitutions from *C.*
adamanteus α and in five substitutions from the β form. The fact that all of
the substitutions involving ionizable residues lead to increased negative charge
in *C. atrox* phospholipase A$_2$ accounts for the observed chromatographic and
electrophoretic properties of the enzyme as contrasted with those of the *C.*
adamanteus α and β forms. The single methionyl residue at position 10 in the *C.*
atrox phospholipase A$_2$ has been alkylated at pH 2.6 by reaction with a 50-fold
molar excess of iodoacetamide. The resulting derivative was separated from
unmodified enzyme by ion-exchange chromatography and, like the native phospholipase, is a fully active dimer. Cleavage with cyanogen bromide at Met-10
yields two fragments, an amino-terminal decapeptide and a 112-residue C-terminal
fragment which contains all of the seven disulfide bonds in the native molecule.
Neither the individual fragments, nor a reconstituted mixture of the two showed
enzyme activity. These findings are discussed relative to those obtained in
studies of enzymes from several sources in an attempt to elucidate structure-
function and evolutionary relationships among phospholipases A$_2$ in general.

INTRODUCTION

There exists in nature a variety of phospholipases, each specific for the
hydrolysis of a particular acyl or phosphodiester bond in the phospholipid molecule. The specificities of phospholipases A$_1$, A$_2$, C and D are indicated in
Fig. 1; phospholipases B, or lysophospholipases can degrade further the products
of hydrolysis with enzymes A$_1$ or A$_2$.

Because of their relative abundance in mammalian pancreas and in the venoms
of snakes and arthropods, phospholipases A$_2$ (EC 3.1.1.4) have been studied in

* Supported by predoctoral Training Grant 5T32 HL-7237 from the United States
Public Health Service.

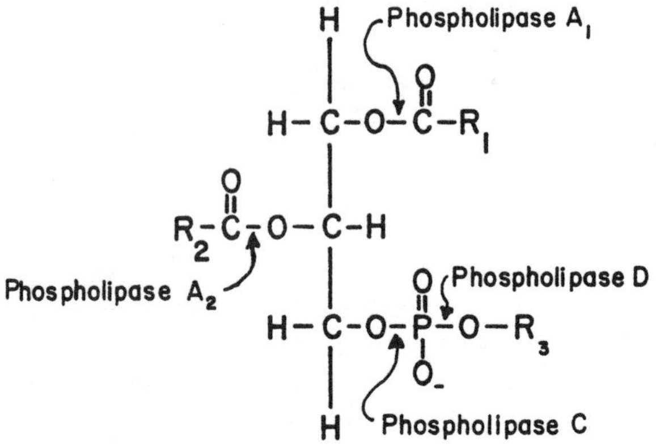

Fig. 1. Specificities of phospholipases in the cleavage of the acyl and phosphodiester bonds of a phospholipid.

greatest detail. These heat-stable, esterolytic enzymes catalyze the selective, calcium-dependent hydrolysis of the 2-acyl groups in **3-sn-phosphoglycerides** (Fig. 1). Phospholipases A_2 play a central role in lipid metabolism[1] and have been applied in probing the structural organization of phospholipids in membranes[2] and in lipoproteins.[3] Since the liberated C-2 acyl substituent may serve as a precursor for prostaglandins and thromboxanes,[4,5] phospholipases A_2 may be involved in processes of thrombosis. Moreover, since these enzymes act at solution-micelle interfaces they serve as useful models for studying heterogeneous catalysis[6] and lipid-protein interactions in general.

Phospholipases A_2 have been the subject of extensive structural investigation. Complete amino acid sequences have been reported for the enzyme from porcine,[7] equine,[8] and bovine[9] pancreas, from honeybee venom,[10,11] and from the venoms of numerous reptiles.[12-20] With the exception of the honeybee enzyme, all of these phospholipases show strong sequence homology.[20] Venom phospholipases A_2 appear to be secreted as the active enzyme, perhaps in the presence of an inhibitor.[21,22] The pancreatic enzymes, however, are secreted as very weakly active zymogens (prophospholipases) with a seven residue extension at the N-terminus that is removed during activation by trypsin. Phospholipases A_2 are relatively small (about 125 amino acid residues) and highly cross-linked with 6 to 7 disulfide bridges. Some venom enzymes have been shown to exist as stable dimers in solution.[23,24] The three-dimensional structures determined by X-ray crystallography have been reported for prophospholipase A_2 from porcine pancreas[25] and for the bovine pancreatic phospholipase A_2.[26] These structures have been useful in interpreting the many results obtained by chemical modification of the phospholipases A_2.

In the present article, we describe our recent studies regarding the primary structural analysis and chemical modification of the venom phospholipase A_2 from *Crotalus atrox* (western diamondback rattlesnake). These findings are considered within the context of our current understanding of structure-function and evolutionary relationships among phospholipases A_2 in general.

STRUCTURAL ANALYSES

Primary structure

An earlier report from this laboratory[20] described the sequence analysis of phospholipase A_2-α from the venom of *Crotalus adamanteus* (eastern diamondback rattlesnake). The venom of this reptile contains two active forms of the enzyme, α and β, which are chromatographically and electrophoretically distinct,[27] but which are nevertheless indistinguishable in terms of specific enzyme activity, molecular weight, and amino acid composition.[19,27] Our studies of the β form of the enzyme indicate that it differs from α only in the amidation state of a single residue, Gln 117, which is Glu in phospholipase A_2-β.

Because of general interest in the crystallographic and kinetic properties of *C. atrox* venom phospholipase A_2 and our interest in homology relationships, we initiated sequence studies of this enzyme based upon strategies utilized with the *C. adamanteus* phospholipases A_2. To summarize briefly, our approach to the sequence analysis was based primarily upon automated Edman degradation of intact reduced and alkylated enzyme and of large fragments obtained therefrom by cleavage with cyanogen bromide and trypsin. Tryptic digestion was restricted to the 4 arginyl residues in the chain by prior citraconylation of lysyl amino groups in the protein derivative. A detailed description of fragment purification and analysis has been reported for *C. adamanteus* phospholipase A_2-α;[20] essentially the same results were obtained in the determination of the *C. atrox* enzyme sequence. The complete amino acid sequence of *C. atrox* venom phospholipase A_2 is given in Fig. 2. It differs from *C. admanteus* A_2-α in six substitutions (residues 12, 14, 35, 56, 74, and 117), four of which lead to an increase in negative charge in the *C. atrox* enzyme (Lys-14 to Gly-14; Arg-35 to Leu-35; Asn-56 to Asp-56; and Gln-117 to Glu-117). Since the *C. atrox* enzyme has a Glu at 117, it differs in only 5 positions from *C. adamanteus* A_2-β. The sequence analyses of *C. atrox* and *C. adamanteus* α and β phospholipases A_2 have thus revealed charge characteristics in accord with the behavior of these enzymes during separation procedures by ion-exchange chromatography[27] and polyacrylamide gel electrophoresis.[19,27]

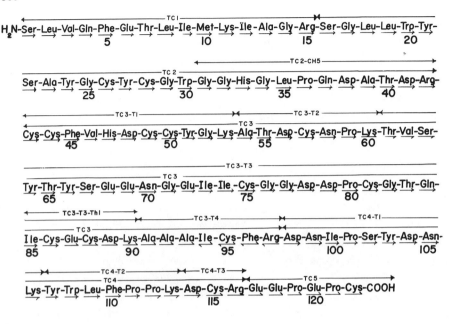

Fig. 2. The amino acid sequence of phospholipase A_2 from the venom of *C. atrox*. Peptide designations are as defined in an earlier publication concerning the sequence analysis of the enzyme for *C. adamanteus* venom.[20]

To date, complete amino acid sequences have been reported for about 20 phospholipases A_2 including those from three mammalian pancreatic sources,[7-9] and for snake venom enzymes from several elapids,[12-16] from a hydrophid,[28] from a viper, *Bitis gabonica*[17] and from three crotalids including two rattlesnake species (20 and the present work), and the Japanese water moccasin, *Agkistrodon halys blomhoffii*.[18] The comparative analysis of representative phospholipase A_2 sequences from the four categories, i.e., pancreatic, elapid (*Naja melanoleuca*), viperid (*B. gabonica*) and crotalid (*C. adamanteus* and A. *halys blomhoffii*) was the subject of relatively lengthy discourse in a recent publication from our laboratory.[20] The sequence of phospholipase A_2 from a member of the *Hydrophidae* (sea snakes), *Laticauda semifasciata*, determined recently by Tamiya et al.[28] provide a fifth category for comparison (Fig. 3) so that we now have phospholipase A_2 sequences representative of all four classes of venomous reptiles (Fig. 4). Several important points have emerged from these comparisons.

1. The sequence of these phospholipases A_2 are all very similar and are clearly homologous, with minimum base changes per codon (mbc/c) ranging from 0.05 to 0.95 (Table 1). Striking identities occur in most of the half-cystine,

TABLE 1

SEQUENCE SIMILARITY MATRIX FOR PHOSPHOLIPASES A_2 FROM VARIOUS SOURCES[a]

	C. atrox	C. adamanteus	A. halys blomhoffii	B. gabonica	N. melanoleuca	L. semifasciata
C. adamanteus	0.05					
A. halys blomhoffii	0.28	0.31				
B. gabonica	0.71	0.69	0.62			
N. melanoleuca	0.86	0.88	0.81	0.85		
L. semifasciata	0.83	0.81	0.77	0.72	0.53	
Bovine pancreas	0.88	0.88	0.89	0.95	0.65	0.70

[a]Sequences are aligned as shown in Fig. 3. The values shown are the minimum number of nucleotide base changes per codon for each comparison (cf. reference 63 for details).

302

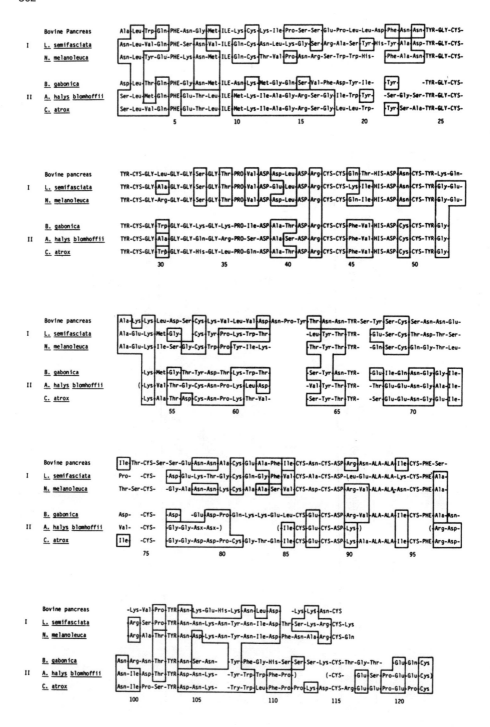

Fig. 3. Comparison of sequences for phospholipases A$_2$ from bovine pancreas and from the venoms of the elapid *N. melanoleuca* DE-II, the hydrophid, *L. semifasciata*, the viper *B. gabonica* and the crotalid *C. atrox*. Gaps are introduced to provide proper alignment of half-cystine residues and the greatest homology. Also included are segments of the sequence of phospholipase A-II from the crotalid *A. halys blomhoffii*, including sequence analysis of the intact chain performed in our laboratory. These segments are enclosed in *brackets* to designate that they are taken from regions of the molecule other than as published originally.[18] Residues invariant in all proteins are capitalized and residues at any position which occur in more than one enzyme sequence are enclosed in *boxes*.

---

and in many of the Gly and Asp positions as well as in apparently invariant lengths of sequence such as Tyr-Gly-Cys-Tyr-Cys-Gly-(——)-Gly-Gly-(——)-Gly-(——)-Pro- (residues 24-36); Cys-Cys-(——)-(——)-His-Asp-(——)-Cys-Tyr (residues 43-51); and Cys-(——)-Cys-Asp-(——)-(——)-Ala-Ala-(——)-Cys-Phe (residues 86-96). The high conservation in the latter three stretches of sequence would strongly imply some functional involvement, if not directly in catalysis, then in substrate binding or maintenance of the active site conformation. We have already commented[20] on the fact that the sequence of the phospholipase A$_2$ (A II) from *A. halys blomhoffii* venom reported by Samejima et al.[18] shows little homology to the structures depicted in Fig. 3. The partial structure for this enzyme given in Fig. 3 is based, in part, upon our own sequence analysis of the *A. halys blomhoffii* protein[29] and some fragments from the published sequence[18] which have been realigned on the basis of homology.

2. Perhaps even more interesting than the sequence similarities were the differences observed. These differences led to our classification[20] of phospholipases A$_2$ into two groups (Fig. 3). Group I comprises the enzymes from elapid and sea snake venom and pancreatic sources while crotalid and viperid venom phospholipases A$_2$ are classified in Group II. The basis for this classification is emphasized by the residues enclosed in boxes in Fig. 3. Moreover, Group I enzymes have a disulfide bond at 11-69 and a loop of structure between residues 52 and 53 (Fig. 3) not seen in Group II phospholipases. The latter enzymes appear to have a unique insertion of Asx-Asn at 98-99, but their most interesting feature is an extension of 6 or 7 residues at the C-terminus of the polypeptide. This extension terminates with Cys-122, a half-cystine residue unique to the Group II phospholipases. The only other residue of this kind is Cys-49 and, presumably, Cys 122 and Cys 49 are bonded in disulfide linkage. As will be mentioned later, His 47 has been implicated in catalysis and the close proximity of the C-terminal extension to this region of the enzyme might play some role in modulating the function of Group II phospholipases.

3. Finally, the results obtained from the sequence comparisons are in accord with phylogenetic relationships inferred from morphology. Hydrophids and elapids are considered to be more primitive than the vipers which have evolved a more complex venom secreting apparatus and hinged fangs. Vipers are, in turn, regarded as the progenitors of pit vipers (crotalids) which possess heat-sensitive facial pits in addition to the other features characteristic of vipers. Rattlesnakes, with their distinctive terminal appendage, represent the most highly evolved of venomous snakes. The tree presented in Fig. 4 was constructed from chemical information (Table 1) and is qualitatively the same as that published by Brattstrom,[30] who reported taxonomy of species within the *Viperidae* and *Crotalidae*. A sequence similarity matrix for phospholipases A_2 from *C. atrox* and *C. adamanteus* (pit vipers with rattles), *A. halys blomhoffii* (pit viper), *B. gabonica* (viper), *N. melanoleuca* (cobra, elapid), *L. semifasciata* sea snake, hydrophid) and bovine pancreas is presented in Table 1 based upon sequences given in Fig. 3. Values of mbc/c less than 1.00 demonstrate the strong homology among all these enzymes. The rattlesnake phospholipases are clearly very similar to one another and, as expected, bear a relationship of decreasing similarity to enzymes from, respectively, another pit viper, a viper, and phospholipases from the hydrophid, the elapid and a pancreatic source.

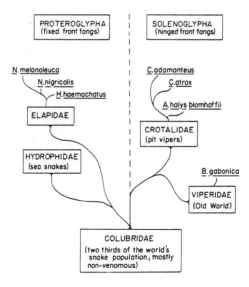

Fig. 4. Evolutionary relationships among elapid, viperid and crotalid venomous snakes as deduced from sequence analysis of venom phospholipases A_2.

Similarly, and in agreement with the taxonomy based upon morphology, one could easily classify the elapids and hydrophids together based upon the sequence similarity observed between their venom phospholipases (Fig. 4). This is all very satisfying in terms of reptilian evolution, but when one calculates the mbc/c in a comparison between the sequences of phospholipases from cobra venom and bovine pancreas, a value of 0.65 is obtained (Table 1). Therefore, one might be led to the amusing conclusion that cobras are more closely related to cows than to rattlesnakes! These findings suggest that although the elapid venom and pancreatic phospholipases A_2 have undergone parallel divergent evolution from the ancestral enzyme, their sequences have not diverged as much from the progenitor as have those of the enzymes from the vipers and crotalids. The more pronounced sequence changes in the Group II phospholipases may reflect some as yet unknown functional attribute of these enzymes. It was of interest to determine whether the rattlesnake with a Group II venom phospholipase might possess a Group I pancreatic enzyme. Analysis of rattlesnake pancreas homogenates failed to detect the presence of phospholipase A_2 activity; it may be that the digestive function is satisfied by the action of the venom enzyme subsequent to envenomation.

It should be noted at this juncture that the covalent structure of another phospholipase A_2, that from honey bee (*Apis mellifica*), has been determined.[10,11] It is an enzyme similar in size to those represented in Fig. 3 but is otherwise considerably different. The *A. mellifica* enzyme sequence shown in Fig. 5 shows strong homology to the corresponding region of the Group I phospholipases. It would seem highly unlikely, therefore, that the honey bee enzyme is an example of convergent evolution to a similar phospholipase A_2 function and we are inclined to believe that this enzyme, too, bears a divergent genetic relationship to those in Fig. 3.

Tertiary structure

Two X-ray crystallographic studies of phospholipases A_2 have been reported recently, both from the laboratory of Dr. Jan Drenth in the Netherlands. The first, that of the proenzyme from porcine pancreas,[25] was fraught with technical difficulties related to the quality of the crystals and diffraction pattern and the fact that the rather poor map was interpreted relative to a covalent structure which has since been revised.[31] More recently, the three-dimensional structure of the active enzyme from bovine pancreas has been reported as determined at 2.4 Å resolution.[26] Stereo diagrams of two views of the molecule are given in Figs. 6 and 7. It should be pointed out that the disulfide bond

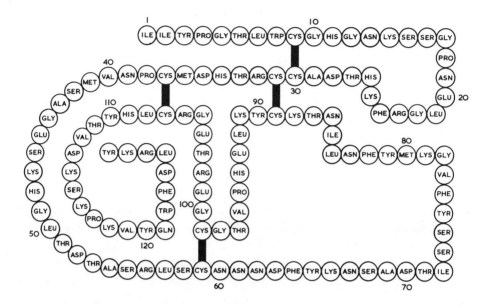

Fig. 5. The amino acid sequence of honey bee venom phospholipase A$_2$ showing the positions of four disulfide bridges. Taken from Shipolini et al.[11] with permission.

assignments in proteins with such a high degree of cross-linking are extremely difficult to determine chemically. Certainly one of the important contributions of the bovine enzyme X-ray analysis was the placement of all seven disulfide bridges in the molecule.

The bovine pancreatic phospholipase molecule is composed of about 50% α-helix (A), followed by a random coil conformation from 14 to 40. One of the two long antiparallel α-helices (C) extends from 40 to 58 and the other (E) comprises residues 90 to 108. These form the backbone of the molecule and could be quite rigid relative to one another due to their attachment by two disulfide bridges. Antiparallel β strands are found between residues 74 to 78 and 80 to 85. It is noteworthy that the disulfide bond between Cys 11 and Cys 77 (11 to 69 in Fig. 3) is missing in Group II phospholipases. A salt bridge may serve this function in the crotalid enzymes since 11 is a Lys and 69 is a Glu (Fig. 3). It also would appear that the Group II phospholipase C-terminal extension of 6 to 7 residues making disulfide bond connection with the half-cystine at 49 (numbering in Fig. 3) could be easily accommodated. It will be of interest in this regard to examine the tertiary structure of *C. atrox* phospholipase A$_2$ currently under analysis in the laboratory of Dr. Paul Sigler at the University of

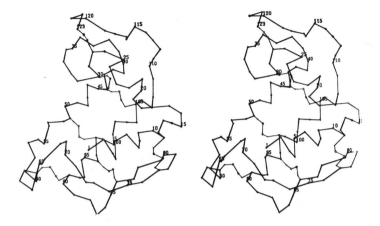

Fig. 6. Stereo diagram showing the $^{\infty}$C-atoms and disulfide bridges of the
bovine pancreatic phospholipase A$_2$ molecule. Reproduced from Dijkstra et al.,[26]
with permission.

Chicago. Unlike the pancreatic phospholipases A$_2$ which are monomeric, the
rattlesnake venom enzymes appear to be isolated and functional as dimers and,
in the case of the C. adamanteus enzyme, the dimer has a K$_d$= 2.0 x 10^{-9}M.[24] The
structural analysis of C. atrox phospholipase A$_2$ at 5 Å resolution reveals a
dimer in the asymmetric unit[32] so it would appear that the chemically identical
monomer chains do not exist in identical conformations in the dimer crystal.
The question as to whether phospholipases are functional as monomers or dimers
(or both) will be addressed in a later section of this paper.

    With this background of structural information relative to the sequence and
conformation of the phospholipases A$_2$ we will proceed to examine and interpret
results accumulated by other lines of investigation concerning the function of
these enzymes. Implicit in these discussions is the assumption that the enzymes
represented in Fig. 3 share a similar tertiary structure. Indeed, as pointed
out by Dijkstra et al.[26] it is easy to accommodate the sequences of homologous
phospholipases A$_2$ into the bovine enzyme structure since the positions of
insertions and deletions are always at the enzyme surface.

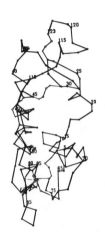 

Fig. 7. Stereo diagram showing C^{α}-carbon atoms of bovine pancreatic phospholipase A_2. The molecule is rotated by 90° about the vertical axis with respect to Fig. 6. Reproduced from Dijkstra et al.,[26] with permission.

STUDIES OF PHOSPHOLIPASE A_2 FUNCTION

Although the catalytic mechanism of phospholipase A_2 hydrolysis remains to be elucidated, some information is available relative to substrate specificity, the nature of the catalytic site, and a region of the molecule that appears to be involved with the recognition of organized lipid-water interfaces.[33] It might at first seem surprising that with so much detailed structural information in hand for the phospholipases A_2, we have only a rudimentary understanding of their mechanism of action. If one considers the well-characterized family of pancreatic serine proteases, however, it will be recalled that the essential features of the mechanism had been correctly deduced by kinetic analysis, prior to determination of the covalent or tertiary structures. In the case of the phospholipases A_2, kinetic studies have been hampered by the fact that the water-soluble enzymes are most efficient in the cleavage of substrates that are insoluble in aqueous media. The enzyme assays employed in many laboratories either do not lend themselves readily to kinetic interpretation or are difficult to manipulate experimentally. Chemical modification studies of phospholipases

A_2 must be interpreted with special care. Loss of activity attending modification of a particular functionality in any given enzyme could be due to the usual reasons, e.g., loss of the ability to bind or cleave the phospholipid substrate or structural alterations in native conformation unrelated to the substrate catalytic or binding sites (hereafter referred to as the active site). However, phospholipases A_2 possess, in addition, some means of recognizing, and modulating their catalytic efficiency with respect to, organized lipid-water interfaces. The residues or mechanisms involved in this interaction must be taken into account when assessing the results of any functional studies of phospholipases A_2. Nevertheless, kinetic analysis, chemical modification studies, comparative sequence analysis and various approaches based upon spectroscopic methods have provided results that can be considered relative to the structural information given in the section on structural analysis.

Substrate requirements

All naturally occurring glycerophosphatides are substrates for phospholipases A_2, regardless of the nature of the polar headgroup (R_3 in Fig. 1). These enzymes may vary considerably, however, in their preference for particular substrates depending upon whether the polar headgroup is choline, ethanolamine or serine. One of the requirements of substrates is that the acyl bond cleaved be on the carbon atom of the glycerol backbone that is adjacent to the carbon bearing the phosphate (or electronegative) group. For example, substrates with the fatty acyl substituents at positions 1 and 3, and the phosphodiester at position 2 are readily cleaved by phospholipases A_2 to yield a 3-acyl-2-sn-phosphoglyceride and fatty acid.[33] Thus, specificity is not necessarily displayed toward ester bonds with secondary alcohols. Phospholipases A_2 are highly stereospecific for 3-sn-phosphoglycerides (e.g., L-lecithins); the stereoisomeric 1-sn analogs (D-lecithins) are pure competitive inhibitors which bind with equal affinity to the enzyme but are not hydrolyzed. The electronegative group may be phosphate or sulfate[34] but it must possess at least one acidic hydroxyl group ionization. Therefore, a negative charge in the polar headgroup is an essential requirement of phospholipases A_2 substrates. Other pure competitive inhibitors in addition to D-lecithins include lysolecithin and L-lecithins in which the R_2 (Fig. 1) ester bond is replaced by an amide or ether linkage or in which R_2 itself contains bulky β substituents.

Phospholipase A_2--interface interaction

Most naturally occurring phosphoglycerides form emulsions composed of particles of varying size because of the presence of two aliphatic side chains of

14 to 18 carbon atoms. However, phosphoglycerides with shorter sidechains are
distinctly soluble due to a reduced hydrophobicity. They yield monomers at low
concentrations and micelles at higher concentrations exceeding the critical
micelle concentration (CMC). One of the dramatic properties of phospholipases
A_2 that is shared with pancreatic lipases and perhaps other enzymes that function in heterogeneous catalysis is that although they are somewhat active
toward monomeric substrates, an exponential increase in activity is observed as
the substrate reaches and exceeds its CMC. This phenomenon is exemplified by
the curves given in Fig. 8[35] which describe the substrate concentration dependency of the rate of hydrolysis of 1,2-diheptanoyl-sn-glycero-3-phosphorylcho-
line by porcine pancreatic phospholipase A_2 (a) and its zymogen (b). The
reactions appear to obey normal Michaelis-Menten kinetics in the concentration
range below the CMC, indicating hydrolysis of monomers at low V_{max}. However, as
the substrate aggregates to form micelles, the rate increases exponentially to
a V_{max} 3-4 orders of magnitude higher than that observed in the monomer range.
The zymogen, although moderately active, does not appear to recognize this
structural transition in the substrate. A number of recent studies have attempted to explain "interface activation", and to determine whether it results
from an enhanced reactivity of the substrate upon aggregation, or from an
activation of the enzyme by a lipid-water interface.

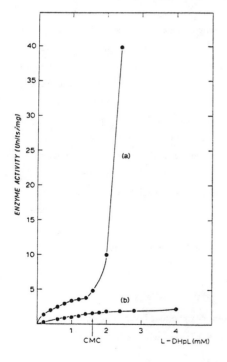

Fig. 8. Concentration dependency of
rate of hydrolysis of 1,2-diheptanoyl-
sn-glycero-3-phosphorylcholine (L-DHpL)
by porcine pancreatic phospholipase A_2
(a) and its zymogen (b). Taken from
Pieterson et al.[35] with permission.

Enhanced substrate reactivity

Several hypotheses have been proposed which implicate enhanced substrate reactivity as an explanation for interface activation. According to Brockerhoff and Jansen,[36] the hydration state of the substrate is a key factor regulating the rate of hydrolysis of lipolytic enzymes. The more soluble the substrate, the thicker the water shell surrounding the ester groups, and the lower the enzymatic activity. For the case of pancreatic lipase, Brockman et al.[37] showed that hydrolysis of a soluble substrate was increased 10^3-fold by the addition of siliconized glass beads. They proposed that the high local concentration of substrate molecules near the active sites of adsorbed enzyme molecules could account for interface activation. This hypothesis would imply that V_{max} for the hydrolysis of a monomeric substrate should be identical to that of the same substrate in aggregated form. However, the V_{max} for aggregated substrates has been shown to be higher than that for monomers with lipase[38] and phospholipase[35] from porcine pancreas. Furthermore, Chapus et al.[39] showed that deacylation of the monoacyl derivative of lipase, an intermediate in the catalytic mechanism, was accelerated by adsorption to siliconized glass beads in the absence of substrate. They concluded that, in addition to the substrate concentration effect proposed by Brockman et al.,[37] the catalytic properties of the enzyme were enhanced upon adsorption.

Enhanced catalytic activity--interfacial recognition site

A model emphasizing the regulation of enzyme catalysis by lipid water interfaces has been proposed by Dr. Gerard de Haas and coworkers at the State University of Utrecht in The Netherlands.[33] Their interfacial recognition site (IRS) hypothesis is based upon studies of pancreatic phospholipases A_2 and their zymogens. This system is nicely suited for chemical modification studies because reactions can be carried out on the zymogen and the functional consequences can be assessed prior to and following activation by tryptic cleavage of the Arg_7-Ala_8 bond in the zymogen. The IRS is defined as an exposed three-dimensional region of the enzyme involved in the formation of the enzyme-interface complex. This site is lacking in the zymogen and is induced following activation by trypsin. It is composed of residues 1-8 in the active enzyme, i.e., [Ala-Leu-Trp-Gln-Phe-Arg-Ser-Met-] and Tyr-69 (the Tyr preceding position 63 in Fig. 3), and is assumed to be topographically and functionally distinct from the active site. It is proposed that this region penetrates lipid water interfaces and anchors the enzyme. This process is facilitated by negative charges in the interface.

The IRS hypothesis is based upon a number of elegant chemical modification studies. The importance of the α-amino group of Ala-1 has been indicated by the pH-dependency of the enzyme-micelle interaction. This interaction is controlled in the absence of Ca^{++} by a group of pK = 8.4 identified as the $\alpha-NH_3^+$.[40] As the Ca^{++} ion concentration is increased, both the pK of the $\alpha-NH_3^+$ group and the pH above which the enzyme fails to bind to the interface appear to shift from 8.4 to 9.3. Slotboom et al.[41] attribute this effect to Ca^{++} binding to a low affinity site close to Trp-3. Pieterson et al.[35] proposed that induction of the IRS was consequent to the formation of a stabilizing salt bridge between the newly liberated $\alpha-NH_3^+$ group of Ala-1 and a buried carboxylate group. Several specific chemical alterations in the N-terminal region of porcine pancreatic phospholipase A_2 have been described by Slotboom and de Haas[42] all of which point to the importance of maintaining the spatial integrity of the charge on the $\alpha-NH_3^+$. Treatment of the proenzyme with methylacetimidate gave a fully ε-amidinated product which, upon hydrolysis with trypsin, yielded *active* ε-amidinated phospholipase A_2 (AMPA). This derivative, AMPA, was then subjected to numerous chemical alterations in structure. Derivatives of AMPA in which Ala-1 is replaced by Gly or β-Ala or in which Trp-3 is replaced by Phe are simi-lar to AMPA in enzymatic activity and show enhanced rates in the presence of micellar substrate. Chain shortening or elongation at the N terminus, blocking the $\alpha-NH_3^+$ group of Ala-1, and replacing L-Ala-1 with D-Ala all yield proteins which no longer show activity towards micellar substrates. Des-Ala-AMPA, Ala-Ala-AMPA, and D-Ala-AMPA are, however, active to varying degrees on substrate monolayers[40] which allow a change in the lipid packing density, but they show a weaker capacity for interaction with the interface than AMPA. The zymogen and the phospholipase A_2 derivatives with a blocked $\alpha-NH_3^+$ group are unable to hydrolyze monolayers even at very low surface pressures.[33] Specific iodination of Tyr-69[43] produces a protein that can hydrolyze monolayers at high surface pressures where native phospholipase is no longer active.[44] Thus, modifications of residues in the so-called IRS can enhance, diminish, or completely abolish the enzyme's ability to interact with interfaces.

The functional importance of the N terminal region has also been demonstrated in studies of a Group II snake venom enzyme. Recently Huang and Law[45] have described the synthesis of a photolabile phosphatidylethanolamine (PE) analog and its use in the inactivation of *C. atrox* phospholipase A_2. Irradiation of the enzyme in the presence of the photolabile PE analog of the pure L-configura-tion resulted in a modification restricted primarily to the N-terminus and sub-sequent analysis of the derivative showed that modification was within the first 10 residues. If racemic reagent was employed, another site of reaction was observed in addition to that described above, this being in the vicinity of the active site His-47. Thus it would appear, in agreement with studies mentioned earlier, that the D-reagent can bind, albeit nonproductively, to a region near the active site. The difference in site modified may reflect this nonproductive binding.

The methionyl residue at position 8 of the pancreatic enzyme appears to be the limit of the IRS as defined thus far in the N-terminal region. Carboxymethylation of this residue leads to complete inactivation of the enzyme towards both monomer and aggregated substrate.[46] Spectroscopic studies indicate that the derivative still binds Ca^{++} and monomeric substrate with affinities equal to those of native enzyme. It is appropriate to consider here experiments carried out in our laboratory that were designed to evaluate the possible functional involvement of the single Met-10 in *C. atrox* phospholipase A_2. Carboxamidomethylation of this residue was accomplished by reaction of the native enzyme (8×10^{-4}M) with a 50-fold molar excess of iodoacetamide in 1% acetic acid, pH 2.6, for 24 to 48 hours. The alkylated derivative was purified by ion-exchange chromatography and was shown to be fully active both in bulk phase and monolayer assay systems. Thus, conversion of the methionyl side chain at 10 to a positively charged sulfonium derivative has no apparent effect upon enzyme activity. Reactions with iodoacetate, in agreement with the findings of Wells[47] gave no loss of activity during alkylation at pH 6.0 or 8.5. Incorporation of reagent was documented, however, and it would appear that Met-10 is the primary site of reaction with these reagents over the whole pH range.

A further study undertaken in our laboratory involving Met-10 concerned the isolation and assay of fragments generated by cleavage of the *C. atrox* enzyme with cyanogen bromide. Both the N-terminal decapeptide and the C-terminal fragment of 112 residues were totally inactive in monolayer assays and activity was not regenerated in mixtures of the separated fragments. Gel filtration of the C-terminal fragment in the concentration range $10^{-5}-10^{-6}$M indicated that it was monomeric and no longer possessed the dimeric character of the native enzyme. Since the disulfide bridges and, presumably most of the native structure of the enzyme are preserved in the C-terminal fragment, one must ascribe some functional role to the N-terminal region of the *C. atrox* phospholipase A_2. It is not clear from the X-ray model (Figs. 6 and 7) which aspect of the enzyme function would be most likely impaired by removal of the N-terminal 10 residues. The lack of activity of the C-terminus could be due solely to its inability to dimerize since it has been postulated that *C. atrox* is only active as a dimer;[24] alternatively, this lack of activity could be attributed to the loss of the N-terminal part of the IRS. Volwerk[48] speculates that the hydrophobic nature of the IRS may be responsible for the dimerization of the snake venom enzymes.

A complete model explaining the action of lipolytic enzymes must account for three main steps in heterogeneous catalysis: 1) the binding or adsorption of the enzyme to the interface, 2) interfacial activation or the ability of the enzyme to hydrolyze aggregated substrates faster than monomeric substrates, and 3) catalysis proper. The IRS hypothesis for pancreatic phospholipase addresses the first two steps. De Haas et al.[3] assume that the IRS is a relatively hydrophobic domain at the surface of the enzyme which is stabilized by an internal salt bridge involving the amino terminal group. However, the X-ray crystallographic analysis of the bovine pancreatic phospholipase[26] failed to discern such a salt bridge. Rather, it was shown that the α-NH$_3^+$ group of Ala-1 is located within H-bond distance of a water molecule possibly liganded to the active site calcium ion. Moreover, the actual penetration and depenetration of a lipid bilayer by the enzyme would be likely to constitute an energetically unfavorable situation. The IRS hypothesis also attributes interfacial activation to a conformational change after penetration of the IRS domain. Any conformational change in the enzyme resulting from interaction with an interface which enhances catalytic efficiency by 3-4 orders of magnitude must involve the active site proper. The IRS, rather than a distinct topographical region, may be a portion of the active site. Such an interpretation would appear to be consistent with the X-ray crystallographic model. Moreover, interfacial activation is probably a combination of effects on the enzyme and on the substrate. In any case, several lines of evidence point to the functional involvement of the N-terminal region in phospholipase catalyzed hydrolysis, and this continues to be an important focus for future research concerning the enzyme mechanism.

Half-site reactivity and the functional dimer hypothesis

In a study of the effects of a variety of reagents toward the activity of
C. adamanteus phospholipase A$_2$, Wells[47] found, among other things, that total loss of activity was observed upon modification of one amino group per dimer. This observation was the first indication of a possible "half-site" reactivity associated with phospholipases A$_2$, and was not so surprising, perhaps, in view of the fact that under the experimental conditions employed in the modification, the enzyme exists as a dimer. Moreover, there is kinetic evidence that the rattlesnake enzymes are functional *only* as dimers.[24] A more recent case of half site reactivity has been reported by Roberts et al.[49] for the phospholipase A$_2$ from the venom of a cobra (*Naja naja naja*). In this instance, nearly complete inactivation by p-bromphenacyl bromide was accomplished following the modification of 0.5 residues of the lone histidine (presumably His-47) per monomer. Although the *N. naja naja* enzyme can dimerize at high concentrations, it is

monomeric under the conditions employed for alkylation and assay. Therefore, if the half-site reactivity is correct, the formation of an asymmetric dimer with nonequivalent histidines must be induced not only by substrate, but by the reagent p-bromophenacyl bromide as well. Moreover, since no activity is regenerated following dilution of the inactive enzyme to monomer concentrations prior to assay, one must assume that unmodified monomers have a much greater affinity for alkylated monomers than for themselves. In the model put forth by Roberts et al.[50] one phospholipase monomer binds Ca^{++}, and then to a phospholipid at the interface. This binding causes a conformational change that leads to dimerization to form the asymmetric dimer with one catalytic subunit. If this would be generally true for all phospholipases, then it would constitute an alternative to the IRS hypothesis to explain the enzyme-interface interaction. A recent publication by Zhelkovskii et al.[51] reports that inactivation of the venom phospholipase from *N. naja oxiana* results from modification of 0.5 mole of aspartic acid per monomer chain, a finding that would serve to reinforce the half-site reactivity model. However, it must be stressed that the half-site reactivity reported for the *N. naja naja* phospholipase A_2 has not been observed for other Group I enzymes including the pancreatic phospholipases[52] (always seen as monomers) and two elapid venom enzymes, notexin, from *N. scutatus scutatus*[14] and the phospholipase from *N. nigricollis*,[53] all of which were inactivated by p-bromophenacyl bromide only when His-47 was *totally* alkylated. Similar results have been reported for the Group II phospholipase A_2 from *B. gabonica*.[54] At present, therefore, the question as to whether phospholipases A_2 are generally, or necessarily, active as monomers or dimers remains unresolved and the case for the dimer hypothesis rests upon evidence from studies of the *N. naja naja*[49,50] and crotalid phospholipases A_2.[23,24,47] The enzyme from *C. atrox* is dimeric in solution at concentrations as low as 10^{-10} M, is apparently active only as a dimer,[24] and crystallizes as a dimer in the asymmetric unit.[32] The complete tertiary structural analysis of this protein may reveal some important insights regarding the possible function of phospholipases A_2 as dimers.

The active site

Initially, it was believed by many that the phospholipases A_2 might be typical serine esterases and that their mechanism would, therefore, involve formation of an intermediate acyl enzyme. Although O-acyl cleavage during the course of hydrolysis has been documented by Wells,[55] no evidence for an acyl enzyme intermediate could be demonstrated. Organophosphorus compounds have generally been found to have no inhibitory effect on phospholipases A_2[47] and the observation that diisopropylphosphofluoridate inhibits the *C. atrox* venom enzyme[56] must be considered highly questionable. One of the strongest arguments against the direct participation of a serine in catalysis is based upon the fact that comparative analysis of the sequences of 20 phospholipases A_2 does not reveal the existence of an invariant serine residue (Fig. 3).

Histidine is always a likely candidate as a component of the machinery involved in hydrolytic catalysis and His-47 (Fig. 3) together with Asp-48 constitutes an invariant couplet present in *all* phospholipases A_2 sequenced to date, including that of the disparate honeybee enzyme (Fig. 5). In fact, alkylation of His-47 by p-bromophenacyl bromide leads to inactivation of phospholipases A_2. These findings, first reported by Volwerk et al.[52] in studies of porcine phospholipase A_2 and its zymogen, have been corroborated in studies of Group I phospholipases A_2 from *N. scutatus scutatus*[14] and *N. nigricollis*[53] venoms, and of a Group II enzyme from *B. gabonica*.[54] Thus it is clear that the C-terminal extension on the Group II phospholipases does not necessarily exert an inhibitory influence on the modification of His-47 by p-bromphenacyl bromide as we speculated earlier.[20] Since neither iodoacetate nor iodoacetamide reacts with His-47 it would seem that the modification by the analogous p-bromophenacyl bromide is facilitated in some way, perhaps by binding in a hydrophobic pocket. Both Ca^{++} ions and competitive inhibitors such as D-lecithin protect both the porcine enzyme and its zymogen from inactivation. These findings from the work of Volwerk, et al.[52] suggest that the molecular architectures of the porcine phospholipase A_2 and its zymogen are essentially the same and that the active site is largely preexistant within the zymogen.

Many chemical modification studies have been undertaken in attempts to assess the possible functional involvement of side chains other than that of His-47. The possible involvement of Trp-30 (numbering in Fig. 3) in the active site of *B. gabonica* was inferred from its selective modification during inactivation by N-bromosuccinimide.[57] Substrate protection studies suggested that Trp-30 is involved in substrate binding. We earlier[20] questioned this interpretation since position 30 is the only variable residue within a highly invariant sequence in phospholipases A_2 (residues 24 to 32, Fig. 3). Our suggestion that inactivation could have been due to a structural perturbation within this highly conserved sequence appears to have been borne out by X-ray crystallographic analysis[26] which indicates binding of the essential Ca^{++} to the peptide backbone in this region.

Comparative sequence analysis provides another approach to the identification of residues that are essential for the proper structure and function of proteins. When the number of sequences compared is large, as is the case for the phospholipases A_2, conservation of residues at particular positions

constitutes strong evidence for some kind of functional requirement and serves
as a basis for testing by chemical modification. Inspection of invariant residues among the nearly 20 phospholipase A_2 sequences determined to date (Fig. 3)
reveals two hydrophobic sidechains within the proposed IRS (Phe-4 and Ile-9),
Pro-36, and numerous residues of tyrosine (positions 24, 27, 51, 66, and 103),
glycine (residues 25, 29, 31, 32, and 34), aspartate (residues 38, 41, 48, and
89) and half-cystine (positions 26, 28, 43, 44, 50, 75, 86, 88, 95 and 115).
It was mentioned earlier that disulfide assignments were established largely
by crystallographic analysis[26] and it might be useful here to note the pairings
relative to the alignments in Fig. 3: 11-69, 26-115, 28-44, 43-95, 49-122, 50-88,
57-81, and 75-86. Invariance of half-cystine residues is the rule among homologous proteins. The high conservation of Gly and Pro residues could be due to
the requirement in proteins such as these which are highly constrained by disulfide cross-links of maintaining certain bend structures in which Gly and Pro
are prominent.

Tyrosyl and aspartyl residues could be functionally as well as structurally
essential. Tyrosine modifications performed with *C. adamanteus* phospholipase
A_2[47] did not appear to alter enzyme activity significantly. Recently, our
laboratory has attempted to assess the role of carboxyl groups in phospholipase
A_2 activity by chemical modification studies. Preliminary results indicate that
reaction of the enzyme from *C. atrox* venom with a carbodiimide-nucleophile
reagent under mild conditions leads to loss of activity. The enzyme was treated
with 1-ethyl-3-(3-dimethylamino-propyl)-carbodiimide·HCl (EDC) and taurine at
pH 6.5. Loss of activity was correlated with taurine incorporation and an
analysis of inactivation kinetics implied the presence of at least one essential
carboxyl group. These findings are consistent with chemical modification
studies which employed a carboxyl group-specific diazo compound.[58] Phospholipase A_2 from *Naja naja oxiana* venom was reacted with N-diazoacetyl-N'-(2,4-
dinitrophenyl)ethylenediamine, leading to complete loss of activity and incorporation of only one-half mole of substituent per mole of enzyme monomer.[51]
Aspartic acid was identified as the site of modification.

Chemical modification of the amino groups in phospholipases A_2 appears almost
always to be accompanied by inactivation of the enzyme; unfortunately the residues modified have usually not been identified. The alignments in Fig. 3 reveal
no lysyl residues invariant among all phospholipases A_2. The experiments of
Slotboom and deHaas[42] cited earlier demonstrated that all of the ε-amino groups
can be amidinated in an active derivative of porcine pancreatic phospholipase
A_2 as long as the α-amino group of Ala-1 is unaltered. Of course, amidination
preserves the charge in the vicinity of the ε-amino groups, so that their

structural or functional involvement may not be impaired. We have found that citraconylation of the C. atrox phospholipase A_2 inactivates the enzyme and that complete reactivation occurs upon exposure of the derivative to acidic conditions. This may offer a useful means of partially and reversibly opening up the native structure in probing the functional consequences of other modes of derivitization. Wells[47] reported inactivation of C. adamanteus phospholipase A_2 with ethoxyformic anhydride by modification of one amino group per dimer. His assumption that the reaction had occurred at an ε-amino group was based upon the incorrect notion that the α-amino group was blocked and therefore it may well be that inactivation followed acylation of Ser-1. Conservation of Lys or positive charge is observed at residues 11, 35, 42, 53, 60 and 90 in Group II enzymes. The inactivation of B. gabonica phospholipase A_2 by incorporation of one mole of pyridoxal phosphate per mole of enzyme[54] is especially interesting in this regard because four sites of modification, i.e., 11, 35, 60 and 114, were documented. Since these modified residues occur at positions that appear to be far removed from the catalytic site it must be that they each are important in perhaps different ways in maintaining enzyme structure or contact with the interface. At the present time, however, there exists no evidence for direct implication of an ε-amino group in the catalytic function.

Proposed mechanisms

Although several hypotheses have been put forth to explain the phospholipase A_2 mechanism, most have been based upon an incomplete or incorrect understanding of the residues involved directly in catalysis. Therefore, we consider here only those proposed recently through inspection of primary and tertiary structures and chemical modification data. The tertiary structural analysis of bovine pancreatic phospholipase A_2[26] shows that His-47 is in the middle of one of the long central α-helices and that it is in a cavity at the molecular surface. Two carboxyl groups, one from Asp-48 and one from Asp-89, point towards the imidazole side chain from above and below, respectively. Some electron density attributed to Ca^{++} is in the vicinity of His-47 and Asp-89, and the phenolic group of Tyr-51 is in contact with this density. All four residues mentioned above are invariant among structures given in Fig. 3, but only His-47 and Asp-48 are observed in the bee venom phospholipase A_2 (Fig. 5).

A mechanism based in part upon the structural analysis of porcine prophospholipase and in part upon consideration of known esterase mechanisms was suggested by Drenth et al.[25] in which Asp-48 is the nucleophile, His-47 serves a stabilizing role similar to that of the amide NH groups in the serine proteinases, and Tyr-27 is the proton donor. Although specific modified residues were not identified, Wells[47] showed that tyrosine modifications did not significantly alter the activity of C. *adamanteus* phospholipase A$_2$, and the replacement of Tyr-27 with a Trp residue in the honeybee enzyme would argue against direct participation of a phenolic side chain in catalysis. Drenth et al.[25] proposed that Arg-90 could be involved in binding the phosphate moiety of the substrate, a function that could be fulfilled by Lys-90 in the Group II enzymes and by His-109 in the bee venom phospholipase (Fig. 5). However, there is not strict conservation of a positive charge at this position (Fig. 3). As of now, none of these speculations is supported by evidence from chemical modification studies.

An intriguing mechanism proposed recently by Drenth[59] based upon the bovine pancreatic structure (Figs. 6 and 7) is that a His---Asp couple reminiscent of that seen in the serine proteases might exist in phospholipases between Asp-89 and His-47. In this model, Ser-195 is replaced by H$_2$O as the nucleophile and calcium ion is liganded through interactions with Asp-48 and the carbonyl groups of the peptide bonds joining residues 30 through 32. A calcium binding site so defined would correlate nicely with the observed inactivation by modification of Trp-30[57] and the spectral perturbations of tryptophan accompanying Ca^{++} binding to C. *adamanteus* phospholipase A$_2$.[60] However, it is noteworthy that in an earlier study,[26] the Ca^{++}-ion was reported to be bound in the vicinity of Asp-89. Limited comparison of the bovine phospholipase A$_2$ sequence in this vicinity with the sequences of Ca^{++} ion binding sites in other proteins indicates similarities. For example, the sequence Asp-Arg-Asn-Ala (residues 89-92) in bovine phospholipase A$_2$ is identical to a tetrapeptide region in Ca^{++} binding site III in rabbit skeletal muscle troponin C[61] and nearly identical to the sequence Asp-Lys-Asn-Ala in site III of bovine cardiac troponin C.[62] Since, at the time of this writing the various catalytic components, including the enzyme side-chains, substrate and Ca^{++} have not been placed definitively with respect to one another, it is an intriguing possibility that Asp-89 not Asp-48 is the site of Ca^{++} binding. Although arguments based upon entropic considerations might be raised against water as the nucleophile, it could well be that the H$_2$O molecule involved is ordered through interactions with the nearby Ca^{++}-Asp-48 complex. In any event, these various proposed mechanisms provide interesting subjects of investigation for evaluation by independent experimental approaches.

CONCLUDING REMARKS

It is clear that, despite an abundance of information concerning the primary and tertiary structures of a considerable number of phospholipases A$_2$, much remains to be done in clarifying the enzyme mechanism. X-ray crystallographic analysis[64] has provided a structural framework for interpreting results from solution studies and, in general, a satisfying agreement exists between the structural model and functional predictions based upon kinetic analysis and chemical modification. Nevertheless, it would be very useful to examine models derived from crystallographic analysis of enzyme complexes with substrates or inhibitors. Assessment should be made of the possible participation in catalysis of aspartyl residues either by general carboxyl-labeling methods or by the development of chromogenic substrates[48] and/or poor substrates that might permit study of intermediate steps or kinetic analysis at low temperatures. Elucidation of the enzyme mechanism is certainly a top priority area of research.

The question must be resolved as to whether the requirement of a dimeric structure in the activity of certain phospholipases A_2 is 1) correct, 2) applies to just these select proteins, or 3) is a property of phospholipases A_2 in general. Unifying concepts are always sought after in science and in the case of these enzymes it is tempting to suppose that, because of the strong structural homology which they all share, they will all function essentially the same. Yet, it is known that phospholipases A_2 may or may not have toxic or anti-coagulant activities in addition to their esterase function. This multiple activity is, at present, impossible to predict based upon comparative sequence analysis and yet these functions must be coded within the primary structure. And so it is not unreasonable to imagine that different phospholipases A_2 will provide unique solutions to the ultimate confrontation with the interface whether it be through an interfacial recognition site or via the generation of an asymmetric dimer with a binding subunit.

ACKNOWLEDGMENTS

This work is supported in part by Grant PCM 75-23506 AOI from the National Science Foundation.

REFERENCES

1. Shen, B.W. and Law, J.H. (1979) in The Biochemistry of Atherosclerosis, Scanu, A.M., Wissler, R.W., and Getz, G.S., eds., Marcel Dekker & Co., New York, p. 275.
2. Dawson, R.M.C. (1973) in Form and Function of Phospholipids, Ansell, G.B., Hawthorne, J.N., and Dawson, R.M.C., eds., Elsevier, Amsterdam, p. 97.
3. Aggerbeck, L.P., Kézdy, F.J., and Scanu, A.M. (1976) J. Biol. Chem., 251, 3283.
4. Kunge, H., and Vogt, W. (1972) Ann. N.Y. Acad. Sci., 180, 123.
5. Schoene, M.W., and Iacono, J.M. (1976) in Advances in Prostaglandin and Thromboxane Research, Vol. 2, Samuelsson, B. and Paoletti, R., eds., Raven Press, New York, p. 763.
6. Bonsen, P.P.M., deHaas, G.H., et al. (1972) Biochim. Biophys. Acta, 270, 364.

7. deHaas, G.H., Slotboom, A.J., et al. (1970) Biochim. Biophys. Acta, 221, 31.
8. Evenberg, A., Meyer, H., et al. (1977) J. Biol. Chem., 252, 1189.
9. Fleer, E.A.M., Verheij, H.M., and deHaas, G.H. (1978) Eur. J. Biochem., 82, 261.
10. Shipolini, R.A., Callewaert, G.L., et al. (1971) FEBS Lett., 17, 39.
11. Shipolini, R.A., Callewaert, G.L., et al. (1974) Eur. J. Biochem., 48, 465.
12. Joubert, F.J. (1975) Biochim. Biophys. Acta, 379, 329.
13. Joubert, F.J. (1975) Biochim. Biophys. Acta, 379, 345.
14. Eaker, D. (1975) in Peptides: Chemistry, Structure, Biology, Walter, R., and Meienhofer, J., eds., Ann Arbor Science Publishers, Inc., Ann Arbor, p. 17.
15. Joubert, F.J. (1975) J. Biochem., 52, 539.
16. Halpert, J., and Eaker, D. (1975) J. Biol. Chem., 250, 6990.
17. Botes, D.P., and Viljoen, C.C. (1974) J. Biol. Chem., 249, 3827.
18. Samejima, Y., Iwanaga, S., and Suzuki, T. (1974) FEBS Lett., 47, 348.
19. Tsao, F.H.C., Keim, P.S., and Heinrikson, R.L. (1975) Arch. Biochem. Biophys., 167, 706.
20. Heinrikson, R.L., Krueger, E.T., and Keim, P.S. (1977) J. Biol. Chem., 252, 4913.
21. Braganca, B.M., Sambray, Y.M., and Sambray, R.Y. (1970) Eur. J. Biochem., 13, 410.
22. Vidal, J.C., and Stoppani, A.O.M. (1971) Arch. Biochem. Biophys., 147, 66.
23. Wells, M.A. (1971) Biochemistry, 10, 4074.
24. Shen, B.W., Tsao, F.H.C., et al. (1975) J. Am. Chem. Soc., 97, 1205.
25. Drenth, J., Enzing, C.M., et al. (1976) Nature, 264, 373.
26. Dijkstra, B.W., Drenth, J., et al. (1978) J. Mol. Biol., 124, 53.
27. Wells, M.A., and Hanahan, D.J. (1969) Biochemistry, 8, 414.
28. Tamiya, N., Nishida, S., and Kim, H.S. (1980) In Frontiers in Protein Chemistry, Liu, D.T., and Yasunobu, K.Y., eds., Elsevier-North Holland, New York, this volume.
29. Keim, P.S., Wells, M.A., and Heinrikson, R.L., unpublished observations.
30. Brattstrom, B.H. (1964) Trans. San Diego Soc. Nat. Hist., 13, 185.
31. Puijck, W.C., Verheij, H.M., and deHaas, G.H. (1977) Biochim. Biophys. Acta, 492, 254.
32. Keith, C.H. (1979) Ph.D. Thesis, University of Chicago.
33. deHaas, G.H., Slotboom, A.J., and Verheij, H.M. (1977) in Cholesterol Metabolism, Lipolytic Enzymes, Proc. Int. Cong. Biochem. Lipids, 19th, p. 191.
34. Bonsen, P.P.M., deHaas, G.H., and van Deenen, L.L.M. (1972) Biochim. Biophys. Acta, 270, 364.
35. Pieterson, W.A., Vidal, J.C., et al. (1974) Biochemistry, 13, 1455.
36. Brockerhoff, H., and Jansen, R.G. (1974) in Lipolytic Enzymes, Academic Press, New York, p. 3.
37. Brockman, H.L., Law, J.H., and Kézdy, F.J. (1973) J. Biol. Chem., 248, 4965.
38. Sarda, L., and Desnuelle, P. (1958) Biochim. Biophys. Acta, 30, 513.
39. Chapus, C., Sémériva, M., et al. (1975) FEBS Lett., 58, 155.
40. Pattus, F., Slotboom, A.J., and deHaas, G.H. (1979) Biochemistry, 18, 2698.
41. Slotboom, A.J., Jansen, E.H.J.M., et al. (1978) Biochemistry, 17, 4593.
42. Slotboom, A.J., and deHaas, G.H. (1975) Biochemistry, 14, 5394.
43. Slotboom, A.J., Verheij, H.M., et al. (1978) FEBS Lett., 92, 361.
44. Pattus, F., Slotboom, A.J., and deHaas, G.H. (1979) Biochemistry, 18, 2703.
45. Huang, K.-S., and Law, J.H. (1978) in Enzymes of Lipid Metabolism, Gatt, S., Freysz, L. and Mandel, P., eds., Planum Publishing Corp., p. 177.
46. van Wezel, F.M., Slotboom, A.J., and deHaas, G.H. (1976) Biochim. Biophys. Acta, 452, 101.
47. Wells, M.A. (1973) Biochemistry, 12, 1086.
48. Volwerk, J.J. (1979) Ph.D. Thesis, State University of Utrecht.

322

49. Roberts, M.F., Deems, R.A., et al. (1977) J. Biol. Chem., 252, 2405.
50. Roberts, M.F., Deems, R.A., and Dennis, E.A. (1977) Proc. Natl. Acad. Sci. USA, 74, 1950.
51. Zhelkovskii, A.M., D'yakov, V.L., et al. (1978) Bioorg. Khim., 4(12), 1665.
52. Volwerk, J.J., Pieterson, W.A., and deHaas, G.H. (1974) Biochemistry, 13, 1446.
53. Yang, C.C., and King, K. (1979) Int. Cong. Biochem., Montreal, Canada, abstract 03-4-S38, p. 185.
54. Viljoen, C.C., Visser, L., and Botes, D.P. (1977) Biochim. Biophys. Acta, 483, 107.
55. Wells, M.A. (1971) Biochim. Biophys. Acta, 248, 80.
56. Wu, T.W., and Tinker, D.O. (1969) Biochemistry, 8, 1558.
57. Viljoen, C.C., Visser, L., and Botes, D.P. (1976) Biochim. Biophys. Acta, 428, 424.
58. Zhelkovskii, A.M., Apsalon, U.R., et al. (1977) Bioorg. Khim., 3(10), 1430.
59. Drenth, J., personal communication.
60. Wells, M.A. (1973) Biochemistry, 12, 1080.
61. Collins, S.M., Potter, J.G., et al. (1973) FEBS Lett., 36, 368.
62. van Eerd, J.P., and Takahashi, K. (1976) Biochemistry, 15, 1171.
63. Dayhoff, M. (1979) in Atlas of Protein Sequence and Structure 5, Suppl. 3, National Biomedical Research Foundation, Georgetown University Medical Center, Washington, D.C.
64. Dijkstra, B.W. (1980) Ph.D. Thesis, University of Groningen, The Netherlands.

CHICKEN MUSCLE MYOSIN LIGHT CHAINS

GENJI MATSUDA, TETSUO MAITA, TOSHIYA UMEGANE AND YUKIO KATA
Department of Biochemistry, Nagasaki University School of Medicine, Nagasaki
852, Japan

INTRODUCTION

Myosin, which is the major component of thick filaments of muscle consists of two heavy chains with molecular weights of about 200,000 and four light chains with molecular weights of about 20,000.[1-6] It is assumed that ATPase activity of actomyosin complex play a major role in muscle contraction. Since the light chains in the myosin were first recognized by Tsao[7] in 1953, many researchers have been studying them. The results of our recent studies on the primary structures of light chains of chicken muscle myosin are presented in this paper.

LIGHT CHAINS OF CHICKEN SKELETAL MUSCLE MYOSIN

Isolation of the L-1, L-2, L-3 and L-4 light chains. There are four components in the light chain fraction in chicken skeletal muscle of myosin, which migrate toward the anode in Cellogel electrophoresis at pH 8.3 (Fig. 1). They are designated light chains L-1, L-2, L-3 and L-4 in the order of their increasing anionic mobilities. Light chains L-1 and L-4 are also called alkali light chain (A1) and alkali light chain 2 (A2), respectively, and the light chains L-2 and L-3, the DTNB light chains. L-3 is a phosphorylated form of light chain L-2. The four components in the light chain fraction were isolated by column chromatography using DE-52 as shown in Fig. 2. The elution was carried out with a gradient between 0.035 M KCl-4M urea-25 mM Tris buffer (pH 7.6) and 0.300 M KCl-4 M urea-25 mM Tris buffer (pH 7.0). The eluted fractions were collected separately and lyophilized.

Amino acid sequence of light chain L-1. The carboxymethylated light chain L-1 was digested with trypsin at 37°C, pH 8.5 for 3 hours. The tryptic peptides thus obtained were fractionated into soluble and insoluble fractions by use of 0.05 M pyridine-acetate buffer (pH 3.1). These peptides were further purified by gel filtration using Sephadex G-50, column chromatography on Chromo Beads P and DE-52, paper chromatography, and paper electrophoresis prior to amino acid sequence determination. On the other hand, carboxymethylated light chain L-1 was digested with pepsin at 37°C, pH 2.0 for 2 hours. The peptic peptides, thus obtained, were purified in almost the same manner, i.e., by gel filtration on

324

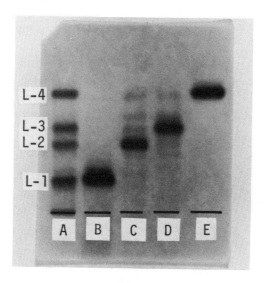

Fig. 1. Cellogel electrophoresis of the light chains of chicken skeletal muscle myosin. The electrophoresis was carried out on a strip (5 x 6 cm) of Cellogel (Chemoteron Co.) at 200 V for 40 mins using 0.02 M Tris-0.15 M glycine buffer, pH 8.3, containing 8 M urea and 2 mM 2-mercaptoethanol. After electrophoresis, the light chains were stained with amino black. The symbols used are: A, light chain fractions; B, light chain L-1; C, light chain L-2; D, light chain L-3 and E, light chain L-4.

Sephadex G-50, column chromatography on Chromo Beads P, paper chromatography and paper electrophoresis prior to amino acid sequence determination. The amino acid sequence of the chicken skeletal muscle myosin light chain L-1 determined here is shown in Fig. 3.

Both the sequences of the chicken and rabbit skeletal muscle myosin light chains L-1, the latter which has already been reported by Frank and Weeds,[8] show that both light chains contain 190 amino acids. Both light chains have blocked N-terminal proline residues and there are 29 amino acid substitutions when the sequences are compared. The physiological function of the light chain L-1 is not clearly understood. However, it has been suggested that light chain L-1 plays a role in the interaction of myosin and ATP.[9-11] The first 33 residues

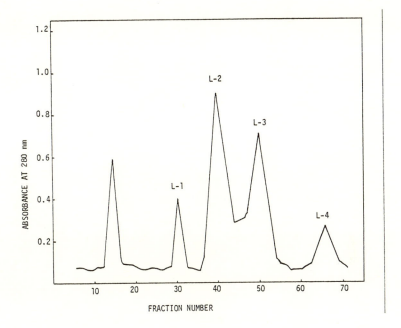

Fig. 2. Column chromatography of the four light chains of chicken skeletal muscle myosin. First, unabsorbed protein and then L-1, L-2, L-3 and L-4 were successively eluted. Conditions used were: column, DE-52, 3.6 x 30 cm; flow rate, 100 ml/hour; and the fraction volume, 12 ml.

from the N-terminal of the light chains are unusually rich in alanine and proline, but the biological significance of this region is not yet well understood.

Amino acid sequence of light chain L-2. The amino acid sequence of the rabbit skeletal muscle myosin light chain L-2 was determined both by Collins[12] and Matsuda et al.[13] Subsequently, Matsuda et al. determined the amino acid sequence of chicken skeletal muscle myosin and reported the sequence in a preliminary communication.[14]

The light chain L-2 of chicken skeletal muscle myosin was digested with trypsin at 37°C, pH 8.0, for 3 hours after it was first carboxymethylated. The tryptic peptides thus obtained were fractionated into soluble and insoluble fractions by the use of 0.05 M pyridine-acetate buffer (pH 3.2). The peptides were further purified by gel filtrations using Sephadex G-50 and Sephadex G-25, column chromatography on Chromo Beads P and DE-52, paper chromatography and paper electrophoresis as reported earlier for the purification of the tryptic peptides of the light chain L-1. The amino acid sequences of the tryptic peptides thus isolated were determined and were placed by sequence homology with the reported sequence of the rabbit skeletal muscle light chain L-2 as shown in Fig. 4.

```
  1                               10                              20
X-Pro-Lys-Lys-Asn-Val-Lys-Lys-Pro-Ala-Ala-Ala-Ala-Ala-Pro-Ala-Pro-Lys-Ala-Pro-Ala-


                                  30                              40
    Pro-Ala-Pro-Ala-Pro-Ala-Pro-Ala-Pro-Lys-Glu-Pro-Ala-Ile-Asp-Leu-Lys-Ser-Ile-Lys-
                                     (Glu-Lys)          (Ser-Ala)


                                  50                              60
    Ile-Glu-Phe-Ser-Lys-Glu-Gln-Gln-Asp-Asp-Phe-Lys-Glu-Ala-Phe-Leu-Leu-Phe-Asp-Arg-
                                  (Glu)                         (Tyr)


                                  70                              80
    Thr-Gly-Asp-Ala-Lys-Ile-Thr-Leu-Ser-Gln-Val-Gly-Asp-Ile-Val-Arg-Ala-Leu-Gly-Glu-
             (Ser)                              (Val-Leu)                    (Thr)


                                  90                              100
    Asn-Pro-Thr-Asn-Ala-Glu-Ile-Asn-Lys-Ile-Leu-Gly-Asn-Pro-Ser-Lys-Glu-Glu-Met-Asn-
                 (Val-Lys)   (Val)                      (Asp)   (Gln)


                                  110                             120
    Ala-Lys-Lys-Ile-Thr-Phe-Glu-Gln-Phe-Leu-Pro-Met-Leu-Gln-Ala-Ala-Ala-Asn-Asn-Lys-
             (Glu)                                          (Ile-Ser)


                                  130                             140
    Asp-Gln-Gly-Thr-Phe-Glu-Asp-Phe-Val-Glu-Gly-Leu-Arg-Val-Phe-Asp-Lys-Glu-Gly-Asp-
                 (Tyr)                                                  (Asp-Gly-


                                  150                             160
    Gly-Thr-Val-Met-Gly-Ala-Glu-Leu-Arg-His-Val-Leu-Ala-Thr-Leu-Gly-Glu-Lys-Met-Thr-
    Thr-Val-Gly)                                                            (Lys)


                                  170                             180
    Glu-Glu-Glu-Val-Glu-Gln-Leu-Met-Lys-Gly-Gln-Glu-Asp-Ser-Asn-Gly-Cys-Ile-Asn-Tyr-
             (Ala)            (Ala)


                                  190
    Glu-Ala-Phe-Val-Lys-His-Ile-Met-Val-Ser
                                 (Ser-Ile)
```

Fig. 3. The amino acid sequence of the L-1 light chain of chicken skeletal muscle myosin. Parenthesized amino acid residues are the amino acids present in the case of light chain L-1 of rabbit skeletal muscle myosin when compared with the L-1 chain of chicken skeletal muscle myosin.

Light chain L-2 of chicken skeletal muscle myosin consisted of 166 amino acids. When this sequence was compared with the light chain L-2 from rabbit skeletal muscle myosin, there were two amino acid deletions and 14 amino acid substitutions.[13] The proline residues in the N-terminal positions of both proteins are blocked.

Since it was recognized that one of the myosin light chains of molluscan muscle might be concerned with the regulation of muscle contraction.[15] the Ca^{2+} binding ability of the L-2 light chain of vertebrate striated muscle myosin has been under scrutiny.[16,17] From the sequence data, Collins[12] suggested that the light chain L-2 of rabbit skeletal muscle myosin might be able to bind with a g atom of Ca^{2+}. Similarly, in the case of the light chain L-2 of chicken

```
 1                                      10                                    20
X-Pro-Lys-Lys-Ala-Lys-Arg-Arg-Ala-Ala-Glu-Gly-Ser-Ser-Asn-Val-Phe-Ser-Met-Phe-Asp-
                               (Ala)   (Gly) Ⓟ


                                       30                                    40
Gln-Thr-Gln-Ile-Gln-Glu-Phe-Lys-Glu-Ala-Phe-Thr-Val-Ile-Asp-Gln-Asn-Arg-Asp-Gly-


                                       50                                    60
Ile-Ile-Asp-Lys-Asp-Asp-Leu-Arg-Glu-Thr-Phe-Ala-Ala-Met-Gly-Arg-Leu-Asn-Val-Lys-
        (Glu)              (Asp)


                                       70                                    80
Asn-Glu-Glu-Leu-Asp-Ala-Met-Ile-Lys-Glu-Ala-Ser-Gly-Pro-Ile-Asn-Phe-Thr-Val-Phe-
                        (Met)


                                       90                                   100
Leu-Thr-Met-Phe-Gly-Glu-Lys-Leu-Lys-Gly-Ala-Asp-Pro-Glu-Asp-Val-Ile-Met-Gly-Ala-
                                                                          (Thr)


                                      110                                   120
Phe-Lys-Val-Leu-Asp-Pro-Asp-Gly-Lys-Gly-Ser-Ile-Lys-Lys-Ser-Phe-Leu-Glu-Glu-Leu-
            (Glu)                     (Thr)              (Gln)


                                      130                                   140
Leu-Thr-Thr-Gln-Cys-Asp-Arg-Phe-Thr-Pro-Glu-Glu-Ile-Lys-Asn-Met-Trp-Ala-Ala-Phe-
                        (Ser-Gln)


                                      150                                   160
Pro-Pro-Asp-Val-Ala-Gly-Asn-Val-Asp-Tyr-Lys-Asn-Ile-Cys-Tyr-Val-Ile-Thr-His-Gly-
                        (Gly)


            166
Glu-Asp-Lys-Glu-Gly-Glu
(Asp-Ala)   (Asp-Gln)
```

Fig. 4. The amino acid sequence of the light chain L-2 of chicken skeletal muscle myosin. Parenthesized amino acid residues are inserted (arrows) or substituted amino acids in the light chain L-2 of rabbit skeletal muscle myosin are compared with the chicken skeletal L-2 chain.

skeletal muscle myosin, the Ca^{2+} binding site was assumed to involve aspartic acid residues 35 and 46 from the N-terminus.[14]

Light chain L-3 of rabbit skeletal myosin is the phosphorylated form of light chain L-2.[18] Light chain L-3 in chicken skeletal muscle myosin is also the phosphorylated light chain L-2. It has been proposed that the serine residue 13 from N-terminal of the light chain L-2 of chicken skeletal muscle myosin is phosphorylated.[14] Moreover, myosin light chain kinase, which phosphorylates specifically this light chain has been reported.[19-21]

Amino acid sequence of light chain L-4. The light chain L-4 of chicken skeletal muscle myosin was digested with trypsin at 37°C, pH 8.0, for 3 hours after the protein had been oxidized with performic acid. The tryptic peptides thus obtained were fractionated into soluble and insoluble fractions by use of 0.05 M pyridine-acetate buffer (pH 3.1), and were further purified by gel filtration using Sephadex G-50, column chromatography on Chromo Beads P and DE-52, paper chromatography and paper electrophoresis, prior to the amino acid sequence determination of the peptides. The amino acid sequence of chicken skeletal muscle myosin light chain L-4 in the present study is shown in Fig. 5 and the tryptic peptides were placed by taking advantage of sequence homology with the light chain L-4 of rabbit skeletal muscle myosin.[8]

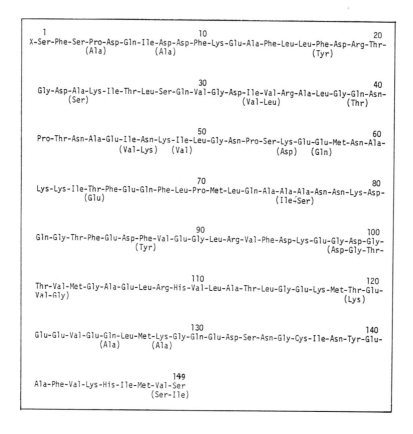

Fig. 5. The amino acid sequence of the light chain L-4 of chicken skeletal muscle myosin. Parenthesized amino acid residues are the amino acids which are substituted in the case of light chain L-4 of rabbit skeletal muscle myosin when compared with the L-4 chain of chicken skeletal muscle.

TABLE 1

AMINO ACID SUBSTITUTIONS OBSERVED AMONG THE

VARIOUS PROTEINS FROM RABBIT AND CHICKEN

Proteins	Total amino acids	Numbers of amino acid[b] substitutions	Number of amino acid substitutions (%)
cytochrome	104	7	6.7%
L-2 light chain of skeletal muscle myosin	168	16	9.5%
L-1 light chain of skeletal muscle myosin	190	29	15.3%
L-4 light chain of skeletal muscle myosin	159	26	16.4%
myoglobin	153	34	22.2%
hemoglobin α^*	141	43	30.5%
hemoglobin β	146	48	32.9%

[a] In the figure, hemoglobin a stands for the AII component of chicken hemoglobin.
[b] In this table, amino acid insertions or deletions are considered to be amino acid substitutions.

When the sequences of the chicken and rabbit skeletal muscle myosin light chains L-4 were compared, both chains contained 149 amino acid residues; the N-terminal residue of both chains were blocked, and there were 26 amino acid sequence differences between the two chains. A comparison of the chicken light chains L-4 with L-1 disclosed that 142 amino acid sequences from the C-terminal of the two chains were identical. However, there were 4 amino acid substitutions in the 7 amino acid segment beyond residue 142 when the L-1 and L-4 chains are aligned starting from the COOH terminal residues. The physiological function of the light chain L-4 is not clearly recognized, but it has been suggested that light chain L-4 like L-1, is concerned with the interaction of myosin with ATP.[9-11] Also, light chain L-4 is present only in fast muscles and not in slow muscles. It should be noted that the ATPase activity is relatively low in slow muscles at least in the case of higher vertebrates. Thus, the difference of the ATPase activity of fast and slow muscle may be due to their content of light chains L-1 and L-4.

Table 1 shows the number of amino acid substitutions observed when various homologous proteins from rabbit and chicken are compared. The number of amino acid substitutions calculated when the various light chains of rabbit and chicken skeletal muscle myosins are compared are greater than when similar calculations are made in the case of cytochrome c, but are less when similar calculations are made for myoglobin and the hemoglobin α and β chains. Furthermore, it is concluded that light chains L-1, L-2 and L-4 have homologous primary structures and they have evolved from a common ancestor.

LIGHT CHAINS OF CHICKEN CARDIAC MUSCLE MYOSIN

Isolation of the light chains L-1, L-2 and L-3. There are three components which migrate toward the anode (as shown in Fig. 6) during Cellogel electrophoresis of the light chain fraction of chicken cardiac muscle myosin at pH 8.3. They are designated L-1, L-2 and L-3 in the order of their increasing anionic mobilities. The amount of L-3 in this experiment, contrary to the case of the skeletal fast muscle myosin, was very small. No light chain L-4 was present in cardiac muscle myosin.

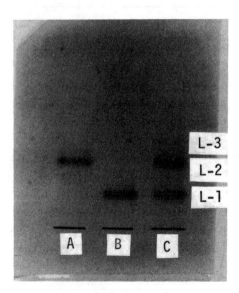

Fig. 6. Cellogel electrophoresis of the light chains of chicken cardiac muscle myosin. Electrophoresis was carried out at 200 V for 45 mins. The conditions are almost the same as reported in the legend to Fig. 1. The symbols used were: A, light chain L-2; B, light chain L-1 and C, light chain fraction.

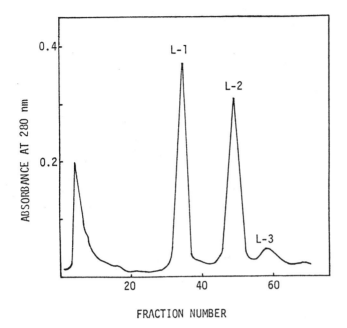

Fig. 7. Column chromatography of the three light chains of cardiac muscle myosin. First unabsorbed protein and then subunits L-1, L-2 and L-3 were successively eluted. The conditions, except for the elution buffer, are almost the same as reported in the legend for Fig. 2.

Of the cardiac myosin components, light chain L-1 corresponds to alkali light chain 1 (A1) of skeletal muscle myosin, and L-2 and L-3 to the DTNB light chain. The L-3 is a phosphorylated form of light chain L-2. The three components were isolated by column chromatography on DE-52 as shown in Fig. 7. The elution was carried out with a gradient between 0.035 M KCl-4M urea-50 mM 2-mercaptoethanol-25 mM Tris buffer (pH 7.6) and 0.300 M KCL-4 M urea-50 mM 2-mercaptoethanol-25 mM Tris buffer (pH 7.0). Each eluted fraction was collected and lyophilized.

Amino acid sequence of light chain L-1. The performic acid oxidized light chain L-1 of chicken cardiac muscle myosin was digested with trypsin at 37°C, pH 8.0, for 4 hours. The tryptic peptides thus obtained were all soluble. They were further purified by column chromatography on Chromo Beads P and then AG 1x2; gel filtration using Sephadex G-50; paper chromatography; and paper electrophoresis prior to amino acid sequence analysis. Also, carboxymethylated light chain L-1 was digested with pepsin at 37°C, pH 2.0, for 2 hours. The peptic peptides, thus obtained, were isolated by gel filtration using Sephadex G-50,

```
                              10                                   20
X-Pro-Lys-Lys-Pro-Glu-Pro-Lys-Lys-Ala-Pro-Glu-Pro-Lys-Lys-Glu-Glu-Pro-Lys-Pro-Ala-
          (Asn-Val-Lys)    (Pro)    (Ala-Ala-Ala-Ala-Pro-Ala-Pro-Lys-Ala)

                              30                                   40
Pro-Lys-Pro-Ala-Glu-Pro-Glu-Pro-Lys-Lys-Glu-Val-Glu-Phe-Asn-Pro-Ala-Ser-Ile-Lys-
     (Ala)          (Pro-Ala-Pro-Ala-Pro)      (Pro-Ala-Ile-Asp-Leu-Lys)

                              50                                   60
Val-Glu-Phe-Thr-Pro-Asp-Gln-Ile-Glu-Glu-Phe-Lys-Glu-Ala-Phe-Ser-Leu-Phe-Asp-Arg-
(Ile)      (Ser-Lys-Glu)     (Gln-Asp)                     (Leu)

                              70                                   80
Thr-Pro-Lys-Ser-Glu-Met-Lys-Ile-Thr-Tyr-Ala-Gln-Cys-Gly-Asp-Val-Leu-Arg-Ala-Leu-
(Gly-Asp-Ala)(—) (—)          (Leu-Ser)    (Val)     (Ile-Val)

                              90                                   100
Gly-Gln-Asn-Pro-Thr-Gln-Ala-Glu-Val-Met-Lys-Val-Leu-Gly-Arg-Pro-Lys-Gln-Glu-Glu-
     (Glu)          (Asn)          (Ile-Asn)  (Ile)     (Asn)   (Ser-Lys)

                              110                                  120
Met-Asn-Ser-Lys-Met-Ile-Asp-Phe-Glu-Thr-Phe-Leu-Pro-Met-Leu-Gln-His-Ile-Ser-Lys-
     (Ala)     (Lys)    (Thr)    (Gln)                         (Ala-Ala-Asn-Asn-

                              130                                  140
Thr-Lys-Asp-Thr-Gly-Thr-Tyr-Glu-Asp-Phe-Val-Glu-Gly-Leu-Arg-Val-Phe-Asp-Lys-Glu-
Asn)      (Gln)     (Phe)

                              150                                  160
Gly-Asp-Gly-Thr-Val-Met-Gly-Ala-Glu-Leu-Arg-His-Val-Leu-Ala-Thr-Leu-Gly-Glu-Arg-
                                                                       (Lys-

                              170                                  180
Leu-Thr-Glu-Glu-Glu-Val-Asp-Lys-Leu-Met-Ala-Gly-Gln-Glu-Asp-Ala-Asn-Gly-Cys-Ile-
Met)              (Glu-Gln)         (Lys)              (Ser)

                    190    192
Asn-Tyr-Glu-Ala-Phe-Val-Lys-His-Ile-Met-Ala-Asn
                    (Val-Ser)
```

Fig. 8. The amino acid sequence of the light chain of chicken cardiac muscle myosin. Amino acid residues parenthesized are the amino acids substituted or deleted in the case of light chain L-1 of chicken skeletal muscle myosin when compared with the cardiac light chain L-1.

column chromatography on Chromo Beads P, paper chromatography and paper electrophoresis prior to amino acid sequence determination. From the results obtained, the amino acid sequence of the chicken cardiac muscle myosin light chain L-1 was determined and is shown in Fig. 8.

A comparison of this amino acid sequence with the light chain L-1 of chicken skeletal muscle myosin indicated the need for two amino acid insertions. The heart light chain L-1 contained 192 amino acids. There were more differences than expected between the two L-1 light chains, especially in the N-terminal region where the alanine and proline contents are unusually high. The ATPase activity of cardiac muscle myosin is reported to be about one-third that of skeletal muscle myosin. Furthermore, the difference in the two ATPase activities

TABLE 2

AMINO ACID SUBSTITUTIONS BETWEEN THE L-1 LIGHT CHAINS OF SKELETAL
AND CARDIAC MUSCLE MYOSIN FROM RABBIT AND CHICKEN

	Total amino acids	Numbers of amino acid[a] substitutions	Number of amino acid substitutions (%)
L-1 light chains from chicken skeletal and cardiac muscle myosins	192	69	35.9%
L-1 light chains from rabbit and chicken skeletal muscle myosins	190	29	15.3%

[a]In this table, amino acid insertions or deletions are classified as amino acid substitutions.

is suspected to come from the difference in the structures of the light chains
which are present in the two muscle myosins.[22] Therefore, the differences in
the sequences of the two kinds of light chain L-1 are of importance.

Table 2 shows the amino acid sequence differences between the L-1 light
chains from chicken skeletal and chicken cardiac muscle myosins (35.9%) and the
number of sequence differences between the L-1 light chains from chicken skeletal and rabbit skeletal muscle myosins (15.3%). The number of sequence differences of skeletal muscle and cardiac muscle myosins from a single animal
specie was greater than was observed for myosin light chain L-1 isolated from
a specific organ or tissue from the rabbit or chicken. This remarkable difference in the primary structures of the L-1 light chains of chicken skeletal
and chicken cardiac myosins may be related to the physiological functions of
these two types of myosin.

REFERENCES

1. Tonomura, Y., Appel, P. and Morales, M. (1966) Biochemistry, 5, 515-521.
2. Gershman, L.C., Stracher, A. and Dreizen, P. (1969) J. Biol. Chem., 244,
   2726-2736.
3. Gazith, J., Himmelfarb, S.H. and Harrington, W.F. (1970) J. Biol. Chem.,
   245, 15-22.
4. Weeds, A.G. and Lowey, S. (1971) J. Biol. Chem., 61, 701-725.
5. Hayashi, Y. (1972) J. Biochem., 72, 83-100.
6. Yagi, K., Okamoto, Y. and Yazawa, Y. (1975) J. Biochem., 77, 333-342.
7. Tsao, T.C. (1953) Biochim. Biophys. Acta., 11, 368-382.
8. Frank, F. and Weeds, A.G. (1974) Euro. J. Biochem., 44, 317-334.
9. Dreizen, P. and Gershman, L.C. (1970) Biochemistry, 9, 1688-1693.
10. Dreizen, P. and Richards, D.H. (1972) Cold Springs Harb. Symp. Quant. Biol.,
    37, 29-45.

11. Wagner, P.D. and Yount, R.G.   (1975) Biochemistry, 14, 1908-1914.
12. Collins, J.H.   (1976) Nature, 259, 699-700.
13. Matsuda, G., Maita, T., et al.   (1977) J. Biochem., 81, 809-811.
14. Matsuda, G., Suzuyama, Y., et al.   (1977) FEBS Letters, 84, 53-56.
15. Kendrick-Jones, T., Szentkiralyi, E.M. and Szent-Grörgyi, A.G.   (1972) Cold Spring Harb. Symp. Quant. Biol., 37, 47-53 (1972).
16. Werber, M.M. and Oplatka, A.   (1974) Biochem. Biophys. Res. Commun., 57, 823-830
17. Morimoto, K. and Harrington, W.F.   (1974) J. Mol. Biol., 88, 693-709.
18. Perrie, W.T., Smillie, L.B. and Perry, S.V.   (1973) Biochem. J., 135, 151-164.
19. Pires, E., Perry, S.V. and Thomas, M.A.W.   (1974) FEBS Letters, 41, 292-296.
20. Yazawa, M., and Yagi, K.   (1977) J. Biochem., 82, 287-289.
21. Yagi, K., Yazawa, M., et al.   (1978) J. Biol. Chem., 253, 1338-1340.
22. Fredricksen, D.W. and Holtzen, A.   (1968) Biochemistry, 7, 3935-3950.

THE SUBUNIT STRUCTURE OF JACK BEAN UREASE

GUNJI MAMIYA, KUNIO TAKISHIMA AND TSUNEJIRO SEKITA[+]
Department of Biochemistry, National Defense Medical College, Tokorozawa,
Saitama, Japan; [+]Clinical Chemistry Laboratory, Kawasaki Municipal Hospital,
Kawasaki, Japan

ABSTRACT

Cyanogen bromide cleavage of urease liberated at least five moles of identical
peptides from one mole urease and cleavage occurred at least after thirteen
different methionine residues. These results suggested the presence of a
minimum of five and a maximum of nine subunits in urease. From one mole of
urease, 38-40 sulfhydryl groups reacted with p-chloromercuribenzoate and N-
ethylmaleimide, and 37 S-methylcysteine residues were present after reaction
with methyliodide. From one mole of reduced urease, 69-70 S-alkylated cysteine
residues were detected after the reaction with methyl-p-nitrobenzenesulfonate
and iodoacetic acid.

INTRODUCTION

Urease is a historically important enzyme because it was the first enzyme to
be crystallized in 1926.[1] It is therefore ironic that fifty years later we know
relatively little about the structure of this enzyme. A partial explanation of
this is the very large molecular weight (483,000) which was determined by
ultracentrifugation by Sumner, Gralen and Eriksson-Quencel.[2] This observation
initially indicated that the structure of urease was very complex and too
formidable to attack. Recently, however, evidence has accumulated that the
structure of urease may be simpler. Several investigators have presented
evidence for the existence of lower molecular weight forms of the enzyme by
ultracentrifugal[3-5] or electrophoretic techniques.[6-8]

We have also tried to find additional evidence for the existence of smaller
subunits from structural studies. The general approach used to solve the
structural problem has been to cleave the urease molecule into smaller fragments
and then to isolate the peptides in high yields. We have chosen to cleave
urease with cyanogen bromide, to isolate the most readily purifiable peptides
and to estimate their yields. Such data should provide us with evidence about
the subunit structure and the preliminary results obtained are presented in this
paper.

MATERIALS AND METHODS

Urease was prepared from Jack bean meal by the procedure previously described[9] and was then freeze-dried.

Cyanogen bromide cleavage: Urease, in 70% formic acid, was treated first with 2-mercaptoethanol under nitrogen at room temperature for 5 hours[10] and then BrCN was added.[11] After 16 hours, the reaction mixture was diluted with water and desalted on a column of Sephadex G-10 with 1 M acetic acid as the solvent.

Gel filtration chromatography: A column (2 x 200 cm) of Sephadex G-50 in 1 M acetic acid was used for the initial fractionation of the desalted BrCN peptides. The pooled fractions were then lyophilized and rechromatographed on a column (2 x 180 cm) of Sephadex G-25 in 1 M acetic acid.

CM-cellulose chromatography: A CM-cellulose column was further used for the fractionation of pooled fractions obtained from the gel filtration chromatography step. The column was eluted with a linear gradient in which the starting buffer was 0.1 M sodium acetic acid, pH 3.7 and the second buffer was 0.2 M sodium acetate, pH 6.0. This column was used for the quantitative analysis of the cyanogen bromide peptides. A linear NaCl gradient was used for the sequential studies instead of a linear pH gradient. The pooled fractions were finally desalted on a column of Sephadex G-10 using 1 M acetic acid. The protein-containing part of the effluent was lyophilized.

Amino acid analysis: The peptide samples were hydrolyzed with 5.7 N HCl at 110°C for either 24 or 72 hours. Homoserine lactone was first converted to homoserine[12] and analyzed as described earlier.[13] The tryptophan content of the peptides were estimated spectrophotometrically.[14] Free SH groups in urease were estimated with p-chloromercuribenzoic acid (PCMB), N-ethylmaleimide (NEM) and methyl iodide (CH_3I) methods and the 1/2 cystine content by the CH_3I, methyl p-nitrobenzenesulfonate (MPNBS), performic acid and iodoacetic acid methods.[15]

Sequence analysis: Sequence determinations were made in the Beckman 890C Sequencer and the Beckman protein quadrol program 122974 was used. Polybrene[16] was also added to peptides prior to sequencing. For the identification of PTH-amino acids, thin layer chromatography[17] and HPLC[18] were used.

RESULTS

The desalted mixture of BrCN peptides was first fractionated on a column of Sephadex G-50 as shown in Fig. 1. Six consecutive pooled fractions were obtained. Rechromatography of pooled fractions E, F, G, and H on separate columns of Sephadex G-25 further resolved the peptides into five pooled fractions, namely, Db (located in between D and E, UV negative), E, F, G and H. Each of the pooled fractions were further purified on separate CM-cellulose columns.

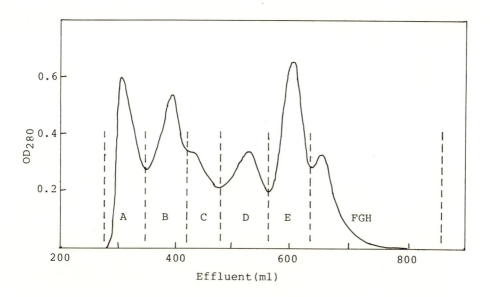

Fig. 1. Sephadex G-50 Chromatography of BrCN peptides from Urease. A column of Sephadex G-50 (medium), 2 x 220 cm, was used; the elutant was 1 M acetic acid, and the flow rate was 23.5 ml per hour. Letters and hatched lines indicate pooled fractions.

Figure 2 shows the chromatogram obtained with pooled fraction E. The purity of each peptide isolated was confirmed by N-terminal sequence and amino acid analyses. It is evident that each large peak contained a pure peptide with a homoserine lactone as C-terminal amino acid and the smaller peak the corresponding peptide with C-terminal homoserine. The amino acid compositions, recoveries and sequences of the peptides isolated are given in Tables 1 and 2, respectively.

Both PCMB and NEM reacted with 38.0 and 40.0 residues of cysteine residues per mole of native urease, respectively. About 10.9 SH groups reacted with NEM after 3 hours and with 38.0% loss of activity; 20.5 SH groups reacted after 20 hours with the complete loss of the activity; and 40.0 SH groups reacted after 40 hours. Amino acid analysis of urease methylated with CH_3I for 5 min at 40°C showed 37.0 residues of S-methylcysteine as well as 30.6 of 1/2 cystine and 1.9 of cysteic acid per molecule. On the other hand, 69.1 residues of S-methylcysteine were present in reduced and methylated (with MPNBS) urease and 69.5 residues of carboxymethylcysteine in reduced and carboxymethylated urease. About 70.0 cysteic acid residues were present in performic acid oxidized urease.

338

TABLE 1

AMINO ACID COMPOSITION AND RECOVERIES OF BrCN PEPTIDES OF UREASE

Amino Acid	LQ	VA	EY	GR	VI	WK	NG	RP	LM	Urease
Asp	4.0(4)	2.9(3)	1.1(1)	1.1(1)			2.0(2)		0.2(-)	494
Thr	2.0(2)	1.1(1)	0.9(1)	2.2(2)		1.0(1)	1.0(1)			289
Ser	3.6(4)	0.9(1)	0.2(-)	0.8(1)		0.9(1)	1.0(1)		0.2(-)	238
Hse	1.0(1)	0.9(1)	0.8(1)	1.0(1)	0.9(1)	0.9(1)	0.9(1)	0.9(1)	1.3(1)	
Glu	2.7(3)	1.1(1)	2.5(3)	1.9(2)			1.0(1)			381
Pro	2.0(2)	4.0(4)				2.1(2)		0.5(1)		248
Gly	4.0(4)	1.0(1)	1.1(1)	2.0(2)	1.9(2)	1.0(1)	1.0(1)		0.4(-)	437
Ala	2.2(2)	2.9(3)	1.8(2)	1.0(1)					0.2(-)	380
Cys(1/4)										68
Val		2.0(2)	1.1(1)	1.8(2)	0.5(1)		1.1(1)		0.1(-)	242
Met										115
Ile	1.4(2)	1.9(2)	0.1(-)	0.7(1)	0.4(1)					281
Leu	3.9(4)	0.1(-)	1.0(1)					0.3(1)	1.0(1)	388
Tyr			0.7(1)							111
Phe	1.9(2)					2.0(2)	1.0(1)			128
Lys	3.0(3)	1.1(1)	1.1(1)	1.1(1)	1.1(1)	2.0(2)				295
His	1.0(1)									131
Arg			1.0(1)	1.9(2)				1.0(1)		191
Trp*		0.9(1)		0.9(1)		1.0(1)				
Total Residue Residue	34	21	14	17	6	12	7	4	2	4453
Total # Recovery	0.9	2.9	2.9	3.7	4.7	2.9	3.6	-	-	

* Determined spectrophotometrically

TABLE 2

N-TERMINAL AMINO ACID SEQUENCES OF UREASE AND SEQUENCES OF ITS BrCN PEPTIDES

Peptide	Pool	Sequence
LQ	Db	Leu-Gln-Ser-Thr-Asp-Asp-Leu-Pro-Leu-Asn-Phe-Gly-Phe-Thr-Gly-Lys-Gly-Ser-Ser-Ser-Lys-Pro-Asp-Glu-Leu-His-Glu-Ile-Ile-Lys-Ala-Gly-Ala-Hse
VA	E	Val-Ala-Trp-Ala-Asp-Ile-Gly-Asp-Pro-Asn-Ala-Ser-Ile-Pro-Thr-Pro-Glu-Pro-Val-Lys-Hse
GR	E	Gly-Arg-Val-Gly-Val-Ile-Ser-Arg-Thr-Trp-Gln-Thr-Ala-Asp-Lys-Hse
EY	E	Glu-Tyr-Ala-Arg-Asp-Gly-Glu-Lys-Thr-Val-Ala-Gln-Leu-Hse
WK	F	Trp-Lys-Pro-Ser-Phe-Phe-Gly-Thr-Lys-Pro-Glu-Hse
VI	F	Val-Ile-Lys-Gly-Gly-Hse
NG	G	Asn-Gly-Val-Phe-Ser-Asx-Hse
RP	G	Arg-Pro-Leu-Hse
LM	H	Leu-Hse
Urease		Met-Lys-Leu-Thr-Pro

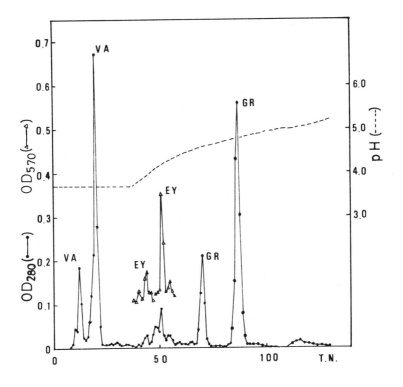

Fig. 2. CM-cellulose chromatography of pooled fraction E. A column of CM 23 (Whatman), 1.0 x 50 cm, was used; fractions of 4.8 ml were collected at a flow rate of 28.8 ml per hour. Linear gradient elution was used with 0.1 M sodium acetate buffer, pH 3.7 in the mixing chamber and 0.2 M sodium acetate, pH 6.0 in the reservoir.

DISCUSSION

It has been generally accepted that the molecular weight of the active form of urease is 483,000. The existence of more highly aggregated forms have also been reported from ultracentrifugal,[9,19] electrophoretic[8,20] or gel filtration[21] studies. In the presence of sulfite[19] or 2-mercaptoethanol[9] these oligomers revert to the 483,000 form. On the other hand several investigators have presented evidence for the existence of lower molecular weight forms. The occurrence of lower molecular weight subunits comes from ultracentrifugal analyses carried out in 6 M guanidine hydrochloride[4] or in sodium dodecyl sulfate (SDS).[5] The molecular weight reported in the former experiment was 83,000 and in the latter case, 60,000. Gel electrophoretic studies of urease in SDS[7] or in SDS and 2-mercaptoethanol[6] indicated the dissociation of urease to subunits with molecular weights of 60,000 or 75,000, respectively.

As shown in Table 1, 2.9 to 4.7 moles of the various BrCN peptides were obtained from one mole urease. (The low yield of peptide LQ during Sephadex G-50 chromatography is due to its presence also in the fraction Db.) This number of BrCN peptides separated indicated that urease consisted of at least five identical subunits. However, the number of subunits present should be greater than five since cyanogen bromide cleavage of urease in 70% formic acid was incomplete. Sekita et al.[10] reported that in pooled fraction D, the purified peptide Lys-Leu-Thr-Pro was contaminated with 10% of the peptide Met-Lys-Leu-Thr-Pro, an NH_2-terminal peptide from urease. Mamiya and Sekita[22] have also reported that urease has six C-terminal phenylalanine residues and also have shown the C-terminal sequence to be -Ser-Tyr-Leu-Phe using carboxypeptidase A and B, and hydrazinolysis analyses. Bailey and Boulter[6] have reported similar results.

The complete amino acid sequences of the nine BrCN peptides which are of lower molecular weights than fraction D are shown in Table 2. Mamiya et al.[23] have also determined the partial amino acid sequences of three other BrCN peptides present in pooled fraction D. These peptides were Lys-Leu-Thr-Pro-Tyr-Gly-Gln-Ile-Gly- and Lys-Ala-Asn-Thr-Gly-. The C-terminal peptide of urease was not included in this list of 12 isolated peptides. Thus, BrCN cleaved after at least 13 different methionine residues in urease. If it is assumed that urease consists of identical subunits, the minimum molecular weight of the subunit was calculated to be 53,000 daltons. Mamiya[14] has determined the tryptophan content (36 residues) of urease by ultraviolet absorption measurements and the resultant data is shown in Table 1. However, Bailey and Boulter[6] have isolated five different tryptophan-containing peptides. Using this figure they calculated the subunit molecular weight to be 75,000. In our studies, four different tryptophan peptides have been isolated. Three of them are shown in Table 2 and the fourth peptide isolated from pooled fraction B was Gly-Leu-Lys-Leu-His-Glu-Thr-Trp-Gly-. The minimum molecular weight calculated from our tryptophan content data was also 53,000. Our results suggest that the number of subunits are five as the lower limit and nine as the upper limit. However, Kobashi[24] from an inhibitor binding study has estimated that there are four active sites per molecule of urease and Staples and Reithel[25] have reported the subunits have a molecular weight of 30,000. Further investigations are under way in order to settle the question of the number of subunits present in urease.

ACKNOWLEDGMENTS

The authors wish to thank the late Dr. P. Edman, Dr. A. Henschen and Dr. K. Sekita for their valuable advice.

REFERENCES

1. Sumner, J.B.   (1926) J. Biol. Chem., 69, 434-441.
2. Sumner, J.B., Gralen, N.G., and Eriksson-Quensel, I.B.   (1938) J. Biol. Chem., 125, 37-44.
3. Blattler, D.P., Contaxis, C.C., and Reithel, F.J.   (1967) Nature, 216, 274-275.
4. Reithel, F.J., Robbins, J.E., and Gorin, G.   (1964) Arch. Biochem. Biophys., 108, 409-413.
5. Gorin, G., Mamiya, G., and Chin, C.C.   (1967) Experientia, 23, 443-445.
6. Bailey, C.J., and Boulter, D.   (1969) Biochem. J., 113, 669-677.
7. Blatter, C.J., and Gorin, G.   (1969) Can. J. Biochem., 47, 989.
8. Fishbein, W.N., Spears, C.C., and Scurzy, W.   (1969) Nature, 223, 191-193.
9. Mamiya, G., and Gorin, G. (1965) Biochim. Biophys. Acta, 105, 382-385.
10. Sekita, T., Mamiya, G., and Sekita, K.   (1975) Keio J. Med., 24, 203-210.
11. Steers, E. Jr., Craven, G.R., and Anfinsen, C.B.   (1965) J. Biol. Chem., 240, 2478-2484.
12. Ambler, R. (1965) Biochem. J., 96, 32p.
13. Spackman, D.H., Stein, W.H., and Moore, S.   (1958) Anal. Chem., 30, 1191-1206.
14. Mamiya, G. (1966) Keio Igaku, 43, 269-278,
15. Oda, T., Ishigami, K., and Mamiya, G.   (1978) Keio Igaku, 55, 527-534.
16. Hunkapiller, M.W., and Hood, L.E.   (1978) Biochemistry, 17, 2124-2133.
17. Edman, P., Henschen, A. in Protein Sequence Determination, Needleman, S.B., ed., Second Edition, Springer-Verlag, Berling, Heidelberg and New York, pp. 232-279.
18. Zimmerman, C.L., Appella, E., and Pisano, J.J.   (1977) Anal. Biochem., 77, 569-573.
19. Creeth, J.M., and Nichol, L.W. (1960) Biochem. J., 77, 230-239.
20. Blattler, D.P., and Reithel, F.J.   (1970) J. Chromatog., 46, 286-292.
21. Sehgal, P.P., Tanis, R.J., and Naylor, A.W.   (1965) Biochem. Biophys. Res. Commun., 20, 550.
22. Mamiya, G., and Sekita, K.   (1967) The 18th Symposium on Protein Structure (in Nagoya), Abstract, 29-32.
23. Mamiya, G., Koizumi, K., et al.   (1971) The 22d Symposium on Protein Structure (in Sendai), Abstract, 105-108.
24. Kobashi, K.   (1972) Seikagaku, 44, 187-204.
25. Staples, S.J., and Reithel, F.J.   (1976) Arch. Biochem. Biophys., 174, 651-657.

INTRACELLULAR Ca^{2+}-DEPENDENT PROTEASE AND ITS INHIBITOR

TAKASHI MURACHI, IWAO NISHIURA[+], KAZUYOSHI TANAKA, TOSHIO MURAKAMI[+],
MICHIYO HATANAKA AND TAKASHI IMANOTO
Department of Clinical Science, Kyoto University Faculty of Medicine, Sakyo-ku,
Kyoto 606, Japan

ABSTRACT

Evidence has been accumulating for the ubiquitous distribution of a Ca^{2+}-dependent neutral protease and its high molecular weight inhibitor(s) among various tissues, including bovine heart muscle, rat liver and brain, and chicken skeletal muscle. We have found that the cytoplasm of human erythrocytes also contained a Ca^{2+}-protease and its specific inhibitor (mol. wt. approx. 280,000). The inhibitor from any one of the tissues and cells tested was heat-stable, but readily inactivated by tryptic digestion. In no case was the inhibition based on sequestering of Ca^{2+} from the medium by the inhibitor. The inhibitor from one tissue could almost equally inhibit the Ca^{2+}-protease from the same and the different tissues. Since the content of the inhibitor relative to that of the Ca^{2+}-protease varies from one tissue to another, the apparent Ca^{2+}-protease activity of the unfractionated tissue homogenate may or may not be detected. From the enzymatically inactive mixture of Ca^{2+}-protease and inhibitor, both components could be at least partly resolved by gel filtration only in the presence of a Ca^{2+}-chelating agent.

The enzymes which catalyze the hydrolytic cleavage of peptide bonds are called proteolytic enzymes or proteases, and they are historically the best known family of enzymes, e.g., pepsin, trypsin, or chymotripsin. These proteases that are best known are the enzymes which are secreted into the digestive fluid or into the blood, and therefore they are termed *extracellular proteases*. In contrast to these extracellular proteases, the proteolytic enzymes, which exist inside the cells and hence called *intracellular proteases*, are much less known. However, lines of evidence have been accumulating to indicate that very important roles are played by these intracellular proteases in the metabolic processes in all types of living organisms.

+Present address: I. Nishiura, Department of Neurosurgery, Kyoto University
Faculty of Medicine, Sakyo-ku, Kyoto 606, Japan; T. Murakami, Department of
Home Economics, Kyoto Kasei College, Uji-shi, Kyoto 611, Japan.

In protein biosynthesis, for example, the initial product of the synthesis has been found, in many cases, to be a larger polypeptide which is called pro-protein or even prepro-protein, and this precursor protein is then transformed into a biologically active protein molecule by clipping off a peptide.[1] This clipping is catalyzed by a protease present in that cell. Furthermore, the active protein once formed, if it is not a secretory protein, should be ultimately degraded within the same cell. This degradation is again to be carried out by intracellular proteases.

An equally important role in such intracellular processing of proteins and peptides must be also played by endogenous inhibitor or inhibitors which may suppress and control these intracellular proteases.

We have been working for several years on the intracellular protease and its endogenous inhibitor(s), particularly of two different types: one is a chymotrypsin-like neutral protease and its poly(ADP-ribose)-like inhibitor, both bound to chromatin of the rat peritoneal macrophages,[2,3] and the other is a Ca^{2+}-dependent neutral protease and its high-molecular-weight proteinous inhibitor from various mammalian tissues and cells.[4-7] We shall present some of our current ideas and experimental results on the latter set of the protease and inhibitor.

Throughout the present report, the protease activity is referred to the hydrolysis of casein per unit time at pH 7.5 and at 37°C, expressed as the increase in absorbance at 750 nm of the acid-soluble fraction using Folin-Ciocalteau reagent.[4] For assaying the inhibitor, the test sample was preincubated with the protease at pH 7.5 and at 37°C for 10 min, and then the remaining activity was determined.

EXPERIMENTS ON RAT BRAIN

We first became interested in characterizing the Ca^{2+}-dependent neutral protease in the soluble fraction of brain tissue, because we had found that such protease activity was significantly increased in the cases of malignant brain tumors.[8] When the 105,000 g supernatant of the brain hemogenate of Wistar strain rats was surveyed for proteolytic activity over a wide range of pH, a small but significant activity at neutral pH values was observed which was insensitive to pepstatin (Fig. 1). The activity was found to be dependent on Ca^{2+}, reconfirming the earlier observation by Guroff.[9] When the 105,000 supernatant was chromatographed on DEAE-cellulose, the Ca^{2+}-dependent neutral protease was eluted at 0.25 M NaCl concentration as is shown in Fig. 2. The enzymatically active fractions were pooled, concentrated and chromatographed on Sephadex G-200, yielding 315-fold purified protease.[8] Using the latter

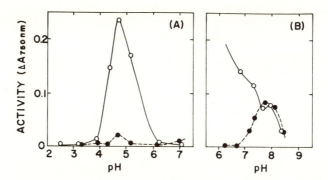

Fig. 1. The effect of pH on the activity of protease in normal rat brain using urea-denatured hemoglobin in (A) and casein in (B) as substrates. Protease activity was assayed in the presence (●) and absence (o) of pepstatin (0.1 mg/ml).[8]

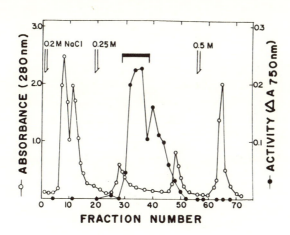

Fig. 2. DEAE-cellulose chromatography of the 105,000 g supernatant fraction from rat brain. The horizontal bar indicates the fractions collected for further purification of the Ca^{2+}-dependent protease.[8]

preparation we studied some of the kinetic and molecular properties of the Ca^{2+}-dependent protease. While we were summarizing our results which are shown in the second column of Table 1, we became aware of the discovery of protein kinase C (then termed as protein kinase M) by Nishizuka and his associates.[10-13] We found that the properties of our protease were very close to those reported by Inoue et al.,[10] who described a novel system of the activation of protein

TABLE 1

PROPERTIES OF Ca^{2+}-DEPENDENT NEUTRAL PROTEASE AS COMPARED
WITH THOSE OF PROTEIN KINASE C-ACTIVATING FACTOR

Properties	Neutral protease[4]	Protein kinase C-activating factor[10]
Elution from DEAE-cellulose	At 0.25 M NaCl	At 0.2 M NaCl
pH optimum	7.0 - 8.0	7.5 - 8.5
Ca^{2+} for 100% activation	5 mM	2 - 5 mM
Activity enhancement with thiol compound	Yes	Yes
Hydrolysis of casein	Yes	Yes
Hydrolysis of synthetic substrates	None	
Leupeptin	Sensitive	
Molecular weight	8.4×10^4	9.4×10^4

kinase in rat brain which was independent on cyclic nucleotides, but which was also dependent on a Ca^{2+}-activated protease. The properties of the protease reported by Inoue et al. are listed in the third column of the table. We then exchanged our preparations with each other, and soon concluded that we had been investigating the same protease, but from different points of research interest.

As shown in Fig. 3, Nishizuka's discovery of protein kinase C and its mode of activation has provided us with a wider view of the regulatory mechanisms controlling the activation of protein kinase. The original scheme has been modified to include our findings of the probable role of the Ca^{2+}-dependent protease in regulating the amount of irreversibly activated protein kinase. We now wish to present data which shows the co-existence of a specific inhibitor of this Ca^{2+}-dependent protease in the brain tissue.

EXPERIMENTS ON RAT LIVER

The discovery of a specific inhibitor was noticed when our studies changed from the brain to the liver tissue. When the supernatant from a homogenate of rat liver was examined for protease activity toward casein, practically no activity was found as shown in the middle column of Table 2.

However, when we chromatographed the same supernatant fluid on DEAE-cellulose

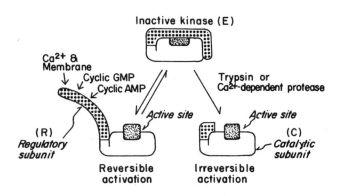

Fig. 3. Reversible and irreversible activation of protein kinases. The original design was made by Dr. Y. Nishizuka. Reversible activations for protein kinase A, C_2R_2 + 2 CyclicAMP $\rightleftharpoons$ 2C + R_2(CyclicAMP)$_2$; for protein kinase G, E_2 = 2 CyclicGMP $\rightleftharpoons$ (E · CyclicGMP)$_2$; for protein kinase C, E + Ca^{2+} + membrane $\rightleftharpoons$ E · Ca^{2+} · membrane.[13]

TABLE 2

Ca^{2+}-DEPENDENT PROTEASE IN RAT LIVER

Addition	Caseinolytic activity (ΔA 750 nm)	
	8,500 g supernatant	Fraction eluted from DEAE-cellulose column at 0.25 M NaCl
None	0	0
5 nM Ca^{2+}	0.020	0.025
5 nM Cysteine	0.010	0.004
5 nM Ca^{2+} + 5 nM Cysteine	0.025	0.647

and assayed it again, we did find in the fraction eluted by 0.25 M NaCl a very strong caseinolytic activity which was dependent on Ca^{2+} and cysteine, as is shown in the right-hand column of Table 2. This was an entirely unexpected observation, and it prompted us to search for an inhibitor in one of the fractions eluted during DEAE-cellulose chromatography.

Figure 4 shows a step-wise elution profile of the rat liver supernatant from a DEAE-cellulose column. The Ca^{2+}-dependent protease was eluted at 0.25 M NaCl.

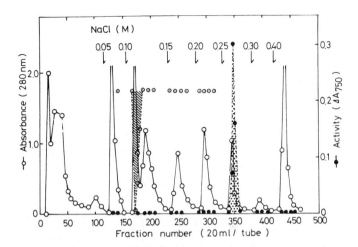

Fig. 4. DEAE-cellulose chromatography of the 105,000 g supernatant fraction from rat liver. For the detection of inhibitory activity, each assay was previously supplemented with a fixed amount of partially purified Ca^{2+}-dependent protease.

We used this protease fraction to test for the presence of protease inhibitor in the other eluted fractions. The results obtained are shown in Fig. 4 by circles with center dots, which indicated that a very strong inhibitor was associated with the fraction eluted by 0.1 M NaCl. Thus, we succeeded in identifying and mutually separating a Ca^{2+}-dependent protease on the one hand, and its inhibitor on the other.

The separation of a protease and its inhibitor could also be achieved by gel filtration through a Sephadex G-200 column (Fig. 5). Each fraction from the column was tested for its caseinolytic activity in the presence of a fixed and known amount of the Ca^{2+}-dependent protease which had been obtained in the preceding experiment. Should the fraction contain a protease, the observed activity would be above the reference level, and conversely, if an inhibitor were present, a negative value would be obtained.[*] As shown in Fig. 5A, the rat liver supernatant was resolved into an enzymatically active fraction and an inhibitory fraction.

We then examined the supernatant of rat brain using identical procedures, and found that the inhibitor was also present in brain tissue (Fig. 5B). A very interesting feature which is common to both liver and brain was that the inhibitor fraction always eluted near the void volume of the Sephadex column

---

[*] Similar method of assay was used throughout the present study (cf. Figs. 4, 5, and 10).

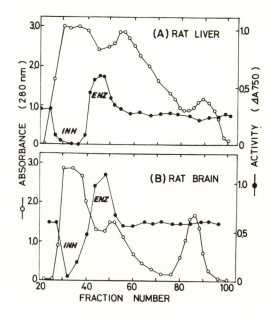

Fig. 5. Resolution of Ca^{2+}-dependent protease and its inhibitor by gel filtration through Sephadex G-200.

(A) rat liver;[4] (B) rat brain.[5]

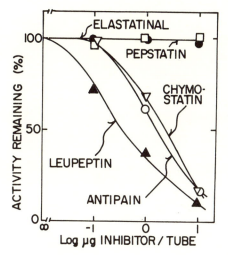

Fig. 6. Effect of microbial protease inhibitors on the Ca^{2+}-dependent protease from rat liver. Casein was used as substrate at pH 7.5.

used, suggesting a high molecular weight for the inhibitor.

Figure 6 illustrates the effects of various microbial protease inhibitors on the Ca^{2+}-dependent protease from rat liver. Leupeption, a trypsin inhibitor,[14] was found to most effective as reported in the case of the brain enzyme.[8] The sensitivity to antipapain seems to be reasonable because the protease requires, in addition to Ca^{2+}, a thiol activator (Table 2).

PROPERTIES AND SPECIFICITY OF THE INHIBITOR

The newly discovered inhibitor of the Ca^{2+}-dependent protease is unique in its large molecular size and extreme heat-stability. When the inhibitor fractions from the Sephadex G-200 column (Fig. 5A) were pooled and concentrated, and the concentrate was heated in a boiling water bath for 20 min, the inhibitory activity was completely retained. Such heat treatment precipitated more than 90% of other proteins present in the concentrate, and thus provided a very efficient procedure for the purification of the inhibitor. Even after the heat treatment, the inhibitor was again eluted in the void volume during a Sephadex G-200 chromatography. The inhibitor was further purified on a hydroxylapatite column, and yielded 370-fold purified inhibitor in 32.6% yield. The preparation thus obtained gave one major band, and at least three fast-moving minor bands during a poly-acrylamide gel electrophoresis in the presence of sodium dodecylsulfate and the calculated molecular weight was 2.8×10^5.

Table 3 summarizes the stability data of the inhibitor. Although it was quite stable against both heat and acid, its proteinous nature was confirmed by rapid digestibility with a catalytic amount of trypsin.

TABLE 3

EFFECTS OF VARIOUS TREATMENTS ON THE POTENCY OF
THE INHIBITOR OF Ca^{2+}-DEPENDENT PROTEASE FROM RAT LIVER

Treatment	Condition	Inhibitory potency remaining (%)
Heating	100°C, 20 min, pH 7.5	100
Trichloroacetic acid	At a final concentration of 10%, 4°C	100
Acetone	At a final concentration of 75%, 4°C	100
Trypsin	1/17.5 w/w, pH 7.5, 37°C 40 min	0
	1/87.5 w/w, pH 7.5, 37°C, 40 min	0
	1/87.5 w/w, pH 7.5, 37°C, 40 min	86
Nuclease P_1	1/1 w/w, pH 7.5, 37°C, 40 min	100

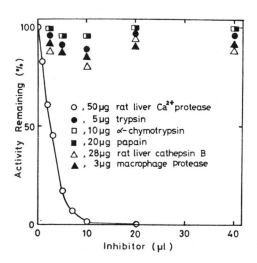

Fig. 7. The effect of the rat liver inhibitor on various neutral proteases. Increasing amounts of the inhibitor (2 mg protein per ml) were pre-incubated with test enzymes.[4]

The specificity of the inhibitor is shown in Fig. 7. It is effective only on the Ca^{2+}-dependent protease, and has no effect on various neutral serine and thiol proteases. Also, it had no effect on rat liver cathepsin D. As mentioned above, the Ca^{2+}-dependent neutral protease was reported to be responsible for irreversible activation of protein kinase C in various mammalian tissues.[13] We, therefore, wanted to know whether our inhibitor was also effective in such a system. When the cyclic nucleotide-independent activation of protein kinase C was assayed by the incorporation of [^{32}P] phosphate from [γ-^{32}P]ATP to the acceptor histone, our preparation of inhibitor clearly inhibited the activation reaction protein kinase to an extent parallel to its inhibitory effect on caseinolysis (Y. Takai and Y. Nishizuka, unpublished observation).

DISTRIBUTION OF THE PROTEASE AND INHIBITOR

In addition to rat brain and liver, a number of different tissues and cells of several species of animals have now been found to contain both the Ca^{2+}-dependent neutral protease and its endogenous inhibitor. Waxman and Krebs[15] reported the identification of two inhibitors of Ca^{2+}-activated neutral protease from bovine cardiac muscle. This report appeared while our first paper on the discovery of the inhibitor from rat liver was in the process of publication.[4] They reported that "no calcium-stimulated proteolytic activity toward casein could be detected in crude extracts, but, after chromatography on DEAE-cellulose, this activity was readily measured." This was exactly what we have independently and concurrently found with rat liver homogenate before and after the purification (Table 2). The most striking similarity between Waxman and Krebs' data and ours was the very large molecular size of the inhibitor, their reported value being 2.7×10^5 and our value was 3×10^5. They reported, however, that their inhibitor lost 35% of its potency upon heat treatment at 100°C for 10 min.

We then found the occurrence of a high molecular weight (2.8×10^5) inhibitor in the cytoplasm of normal adult human erythrocytes, together with its own Ca^{2+}-dependent neutral protease. The erythrocyte inhibitor was also found to be quite heat-stable and resembled the liver and brain inhibitors in many other respects (see below). Similarly, monkey erythrocytes and rat spleen cells were found to contain their own respective pairs of the Ca^{2+}-protease and its inhibitor. Meanwhile, Waxman has also surveyed the distribution of the protease and inhibitor in various tissues.[16] Table 4 summarizes the data available at the present time.

It is interesting to note that the distribution of the Ca^{2+}-dependent neutral protease and its high molecular weight inhibitor is ubiquitous, but the relative abundance of the enzyme to inhibitor varies from one tissue to another. However, the data from the two different laboratories are not in complete agreement. Nevertheless, it is hoped that further experimentation will yield accurate data and will also elucidate the biological significance of such differences in various tissues.

Ishiura et al.[18] reported that the Ca^{2+}-activated protease they purified from chicken skeletal muscle cleaved several muscle proteins and spectrum. Waxman[16] described Ca^{2+}-activated proteases of muscular and nonmuscular origins which cleaved tropomyosin and troponin. We have found that the Ca^{2+}-dependent protease partially purified from human erythrocytes digested rather preferentially the erythrocyte membrane proteins, bands 3 and 4.1, with little or no cleavage of spectrum (unpublished observation). Apart from the action on these structural proteins, the ubiquitous distribution of this protease and its endogenous inhibitor lends more attractiveness to the hypothesis that this protease and inhibitor system are involved in some way in the irreversible activation of protein kinase C (Fig. 3), which also has recently been found to be distributed ubiquitously.[13]

Figures 8A and 8B show the mutual effectiveness of the high molecular weight inhibitors from rat liver and brain on the proteases from both sources. The data suggests great similarity in the properties of the two inhibitors as far

TABLE 4

DISTRIBUTION OF Ca^{2+}-DEPENDENT PROTEASE AND ITS INHIBITOR

Species	Organ	Relative abundance[a]	Reference
Rat	Liver	E << I	4
	Liver, kidney	E > I	16
	Brain	E > I	5
	Thymus cells	E > I	Present paper
	Spleen cells: whole cells	E < I	Present paper
	nylon wool column- passed cells	E << I	Present paper
	Erythrocytes	E >>> (I)[b]	Present paper
Monkey	Erythrocytes	E << I	Present paper
Human	Erythrocytes[c]	E < I	7
Rabbit	Reticulocytes	E > I	16
	Brain, Lung	E > I	16
	Skeletal muscle	E > I	16
Bovine	Heart muscle	E < I	15
Calf	Thymus	E > I	16
Chicken	Breast muscle	E > I	16
	Gizzard	E > I	16
	Skeletal muscle	E	18

[a] E, Ca^{2+}-dependent protease; I, inhibitor. E > I denotes that the protease activity could be detected using the crude extract, while E < I means that the protease activity became measurable only after its removal from the inhibitor.
[b] The existence of the inhibitor was hardly detectable, unless the fractions were highly concentrated.
[c] The presence of a Ca^{2+}-dependent protease was recently reported by Allen and Cadman,[17] but they did not report the presence of its inhibitor.

as their mode of inhibition is concerned. Furthermore, the inhibitor from human erythrycytes was found to be effective, though to lesser extents, on the rat enzymes (Fig. 8C). These findings are very suggestive, in that some general, rather than organ-specific roles for the protease and its inhibitor exist.

MECHANISM OF INHIBITION

When we discovered the inhibitors in rat tissue homogenates, the question

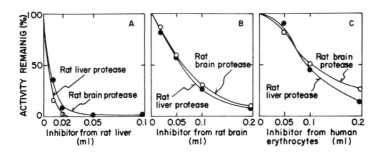

Fig. 8. Mutual effectiveness of Ca^{2+}-dependent proteases and inhibitors of different origins.

was asked if the inhibitor exhibited its effect simply by binding and thus removing Ca^{2+} from the medium, which is absolutely essential for the protease. We carried out a series of inhibition experiments in the presence of varying concentrations of Ca^{2+}. The results obtained are shown in Fig. 9, which clearly indicated that the inhibition cannot be reversed by adding even excess Ca^{2+} to the medium.[4,5] It was therefore concluded that the inhibition was not due to sequestering of Ca^{2+} by the inhibitor protein. Exactly the same conclusion was independently reached by Waxman and Krebs,[15] who used the protease and inhibitor from bovine heart muscle.

The fact that the protease and the inhibitor in the crude supernatant could be resolved simply by applying it on a Sephadex G-200 column in the presence of

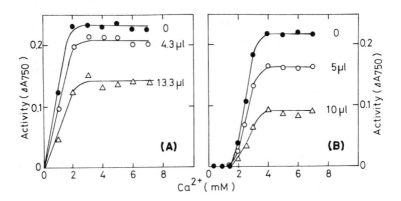

Fig. 9. Inhibition of Ca^{2+}-dependent protease as a function of Ca^{2+} concentration in the medium. (A) rat liver;[4] (B) rat brain.[5]

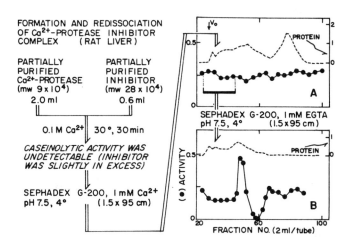

Fig. 10. Evidence for the formation of the protease-inhibitor complex in the presence of Ca^{2+} (A) and for the redissociation of the complex in the presence of a Ca^{2+}-chelating agent (B).

5 mM EGTA[*4,5] seemed to be rather puzzling in view of the specific and strong inhibition between the two components as shown in Fig. 6. The pattern of inhibition suggested a high affinity of the inhibitor for the enzyme and an almost undissociable complex under the conditions employed, i.e., in the presence of Ca^{2+} is expected. In fact, when the Ca^{2+}-dependent protease and the inhibitor, both partially purified from rat liver, were mixed in the absence of Ca^{2+} and then chromatographed on Sephadex G-200 in the presence of 5 mM EGTA, a perfect separation into both components could be demonstrated.[4] To the contrary, when the protease was first mixed with a slight excess of the inhibitor in the presence of Ca^{2+} and after confirming that no detectable caseinolytic activity was present, the mixture was applied to a Sephadex G-200 column still in the presence of Ca^{2+}, no protease or inhibitor activity was detected in any of the fractions eluted (Fig. 10A). The void volume fractions were pooled, concentrated and dialyzed against the buffer which now contained EGTA in place of Ca^{2+}, and the concentrate was again chromatographed on an identical Sephadex G-200 column but now in the presence of EGTA. The elution profile shown in Fig. 10B clearly indicated recovery of both the proteolytic and inhibitory activities. It was thus concluded that the protease and inhibitor formed an inactive complex when the Ca^{2+} concentration was sufficiently high, while removing Ca^{2+} from the

* Abbreviation: EGTA, ethylene glycol bis (ß-aminoethyl ether)-N,N,N',N'-tetraacetic acid.

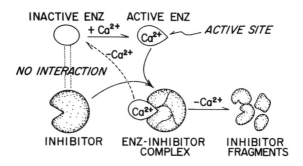

Fig. 11. Hypothetical mode of interaction between Ca^{2+}-dependent protease and its high molecular weight inhibitor.

medium redissociated the complex into the enzyme and inhibitor.

Although the stoichiometry of the formation and redissociation of the enzyme-inhibitor complex has not been established, we wish to present a speculative scheme which explains the events observed (Fig. 11). The protease recovered from the complex did not seem to have undergone drastic changes in its molecular and catalytic properties, while, as is apparent from Fig. 10B, the inhibitor recovered behaved as a molecular entity of much smaller size as compared with the original inhibitor. This is the reason for assuming proteolytic fragmentation of the inhibitor during inhibitor-protease complex formation. The lower recovery of the total inhibitor potency compared with the recovery of the total protease activity after redissociation of the complex (data not shown) implies that the inhibitor fragments only partially cross-reacted with the protease. The latter implication is symbolized in Fig. 11 by showing a couple of small binding clefts in several of the inhibitor fragments, while assigning a full large cleft to the original inhibitor. Furthermore, it is postulated that only the Ca-protease complex which has a different conformation from the free protease, is capable of reacting with the inhibitor.

CONCLUSION

As is apparent from the foregoing discussions, it is very premature to draw any definite conclusion about the biological roles of these Ca^{2+}-dependent proteases and their inhibitors or the mechanism of their interactions. Nonetheless, the strikingly ubiquitous distribution and very similar natures of these proteases and inhibitors in wide varieties of tissues and cells seem to provide us with plenty of room for future challenges from diverse viewpoints.

In addition, a unique large molecular size and the extreme heat stability of the newly found inhibitor may also elicit further active investigations with both physical and biological interests.

ACKNOWLEDGMENTS

This work was supported in part by Grants-in-Aid for Scientific Research and Cancer Research from the Ministry of Education, Science and Culture, Japan, and by a grant from the Yamanouchi Foundation for Research on Metabolic Disorders.

REFERENCES

1. Blobel, G., and Dobberstein, B.   (1975) J. Cell Biol., 67, 835-851.
2. Suzuki, Y., and Murachi, T.   (1978) J. Biochem. (Tokyo), 84, 977-984.
3. Murachi, T., and Suzuki, Y.   (1978) in Versatility of Proteins, Li, C.H., ed., Academic Press, New York, pp. 183-202.
4. Nishiura, I., et al.   (1978) J. Biochem. (Tokyo), 84, 1657-1659.
5. Nishiura, I., Tanaka, K., and Murachi, T.   (1979) Experientia, in press.
6. Murachi, T., et al.   (1979) Abstr., ACS/CSJ Chem. Congr., Honolulu, Hawaii, April 2-5, Biol. Chem., No. 6.
7. Murachi, T., et al.   (1979) Abstr. 11th Internatl. Congr. Biochem., Toronto, July 8-13.
8. Nishiura, I., Handa, H., and Murachi, T.   (1979) Neurol. Med. Chirur., 19, 1-7.
9. Guroff, G.   (1964) J. Biol. Chem., 239, 149-155.
10. Inoue, M., et al.   (1977) J. Biol. Chem., 252, 7610-7616.
11. Takai, Y., et al.   (1977) Biochem. Biophys. Res. Commun., 77, 542-550.
12. Nishizuka, Y., et al.   (1978) in Advan. in Cyclic Nucleotide Res., George, W.J., and Ignarro, L.J., eds., Raven Press, New York, Vol. 9, pp. 209-220.
13. Nishizuka, Y., et al.   (1979) Mol. Cell. Biochem., 23, 153-165.
14. Aoyagi, T., and Umezawa, H., (1975) in Proteases and Biological Control, Reich, H., Rifkin, D., and Shaw, E., eds., Cold Spring Harbor Laboratory, pp. 429-454.
15. Waxman, L., and Krebs, E.G.   (1978) J. Biol. Chem., 253, 5888-5891.
16. Waxman, L.   (1979) Fed. Proc., 38, 479.
17. Allen, D.W., and Cadman, S.   (1979) Biochim. Biophys. Acta, 551, 1-9.
18. Ishiura, S., et al.   (1978) J. Biochem. (Tokyo), 84, 225-230.

STRUCTURE AND FUNCTION OF CARBOXYL PROTEASES: A COMPARISON

OF THE INTRACELLULAR AND EXTRACELLULAR ENZYMES

JUNG SAN HUANG, SHUAN SHIAN HUANG, AND JORDAN TANG[*],
Laboratory of Protein Studies, Oklahoma Medical Research Foundation;
*Department of Biochemistry and Molecular Biology, University of Oklahoma Health
Sciences Center, Oklahoma City, Oklahoma 73104

ABSTRACT

The structure and function of intracellular and extracellular carboxyl proteases are compared. Since the primary and tertiary structures of several of
these enzymes are very homologous, a single structure-function model representative of the extracellular carboxyl proteases can be summarized. The main
features are similar polypeptide chain foldings, identical active center aspartyl
residues, and an extended substrate binding cleft.

Recent results obtained from bovine and porcine spleen cathepsin D indicate
that the active center residues are nearly identical to those of the extracellular enzymes. The partial amino acid sequence is also homologous to the
extracellular carboxyl proteases. These results suggest that the active center,
catalytic mechanism and evolutionary origin of cathepsin D, and the extracellular enzymes are related. However, special structural features unique to
cathepsin D were observed. These include 4 carbohydrate attachment sites, a
"hydrophobic tail," and a high-molecular-weight precursor.

Original data have been presented for the presence of a cathepsin D "activator." It is a protein of about 27,500 daltons and is obviously a SH-protease.
Preliminary studies of the characteristics of this synergism have been made.

INTRODUCTION

Carboxyl proteases represent a group of proteolytic enzymes with diverse
species origins and biological functions.[1] The best known extracellular carboxyl
proteases are the mammalian gastric digestive proteases: pepsin, gastricsin,
and chymosin. Similar enzymes are secreted extracellularly by microorganisms,
especially fungi (e.g., penicillopepsin, rhizopuspepsin, endothiapepsin, etc.).
Other samples of extracellular carboxyl proteases are plasma renin, which converts angiotensinogen to angiotensin I, and an acid protease from seminal plasma.
Intracellularly, carboxyl proteases are found in mammalian lysosomes (cathepsin
D and E), vacuoles of the yeast cells,[2] and in the protozoan cells.[3,4]

In spite of the diversity in the function and origin, it is now quite clear
that the carboxyl proteases are a group of proteolytic enzymes with common

structural features and catalytic mechanisms. In this article, we wish to
first establish a structure and function model for the extracellular carboxyl
proteases. Then we shall compare this model to the structure and function of
lysosomal cathepsin D, based on some recent results. Finally, we shall describe
some new data on possible cooperative aspects of cathepsin D with another protease.

STRUCTURE AND FUNCTION OF EXTRACELLULAR CARBOXYL PROTEASES

The amino acid sequences of three extracellular carboxyl proteases are very
homologous. Pepsin,[5] chymosin,[6] and penicillopepsin[7] are not only about the
same size, but they also contain many identical residues in the alignment of the
three amino acid sequences.[8] Partial sequence of other extracellular carboxyl
proteases are also very homologous.[8] The X-ray crystallographic studies so far
have been carried out to atomic resolutions for four carboxyl proteases: pepsin,[9] penicillopepsin,[10] rhizopuspepsin,[11] and endothiapepsin.[11] It is clear
that the overall folding of the polypeptide chains in four enzymes are nearly
identical. All four extracellular carboxyl proteases contain two distinct lobes
which are divided by an apparent substrate-binding cleft. The size of this
cleft is sufficient to accommodate seven or eight amino acid residues in a polypeptide substrate, as was predicted from the specificity studies of pepsin (for
review, see reference 1). There are two active sites, Asp-32 and Asp-215, which
had previously been identified in pepsin using chemical modification with
active-center-directed reagents, a substrate-like epoxide,[12] and a diazo compound.[13] In the crystal structures these two aspartyl groups are located in
the center of the binding cleft and are probably hydrogen-bonded to each other.
In addition, Asp-32 appears to be hydrogen-bonded to Ser-35, which is found in
all carboxyl proteases.[10]

Even though the precise substrate binding positions in the crystal structures
are still under progress (M. James and T.L. Blundell, personal communications),
a reasonable catalytic mechanism for these extracellular enzymes can be proposed. The results in the pH dependence of epoxide inhibition of pepsin had
revealed that Asp-32 is the low pKa (~ 2.8) carboxyl group suggested in previous
kinetic studies.[15] This downshift in pKa can be readily explained by its hydrogen bonding to Ser-35. It is generally agreed among the investigators that
Asp-32 is the probable nucleophile involving nucleophilic attack to the carbon
atom of the substrate carbonyl group, either directly[14] or through a hydrogen-
bonded water.[16] The roles of Asp-215 and another nearby Tyr-75 are less certain.
It is likely that one of the two serves as a proton donor to the leaving amide
group of the substrate, and the other may be hydrogen-bonded to the oxygen atom

of the substrate carbonyl group.[14,16] In spite of these uncertainties on the detailed catalytic mechanism, it is clear that all the extracellular carboxyl proteases, including renin, have similar active center structures and catalytic mechanisms. This point is indicated from the universality of the active-site directed inactivators mentioned above and the universality of the inhibition by pepstatin, a transition-state inhibitor.[17] The differences in the specificities, e.g., between pepsin, chymosin, and renin, can be easily explained on the basis of topographical differences in the substrate binding clefts.

CATHEPSIN D--AN INTRACELLULAR CARBOXYL PROTEASE

Because the intracellular proteases are potentially destructive to the cellular structures and contents, the need for the regulation of their synthesis, packaging, transport, and finally, activity is apparent. One might anticipate, therefore, that the intracellular carboxyl proteases are more complicated in their structure, function, and genesis. For this reason, we have chosen lysosomal cathepsin D as a model for structure-function studies. In the following, we wish to discuss three areas of our recent findings.

Structure and function of cathepsin D

We have recently isolated six cathepsin D isozymes from porcine spleen[18] and two isozymes from bovine spleen.[19] The characteristics of these isozymes are summarized in Table 1. Five porcine isozymes, I-V, which are different is isoelectric points, are 50,000 daltons in molecular weight. Four of these isozymes, I-IV, contain in each two polypeptide chains. The heavy chain is 35,000 daltons, which can be separated from the light chain, 15,000 daltons, on a column of Sephadex G-75 in the presence of 6 M urea.[18] Isozyme V, however, is a single-chain enzyme. These five isozymes have similar specific activities in proteolysis.

Bovine spleen cathepsin D isozymes, A and B, are 46,000 daltons.[19] Like the porcine cathepsin D isozymes, the bovine isozymes contain both the one-chain and the two-chain structures. The bovine heavy chain is 34,000 daltons, and the light chain is 12,000 daltons. Unlike the porcine isozymes, however, each bovine isozyme contains about 60% of the single-chain cathepsin D and 40% of the two-chain species.

Cathepsin D is a glycoprotein which can be retained and eluted from concanavalin A-sepharose column.[18] The carbohydrate contents are identical in all the isozymes. As summarized in Table 1, each enzyme molecule contains 8 residues of mannose and 4 residues of glucosamine. Since the light chain contains only 2 mannose and 1 glucosamine, it probably has only a single carbohydrate

TABLE 1

SUMMARY ON THE PROPERTIES OF CATHEPSIN D ISOZYMES FROM BOVINE
SPLEEN AND COMPARISON WITH THE ISOZYMES FROM PORCINE SPLEEN[a]

		pI	Molecular weight	Number of subunits	Norleucine incorporation[a] res/mol	Carbohydrate[b] Man res/mol	Carbohydrate[b] GlcN res/mol	Heavy chain antiserum
Porcine isozyme	I	7.54	50,000	2	1	8	4	+
	II	7.09	50,000	2				+
	III	6.60	50,000	2				+
	IV	6.19	50,000	2				+
	V	5.51	50,000 } 100,000	1				+
Light chain			15,000		0	2	1	-/+[c]
Heavy chain			35,000		1	6	3	+
Bovine isozyme	A	6.49	46,000	1 + 2		8	4	+
	B	6.04	46,000	1 + 2		8	4	+
Light chain			12,000					
Heavy chain			34,000					

[a] Data are that of amino acid analyses after reacting the native enzyme with diazoacetyl norleucine methyl ester.

[b] Man and GlcN are mannose and glucosamine, respectively.

[c] Only trace reactivity observed.

Residue numbers	-2	-1	1	2	3	4	5	6	7	8	9	10	11	12	13	14	15	16	17	18	19	20
Bovine cathepsin D	Gly-	Pro-	Ile-	Pro-	Glu-	Leu-	Leu-	Lys-	Asn-	Tyr-	Met ──		Asp-	Ala-	Gln-	Tyr-	Tyr-	Gly-	Glu-	Ile-	Gly-	Ile-
Gastric Proteases:																						
Porcine pepsin	-Ala-	Leu/Ile-	Gly-	Asp-	Glu-	Pro-	Leu-	Glu-	Asn-	Tyr-	Leu ──		Asp-	Thr-	Glu-	Tyr-	Phe-	Gly-	Thr-	Ile-	Gly-	Ile-
Bovine chymosin	Gly-	Glu-	Val-	Ala-	Ser-	Val-	Pro-	Leu-	Thr-	Asn-	Tyr-	Leu ──		Asp-	Ser-	Gln-	Tyr-	Phe-	Gly-	Lys-	Ile-	Tyr-Leu-
Microbial Proteases:																						
R. chinensis	Ala-	Gly-	Val-	Gly-	Thr-	Val-	Pro-	Met-	Thr-	Asp-	Tyr-	Gly-	Asn-	Asp-	Val/Ile-	Glu-	Tyr-	Tyr-	Gly-	Gln-	Val-	Thr-Ile-
Penicillopepsin	Ala-	Ala-	Ser-	Gly-	Val-	Ala-	Thr-	Asn-	Thr-	Pro-	Thr-	Ala-	Asn-	Asp-	Glu-	Glu-	Tyr-	Ile-	Thr-	Pro-	Val-	Thr-Ile-

Residue numbers	21	22	23	24	25	26	27	28	29	30	31	32	33	34	35	36	37
Bovine cathepsin D	Gly-	Thr-	Pro-	Pro-	Gln-	Ser-	Phe-	Thr-	Val-	Val-	Phe-	Asp-	Thr-	(Gly-	Ser-	Ser-	Asn)
Gastric Proteases:																	
Porcine pepsin	Gly-	Thr-	Pro-	Ala-	Gln-	Asp-	Phe-	Thr-	Val-	Ile-	Phe-	Asp-	Thr-	Gly-	Ser-	Ser-	Asn-
Bovine chymosin	Gly-	Thr-	Pro-	Pro-	Gln-	Glu-	Phe-	Thr-	Val-	Leu-	Phe-	Asp-	Thr-	Gly-	Ser-	Ser-	Asp-
Microbial Proteases:																	
R. chinensis	Gly-	Thr-	Pro-	Gly-	Lys-	Ser-	Phe-	Asn-	Leu-	Asn-	Phe-	Asp-	Thr-	Gly-	Ser-	Thr	
Penicillopepsin	Gly	---	---	Gly-	Thr-	Thr-	Leu-	Asn-	Leu-	Asn-	Phe-	Asp-	Thr-	Gly-	Ser-	Ala-	Asp-

Fig. 1. Amino-terminal sequences of bovine and porcine cathepsin D and their homology to extracellular acid proteases.

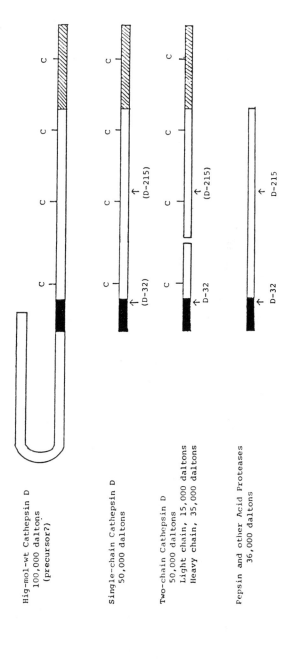

Fig. 2. Schematic presentation of polypeptide chain structures of cathepsin D and their structure relationship with other acid proteases.

attachment site. The heavy chain, which contains 6 mannose and 2 glucosamine, probably has three carbohydrate sites.

The N-terminal sequences of the light chains from both porcine and bovine enzymes are summarized in Fig. 1.[18,19] The extensive structural homology of these sequences with the sequence of extracellular carboxyl proteases is apparent. Particularly striking is the nearly identical sequence surrounding the active-site residue, Asp-32. We had previously shown that cathepsin D is inhibited by the active-site directed epoxide.[20] The modification site of cathepsin D by this reagent is probably at Asp-32, as in the case of pepsin inactivation. The N-terminal sequence of the single-chain cathepsin D is the same as that for the light chain, while that for the heavy chain showed considerable microheterogeneity.[18]

Based on these data, the overall chain structure of cathepsin D can be aligned as illustrated in Fig. 2. The presence of a second active-site carboxyl group equivalent to Asp-215 was verified by the inhibition of cathepsin D by diazoacetyl-norleucine methyl ester. We observed that after the inactivation, one residue of norleucine was incorporated into the amino acid composition of the heavy chain.[18] It is quite clear, then, that the active center structure of cathepsin D is very similar to that of the extracellular carboxyl proteases. The catalytic mechanism of these enzymes must also be very similar. The sequence homology shown in Fig. 1 also suggests a divergent evolutionary path for cathepsin D and the extracellular carboxyl proteases.[21]

An interesting aspect of the cathepsin D structure is the extra length at the C-terminal region when aligned against the extracellular enzymes (Fig. 2). We have been able to estimate the approximate amino acid composition of this "tail" region of the molecule. It appears that the "tail" region contains a large number of hydrophobic amino acids and may have a function of membrane binding.[18]

A possible precursor of cathepsin D

From porcine spleen we have also isolated a 100,000-dalton cathepsin D which has only 5% of the specific activity as compared to those of the 50,000-dalton species.[18] This high-molecular-weight cathepsin D is apparently a single polypeptide chain, since no mobility change was observed on SDS-gel electrophoresis after the reduction of disulfide bonds. Like the 50,000-dalton isozymes, however, the high-molecular-weight cathepsin D was retained on a column of pepstatin-Sepharose. The structural relationship between the high-molecular-weight cathepsin D and other isozyme species was demonstrated by using immunochemical techniques. The antiserum raised against porcine cathepsin D heavy chain cross-reacted with urea-denatured high-molecular-weight cathepsin D in Ouchterlony

TABLE 2

RESTORATION OF TOTAL CATHEPSIN D ACTIVITY BY
RECOMBINATION OF THE FRACTIONS AT STEP 4

Purification step	Cathepsin D recovery %
1. Extract	100
2. Acid supernatant	80
3. DEAE-Sephadex unretained	60
4. Pepstatin-Sepharose column	
(a) Unretained fraction	1.5
(b) Retained fraction	30
(a) + (b)	57

diffusion. The precipitin line for the high-molecular-weight cathepsin D
showed identity with the 50,000 dalton isozymes or with the heavy chain itself.[18]

The hypothesis that the high-molecular-weight cathepsin D is a precursor is
depicted in Fig. 2. The preliminary results on the polypeptide chain relationships suggest that the 100,000 dalton precursor is converted to the 50,000-
dalton single chain, then to the two-chain cathepsin D by limited proteolysis.

Synergism of cathepsin D and lysosomal sulfhydryl proteases

The synergism of the proteases in tissue protein catabolism has long been
implied but has received only a little experimentation.[22] We first encountered
the phenomenon during the purification of cathepsin D. As shown in Table 2,
the total activity of cathepsin D recovered after the pepstatin-Sepharose chromatography (b, step 4) was about one-half, from 60% to 30%. The unretained
fraction (a, step 4) had very little proteolytic activity. However, when the
recovered enzyme was reconstituted with the breakthrough fraction, the original
total enzyme activity was almost completely restored. Apparently, the breakthrough fraction contained an "activator" which can enhance the proteolytic
activity of cathepsin D. The "activator," which was nondialyzable, heat labile,
and inactivated by subsilisin digestion, was evidently a protein.

The dose respone of the "activation" on cathepsin D activity (Fig. 3) showed
an expected saturation curve. The effect of "activator" on cathepsin D activity at different pH is shown in Fig. 4. The elevation of activity was observed
within the whole range of the cathepsin D activity from pH 2 to 6. The shift
of the optimum pH caused by "activator" was only slight. The enhancement in

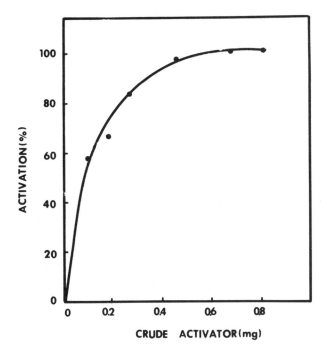

Fig. 3. Dependence of activity of porcine cathepsin D on "activator" concentration. The reaction mixture contained various amounts of "activator," 5 µg of porcine spleen cathepsin D, and 1% of bovine hemoglobin in 2.5 ml 0.25 M sodium formate buffer, pH 3.2. After incubation at 37°C for 20 min, 2 ml of 10% trichloroacetic acid was added to stop the reaction. The O.D.$_{280}$ of the filtrates were measured. All appropriate controls were carried out.

enzyme activity by the "activator" was observed using various protein substrates. In addition to bovine hemoglobin, which was the standard substrate for cahtepsin D, the hydrolysis of bovine serum albumin and rat liver soluble proteins were also increased by the "activator."

Although the "activator" did not increase the activity of trypsin, α-chymotrypsin, and pepsin, it was nevertheless effective for all carboxyl proteases tested. As shown in Fig. 5, the activities of chymosin and microbial carboxyl proteases were increased between 2 and 3 fold. A 3.5-fold increase of activity for A. *saitoi* carboxyl protease was particularly advantageous for using this enzyme in the assay of "activator." The enhancement of the activity of pepsin was not observed at pH 3.2 using hemoglobin as substrate. However, at pH 5 the milk-clotting activity of pepsin was greatly increased in the presence of the activator (Fig. 6). It is possible that the "activator" is digested by

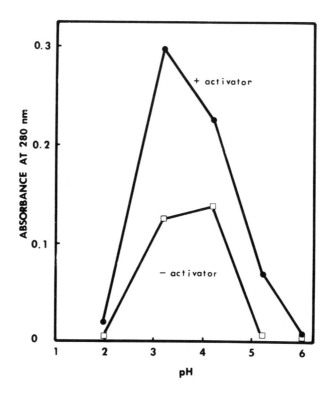

Fig. 4. The pH profile of cathepsin D activity in the presence and absence of the activator. The assay conditions were essentially the same as described in the legend of Fig. 3. The buffers for different pH were 0.25 M glycine-HCl, pH 2.0; 0.25 M sodium formate, pH 3.2; 0.25 M sodium acetate, pH 4.2; 0.25 M sodium acetate, pH 5.2; and 0.25 M sodium phosphate, pH 6.0.

pepsin in pH 3.2, thus causing the loss of "activator" activity.

When the crude "activator" from step 4, Table 1, was fractionated on a column of Sephadex G-100, only a single "activator activity" was observed which corresponded to a molecular weight of 27,500 daltons (Fig. 7). The effects of various compounds on the activity of the "activator" was tested. It was found that the activation was partially abolished by known sulfhydryl reagents (Table 3). Therefore, several known protease inhibitors were tested. Leupeptin and N-tosyl-L-lysine chloromethyl ketone (TLCK) were found to strongly reduce the activity of the activator (Table 3). These results indicate that a sulfhydryl (SH-) protease is responsible for the activity of the "activator."

We have carried out some preliminary studies on the nature of the SH- protease with the "activator" properties. As shown in Table 2, the SH- protease

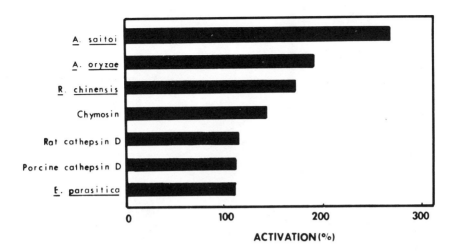

Fig. 5. Activation of carboxyl proteases by "activator." The reaction mixtures included various amounts of acid proteases, from 0.1 µg to 10 µg, either with or without 1 mg of crude activator. The assay was carried out as described for Fig. 3. The "activator" increased the activities from 2 to 3.5 fold.

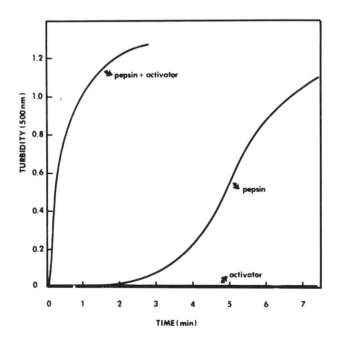

Fig. 6. Effect of activator on the milk-clotting activity of pepsin. The milk-clotting assay was carried out using a spectrophotometric method.[23] The amounts of pepsin and activator in the reaction mixture were 2.5 µg and 10 µg, respectively.

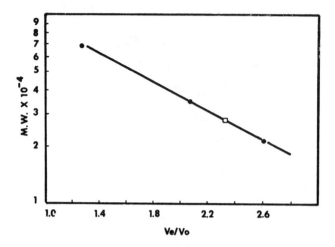

Fig. 7. Molecular weight determination of activator by molecular sieve chromatography. The ratio of the elution volume (Ve) and void volume (Vo) is plotted against the log molecular weight for sephadex G-100 column chromatography. The standards are bovine serum albumin (68,000), *Aspergilus saitoi* acid protease (35,000), and α-chymotrypsin (21,600). The molecular weight of "activator" (□) was estimated as 27,500.

TABLE 3

EFFECT OF VARIOUS COMPOUNDS ON THE ACTIVATOR ACTIVITY

Additions	Concentrations	% Activation	
		With Cathepsin D	With A. *saitoi* protease
None	–	100	100
TLCK	0.1 mM	60	15
TPCK	0.1 mM	98	81
Leupeptin	10 μg/ml	--	10
Iodoacetic acid	1 mM	70	--
p-hydroxy mercuric phenyl-sulfonic acid (sodium salt)	0.1 mM	60	28
EDTA	1 mM	103	--
Heparin	0.2 mg	111	--
Phospholipids	(0.1 ml)	105	--

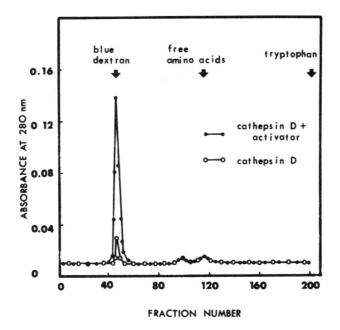

Fig. 8. Column chromatography of the trichloroacetic acid-filtrates of bovine hemoglobin digested by cathepsin D in the presence or absence of activator. The experimental conditions were the same as previously described. After trichloroacetic acid precipitation the filtrates were passed through a column of sephadex G-10 (1.2 x 40 cm) previously equilibrated with 0.1 N acetic acid. The peptide concentration in each fraction (1.35 ml) was monitored by the absorbances at 280 nm and 230 nm. The profile at O.D.$_{230}$ gave the same pattern as that at O.D.$_{280}$. The elution positions of free amino acids were determined with authentic samples in different runs.

has little proteolytic activity toward hemoglobin or other protein substrates. Since the acid exopeptidases are known to be present in various tissues (e.g., cathepsin A, B$_2$, and C), we tested the size of proteolytic products of hemoglobin by cathepsin D in the presence and absence of "activator" SH-protease. Figure 8 shows that the "activation" did not produce small hydrolytic products such as free amino acid. Instead, the enhanced peak by the SH-protease on the chromatogram of a Sephadex G-10 column appeared exclusively at the breakthrough position. These results indicated that the "activator" is not an exopeptidase.

Since the A. *saitoi* carboxyl protease can activate trypsinogen[24] to trypsin, we tested the effect of the "activator" SH-protease on the rate of trypsinogen-

TABLE 4

EFFECT OF ACTIVATOR ON THE ACTIVATION OF TRYPSINOGEN
BY *ASPERGILLUS SAITOI* ACID PROTEASE[a]

	Aspergillus saitoi acid protease		
	50 ng	100 ng	200 ng
	O.D.$_{405}$		
+ activator	0.900	1.415	1.869
- activator	0.329	1.007	1.520

[a]The activation of trypsinogen by *Aspergillus saitoi* acid protease was carried out in a reaction mixture containing 0.1 ml of trypsinogen solution (5 mg in 10 ml of 0.0025 M HCl), 0.2 ml of 0.1 M sodium citrate buffer at pH 3.4, various amounts of *Aspergillus saitoi* acid protease and 50 µg or none of activator. After incubation at room temperature for 10 min, 2 ml of α-N-benzoyl-DL-arginine p-nitroanilide solution (45 mg in 100 ml of 0.05 M·Tris-HCl buffer at pH 8.2 containing 0.002 M CaCl$_2$) was added. After 10 min the reaction was stopped by addition of 1 ml of 30% acetic acid. The absorbance at 405 nm was then measured.

trypsin conversion. As shown in Table 4, about 3-fold increase in the trypsin production was observed at an appropriate SH-protease concentration. Since the conversion of trypsinogen to trypsin involves primarily the cleavage of a single bond at position 6 to 7, the enhancement in this conversion is particularly interesting. We have also compared the polyacrylamide gel electrophoretic patterns of cathepsin D-digested hemoglobin, with and without the SH-protease "activator." The bands, which represent the products, are visibly increased, due to the presence of the "activator." However, the overall positions of the bands were the same. Even though the detailed mechanism of "activation" would require further studies with purified responsible SH-protease, we can tentatively conclude that the "activation" phenomena described here are a result of synergism which requires the proteolytic activity of cathepsin D and at least one SH-protease. Since the "activation" takes place in the acidic pH range, the "activator" SH-protease is probably a lysosomal enzyme.

It is not certain at the present time whether the observed synergism between cathepsin D and the SH-protease has any physiological significance. It could be an effective way to carry out the lysosomal degradation of proteins. However, other biological effects in the regulatory functions are conceivable. Some of these uncertainties will hopefully be clarified when the mechanism of this synergism is studied with purified SH-protease.

ACKNOWLEDGMENT

The original data and the studies cited from our laboratory were supported in part by Research Grants AM-01107 and AM-20212 from the National Institutes of Health. The authors would like to thank Mrs. Azar Fesmire for skillfull technical assistance.

REFERENCES

1. Tang, J. (1977) Acid Proteases, Structure, Function, and Biology, Plenum Press, New York.
2. Hata, T., Hayashi, R., and Doi, E. (1967) Agric. Biol. Chem., 31, 357-367.
3. Levy, M.R., Siddiqui, W.A., and Chou, S.C. (1974) Nature, 247, 546-549.
4. Dickie, N., and Liener, I.E. (1962) Biochim. Biophys. Acta., 64, 41-51.
5. Tang, J., et al. (1973) Proc. Nat. Acad. Sci. USA, 70, 3437-3439.
6. Foltmann, B., et al. (1977) Proc. Nat. Acad. Sci. USA, 74, 2321-2324.
7. Hsu, I.-N., et al. (1977) Nature, 266, 140-145.
8. Foltmann, B., and Pedersen, V.B. (1977) in Acid Proteases, Structure, Function, and Biology, Tang, J., ed., Plenum Press, New York, pp. 3-22.
9. Andreeva, N., et al. (1978) J. Mol. Biol. (Russian), 12, 922-927.
10. Hsu, I.-N, et al. (1977) in Acid Protease, Structure, Function, and Biology, Tang, J., ed., Plenum Press, New York, pp. 61-81.
11. Subramanian, E., et al. (1977) Proc. Nat. Acad. Sci. USA, 74, 556-559.
12. Chen, K.C.S., and Tang, J. (1972) J. Biol. Chem., 247, 2566-2574.
13. Rajagopolan, T.G., et al. (1966) J. Biol. Chem., 241, 4294-4297.
14. Hartsuck, J.A., and Tang, J. (1972) J. Biol. Chem., 242, 2575-2580.
15. Clement, G.E. (1973) Prog. in Bioorg. Chem., 2, 177-238.
16. James, M.N.G., Hsu, I.-N., and Delbaere, T.J. (1977) Nature, 267, 808-813.
17. Marciniszyn, J., Jr., Harsuck, J.A., and Tang, J. (1976) J. Biol. Chem., 251, 7088-7094.
18. Huang, J.S., Huang, S.S., and Tang, J. (1978) J. Biol. Chem., 254, 11405-11417.
19. Huang, J.S., Huang, S.S., and Tang, J. (1979) in Enzyme Regulation and Mechanism of Action, Mildner, P. and Ries, B., eds., Pergamon Press, 289-306.
20. Cunningham, M., and Tang, J. (1976) J. Biol. Chem., 215, 55-66.
21. Tang, J. (1979) Mol. Cell. Biochem., 26, 93-109.
22. McQuillan, M.T., Mathews, J.D., and Trikojus, V.M. (1961) Nature, 192, 333-336.
23. McPhie, P. (1976) Anal. Biochem., 73, 258-261.
24. Gabeloteau, C., and Desnuelle, P. (1960) Biochim. Biophys. Acta., 42, 230.

IV
Oxydases, Dehydrogenases and Hemoglobin

AMINO ACID SEQUENCE STUDIES OF BOVINE HEART
CYTOCHROME OXIDASE: SEQUENCE OF SUBUNIT VI

MASARU TANAKA, TATSUO ETO, KERRY T. YASUNOBU, YAU-HUEI WEI[+] AND TSOO E. KING[+]
Department of Biochemistry-Biophysics, University of Hawaii, Honolulu, Hawaii
96822; [+]Laboratory of Bioenergetics, State University of New York at Albany,
Albany, New York 12222

ABSTRACT

Studies are in progress to determine the amino acid sequences of the various
subunits of bovine heart cytochrome oxidase. Thus far, the amino acid sequence
of subunit V, one of the heme a-binding subunit, has been completed in our
laboratory[1] and the partial sequence of subunit II, one of the possible copper-
containing subunits, has been reported.[2]

The amino acid sequence of subunit VI, the 9.5 K subunit, now has been under
investigation in our laboratory and the results of these investigations will be
presented. According to Yu and Yu,[3] subunit VI is disulfide bonded to subunit
IV. In yeast and probably in mammals, subunit VI is synthesized under the
control of nuclear DNA. Subunit VI has been proposed to face the cytosol-side
of the inner membrane of the mitochondrion.

INTRODUCTION

The importance of cytochrome oxidase, the terminal oxidase of the mitochondrial electron transport system, is illustrated by the dedication of an entire
symposium to cytochrome oxidase at the JASCO Meeting in Kobe, Japan in 1978.[4]
At this meeting our goals and objectives for the sequence determination were
presented.[2] During the past year, new sequence data has been obtained in our
laboratory and this data are presented in this report. Past sequence progress
includes the complete determination of the amino acid sequence of subunit V,[1]
the heme a-containing subunit, and partial sequence of subunit II.[2] Sequence
study of subunit VI has also been now under active investigation and the current
status of our sequence studies of bovine heart cytochrome oxidase subunit VI
are presented in this report.

EXPERIMENTAL PROCEDURES

Materials. Bovine heart cytochrome oxidase subunits were isolated from
lipid-depleted cytochrome oxidase as reported previously.[3,5] The isolated subunits were in pure state by polyacrylamide gel electrophoresis in sodium
dodecyl sulfate. Trypsin, α-chymotrypsin, carboxypeptidase A-DFP (41 U/mg) and

carboxypeptidase B-DFP (83 U/mg) were purchased from the Worthington Biochemical
Corporation. Prior to use, trypsin and α-chymotrypsin were treated with L-l-
tosylamido-2-phenylethyl chloromethyl ketone and L-l-tosylamido-2-lysylethyl
chloromethyl ketone, respectively. All of the reagents used for sequence
determinations were of sequanal quality and were purchased from Pierce Chemical
Company.

Methods. Methods used for sequencing have been described in previous reports
from our laboratory.[1,6] Amino acid analyses were carried out in the Beckman-
Spinco Model 120C and 121MB automated amino acid analyzers as described by
Spackman et al.[7] Sequence determinations were performed manually by a slight
modification of the Edman procedure.[8]

RESULTS

Amino acid sequences of NH$_2$- and COOH-terminal regions of bovine heart
cytochrome oxidase subunits. Table 1 shows amino acid sequences of the amino-
and carboxyl-terminal regions of bovine heart cytochrome oxidase subunits.
Subunit I has an N-formyl blocking group on the amino terminal residue. Pronase
digestion of carboxymethylcysteinyl-subunit I yielded a long size of fragment
consisting of about 23 amino acid residues. This fragment was passed through
a Dowex 50 column and contained no basic amino acids. Carboxypeptidase A
digestion, column purification of the liberated amino acids and amino acid
analyses with and without prior acid hydrolysis of the amino acids established
the presence of a carboxyl-terminal phenylalanine residue and a penultimate
asparagine residue. Carboxypeptidase B and A digestion plus hydrazinolysis
experiments of the carboxymethyl cysteinyl-subunit I showed that carboxyl-
terminal sequence of subunit I was Val-Tyr-Leu-(Asn or Gln)-Lys-COOH. Of the
seven subunits, subunits I and II have amino-terminal formyl groups, while
subunit VI contains an acetyl group, and the three subunits, III, IV and V,
start with amino-terminal alanine, alanine and serine, respectively. Subunit VII
has a blocked amino terminal residue of an unknown nature. The carboxyl-terminal
amino acids of subunits I, III and IV are respectively lysine, lysine and histidine, namely basic amino acids. Subunits II, VI, V and VII contain respectively
leucine, isoleucine, valine and valine, namely hydrophobic amino acids.

Sequence studies of bovine heart cytochrome oxidase subunits. Amino acid
sequence studies of the seven subunits of bovine heart cytochrome oxidase are
in progress. The complete amino acid sequence of subunit V has already been
published.[1] About 93% of partial amino acid sequence of subunit II has also
been published.[2] Sequence studies of subunit VI are now under active investigation in our laboratory.

TABLE 1

AMINO ACID SEQUENCES OF NH_2- AND COOH-TERMINAL REGIONS
OF BOVINE HEART CYTOCHROME OXIDASE SUBUNITS

Subunit	NH_2-terminal Region	COOH-terminal Region
I	Formyl-(Asx$_2$Thr$_2$Ser$_1$Glx$_3$Gly$_2$Ala$_1$Val$_2$- Met$_2$Ile$_1$Leu$_2$Tyr$_2$Phe)-Asn-Phe.	Asn Val-Tyr-Leu- or -Lys Gln
II	Formyl-Met-Ala-Tyr-Pro-Met-Gln-Leu-Gly- Phe-Gln	Ser-Ala-Ser-Met-Leu
III	Ala-His-Gly-Ser-Val-Val-Lys-Ser-Glu-Asp . . .	Asn Leu-Ala- or -Lys-Lys Gln
IV	Ala-Ser-Gly-Gly-Gly-Val-Pro	His
V	Ser-His-Gly-Ser-His-Glu-Thr-Asp-Glu-Glu . . .	Gly-Leu-Asp-Lys-Val
VI	Acetyl-Ala-Glu-Asp-Ile-Gln-Ala-Lys	Phe-Pro-Gly-Lys-Ile
VII	X	Asn (Leu,Tyr,Lys)- or - Val Gln

Amino acid analyses of carboxymethylcysteinyl-subunit VI showed that the
protein consisted of about 85 amino acid residues including six lysine and six
arginine residues. Thus far, from sequence studies of the tryptic peptides of
carboxymethylcysteinyl-subunit VI, six lysine-containing peptides (peptides T2a,
T6a, T7b, T9b, T11b3, and T12b); three arginine-containing peptides (T10b1,
T10c, and T11a); and two free arginine residues (T10a in approximately 200%
yield) have been isolated. Also, free isoleucine (T3a) was obtained, obviously
from carboxyl-terminus as shown in Fig. 1.

In order to obtain overlapping sequences, carboxymethylcysteinyl-subunit VI
was next subjected to chymotrypsin digestion. Six chymotryptic peptides (C1b,
C2a, C3d, C5d, C5e, and C5f) have been isolated thus far as indicated in Fig. 1.

Carboxymethylcysteinyl-subunit VI was cleaved with cyanogen bromide to
obtain overlap data and a useful peptide fragment, B3bA4, was obtained as shown
in Fig. 1.

Furthermore, end group analyses of the carboxymethylcysteinyl-subunit VI
showed that amino-terminus was blocked by an acetyl group and the carboxyl-
terminal amino acid was isoleucine. The tentative partial amino acid sequence
of subunit VI is shown in Fig. 1.

Throughout the sequence studies of subunit VI, we often encountered many difficulties in peptide purification of the tryptic, chymotryptic or CNBr peptides.

DISCUSSION

At the present time, about 93% of amino acid sequence of subunit VI may be tentatively assigned as shown in Fig. 1. The NH_2-terminus of subunit VI was shown to be blocked by an acetyl group while the NH_2 group of subunits I and II are blocked by formyl groups, and subunits III, IV and V have alanine, alanine and serine, respectively, in their amino-terminal positions (Table 1). Subunit VII has a blocked N-terminus of unknown structure.

A tryptic peptide, T10c, obtained from subunit VI has also been isolated and sequenced by Buse et al.[9] from what they designate as subunit VII. The reported sequence for this peptide was H_2N-Gly-Gly-Asp-Val-Ser-Val-Cys-Glu-Glu-Tyr-Arg-COOH. From our sequence studies of peptide T10c, the sequence of the peptide was found to be H_2N-Gly-Gly-Asp-Val-Ser-Val-Cys-Glu-Trp-Tyr-Arg-COOH. Thus, the glutamic acid residue at the position 9 in the peptide reported by the German group should be changed to a tryptophan residue according to our sequence analysis of peptide T10c.

Yu and Yu[3] have reported that subunit VI is -S-S- bonded to subunit IV. As shown in Fig. 1, subunit VI contains three cysteine residues in its sequence. Therefore, at least one cysteine of these three cysteines must be involved in disulfide bonding to subunit IV.

A comment concerning the sequence determination of the cytochrome oxidase subunits is warranted here. Firstly, the insolubility of some of the subunits and the derived peptide fragments caused great difficulties. Secondly, lipids or other membrane components adhere tightly to the peptides and it was difficult to purify the peptides. Thirdly, the strong hydrophobic interactions between the subunits and between the peptide fragments made the purification of the protein and peptides quite difficult. During molecular sieving of the peptide fragments, separate peaks are obtained but without resolution of the peptides. In some cases, the elution patterns were not reproducible. All of these factors make the sequence determination of membrane proteins difficult.

ACKNOWLEDGMENTS

This research was supported in part from Grants GM 22556 from the National

Fig. 1. Tentatively overlapped amino acid sequence of bovine heart cytochrome oxidase subunit VI.

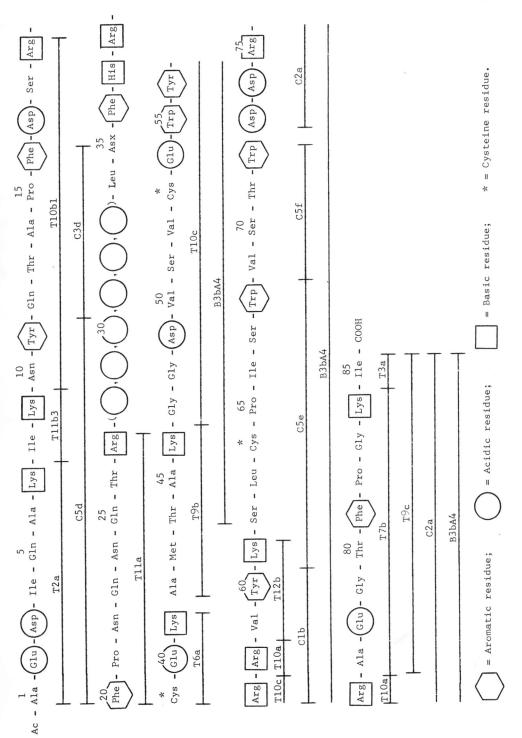

Institutes of Health, RIAS-SER 77-06923 from the National Science Foundation and United States Public Health Service Grants to T. E. King.

REFERENCES

1. Tanaka, M., Haniu, M., Yasunobu, K.T., et al. (1979) J. Biol. Chem., 254, 3879-3885.
2. Yasunobu, K.T., Tanaka, M., Haniu, M., et al. (1979) in Cytochrome Oxidase, King, T.E., et al., eds., Elsevier-North Holland Biomedical Press, Amsterdam, New York, pp. 91-101.
3. Yu, C.A., and Yu, L. (1977) Biochim. Biophys. Acta, 495, 248-259.
4. King, T.E., Orii, Y., Chance, B., and Okunuki, K., eds. Cytochrome Oxidase, (1979) Elsevier-North Holland Biomedical Press, Amsterdam, New York.
5. King, T.E., Yu, L., Yu, C.A., and Wei, Y.H. (1979) in Cytochrome Oxidase, King, T.E., et al., eds., Elsevier-North Holland Biomedical Press, Amsterdam, New York, pp. 53-65.
6. Tanaka, M., Haniu, M., Yasunobu, K.T., and Yoch, D.C. (1977) Biochemistry, 16, 3525-3537.
7. Spackman, D.H., Moore, S., and Stein, W.H. (1958) Anal. Chem., 30, 1190-1206.
8. Tanaka, M., Haniu, M., Yasunobu, K.T., and Mortenson, L.E. (1977) J. Biol. Chem., 252, 7081-7088.
9. Buse, G., Steffens, G.J., and Steffens, G.C.M. (1978) Hoppe-Seyler's Z. Physiol. Chem., 359, 1011-1013.

Published 1980 by Elsevier North Holland, Inc.
Liu/Mamiya/Yasunobu, eds. Frontiers in Protein Chemistry

OPTICAL PROPERTIES, SH-REACTIVITY AND FUNCTION OF HEMOGLOBINS

SHIGERU MATSUKAWA, KAZUHIRO MAWATARI, YOSHIMASA YONEYAMA AND MASAKO NAGAI[*]
Department of Biochemistry, Kanazawa University School of Medicine, and
[*]Biological Laboratory, Kanazawa University School of Paramedicine,
Kanazawa 920, Japan

SUMMARY

The absorption and circular dichroism spectra as well as the reactivity of
4,4'-dithiopyridine with the β-chain cysteine residue-93 of the deoxy-hemoglobin
forms of Hb A, Hb Kempsey, Hb Yakima, Hb Chesabeake and Hb J Capetown were
determined in the presence and absence of inositol hexaphosphate. These properties were correlated with their oxygen binding capabilities. The conclusions
reached were: 1) the absorption and CD spectra could be analyzed according to
the two state theory of Monod, Wyman and Changeux in which there are two extreme
states which are designated as the R- and T-states and ratio of the two forms
is controlled by the equilibrium constant, L. 2) The apparent first order rate
constants for the reaction of the individual hemoglobin chains with 4,4'-
dithiopyridine are qualitatively correlated to L but not when inositol hexaphosphate is added. In the presence of inositol hexaphosphate, the apparent first
order rate constants are instead related to c, which is a ratio of affinity
constants in the equation derived by Monod, Wymand and Changeux. These results
suggested that the T-states of these hemoglobin were heterogeneous.

INTRODUCTION

In order to investigate structure-function relationship of hemoglobin, the
optical properties of the various hemoglobins and β-93 SH reactivities were
studied. The absorption (AB) and circular dichroic (CD) spectra in the Soret
and visible regions represent electronic states of heme, while the β-93 SH
reactivity provides information related to the globin conformation near the
heme. Four mutant hemoglobins i.e., Hb Kempsey, Hb Yakima, Hb Chesapeake and
Hb J Capetown, with $\alpha_1\beta_2$ contract region anomalies, were studied.

EXPERIMENTAL

Materials. Inositol hexaphosphate (IHP) and [bis(2-hydroxyethyl)amino]tris
(hydroxymethyl)methane (bis-Tris) were obtained from Sigma and 4-PDS from
Aldrich.

Hemoglobins. Four abnormal hemoglobins were purified by column chromatography on DEAE cellulose (DE 32, Whatman). Their purities were checked by

384

electrophoresis on polyacrylamide slab gels. Hb A was obtained from fresh
human erythrocytes after lysis with deionized water. Stripped hemoglobins were
prepared by using Sephadex G-25 which was equilibrated with 0.1 M NaCl in 0.05 M
bis-Tris buffer, pH 7.0.

Spectrophotometric measurements. The AB spectra were taken in the Cary
Model 14 spectrophotometer or a Union model SM 401 spectrophotometer (Union
Giken Co., Hirakata, Japan). CD spectra were taken in the JASCO J-20 recording
spectropolarimeter (Japan Spectroscopic Co., Tokyo). The CD measurements were
expressed in degrees square centimeters per decimole of heme. If necessary,
deoxygenation was carried out by repeated alternate evacuation and flushing with
Q gas (helium/isobutane, 99.05: 0.95) in a Thunberg-type cell. Small quantities
of sodium borohydride were added to ensure complete deoxygenation of the sample.
The addition of borohydride had no effect on AB and CD spectra in the Soret and
visible regions. The AB and CD spectra were measured immediately after
deoxygenation. Then the specimens were oxygenated by air and the spectra were
again recorded. Since Hb A and Hb J Capetown could not completely be oxygenated
by air in the presence of IHP (at pH 7.0), 100% oxygenation of the hemoglobin
was performed by flushing with pure oxygen.

Reactivity of the β-93 Sulfhydryl groups with 4-PDS. The reaction of the
reactive sulfhydryl group (β-93 CySH) with 4-PDS was determined as described
by Ampulski et al.[1] with minor modifications. Three ml samples which contained
hemoglobin in 0.1 M NaCl-0.05 M bis-Tris buffer, pH 7.0, with or without 1 mM
IHP, and 0.4 ml of 1.7×10^{-4} M 4-PDS were put into the cuvette and the side
arm of a Thunberg-type cell, respectively. For the complete deoxygenation of
the abnormal hemoglobins with high oxygen affinity, e.g., Hb Yakima and Hb
Kempsey, a small amount of freshly prepared sodium borohydride solution in cold
bis-Tris-NaCl solution was injected through a rubber cap into the evacuated
hemoglobin solution. The contents were allowed to stand at room temperature in
order to further deoxygenate it and also to decompose the excess reductant. The
reaction was started by mixing vigorously the two solutions. The change in
absorbance at 324 nm as a function of time was recorded in a Hitachi model 124
recording spectrophotometer or a Union Model SM 401 spectrophotometer for at
least 60 min at 25°C. The reaction of the SH group in deoxyhemoglobin could
not be followed to the end of reaction because of the slowness of the reaction
and therefore, k_{app} was obtained from Guggenheim plot of the data.

Oxygen equilibrium. The oxygen equilibrium curves of hemoglobins in the
presence and absence of IHP were obtained spectrophotometrically as described
by Sugita and Yoneyama.[2] Deoxygenation was carried out by the same method used
in the spectrophotometric measurements. The pH of hemoglobin solution was

checked by a Hitachi-Horiba pH meter after the oxygen equilibrium measurements
were completed. The Hill's constant, n, was calculated from the slope of the
Hill's plot at 50% oxygen saturation. The content of methemoglobin in the
samples was negligible since the titration spectra showed a sharp isosbestic
point and since no increase of absorbance at 630 nm was observed during the
oxygen equilibrium measurements. In some cases the oxygen equilibrium curve
of hemoglobins was precisely determined using the automatic recording apparatus
similar to the apparatus developed by Imai et al.[3]

Data analysis. The degree of correlation between the intensity of the CD
band and the allosteric constant (L) of hemoglobins was examined by use of a
FACOM 230-35 computer (Fujitsu, Co., Tokyo) by the least square method for a
nonlinear function.

RESULTS

AB and CD spectra of hemoglobins. The AB spectra from 375 nm to 625 nm of
the oxy- and deoxy-forms of four abnormal hemoglobins were measured in 0.05 M
bis-Tris buffer, pH 7.0, containing 0.1 M NaCl. The AB spectra of the oxy-forms
of the four abnormal hemoglobins did not show any large difference from those
of Hb A. However, the spectra of the deoxy-forms of Hb Yakima and Hb Kempsey
were markedly different (Fig. 1A) and those of the deoxy-forms of Hb Chesapeake
and J Capetown were slightly different from that of Hb A.

The CD spectra of the oxygenated forms of four abnormal hemoglobins were
nearly identical to the spectrum of Hb A. The spectra of the deoxygenated forms
of Hb Yakima and Hb Kempsey were, however, greatly different from those of Hb A
(Fig. 1B). The intensity of the positive Soret extremum was less and its
position was red-shifted. The intensity of the positive visible extremum also
was less. The CD spectra of deoxy-Hb Chesapeake and J Capetown, were similar
but yet slightly different from the spectra of Hb A.

Upon the addition of IHP, small changes were observed in the AB and CD
spectra of oxy-forms of the four abnormal hemoglobins. Large changes were
observed in the AB and especially the CD spectra of the abnormal deoxy-
hemoglobins, especially in the cases of Hb Yakima and Hb Kempsey. The resultant
AB and CD spectra of deoxy-Hbs Yakima and Kempsey in the presence of IHP were
almost similar to those for Hb A (Fig. 2A,B). A summary of the AB and CD
spectral properties of Hb A and the four abnormal hemoglobins (deoxygenated
forms) is shown in Table 1.

Reactivity of the SH groups in oxy- and deoxy-hemoglobins with 4-PDS.
Apparent first order rate constants for the reaction of Hb A and four abnormal
hemoglobins with 4-PDS in the presence and absence of IHP, are shown in Table 2.

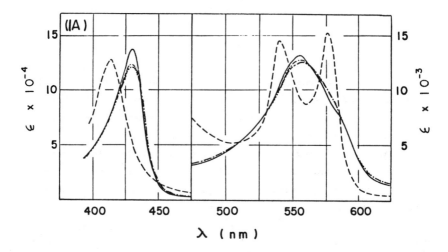

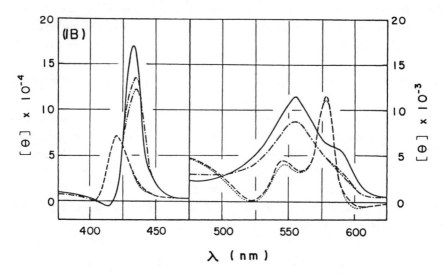

Fig. 1A and B. The AB and CD spectra of some hemoglobins in the Soret and visible wavelength regions shown in Figs. 1A and 1B are the AB and CD spectra, respectively, of the oxygenated forms of Hb A, Hb Yakima, Hb Kempsey (----), and the deoxygenated forms of Hb A(——), Hb Yakima, (-·--·-), and Hb Kempsey (-··--··-). The heme concentration was 50 μM and the samples were dissolved in 0.05 M bis-Tris buffer, pH 7.0, which contained 0.1 M NaCl. The extinction coefficient (ε) ($M^{-1}cm^{-1}$) and the ellipticity (θ) (degrees square centimeters per decimole) values were calculated on the basis of the heme concentration of the samples.

The rate constants (k_{app}^{oxy}) of Hb A and four abnormal hemoglobin were not greatly different. However, larger values of k_{app}^{deoxy} were calculated for the abnormal hemoglobins than for Hb A, except for the case of Hb J Capetown. In the case of Hb Kempsey, the k_{app}^{deoxy} value was almost similar to that for the k_{app}^{oxy} value

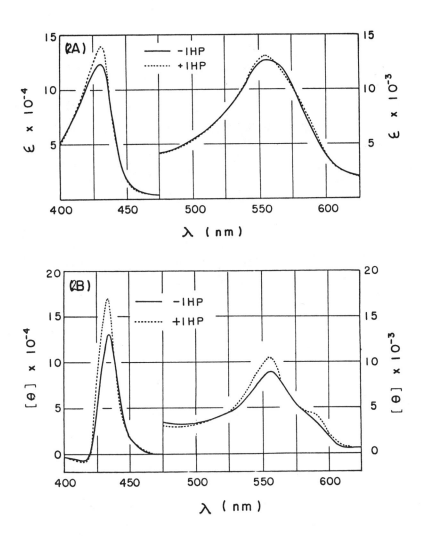

Fig. 2A and B. The AB (Fig. 2A) and CD (Fig. 2B) spectra of deoxygenated Hb Yakima in the presence and absence of IHP. The heme concentration was 50 µM and was dissolved in 0.05 M bis-Tris buffer, pH 7.0, containing 0.1 M NaCl. The IHP concentration when present was 500 µM.

TABLE 1

SUMMARY OF THE SPECTROPHOTOMETRIC PROPERTIES OF THE ABNORMAL
HEMOGLOBIN AND HB A IN THE PRESENCE AND ABSENCE OF IHP

Hemoglobin	IHP	Absorption band Max (nm) $\varepsilon \times 10^{-3}$		CD band Max (nm) $\theta \times 10^{-3}$	
A	−	430	141	433	173
		555	13.3	555	11.1
	+	430	140	433	175
		555	13.4	555	11.5
J Capetown	−	430	140	433	171
		555	13.1	555	10.9
	+	430	140	433	170
		555	13.4	555	11.0
Chesapeake	−	430	140	433	173
		555	13.4	555	11.3
	+	430	142	433	170
		555	13.4	555	10.7
Yakima	−	430	125	435	135
		556	12.9	555	8.8
	+	430	141	433	168
		555	13.4	555	10.5
Kempsey	−	430	122	435	128
		557	12.8	555	8.7
	+	430	140	433	162
		555	13.3	555	10.9

The experimental conditions used are described in the legends for Figs. 1 and
2.

suggesting that the local conformation about the β-93 Cys in this hemoglobin did
not change upon reaction of the -SH group with 4-PDS to the same extent observed
for Hb A. The k_{app}^{deoxy} value for Hb Yakima was about half of its k_{app}^{oxy} value. It
can be seen in Table 2 that the SH reactivities of the various deoxyhemoglobins,
i.e., the k_{app}^{deoxy} values for the various hemoglobins in the presence and absence
of IHP, decreased in the following order: Hb Kempsey, Hb Yakima, Hb Chesapeake,
Hb J Capetown and Hb A.

Oxygen equilibrium of hemoglobins. The oxygen equilibrium properties of Hb A
and the abnormal hemoglobins in the presence and absence of IHP are shown in

TABLE 2

EXPERIMENTAL VALUES OF THE SH REACTIVITY WITH 4-PDS OF THE
VARIOUS HEMOGLOBINS IN THE PRESENCE AND ABSENCE OF IHP

Hemoglobin	IHP	k_{app} ($\times 10^{-4} sec^{-1}$)	
		oxy	deoxy
A	-	61.7	1.90
	+	41.7	1.34
J Capetown	-	76.3	5.83
	+	57.6	3.33
Chesapeake	-	50.0	11.3
	+	46.7	6.17
Yakima	-	62.5	41.7
	+	42.0	7.33
Kempsey	-	78.3	80.0
	+	66.7	23.3

The reaction mixture contained 50 μM hemoglobin and 0.1 μM of 4,4'-dithiopyri-
dine (4-PDS) in 0.05 M bis-Tris buffer-0.1 M NaCl, pH 7.0. IHP (1 mM final
concentration) was added to the hemoglobin solution in the cuvette prior to the
reaction with 4-PDS present in the side arm.

Table 3. The P_{50} value which is the oxygen pressure at 50% saturation for the
various types of hemoglobin in the presence and absence of IHP decreased in the
following order: Hb A, Hb J Capetown, Hb Chesapeake, Hb Yakima and Hb Kempsey.
The Hill's constant, n, an indicator of cooperativity determined for the various
types of hemoglobins changes in a similar way as did the P_{50} values reported
above with minor exceptions. The addition of IHP raised the values of both the
values of P_{50} and n.

DISCUSSION

Optical properties and oxygen equilibria. In order to correlate the optical
properties of hemoglobin A and the four abnormal hemoglobins to their oxygen
equilibria in the presence and absence of IHP, the allosteric parameters L and c,
of Monod, Wyman and Changeux[4] were calculated from the P_{50} and n values as
described by Edelstein[5] and Bunn and Guidotti[6] using the following equation:

$$n = 1 + 3 \frac{(\gamma - 1)(c\gamma - 1)}{(1 + \gamma)(c\gamma + 1)} \qquad (1)$$

TABLE 3

OXYGEN BINDING PROPERTIES AND MWC PARAMETERS OF NORMAL
AND ABNORMAL HEMOGLOBINS IN THE PRESENCE AND ABSENCE OF IHP

Hemoglobin	IHP	Hill's $\underline{n}$	P_{50} (mmHg)	$\underline{L}$	$\underline{c}$
A	−	2.7	5.4	3.3×10^{4}	0.01
	+	2.6	54.0	3.3×10^{8}	0.004
J Capetown	−	2.2	3.0	4.1×10^{3}	0.042
	+	2.4	19.5	5.7×10^{6}	0.017
Chesapeake	−	1.2	0.72	1.0×10^{1}	0.35
	+	1.9	5.6	3.8×10^{4}	0.035
Yakima	−	1.0	0.36	1.0	1.0
	+	1.8	2.25	1.0×10^{3}	0.098
Kempsey	−	1.0	0.28	2.5×10^{-1}	1.43
	+	1.6	1.4	1.5×10^{2}	0.13

The oxygen binding curves were measured at 25° in 0.05 M bis-Tris buffer pH 7.0,
containing 0.1 M NaCl, with or without 1 mM IHP. The hemoglobin concentration
was 50 μM (heme basis). The MWC parameters, $\underline{L}$ and $\underline{c}$ were determined using the
equations shown in the text.

In this equation n is Hill's constant and γ is equal to $P_{50}/0.4$. The $\underline{L}$ and $\underline{c}$
values obtained for the various hemoglobins studied are summarized in Table 3.

The CD spectra of Hb A and the abnormal hemoglobins were analyzed in terms of
the two state MWC model as follows. The intrinsic ellipticities θ_{R} and θ_{T} were
assigned to the R and T state, respectively. Then the observed ellipticity (θ)
for each type of hemoglobin was related to the R and T states by the following
equation:

$$(\underline{\theta}) = \frac{R}{T + R} \ (\theta_{R}) + \frac{T}{T + R} \ (\theta_{T}) \tag{2}$$

where T/R is the allosteric constant, $\underline{L}$. Substituting for $\underline{L}$ in Equation 2,
Equation 3 is obtained:

$$(\underline{\theta}) = \frac{1}{1 + L} \ [(\theta_{R}) + \underline{L} \ (\theta_{T})] \tag{3}$$

Among the various extrema present in the CD spectrum of the various deoxy hemoglobin samples, the positive CD band in the Soret region, the intensity of which

is given in Table 1, were used for initial analysis. The logarithmic values of θ were plotted against log $(1 + \underline{L})$ as shown in Fig. 3. From the observed ellipticities of deoxy-Hb A and the abnormal hemoglobins, with and without IHP, the most probable values of (θ_R) and (θ_T) which satisfy the Equation 3, were estimated by computer curve fitting analysis. The values of (θ_R) and (θ_T) thus calculated were 1.1×10^5 and 1.60×10^5, respectively (Fig. 3). The preliminary results strongly suggested that the positive CD band in the Soret region could be used for the two state model analysis, in which the intrinsic (θ_R) and (θ_T) values for the R and T states are used for the analysis. The ellipticity value of the hemoglobin fixed in the R state (θ_R) was calculated to be 1.1×10^5 which was significantly larger than the mean value for individual α and β chains $[(\theta) = 0.8 \times 10^5]$ which indicated the structure of the hemoglobin fixed in the R state were not simply a sum of values for the separate α and β chains but indicated some different interactions from the T state also exist among four subunits.

Expansion of these analyses to all spectral portions of the Soret band led to the theoretical AB and CD spectra of the deoxyhemoglobin samples in the intrinsic R and T states. Calculated AB spectra for hemoglobin fixed in the R

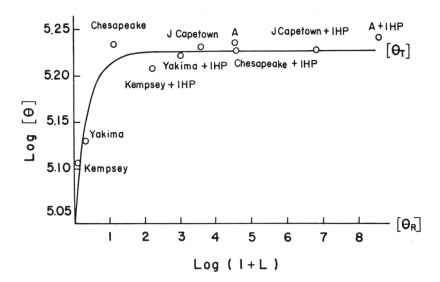

Fig. 3. Correlation between molar ellipticity at 433 nm and the allosteric constant $\underline{L}$ of the deoxygenated hemoglobins. The logarithmic values of θ of deoxygenated hemoglobins shown in Table 1 were plotted against log $(1 + \underline{L}$. The values of $\underline{L}$ shown in Table 3 were used in this figure.

state and in the T state, together with the difference spectra are shown in Fig. 4. In these calculations, the $\underline{\theta}$, $\underline{\theta}_R$, and $\underline{\theta}_T$ values of Equation 3 were replaced by $\underline{\varepsilon}$, $\underline{\varepsilon}_R$, and $\underline{\varepsilon}_T$, respectively, using the least square method. The difference spectra thus obtained, were very similar to those reported by Gibson[7] for the deoxy-T minus the deoxy-R hemoglobin, a spectrum formed immediately after flash photolysis of Hb A-CO and Hb A-O_2.

SH reactivities and oxygen equilibria. Like the optical properties, the SH reactivities of Hb A and the four abnormal hemoglobins were analyzed according to the two state model. The apparent first order rate constant, k_{app}^{deoxy}, experimentally determined for the deoxy-hemoglobins, would be calculated by

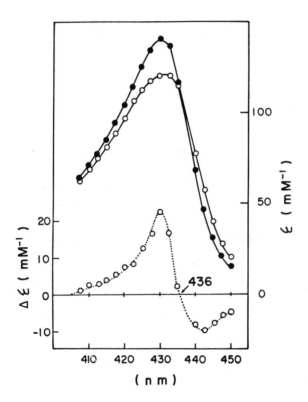

Fig. 4. Calculated AB spectra of the R and T states of deoxygenated hemoglobin in the Soret region and difference spectrum between them. Symbols used are (●——●), deoxy-T state hemoglobin; (o——o) deoxy-R state hemoglobin; and (----), the difference spectrum (T minus R states). The arrow points to the isosbestic point at 436 nm.

K_{app}^{T}, an apparent first order rate constant for hemoglobin in the T state, and k_{app}^{R} that for the R state hemoglobin, and by the relative population ratio of the hemoglobin in the T and R states according to the following equation:

$$k_{app}^{deoxy} = \frac{T}{T + R} \; k_{app}^{T} + \frac{R}{T + R} \; k_{app}^{R} \qquad (4)$$

In the equation, T/R is equal to $\underline{L}$, the allosteric constant for the two state model (MWC). Substitution of L into Equation 4 gave rise to Equations 5 and 5a shown below.

$$k_{app}^{deoxy} = \frac{1}{1 + \underline{L}} \; (k_{app}^{T} \; \underline{L} + k_{app}^{R}) \qquad (5)$$

$$= \frac{k_{app}^{R}}{1 + \underline{L}} \; (A \cdot L + L) \qquad (5a)$$

In Equation 5a, A is equal to k_{app}^{T}/k_{app}^{R}. If the assumption is made that k_{app}^{R} abd k_{app}^{T} values correspond to the k_{app}^{oxy} and k_{app}^{deoxy} values for Hb A since it is generally accepted that the oxy-and deoxy Hb A have typical R and T structures, respectively, we can calculate the value of k_{app}^{deoxy}, the intrinsic rate constant, easily using the $\underline{L}$ values obtained for a series of hemoglobins. As shown in Table 4, some of the calculated values were approximately the same as the experimental values in the absence of IHP, except in the case of Hb J Capetown. IHP could possibly displace the R-T equilibrium state of the hemoglobins, and thereby cause an increase in their $\underline{L}$ values.[8] Consequently, the k_{app}^{deoxy} value, which was estimated by introducing $\underline{L}$ value for each hemoglobin in the presence of IHP into Equation 5 should approach a constant value of approximately 1.9 which is the value found for Hb A. However to the contrary, as shown in Table 2, the experimental values of k_{app}^{deoxy} for hemoglobins in the presence of IHP are different. It is suggested that these discrepancies in the SH reactivity of all the unliganded hemoglobins tested in the present investigation cannot be explained in terms of the simple two state model, in which the k_{app}^{R} and k_{app}^{T} values must be fixed to the same values for the various hemoglobins tested.

The above argument led to the concept that the rate constant for the SH reactivity of the hemoglobins in the T state differ from one another, while the oxy-forms of these hemoglobins which were predominantly in the R state had slightly different values of k_{app}^{oxy} (see Table 2). In order to check this idea, the parameter $\underline{A}$ (which is equal to k_{app}^{T}/k_{app}^{R}) for each hemoglobin was calculated by substituting the $\underline{L}$ and the experimental values of k_{app}^{deoxy} and k_{app}^{oxy} in place of k_{app}^{R} value into Equation 5a and the results are summarized in Table 4. None

TABLE 4

CALCULATED APPARENT FIRST ORDER RATE CONSTANTS FOR THE REACTION OF THE
SH GROUPS WITH 4-PDS (k_{app}^{deoxy}) AND PARAMETER $\underline{A}$ OF VARIOUS HEMOGLOBINS IN
THE PRESENCE AND ABSENCE OF IHP

Hemoglobin	IHP	k_{app}^{deoxy} (x $10^{-4}sec^{-1}$)[a] calculated	Parameter[b] $\underline{A}$
A	−	1.90	0.030
J Capetown	−	1.90	0.076
Chesapeake	−	7.34	0.149
Yakima	−	31.8	0.334
Kempsey		49.8	1.1
A	+	1.90	0.032
J Capetown	+	1.90	0.058
Chesapeake	+	1.90	0.132
Yakima	+	1.90	0.174
Kempsey	+	1.90	0.35

[a]It is assumed that k_{app}^{R} and k_{app}^{T} values correspond to k_{app}^{oxy} and k_{app}^{deoxy} values,
respectively for dissociated Hb A, and were calculated by varying only the $\underline{L}$
values in Eq. (5) irrespective of the presence or absence of IHP.

[b]Calculated from Eq. (5a) using $\underline{L}$ values and the experimental k_{app}^{deoxy} and k_{app}^{oxy}
values which were assumed to be equal to k_{app}^{R}. The k_{app}^{deoxy} and k_{app}^{oxy} values are
presented in Table 2. The use of the k_{app}^{oxy} value instead of the k_{app}^{R} value is
made on the assumption that the R state quaternary structure is the same for
all hemoglobins.

of the $\underline{A}$ values calculated for these hemoglobins were found to be constant
either in the presence or absence of IHP. This finding suggested that these
hemoglobins have their own intrinsic values of k_{app}^{T} as well as $\underline{c}$, the MWC
parameters for binding of oxygen.

Inspection of $\underline{A}$ and $\underline{c}$ values have suggested that some sort of relationship
existed between these parameters. When the $\underline{A}$ and $\underline{c}$ values of each hemoglobin
shown in Table 4 were plotted on a log-log graph as shown in Fig. 5, a linear
relationship was noted. Although $\underline{c}$ is a ratio of dissociation constants and
$\underline{A}$ is the ratio of kinetic constants, the linear relationship observed indicated

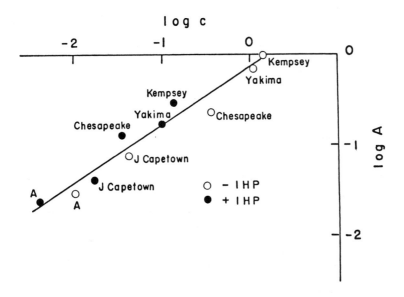

Fig. 5. Relation between log $\underline{A}$ (equal to k^{T}_{app}/k^{R}_{app}) and log $\underline{c}$ equal to (k_{R}/k_{T}), the MWC model parameters. The values of $\underline{A}$ and $\underline{c}$ used were obtained from Tables 4 and 3, respectively.

that the conformational factors governing oxygen affinity of the hemoglobin in the T state is closely related to the factors which determine the SH reactivity of each hemoglobin. Furthermore, from structural considerations, the results supported the concept that there are real quantitative differences in the T state quaternary structures but not in the R structure for these hemoglobins, both in the absence and presence of IHP.

What is the reason for the linear relationship between the values of $\underline{A}$ and $\underline{c}$? In the case of normal hemoglobin, the proximal His, which is present in helix F in both the α and β chains are directly coordinated to heme iron. The position of helix F, which anchors the heme iron, may be capable of finely regulating the oxygen affinity of hemoglobin in the T state. The β-chain CySH-93 which is an immediate neighbor of β-chain His 92, the proximal base of β chains, must move together with helix F. It is likely that the SH reactivity of deoxy-hemoglobin is influenced by the flexibility of helix F present in the β-chains. The high affinity for oxygen of the T state of some of the abnormal hemoglobins relative to normal hemoglobin may be explained by making the assumption that helix F is not restricted as it is in the case of Hb J Capetown and Hb A but is more flexible with regards to the movement towards the heme in case of Hb Kempsey, Hb Yakima, and Hb Chesapeake. The difference is thought to provide an explanation for the linear relationship observed between the kinetic parameter $\underline{A}$ and MWC equilibrium parameter $\underline{c}$.

ACKNOWLEDGMENTS

Thanks are due to Drs. R.T. Jones, R.D. Koler (Hb Yakima), S. Gordon (Hb Kempsey), S. Charache (Hb Chesapeake), and M.C. Botha for the shipment of blood samples containing the various hemoglobins and to Dr. T. Saito for the purification of hemoglobin from the various blood specimens. The study was partly supported by funds from the Japanese Education Ministry, the Japan-U.S. Science Cooperative Program, the Mistubishi Foundation and the Naito Foundation.

REFERENCES

1. Ampulski, R.S., Ayers, V.E., and Morell, A.S. (1969) Anal. Biochem., 32, 163-169.
2. Sugita, Y. and Yoneyama, Y. (1971) J. Biol. Chem., 246, 389-394.
3. Imai, K., Morimoto, H., et al. (1970) Biochim. Biophys. Acta., 200, 189-196.
4. Monod, J., Wyman, J. and Changeux, J.P. (1969) J. Mol. Biol., 12, 88-118.
5. Edelstein, S.J. (1971) Nature, 230, 224-227.
6. Bunn, H.E. and Guidotti, G. (1972) J. Biol. Chem., 247, 2345-2350.
7. Gibson, Q.H. (1959) Biochem. J., 71, 293-303.
8. Matsukawa, S., et al. (1979) J. Biol. Chem., 245, 2358-2363.

SEQUENCE STUDIES OF THE *PSEUDOMONAS PUTIDA* CYTOCHROME P-450: A PROGRESS REPORT ON THE PARTIAL SEQUENCE

MITSURU HANIU, LYMAN G. ARMES, KERRY T. YASUNOBU, AND IRWIN C. GUNSALUS[+]
Department of Biochemistry-Biophysics, University of Hawaii, Honolulu, Hawaii, 96822; [+]Department of Biochemistry, University of Illinois, Urbana, Illinois, 61801

ABSTRACT

Pseudomonas putida contains a camphor inducible mixed-function oxidase system which catalyzes the hydroxylation of the 2-boranone skeleton at the 5-exo position. The system consists of three protein components; putidaredoxin reductase, a flavoprotein; putidaredoxin, an iron-sulfur protein; and cytochrome P-450$_{cam}$, the heme containing oxidase. Our group is engaged in determining the primary structures of the various components of this system. The amino acid sequence of putidaredoxin has been reported by Tanaka et al. (J. Biol. Chem., 249, 3680, 1974). Present investigations are centered on the heme containing oxidase. The procedures studied thus far include: amino terminal sequence determination, isolation of the tryptic and the clostripain peptides, isolation of the cyanogen bromide fragments, and isolation of the limited acid cleavage products. The current status of the sequence research is presented.

INTRODUCTION

Cytochrome P-450 coupled oxidation-reduction activity has been observed in numerous living systems.[1-4] In humans, P-450 systems participate in steroidogenesis, and the metabolism of fatty acids and xenobiotics. The physical and chemical properties of the catalytic centers of all such P-450 systems show considerable analogy.

Pseudomonas putida, cultured on a medium containing (+)-camphor as sole carbon source, possesses two parallel metabolic pathways for the oxidative cleavage of the 2-boranone skeleton.[5] The inducible enzymes of these pathways include a mixed-function oxidase system which catalyzes the sterospecific hydroxylation of the 5-exo position of the 2-boranone molecule. This system consists of three components: a flavoprotein, putidaredoxin reductase; an iron-sulfur protein of the $Fe_2S_2Cys_4$ class, putidaredoxin; and a heme containing b-variant cytochrome, P-450$_{cam}$.[3,6] The homogeneous constituents of this system are easily obtained in substantial quantities and, hence, are particularly amenable to investigation.

We are presently determining the primary structure of cytochrome P-450$_{cam}$.
It is hoped that this data in conjunction with X-ray crystallographic data will
elucidate the nature of the reductive cleavage of oxygen as well as provide a
model for other enzymes of the P-450 class.

EXPERIMENTAL PROCEDURES

Materials. Cytochrome P-450 was isolated and purified from *Pseudomonas*
putida P$_p$G786 (ATCC 29607) as described previously.[7] The crystalline preparations used were pure by SDS-disc electrophoretic analyses. The various
sources of materials used for this study have been described in a previous
publication from our laboratory.[8] Sources of chemicals not described previously
include clostripain (500 units per mg) which was purchased from Boehringer
Mannheim Co. and 3-hydroxy-propane-sulfonic acid γ-sultone from Aldrich
Chemical Co.

Preparation of enzyme derivatives. The CysCm-derivative was prepared as
described by Crestfield et al.[9] and the sulfopropyl-cysteine derivative was
prepared by an adaptation of the procedure described by Rüegg and Rudinger.[10]

Amino acid analyses were run on the Beckman Model 121 MB automated amino
acid analyzer which is a commercial model of the instrument developed by
Spackman et al.[11] The NH$_2$-terminal sequence of the protein was run in the
Beckman Model 890 Protein Sequencer which is the commercial version of the
instrument developed by Edman and Begg.[12] Manual Edman degradation was performed by the procedure of Edman.[13] Carboxypeptidase digestions were carried
out as described by Ambler.[14]

Tryptic cleavage of the CysCm-cytochrome P-450 were carried out at 25° for
24 hours and the reactant mixture applied to Sephadex G-50 column (3.0 x 55 cm)
after lyophilization.

Clostripain digestion was carried out as described by Mitchell.[15] The CysCm-
cytochrome P-450 was dissolved in 0.1 M sodium phosphate buffer, pH 7.8, containing 5 mM dithiothreitol. The clostripain protein ratio was 1 : 30 (w/w)
and the digestion was carried out at 37°C for 64 hours. The digestion mixture
was applied to the column of Sephadex G-100 (3.6 x 60 cm) after lyophilization.

Limited acid cleavage of P-450$_{cam}$ was effected by placing the protein in
7.0 M guanidine hydrochloride brought up in 70% formic acid for 48 hours at
37°C.[16]

The cyanogen bromide digestion of P-450$_{cam}$ was carried out in the dark at
room temperature under a nitrogen atmosphere for a period of 24 hours in 70%
formic acid and a 50-fold excess of cyanogen bromide over methionine.[17]

TABLE 1

AMINO ACID COMPOSITION OF THE *P. Putida* Cytochrome-P-450

Amino Acid	Residues/mole
Lysine	13
Histidine	12
Arginine	24
Aspartic Acid	27
Asparagine	9
Threonine	19
Serine	21
Glutamic Acid	42
Glutamine	13
Proline	27
Glycine	26
Alanine	34
Half Cystine	6
Valine	24
Methionine	9
Isoleucine	24
Leucine	40
Tyrosine	9
Phenylalanine	17
Tryptophan	1
Total Amino Acids	397

RESULTS AND DISCUSSIONS

Amino acid composition and end groups. The amino acid composition of cytochrome P-450$_{cam}$ as reported by Dus et al.[18] is summarized in Table 1. In contrast to these findings, seven cysteine and 2-3 tryptophan residues have been determined by sequence analysis. The end groups of the P-450 have been previously reported; the amino terminus being threonine and the carboxyl terminus valine. The amino terminal sequence is shown in Fig. 1. The single polypeptide chain contains no disulfide bridges.[19] Evidence suggests that the heme iron is ligated through a mercaptide linkage to a cysteinyl residue.[20]

Strategy of protein sequencing. Four methods of fragmentation have been performed thus far. These include: 1) tryptic cleavage of the CysCm derivative, 2) clostripain fragmentation of the CysCm derivative, 3) cyanogen bromide

TABLE 2

AMINO ACID SEQUENCES OF SOME TRYPTIC PEPTIDES

Peptide No.	Sequence
1B1e	Ile-Gln-Glu-Leu-Ala-Cys-Ser-Leu-Ile-Glu-Ser-Leu-Arg-Pro-Gln- Gly-Gln-Cys-Asn-Phe-Thr-Glu-Asp-Tyr-Ala-Glu-Pro-Phe-Pro-Ile- (Arg)
1B1g	Thr-Thr-Glx-Thr-Ile-Glx(Ser)-Asx-Ala-Asx-Leu-Ala-Pro-Leu-Pro- Pro( )Val-Pro-Glx( )Leu------
2cd	Ile-Pro-Ala-Ala-Cys-Glu-Glu-Leu---------
2F	Met-Cys-Gly-Leu-Leu-Leu-Val-Gly-Gly-Leu-Asp-Thr-Val-Val-Asn- Phe-Leu-Ser-Phe-Ser-Met-Glu-Phe-Leu-Ala(Lys)
3Ba	Glx(Asx)Ile-Val(Thr)-Leu-Lys
3Bb	Gln-Asn-Ala-Cys-Pro-Met(His)-Val-Asx-Phe-Ser(Arg)
3Bd	Ile-Pro-Asx-Phe(Ser)Ile-Ala-Pro-Gly-Ala-Glx-Ile-Glx--------- (Lys)
3Bf	Glx-Ala-Leu-Tyr-Ile-Tyr-Leu-Ile-Pro-Ile-Ile-Glx------
3Dc	Cys-Asp-Gly-Gly-His-Trp-Ile-Ala-Thr-Arg
3De	His-Phe-Ser(Ser)Glu-Cys-Pro-Phe-Ile-Pro(Arg)
4Fa	Ser-Pro-Glx-His-Arg
4G	Glx(Trp)-Leu-Thr-Arg

digestion of the sulfopropylcysteine derivative, and 4) limited acid cleavage of the sulfopropylcysteine derivative. The sequence data obtained by these procedures are presented in Tables 2 and 3.

Isolation and sequence determination of the clostripain peptides. Since clostripain cleaves after the arginine residues only, it was used to isolate larger fragments with the lysine peptide bonds still intact. Since peptides are still being isolated, sequence data of the peptides purified to date are summarized in Table 3. Note that peptide C-2C is the NH_2-terminal peptide while peptide C3Af is the COOH-terminal peptide of cytochrome P-450.

Cyanogen bromide digestion of the sulfopropylcysteine derivative. As shown in Table 1, Dus et al.[18] have reported nine methionine residues in P-450$_{cam}$, hence, ten fragments are expected assuming quantitative cleavage. The cyanogen bromide fragments are currently being purified. The gel filtration profile of the fragments on Sephadex G-75 (2.5 x 92.5 cm) is given in Fig. 2. The amino terminal sequence of the peptide denoted by Roman numeral IV is shown in Fig. 3.

Limited acid cleavage of the sulfopropylcysteine derivative. Both the gel

TABLE 3

SEQUENCES OF SOME CLOSTRIPAIN PEPTIDES

Peptide No.	Sequence
C3Aa	Ala-Leu-Ala-Asn-Gln-Val-Val-Gly-Met-Pro-Val-Val-Asp-Lys- Leu-Glu-Asn-Arg
C3Af	Ile-Pro-Asp-Phe-Ser-Ile-Ala-Pro-Gly-Ala-Gln-Ile-Gln-His- Lys-Ser-Gly-Ile-Val-Ser-Gly-Val-Gln-Ala-Leu-Pro-Leu-Val- Ser-Asn-Pro-Ala-Thr-Thr-Lys-Ala-Val
C3Eb	Gly-Gln-Leu-Ile-Arg-Glu-Ala-Tyr-Glu-Asp-Tyr-Arg
C3Ec	Ile-Pro-Ala-Ala-Cys-Glu-Glu-Leu-Leu-Arg
C3Fc	His-Phe-Ser-Ser-Glu-Cys-Pro-Phe-Ile-Pro-Arg-Glu---------- -(Arg)
C3Gd	Ile-Gln-Glu-Leu-Ala-Cys-Ser-Leu-Ile-Glu-Ser-------(Arg)
C2A	Ile-Leu-Thr-Ser-Asp-Tyr-Glu-Phe-His-Gly-Val-Gln-Leu- (Lys)(Lys)Gly(Asx)(Glx)Ile-Leu-Leu-Pro(Glx)(Met)-Leu-Ser- Gly-Leu(Asx,Glx,Arg)
C2C	Thr-Thr-Glu-Thr-Ile-Gln-Ser-Asn-Ala-Asn-Leu-Ala-Pro-Leu- Pro-Pro(His)Val-Pro-Glu(His)-Leu-Val-Phe-Asx-Phe-Asx- Met-Tyr--------------------------(Arg)

10
Thr-Thr-Glu-Thr-Ile-Gln-Ser-Asn-Ala-Asn-Leu-Ala-Pro-
20
Leu-Pro-Pro-His-Val-Pro-Glu-His-Leu-Val-Phe-Asx-Phe-
39
Asx-Met-Tyr-Asx-Pro-Ser-Asx-Leu-Ser-Ala-Gly-Val-Glx-

Fig. 1. NH_2-terminal sequence of cytochrome P-450$_{cam}$.

filtration profile shown in Fig. 4 and SDS-urea gel electrophoretic data indicate the presence of two components in the limited acid cleavage mixture. Dansylation data indicates the lower molecular weight constituent to be the amino terminal portion of the protein. The partial amino terminal sequence of the higher molecular weight component is given in Fig. 5.

From the various enzymatic and chemical digests of cytochrome P-450, six of the nine residues preceding and following the methionine residues and this data is summarized in Table 4. In addition, there is a methionine residue at

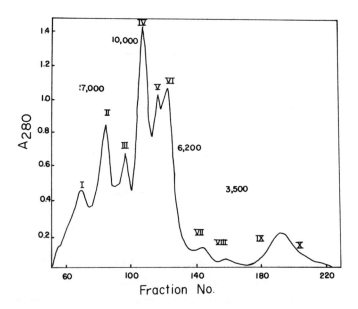

Fig. 2. Gel filtration profile of cyanogen bromide digest of cytochrome P-450_{cam} on sephadex G-75 superfine column. The digest was dissolved and chromatographed in 0.2 M sodium formate buffer, pH 4.0, 5.0 M guanidine hydrochloride, and 0.001 M dithiothreitol. Arabic numbers on the graph represent molecular weight estimates determined from a previous chromatography of protein standards.

```
1                                    10
Leu-Ser-Gly-Leu-Asx-Glx-Arg-Glx-Asx-Ala-Cys-Pro-Met
```

Fig. 3. NH$_2$-terminal sequences of CNBr-IV.

position 28 from the NH$_2$-terminal end of the protein.

SUMMARY

Cytochrome P-450$_{cam}$ is a large single chain protein consisting of approximately 400 residues. The NH$_2$-terminal and COOH-terminal sequences were determined as shown in Fig. 6. In order to produce the overlaps necessary for the complete primary structural determination, four methods of fragmentation are being investigated. Data gathered from the sequencing of the amino terminus of the protein, the tryptic and clostripain peptides, the cyanogen

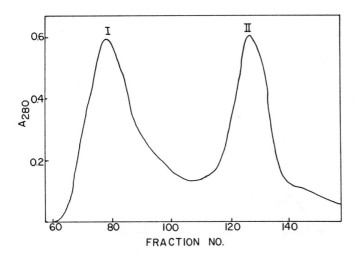

Fig. 4. Gel filtration profile of limited acid cleavage products of cytochrome P-450$_{cam}$ on sephadex G-75 superfine column. The digest was dissolved and chromatographed in 0.2 M sodium formate buffer, pH 4.0, 5.0 M guanidine hydrochloride, and 0.001 M dithiothreitol.

```
      1                                          10
Pro-Pro-Glx-Glx-Arg-Glx-Phe-Arg-Ala-Leu-Ala-Asx-Glx-Val-Val-

        17
Gly-Met------------------------------------------------------
```

Fig. 5. Partial NH$_2$-terminal sequence of the large limited acid cleavage fragment.

TABLE 4

MET-X LINKAGES IN CYTOCHROME P-450

	Peptide	Met-X
1.	C2C and 1Blg	Asx-Met-Tyr
2.	3Bb	Pro-Met-His
3.	2F	(Lys/Arg)-Met-Cys
4.	C3Aa	Gly-Met-Pro
5.	C2A	Glx-Met-Leu
6.	2F	Ser-Met-Glu

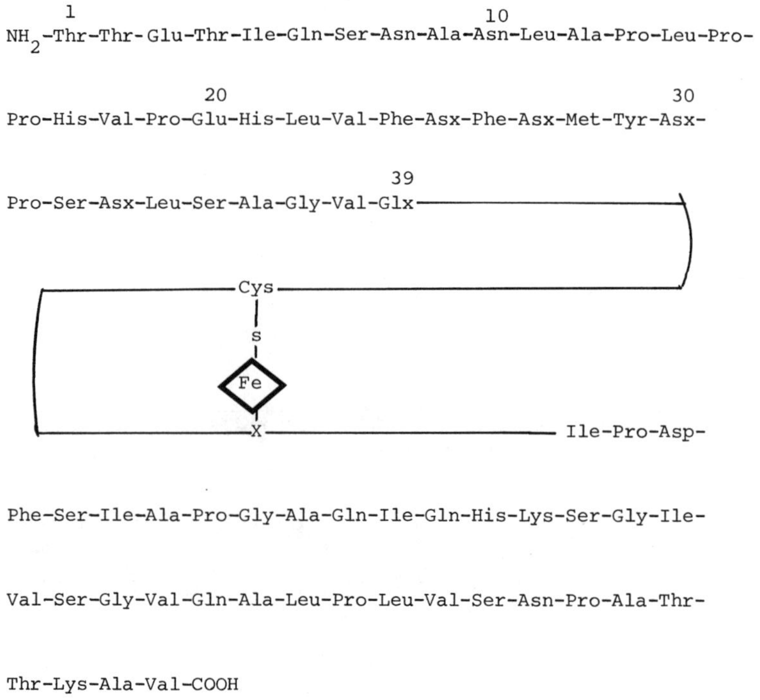
1
10
NH2-Thr-Thr-Glu-Thr-Ile-Gln-Ser-Asn-Ala-Asn-Leu-Ala-Pro-Leu-Pro-
20
30
Pro-His-Val-Pro-Glu-His-Leu-Val-Phe-Asx-Phe-Asx-Met-Tyr-Asx-
39
Pro-Ser-Asx-Leu-Ser-Ala-Gly-Val-Glx
Cys
S
Fe
X
Ile-Pro-Asp-
Phe-Ser-Ile-Ala-Pro-Gly-Ala-Gln-Ile-Gln-His-Lys-Ser-Gly-Ile-
Val-Ser-Gly-Val-Gln-Ala-Leu-Pro-Leu-Val-Ser-Asn-Pro-Ala-Thr-
Thr-Lys-Ala-Val-COOH

Fig. 6. NH_2-terminal and COOH-terminal sequences of cytochrome P-450_{cam}.

bromide fragments, and the limited acid cleavage products has led to the tentative assignment of approximately seventy percent of the primary structure. This chemical sequence data will complement the X-ray crystallographic studies of P-450_{cam} now in progress and aid in the elucidation of the three dimensional structure of this important enzyme.

ACKNOWLEDGMENTS

This work is supported in part by research grants NIH GM 22556 and NSF RIAS-SER 77-06923.

REFERENCES

1. Omura, T., and Sato, R. (1964) J. Biol. Chem., 239, 2370.
2. Madyastha, K.M., Meehan, T.D., and Coscia, C.J. (1976) Biochemistry 15, 1097.
3. Katagiri, M., Ganguli, B., and Gunsalus, I.C. (1968) J. Biol. Chem., 243, 3543.
4. Haugen, D.A., and Coon, M.J. (1976) J. Biol. Chem., 251, 7929.

5. Hedegaard, J., and Gunsalus, I.C. (1965) J. Biol. Chem., 240, 4038.

6. Cushman, D.W., Tsai, R.L., and Gunsalus, I.C. (1967) Biochem. Biophys. Res. Commun., 26, 577.

7. Tsai, R.L., Gunsalus, I.C. and Dus, K. (1971) Biochem. Biophys. Res. Commun., 45, 1300.

8. Tanaka, M., Haniu, M., et al. (1977) J. Biol. Chem., 252, 7081.

9. Crestfield, A.M., Moore, S., and Stein, W.H. (1963) J. Biol. Chem., 238, 622.

10. Rüegg, U.T., and Rudinger, J. (1977) Methods Enzymol., 47, 116.

11. Spackman, D.H., Moore, S., and Stein, W.H. (1963) J. Biol. Chem., 238, 618.

12. Edman, P., and Begg, G. (1967) Eur. J. Biochem., 1, 80.

13. Edman, P., and Henshen, A. (1975) in Protein Sequence Determination, Needleman, S.B., ed., Springer-Verlag, Berlin, p. 232.

14. Ambler, R.B. (1967) Methods Enzymol., 11, 436.

15. Mitchell, W.M. (1977) Methods Enzymol., 47, 165.

16. Jauregui-Adell, J., and Marti, J., (1975) Anal. Biochem., 69, 468.

17. Weber, K., Notani, G., and Zinder, N. (1966) J. Biol. Chem., 241, 2379.

18. Dus, K., Katagiri, M., et al. (1970) Biochem. Biophys. Res. Commun., 40, 1423.

19. Dus, K. (1975) in Cytochromes P-450 and b_5, Cooper, D.Y., et al., eds., Plenum Press, New York, p. 287.

20. Stern, J.O., and Peisach, J. (1974) J. Biol. Chem., 249, 7495.

MECHANISM OF HEMICHROME FORMATION FROM

HEMOGLOBIN SUBUNITS BY HYDROGEN PEROXIDE

AKIO TOMODA AND YOSHIMASA YONEYAMA
Department of Biochemistry, Kanazawa University School of Medicine,
Kanazawa 920, Japan

SUMMARY

Hydrogen peroxide oxidation of ferrous hemoglobin subunits such as the α_{SH}, α_{PMB} and β_{PBM} chains were studied under various conditions. These chains were converted to hemichrome in the presence of hydrogen peroxide generating systems. Absorption and ESR spectral studies showed that the oxidation of these hemoglobin subunits was not inhibited by superoxide dismutase, but was decreased by catalase. The rate of oxidation of the hemoglobin subunits by hydrogen peroxide was found to be dependent on the extent of masking of sulfhydryl group of the chains by p-chloromercuribenzoate. Hemichrome was formed directly from these ferrous chains upon attack by hydrogen peroxide without the formation of methemoglobin. The significance of hydrogen peroxide in relation to the hemichrome formation in the disorders of erythrocytes such as thalassemia is discussed.

INTRODUCTION

Hemichrome is an intermediate in the denaturation of hemoglobin. However, the detailed mechanism for the formation of this oxidized hemoglobin derivative remains to be clarified. Hemichrome and methemoglobin are ferric heme compounds which show characteristic absorption[1] and low spin ESR spectra.[2] The heme iron of hemichrome is bonded directly to the distal histidine E7, while the heme iron of methemoglobin is liganded with water.[1]

Rachmilewitz et al.[3] suggested when the hemoglobin tetramer is dissociated into its free α and β subunits, the subunits are more susceptible to hemichrome formation than is the native hemoglobin tetramer. However, the oxidant of hemoglobin in the red cells was not identified or specified. Judging from the recent advances on the effects of superoxide on cells and proteins, it is most probable that oxidants such as superoxide and peroxide will exert some influences on the oxidation of hemoglobin to hemichrome in the red cells. Brunori et al.[4] mentioned the possibility that these oxidants are concerned with the formation of hemichrome, based on autoxidation studies of the hemoglobin α and β chains. Winterbourn et al.,[5] however, indicated that the effects of these oxidants might be small. However, direct evidence for hemichrome formation during the reaction between hemoglobin and superoxide or peroxide is lacking.

In the present study, hemichrome formation during the oxidation of the hemoglobin α and β chains by hydrogen peroxide was investigated under various experimental conditions, and a mechanism of hemichrome formation by the oxidant is presented.

EXPERIMENTAL PROCEDURE

Preparation of hemoglobin subunits. Solutions of hemoglobin were obtained as described previously,[6] and were further purified by DEAE-Sephadex chromatography by the procedure reported by Huisman and Dozy[7] in which the column is eluted with 50 mM Tris-HCl buffer, pH 8.0. The purified hemoglobin was further treated with p-chloromercuribenzoate (PCMB) at 4°C for 16 hours. Free PCMB was removed by passing the solution through a column of Sephadex G-25 (coarse) equilibrated with 10 mM potassium phosphate buffer, pH 7.0. The PCMB derivatized (α_{PMB} and β_{PMB}) chains were separately collected from the CM-cellulose chromatography and the underivatized (α_{PMB} and β_{SH}) chains were obtained by removal of PCMB from chromatography on thiolated Sephadex G-25.[8]

The oxidation of hemoglobin subunits by H_2O_2 or H_2O_2-generating systems. The oxidation of hemoglobin subunits by H_2O_2 and H_2O_2-generating systems was carried out as follows. For each 0.2 ml of solution of the α_{SH}, β_{SH}, α_{PMB} or β_{PMB} chains of hemoglobin, 1.3 ml of 0.2 M potassium phosphate buffer, pH 7.0 was added. Changes in the absorption spectra of the mixture were measured at 578 nm, or between 500 and 650 nm, or between 240 and 300 nm at 25°C after the addition of up to final concentrations of up to 120 µM of H_2O_2 or after the addition of the H_2O_2-generating systems, i.e. glucose-glucose oxidase or xanthine-xanthine oxidase. The H_2O_2-generating rates for these enzyme systems were 15 and 30 µM/min, respectively.

Measurement of ESR spectra. The ESR spectra of the H_2O_2-oxidized α_{SH} (633 µM) and β_{SH} (733 µM) subunits were recorded at 123°C in a JEOL JEX PE-3X X-band spectrometer equipped with a 100 kHz field modulation unit.

RESULTS AND DISCUSSION

Absorption spectra of the H_2O_2-oxidized hemoglobin subunits. Figures 1A and 1B show the changes in absorption spectra between 500 and 650 nm during the oxidation of the α_{SH} and α_{PMB} chains by H_2O_2. In both cases, the absorption at 542 and 578 nm decreased gradually as the reaction proceeded. Isosbestic points were initially observed at 526 and 587 nm but then disappeared as the reaction reached completion. The final absorption spectrum obtained corresponded to spectrum of hemichrome published by Rachmilewitz.[2]

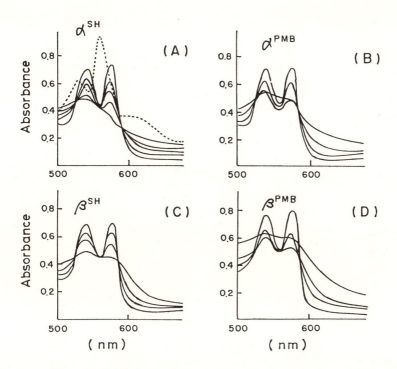

Fig. 1. Changes in the absorption spectra during oxidation by H_2O_2 of the separate hemoglobin subunits. The changes in absorption spectra between 500 and 650 nm were measured after the addition of H_2O_2 to the solutions of separate hemoglobin subunit. The concentrations of H_2O_2 were 120 μM. A, α_{SH} chain (48 μM) and the broken lines show the same spectra after addition of dithionite to the H_2O_2-oxidized α_{SH} chain; B, β_{PMB} (47.9 μM); C, β_{SH} chain (46.7 μM); and D, β_{PMB} chain (52.7 μM).

Upon the addition of dithionite (Fig. 1A), the final absorption spectrum changed to one that is typical for a hemochrome (dotted lines). These results suggest that hemichrome is directly formed during H_2O_2 oxidation of the hemoglobin α_{SH} and α_{PMB} chains.

Similar spectral changes were also observed with the β_{SH} and β_{PMB} chains (Figs. 1C and 1D). The final absorption spectra observed indicated hemichrome was the product of the reactions.

Since a perturbation of the spectra in the UV region occurs when hemichrome is formed,[1] changes in the absorption spectra of the α_{SH} and β_{SH} chains between 240 and 32 nm were measured (Figs. 2A and 2B). A decrease in the absorbance at 264 nm was typically observed with both the α_{SH} and β_{SH} chains. Isosbestic

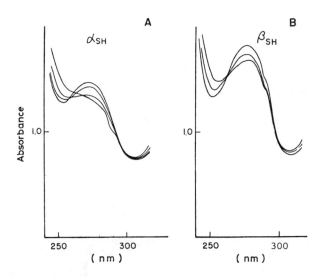

Fig. 2. Changes in UV spectra during the oxidation of the separate hemoglobin subunits. Symbols adopted are: A, α_{SH}-chain (47.9 μM); and B, β_{SH}-chain (47.0 μM).

points were observed at 260 and 290 nm. Since the absorbance at 264 nm arises probably from the tyrosine and tryptophan residues present in the hemoglobin subunits, it is likely that oxidation of these amino acid residues by H_2O_2 occurs during the hemichrome formation.

ESR spectra of the H_2O_2-oxidized hemoglobin. Rachmilewitz et al.[2] studied the ESR spectra of hemichrome and showed presence of a typical low spin compound. In order to see whether the oxidized compounds derived from H_2O_2-oxidized hemoglobin subunits were hemichrome derivatives, the ESR spectra were taken (Fig. 3). The ESR spectra of the H_2O_2-oxidized α_{SH} and β_{SH} chains were typically that of low spin compounds (g = 2.45, 2.29 and 2.05) and therefore the spectrum of a hemichrome. The sharp signal near g = 2, which was observed during the oxidation of the α_{SH} chains, is probably due to protein free radicals produced by H_2O_2.

Oxidation of hemoglobin subunits by H_2O_2. Figure 4 shows the changes in absorbance at 578 nm when the α_{SH}, α_{PMB}, β_{SH} and β_{PMB} chains of hemoglobin were exposed to H_2O_2. A rapid decrease in the absorbance at 578 nm was observed within 12 min at 25°C and were found to follow first order kinetics. The absorbance changes were completely inhibited in the presence of catalase, as shown by the broken lines in the figure.

Next, the initial rates of the oxidation of α_{SH} and β_{SH} chains by H_2O_2 were

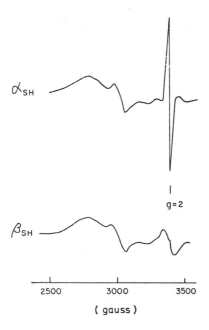

α_{SH}

β_{SH}

g = 2

2500   3000   3500

(gauss)

Fig. 3. ESR spectra of the H_2O_2 oxidized individual hemoglobin subunits. The ESR spectra of the H_2O_2-oxidized α_{SH}-chain (653 μM) and β_{SH}-chain (733 μM) were measured at a magnetic field between 250 and 350 mT at 123°K. The final concentrations of H_2O_2 used for the oxidation of the chains was 1.3 mM.

studied at different concentrations of H_2O_2 (Fig. 5). When the concentration of H_2O_2 was changed from 160 to 1200 μM, the initial rates of oxidation of the α_{SH} and β_{SH} chains were dependent on H_2O_2 concentrations. The reaction rate constants, at the higher H_2O_2 concentrations were estimated to be 15 M^{-1} s^{-1} for the chain α_{SH} and 6.4 M^{-1} s^{-1} for the β_{SH} chain.

Effect of PCMB binding to the hemoglobin subunits of the oxidation rates of hemoglobin subunits by H_2O_2. Figure 6 shows the plot of the initial rates of hemichrome formation from the α_{SH} and β_{SH} chains by H_2O_2 when the concentrations of PCMB were varied. Since the oxidation rates reached a maximum value when the α_{SH} chains (25 μM) were saturated with PCMB, one mole of PCMB binds to one mole of the α_{SH} chains. This is consistent with the fact that the α_{SH} chains have one cysteine residue.

On the other hand, the maximum rates of β_{SH} chain oxidation were obtained when a 2-fold excess of PCMB (50 μM) were present per β_{SH} chain (25 μM). This end point is in good agreement with the fact that β_{SH} chains have two cysteine

412

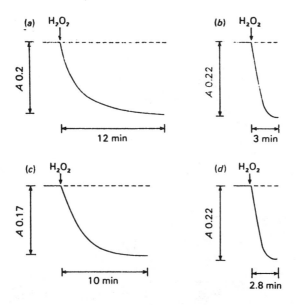

Fig. 4. Effect of H_2O_2 on the separated hemoglobin subunits. The changes in the absorbance at 578 nm for the a, the α_{SH}-subunit (42.7 μM); b, the α_{PMB}-subunit (37.8 μM); c, the β_{SH}-subunit (42 μM); and d, the β_{PMB}-subunit (37.8 μM) chains were measured after the addition of 120 μM of H_2O_2. The broken lines show the results when catalase was added.

residues. Thus, it is possible to say that hemoglobin subunits were more susceptible to hemichrome formation when the sulfhydryl group of these chains were blocked by PCMB.

Effect of H_2O_2-generating systems on oxidation of the hemoglobin subunits. Since both the glucose-glucose oxidase and xanthine-xanthine oxidase systems are known to produce H_2O_2, the effect of these H_2O_2 generating systems on the oxidation of hemoglobin subunits were studied (Figs. 7 and 8).

Figure 7 shows the effect of H_2O_2 produced by glucose oxidase or xanthine oxidase H_2O_2 generating systems on the oxidation of α_{SH} chains. Decreases in the absorbance at 578 nm were observed in the presence of these enzyme systems. These absorption spectral changes observed with the oxidized α_{SH} chains were similar to those observed when H_2O_2 was added, which showed that hemichrome was formed.

When catalase was added, the changes in the absorption spectra of the α_{SH} chains were largely inhibited (about 60%), although not as completely as observed when H_2O_2 was the oxidant. This result further confirmed our finding

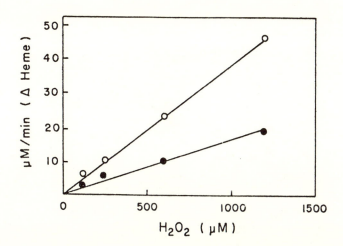

Fig. 5. Oxidation rates of the α_{SH}- and β_{SH}-subunits in the presence of various concentrations of H_2O_2. The oxidation rates of the α_{SH} (o) and the β_{SH}-subunits (•) (42.7 and 24.0 µM, respectively, were measured from the decrease in the absorbance at 578 nm in the presence of various concentrations of H_2O_2 (160–1200 µM).

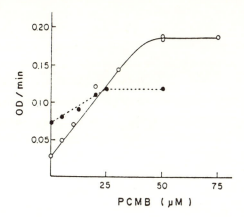

Fig. 6. Effect of p-chloromercuribenzoate on the oxidation rate of individual hemoglobin subunits by H_2O_2. The rate of oxidation of separate hemoglobin subunits by H_2O_2 were studied at various concentrations of p-chloromercuribenzoate. The concentrations of the α_{SH}-chain, β_{SH}-chain and H_2O_2 were 25 µM, 25 µM and 120 µM, respectively. The symbols used were: (•) the α_{SH}-subunit and (o), β_{SH}-subunit.

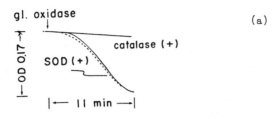

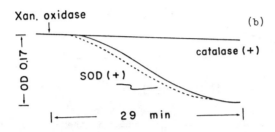

Fig. 7. Effect of the use of H_2O_2 generating systems on the oxidation of the α_{SH}-chain. The changes in absorbance of the α_{SH}-chain at 578 nm were measured after the addition of glucose oxidase (a) or xanthine oxidase (b). The effects of catalase and superoxide dismutase (SOD) additions were also studied. H_2O_2 was generated by glucose oxidase at a rate of 15 µM/min and by xanthine oxidase at a rate of 30 µM/min.

that the hemichrome was produced from the α_{SH} chains by a direct attack by H_2O_2. If superoxide anion was involved in hemichrome formation, the oxidation rate of α_{SH} chains should decrease upon the addition of superoxide dismutase when the xanthine oxidase H_2O_2-generating system was used, since xanthine oxidase is known to produce superoxide as well as H_2O_2. However, the oxidation rate was not inhibited by superoxide dismutase and indicated that superoxide anion was not involved in hemichrome formation, a finding which is in good agreement with the report of Winterbourn et al.[5]

Similar results were also observed with the β_{SH} chains of hemoglobin (Figs. 8A and 8B). When H_2O_2 was generated in the reaction mixture by the addition of glucose-glucose oxidase or xanthine-xanthine oxidase, the oxidation of β_{SH} chains to hemichrome was noted spectrally. The inhibition of the oxidation of β_{SH} chains to hemichrome was noted spectrally. The inhibition of the oxidation of β_{SH} chains in the presence of catalase was 60%. However, the addition of superoxide dismutase to this system increased slightly the rate of hemichrome formation when H_2O_2 was generated by the oxidase system.

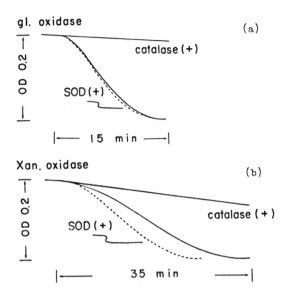

Fig. 8. Effect of the use of H_2O_2 generating systems on the oxidation of the β_{SH}-chains. Experimental conditions and procedures adopted are the same as those mentioned in the legend to Fig. 7.

Mechanism of hemichrome formation. It is generally accepted that hemichrome is formed from methemoglobin.[1,5] However, the results obtained in this study suggest that hemichrome is formed directly from the ferrous chains of hemoglobin. This finding was further supported by the following results. When hemichrome formation was investigated in the presence of large amounts of cyanide, it is expected that met-form of the hemoglobin chains would be trapped by cyanide complex and an inhibition of the rate of oxidation by H_2O_2 would occur. However, the rate of hemichrome formation was the same in the presence or absence of cyanide (100 fold excess of cyanide over heme) and ruled out methemoglobin as an intermediate.

In the case hemoglobin tetramer (free from catalase and superoxide dismutase), methemoglobin is readily produced by the oxidizing action of H_2O_2.[9] This finding suggests that when the α and β chains are associated to form the tetramer, hemoglobin is protected against attack by H_2O_2.

The present results obtained demonstrated that a small amount of H_2O_2 added exogenously or generated by the H_2O_2 generating enzyme systems can easily attack the individual hemoglobin α and β chains. Furthermore, the inhibitory effect of catalase on hemichrome formation from the dissociated hemoglobin chains is incomplete when H_2O_2 is generated by the H_2O_2 generating systems. Taking these findings into consideration and the fact that H_2O_2 is a physiologically produced substance, it is possible that in the red cells which contain an excess of hemoglobin subunits, hemichrome formation from the dissociated chains may be accelerated. In Thalessemic red cells, which contain an excess of either the α or β chains, the accumulation of hemichrome has been reported by Rachmilewitz et al.[2] Recently, Beutler et al.[10] found that the content of glutathione peroxidase is greatly increased in thalassemic red cells. The presence of an excess of the peroxidase seems fortuitous because the enzyme detoxifies the intracellular H_2O_2 in the red cells and thereby eliminates the hemoglobin subunit-H_2O_2 reaction.

Production of H_2O_2 under physiological conditions. The generation of H_2O_2, by various *in vivo* routes, is considered to be very low in the red cells. One possible source of this H_2O_2 comes from the dismutation of superoxide anion by superoxide dismutase in the intracellular space of the red blood cells or in the extracellular space, e.g., in the blood. Weber et al.[11] showed that superoxide anion is produced during the autoxidation of the hemoglobin tetramer. However, the autoxidation rate of individual hemoglobin subunits are much greater than the rate of autoxidation of the hemoglobin tetramer,[4] and therefore, the generation of superoxide anion will be greater during the autoxidation of the α and β chains. Superoxide anion, thus generated, may revert to H_2O_2 spontaneously from the action superoxide dismutase present in the red blood cells. The H_2O_2 produced can then induce the formation of hemichrome from the unassociated hemoglobin subunits as shown in Scheme 1.

Another possible source of H_2O_2 in the red blood cell is from the superoxide anion released by the leucocytes. Recent studies have revealed that a large amount of superoxide anion is generated by leucocytes during phagocytosis or during pseudo-phagocytosis induced by chemical substances.[12,13] Since superoxide anion is permeable to red cell membrane,[14] this anion can be converted by superoxide dismutase to H_2O_2, which can then attack the unassociated hemoglobin subunits and form hemichrome. This proposal is supported by the observation that when infection occurs the leucocytes are activated, the formation of Heinz bodies, as a result of hemichrome formation, are accelerated in the thalassemic red cells.

REFERENCES

1. Rachmilewitz, E.A. (1969) in Red Cell Structure and Metabolism, Ramot, B., ed., Academic Press, New York, 94-102.
2. Rachmilewitz, E.A., et al. (1969) Nature (London), 222, 248-250

(1) Autoxidation of hemoglobin subunits

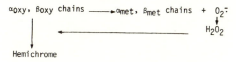

Hemichrome

(2) Release of superoxide anion from leucocytes

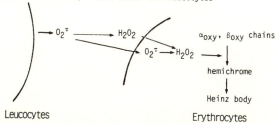

Leucocytes   Erythrocytes

Scheme 1. Possible mechanism of hemichrome formation from the oxidation of the hemoglobin subunits in the red blood cells by H_2O_2.

3. Rachmilewitz, E.A., Peissach, J. and Blumberg, W.E. (1971) J. Biol. Chem., 246, 3356-3366.
4. Brunori, M., et al. (1975) Eur. J. Biochem., 53, 99-104.
5. Winterbourn, C., McGrath, B.M. and Carrell, R.W. (1976) Biochem. J., 155, 493-502.
6. Tomada, A., et al. (1978) Biochem. J., 171, 329-335.
7. Huisman, R.H.J. and Dozy, A.M. (1965) J. Chromatogr., 19, 160-169.
8. Tyuma, I., Benesch, R.E. and Benesch, R. (1966) Biochemistry, 5, 2957-2962.
9. Eyer, P., Hertle, H., Kiese, M. and Klein, G. (1975) G. Mol. Pharmacol., 11, 326-334.
10. Beutler, E., et al. (1977) Blood, 50, 647-655.
11. Wever, R., Oudega, B., and Van Gelder, B.F. (1973) Biochim. Biophys. Acta, 302, 475-478.
12. Nakagawara, A. and Minakami, S. (1975) Biochem. Biophys. Res. Commun., 64, 760-767.
13. Nakagawara, A., Nabi, B.Z.F. and Minakami, S. (1976) Clin. Chim. Acta, 74, 173-176.
14. Lynch, R.E. and Fridovich, I. (1978) J. Biol. Chem., 253, 4697-4699.

V
Regulatory Proteins and Related Substances

CHEMISTRY AND FUNCTION OF LONG CHAIN CARDIOTOXINS FROM *BUNGARUS FASCIATUS* VENOM

TUNG-BIN LO AND HSIENG-SEN LU
Institute of Biological Chemistry, Academia Sinica, Taipei, Taiwan (ROC)

SUMMARY

Three cardiotoxin-like components (long chain cardiotoxins), Toxins V-2, V-3 and VI were isolated and purified from the venom of *Bungarus fasciatus*. They were less basic (pI 9.2-9.7) than the cobra cardiotoxin ($\geq$ 11.0) and their molecular weights were about double (117-118 amino acid residues) that of known cardiotoxins (58-61 amino acid residues). The amino acid sequences of the three toxins were determined and there was no homology of the amino acid sequence nor of the conformation as estimated by their CD spectra and by Chou-Fasman calculations, with the cobra cardiotoxin. While these toxins showed about 30% homology in amino acid sequence, and rather similar CD spectra to phospholipase A_2, neither of the three toxins exhibited any phospholipase A_2 activity. Toxin VI stimulated skeletal muscle contraction (chicken cervicis muscle assay) like certain other known cardiotoxins and some basic phospholipase A_2. These three toxins showed extraordinary stable conformations (stable in 8M urea or in 7M quanidine hydrochloride) and were predicted to have $\leq$ 16% α-helix and $\leq$ 24.5% β-pleated sheet structures in their native conformations.

INTRODUCTION

Major toxic principles of the Elapidae venoms are classified into two kinds of protein toxins, i.e., neurotoxins and cardiotoxins which can be easily differentiated from each other by the simple chicken biventer cervical muscle assay. Neurotoxins cause nerve blockage, while cardiotoxins cause heart contraction followed by paralysis.[1] Neurotoxins are divided into postsynaptic and presynaptic neurotoxins and postsynaptic toxins are subdivided to Type I (61-62 amino acid residues) and Type II (71-74 amino acid residues), i.e., by the length of the polypeptide chains. Cardiotoxins are usually less toxic than neurotoxins and are also designated cytotoxins or direct lytic factor (DLF) by different authors. They are very basic proteins with pI $\geq$ 11 and more than 50 toxins of this group have been isolated and sequenced. They are all single polypeptide chain proteins which contain 48-61 amino acid residues with four pairs of intramolecular disulfide linkages.

Recently, three cardiotoxin-like protein toxins have been isolated and purified from the *Bungarus fasciatus* venom.[2] They show some pharmacological

action of cardiotoxins[3] but their molecular sizes and amino acid compositions
differ considerably from those of known cardiotoxins. These three cardiotoxin-
like proteins were tentatively designated as long chain cardiotoxin, Toxins V-2,
V-3 and VI.[4] This communication describes the chemical studies which led to
the amino acid sequence determinations of these three toxins and preliminary
comparative observations of their pharmacological actions with known cardiotoxins and with venom phospholipase A_2.

AMINO ACID SEQUENCE DETERMINATIONS

Three cardiotoxin-like protein toxins, Toxins V-2, V-3 and VI were isolated
and purified from *Bungarus fasciatus* venom (obtained from Miami Serpentarium
Laboratories, Miami, Florida, USA, Lot No. BF 101) by repeated CM-cellulose
column chromatography[4] and their amino acid compositions are shown in Table 1.

The native Toxin VI was reduced with dithiothreitol and alkylated by
iodacetate (RCM-Toxin VI). RCM-Toxin VI was then digested with trypsin and the
tryptic peptides were isolated and purified by DEAE-cellulose chromatography,
gel filtration and paper electrophoresis. RCM-Toxin VI and all its tryptic
peptides were sequenced either in the automatic sequence analyzer (JEOL JAS 47K)
or by the modified manual Edman degradation procedure.[5] The liberated PTH-amino
acids were identified by gas liquid chromatography on the SP 400 column of a
Beckman GC 65 Gas Chromatograph,[6] by thin layer chromatography on polyamide sheet[7]
or by back hydrolysis either with 4M methane sulfonic acid[8] or with 6N-HC1-
mercaptoethanol-phenol.[9] The alignment of all tryptic peptides was achieved by
the analysis of all tryptic peptides of maleylated RCM-Toxin VI and the chymotryptic peptides of RCM-Toxin VI. The complete amino acid sequence is shown in
Fig. 1.[10] The native Toxin VI consists of 118 amino acid residues in a single
polypeptide chain with six pairs of intramolecular disulfide bonds.

For determination of amino acid sequences of Toxins V-2 and V-3, both toxins
were also reduced and carboxymethylated, then RCM-Toxins V-2 and V-3 were
separately digested with trypsin. The two sets of tryptic peptides were
isolated by the peptide mapping technique on Whatman 3MM paper (45 x 57 cm).
The results are shown in Fig. 2 along with that of RCM-Toxin VI. They show
very similar patterns with only slight differences. All peptides were recovered
from paper and amino acid composition and N-terminal residues were determined.
Finally, the peptide fragments with different composition were sequenced by a
modified manual Edman technique.[5] The complete amino acid sequences of Toxins
V-2 and V-3 are shown in Fig. 3 which emphasizes the sequence differences of
these toxins. Toxin V-2 also consists of 118 amino acid residues but four
residues in the amino acid sequence are different: Thr[14] is replaced with Ser,

423

TABLE 1

AMINO ACID COMPOSITION OF *BUNGARUS FASCIATUS* CARDIOTOXINS

Amino Acids	Toxin V-2		Toxin V-3		Toxin VI	
Lys	8.2	(8)	8.1	(8)	8.7	(9)
His	3.1	(3)	2.8	(3)	3.0	(3)
Arg	3.8	(4)	3.7	(4)	3.8	(4)
Asx	16.2	(16)	16.1	(16)	15.9	(16)
Thr	8.6	(9)	9.5	(10)	9.7	(10)
Ser	1.6	(2)	0.8	(1)	0.8	(1)
Glx	6.8	(7)	6.7	(7)	6.8	(7)
Pro	5.5	(6)	5.7	(6)	4.8	(5)
Gly	11.0	(11)	9.8	(10)	11.1	(11)
Ala	11.0	(11)	11.0	(11)	11.0	(11)
1/2 Cys[a]	11.6	(12)	11.8	(12)	11.6	(12)
Val	2.8	(3)	2.7	(3)	2.9	(3)
Met	1.1	(1)	1.3	(1)	0.8	(1)
Ile	4.7	(5)	5.0	(5)	4.8	(5)
Leu	5.8	(6)	6.2	(6)	5.8	(6)
Tyr	7.6	(8)	7.8	(8)	9.1	(9)
Phe	5.2	(5)	5.1	(5)	4.2	(4)
Trp[b]	1.1	(1)	0.9	(1)	0.8	(1)
Total	118		117		118	

[a]Quantitative results were obtained from the analysis of the hydrolysates of the reduced and S-carboxymethylated toxins.
[b]Using 4M methanesulphonic acid hydrolysis.

Lys^{29} is missing, Tyr^{60} is replaced by Phe and Pro is inserted between Asn^{61} and Leu^{62}. Toxin V-3 contains only 117 amino acid residues and five amino acid sequences are different from Toxin VI: Glu^{10} is replaced with Gln, both Lys^{29} and Gly^{52} are missing, Tyr^{60} is replaced with Phe and Pro is inserted between Asn^{61} and Leu^{62}.

These three cardiotoxin-like proteins show no amino acid sequence homology to known cardiotoxins while there is about 30% homology with snake venom phospholipase A_2 as shown in Fig. 4. As described later, however, these toxins are devoid of any enzymic and immunological properties of phospholipase A_2.

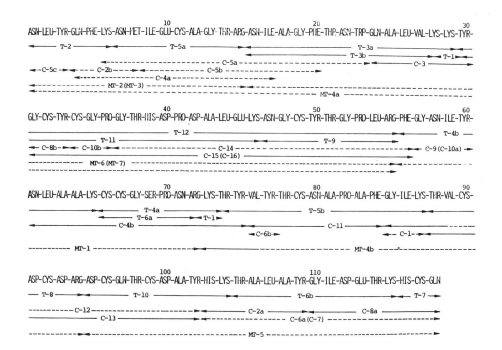

Fig. 1. Amino acid sequence of Toxin VI.

PHYSICOCHEMICAL AND CONFORMATIONAL STUDIES

The isoelectric point (pI) of Toxins V-2, V-3 and VI were determined by the isoelectric focusing technique using pH 9-11 Ampholine in a 110 ml LKB cell. As shown in Table 2, their pI values determined were 9.16, 9.32 and 9.70, respectively[4] and were much lower than the corresponding value for the cobra cardiotoxin ($\geq$ 11.0).

The gross conformation of these three toxins were investigated by circular dichroism spectral (CD spectra) analysis in water and in neutral buffer solution in a JASCO J-20 Spectropolarimeter. Conformational stabilities were also studied by CD spectra change caused by various denaturants, pertubants and pH of the solutions. The CD spectra in aqueous solutions are shown in Fig. 5. Toxins V-2, V-3 and VI showed almost identical spectra. Their spectra are quite different from that of cobra cardiotoxin but show some similarity to cobra phospholipase A_2. A characteristic double minimum at 222nm and 208-210nm and a maximum near 191-194nm as observed in the spectrum of phospholipase A_2 are well-known indicators of the presence of α-helix.[11] The spectra of Toxins V-2, V-3 and VI do show those extrema but the minimum of 208-210 nm is not clear.

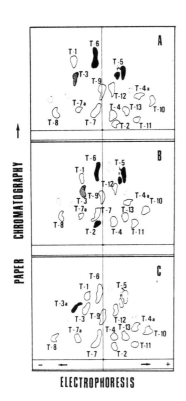

PAPER CHROMATOGRAPHY

ELECTROPHORESIS

Fig. 2. Peptide maps of tryptic digests, A: RCM-Toxin V-2, B: RCM-toxin V-3, C: RCM-Toxin VI paper chromatography; n-butanol:acetic acid:water (60:15:75, v/v), paper electrophoresis; pH 6.65 collidine: acetic:water (8.9:3.1:988 v/v) at 2,000 V, 1h.

The α-helical content of Toxin VI was calculated by the method of Chen et al. (based on the value of θ_{222nm})[12] and by the procedure of Greenfield and Fasman (based on the value of θ_{210})[11] and the values were 36.8% and 16.0% α-helix; respectively. This discrepancy can be explained by the interferences due to the high disulfide content and hydrophobic nature of Toxin VI in the 222nm region. Thus, the high negative value of θ_{222nm} was not exclusively contributed by the helix present in Toxin VI. This explanation was verified in the case of porcine phospholipase A_2 in which case about 50% helix was predicted from the CD spectra[13] but X-ray crystallographic analysis showed only 12% helix.[14] For this reason, it is rather reasonable to predict 16% or less α-helix in Toxin VI. The secondary structure of Toxin VI was predicted by the Chou-Fasman[15] method and α-helical content certainly cannot be more than 24.5% as shown in Fig. 6. The same predictive procedure indicated that as much as 22% of β-sheet structure can exist in Toxin VI.

Conformational stability was investigated by the changes of the CD spectra

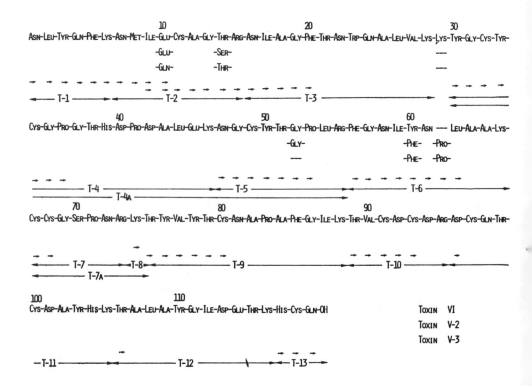

Fig. 3. Amino acid sequence of Toxins VI, V-2 and V-3.

in 8M urea, in 7M guanidine hydrochloride, in ethylene glycol and in ethylene
glycol/isopropyl alcohol (1/1). As shown in Fig. 7, the two denaturants do not
drastically change the conformation and organic solvents also do not perturb the
CD spectra at far UV region but a broad negative band appears between 300nm and
250nm. This change may be attributed to disulfide bonds.[16] The effects of
acid and anionic surfactant (sodium dodecyl sulfate) were also investigated.
The results are shown in Fig. 8. At pH 2.6, the CD spectra remains essentially
the same as in neutral solution. Prolonged exposure to pH 1 (about 6 h) led to
precipitation of the protein but the precipitates could be solubilized by SDS.
The most characteristic features of the effect of SDS in the acidic pH region
are the enhancement of helical content (double minimum becomes more evident and
both of their magnitudes increase) and concomitant disruption of the local
conformation (drastic change in 250-300nm region below pH 2.7) occurred.

Toxin VI contains 9 tyrosine residues. By employing the spectrophotometric

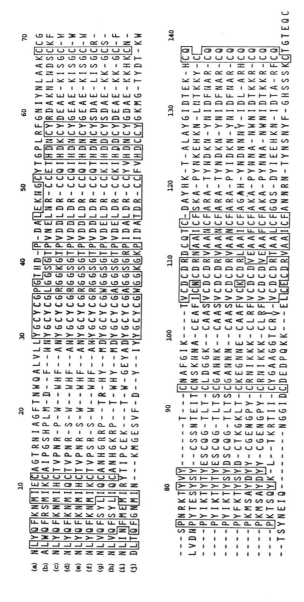

Fig. 4. Comparison of the amino acid sequence of Toxin VI (a) to related toxins; (b) phospholipase A from porceine pancreas, (c) DE-I of *Naja melanoleuca*, (d) DE-II of *Naja melanoleuca*, (e) DE-III of *Naja melanoleuca*, (f) phospholipase A of *Hemachatus hemachatus*, (g) Notexin, (h) phospholipase A (neurotoxic) of *Notechis scutatus*, (i) A chain of β-Bungarotoxin of *Bungarus multicinctus*, (j) phospholipase A of *Bitis gabonica*.

TABLE 2

SOME PROPERTIES OF TOXINS

	Toxin V-2	Toxin V-3	Toxin VI	Cobra Cardio-toxin	Cobra Phospholi-phase A$_2$	Notexin
Amino acid residue	118	117	118	60	119	119
pI	9.16	9.32	9.70	$\geq$ 11.0	5.32	Basic
LD$_{50}$ (µG/G of mice)	4.8	3.8	3.1	1.48	8	0.017
Skeletal muscle contracture	+	+	+	++	-	++
Smooth muscle contracture	(-)	(-)	(-)	+	++	+
Heart muscle depression	±	±	±	++	positive inotropic effect (at low conc.)	positive inotropic effect (at low conc.)
Local irritation	+	+	+	==		
Direct hemolytic action	-	-	-	++	-	(+)
Anticholinesterase	+	+	+	++	-	
Arterial blood pressure	-	-	-		fall	fall

titration technique, the titration curve shown in Fig. 9 was obtained. Only one Tyr was normally titratable with an apparent pk of 10.2, five residues with pk of 11.8 and the three remaining Tyrosine residues began to ionize at pH 12.3. The result agrees well with the result from nitration of the tyrosine residues with tetranitromethane (TNM). Only one Tyr was easily nitrated when the molar ratio was 3:1 (TNM:Toxin) and 6 Tyr were modified when TNM was increased to 12 times the concentration of protein. Conformational stability in alkaline solution was also studied. The results are shown in Fig. 10. At pH 11.02 where only one Tyr was ionized, the shape and magnitude of CD bands were virtually unchanged except for a 1-2nm red-shift. At pH 12.23, all of CD bands were completely changed to the unordered form, like the CD spectrum of RCM-toxin.

From these results, it is quite obvious that the conformation of Toxin-VI is

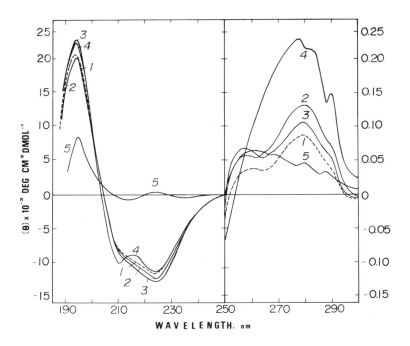

Fig. 5. CD spectra in aqueous solution, L: Toxin V-2, 2: Toxin V-3, 3: Toxin VI, 4: cobra phospholipase A_2, 5: cobra cardiotoxin. The 10 mm, 0.5 mm and 0.1 mm cells were used for the wave length regions of 300-250 mmn, 250-205 mm and below 220 mm, respectively, at a speed of 0.4 mm/min with a time constant of 4 sec.

different from cobra cardiotoxin and shows some resemblance to phospholipase A_2. It may be suggested that native conformation of Toxin-VI contains about 16% (or less) of α-helix and is rich ($\leq$ 24.5%) in β-structure. Its conformation was found to be extraordinarily stable in pH range 2.7-11.02 and even in 8M urea or in 7M guanidine hydrochloride. Cobra cardiotoxin has been reported to be denatured in 6.0M guanidine hydrochloride and the native conformation to be rich in β-structure.[17]

PHARMACOLOGICAL CHARACTERIZATION

Some properties of Toxins V-2, V-3 and VI are listed in Table 2. Toxicity (LD_{50}) was tested in mice by intraperitoneal injection. They are 2 to 3 times less toxic than cobra cardiotoxin. Three toxins showed several actions characteristic of cobra cardiotoxin, i.e., such as skeletal muscle (cervicis muscle of baby chicks) contracture, depression of cardiac muscle, blockage of neuromuscular transmission, local irritation and anticholinesterase activity

430

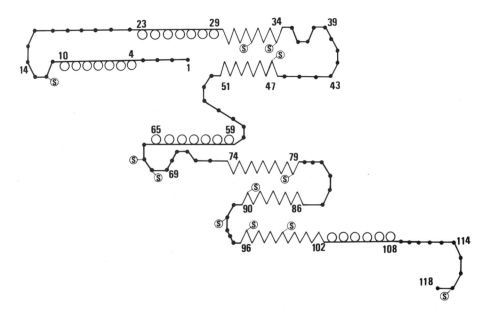

Fig. 6. Predicted model of secondary structure of Toxin VI by Chou-Fasman method.

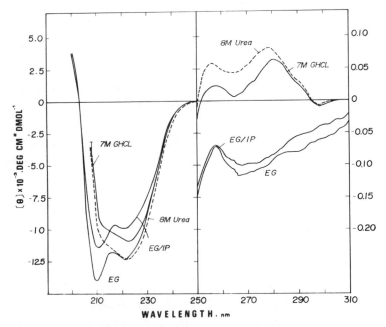

Fig. 7. CD spectra of Toxin VI in different denaturants and perturbants solution as indicated. Conditions were the same as in Fig. 5.

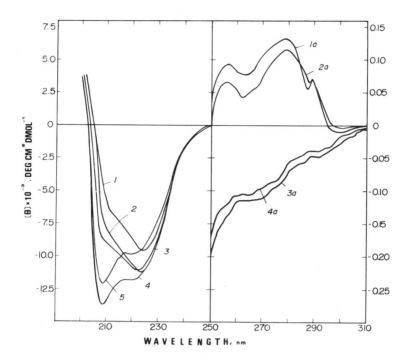

Fig. 8. CD spectra of Toxin VI in acidic and in acid-SDS solution. 1: pH 1, 2: pH 2.7, 3: SDS (pH 6.4), 4: SDS (pH 1), 5: SDS (pH 2.7). Conditions were the same as in Fig. 5.

but their actions were mostly weaker than those of cobra cardiotoxin.[3] They did not affect smooth muscle (ileum of guinea pig) contracture and direct hemolysis which were also characteristic of cobra cardiotoxin. Purified cobra cardiotoxin antibody did not show any cross reaction with Toxins V-2, V-3 and VI, as shown in Fig. 11.

As described before, Toxins V-2, V-3 and VI showed some amino acid sequence homology and in the CD spectra with phospholipase A_2 but three toxins are devoid of any phospholipase A_2 activity irregardless of the assay method used. At a concentration of 10^{-7} g/ml, cobra phospholipase A_2 caused a stimulation of a guinea pig atria.[18] Intravenous injection of phospholipase A_2 at a dose of 0.05 mg/kg induced a fall in arterial blood pressure in the rat.[18] However, Toxins V-2, V-3 and VI caused neither a stimulation of the guinea pig ileum (or atria) nor showed a depressor effect on the rat, even when the doses were 10 to 100 times that used for phospholipase A_2 (at a concentration which evoked a response). The induction of contraction of the biventer cervicis muscle of

432

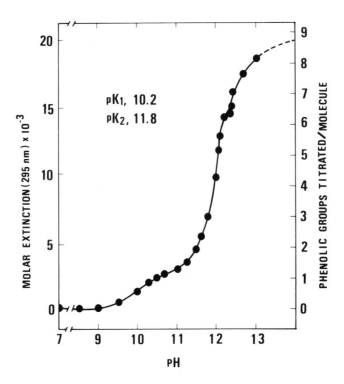

Fig. 9. Titration curve of Toxin VI obtained by spectrophotometric method, 2300 was used as the differential molar extinction coefficient at 295 nm.

baby chicks was observed only with Toxin VI and the basic phospholipase A_2 of *Naja nigricollis* (not for acidic phospholipase A_2) and *Notexin*.[18] As shown in Fig. 11, purified antibody of Toxin VI showed equivalent cross reaction with Toxins V-2, V-3 and VI but no reaction with cobra phospholipase A_2 and cobra cardiotoxin.

ACKNOWLEDGMENT

This work was supported in part by grants NSC-67B-0203-03(01) and NSC-68B-0203-03(03) from the National Science Council, Republic of China.

REFERENCES

1. Lee, C.Y., et al. (1968) Naunyn. Schmiedeberg Arch. Exp. Phathol. Pharmakol., 259, 360.
2. Lu, H.S. and Lo, T.B. (1974) J. Chinese Biochem. Soc., 3, 57.
3. Shiau Lin, S.Y., Huang, M.C., and Lee, C.Y. (1975) Toxicon, 13, 189
4. Lo, T.B., and Lu, H.S. (1978) in Toxins: Animal, Plant and Microbial, Rosenberg, P., ed., Pergaman Press, Oxford and New York, pp 161-181.

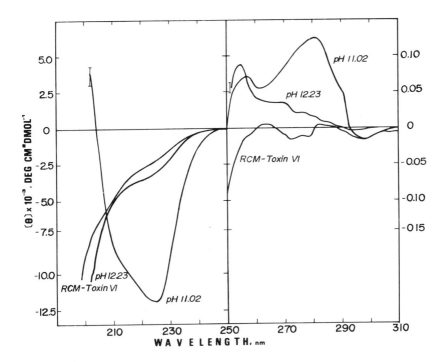

Fig. 10. CD spectra of Toxin VI in alkaline pH and of RCM-Toxin VI in aqueous solution. Conditions were the same as in Fig. 5.

5. Edman, P., (1970) in Protein Sequence Determination, Needleman, S.B., ed., Springer Verlage, Berlin, pp 211-255.
6. Pisano, J.J., Bronzert, T.J., and Brewer, H.B., Jr., (1972) Anal. Biochem., 45, 43.
7. Summers, M.R., Smythers, G.W., and Oroszlan, S., (1973) Anal. Biochem., 53, 624.
8. Kortt, A.A., Wysocki, J.R., and Liu, T.Y. (1976) J. Biol. Chem., 251, 1941
9. Lu, H.S., Sun, W.M., Chen, S.W., and Lo, T.B. (1978) Proc. Natl. Sci. Counl. (Taiwan), 2, 352.
10. Lu, H.S. and Lo, T.B. (1978) Int. J. Peptide Protein Res., 12, 181.
11. Greenfield, N. and Fasman, G.D., (1969) Biochemistry, 8, 4108.
12. Chen, Y.H., Yan, J.T., and Martinez, H.M. (1972) Biochemistry, 11, 4120.
13. Tigensons, B. and de Haas, G.H. (1977) Biochim. Biophys. Acta, 494, 285.
14. Drenth, J., et al. (1976) Nature, 264, 373.
15. Chou, P.Y., and Fasman, G.D., (1974) Biochemistry, 13, 222; (1977) J. Mol. Biol., 115, 135; and Fasman, G.D., Chou, P.Y., and Adler, A.J. (1976) Biophys. J., 16, 1201.
16. Yamashiro, D., Rigbi, M., et al. (1975) Intl. J. Peptide Protein Res., 7, 385.
17. Hung, M.C., and Chen, Y.H., (1977) Int. J. Peptide Protein Res., 10, 277.
18. Ho, C.L., and Lee, C.Y., unpublished data.

434

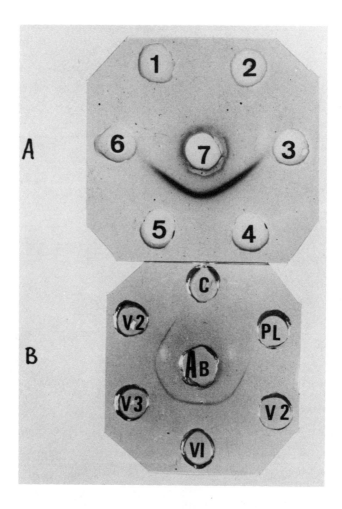

Fig. 11. Ouchterlony immodiffusion pattern. (A) Purified antibody of cobra cardiotoxin (7) was challenged with (1) Toxin V-2 (25 μg), (2) Toxin V-3 (25 μg), (3) Toxin VI (25 μg), (4) cobra cardiotoxin (5 μg), (5) cobra cardiotoxin (25 μg), (6) bovine serum albumin (25 μg); (B) Purified antibody of Toxin VI was challenged with Toxin V-2, Toxin V-3, Toxin VI, PL: cobra phospholipase A$_2$; (C) cobra cardiotoxin.

Published 1980 by Elsevier North Holland, Inc.
Liu/Mamiya/Yasunobu, eds. Frontiers in Protein Chemistry

TETANUS TOXIN: STRUCTURE-FUNCTION RELATIONSHIPS

WILLIAM H. HABIG, M. CAROLYN HARDEGREE AND L. D. KOHN[*]
Food and Drug Administration, Bureau of Biologics, Bethesda, Maryland 20205;
*National Institute of Arthritis Metabolism and Digestive Diseases, National
Institutes of Health, Bethesda, Maryland 20205

ABSTRACT

Recent studies suggest structural relationships between tetanus toxin and the
glycoprotein hormone, thyrotropin (TSH). Both proteins interact with gangliosides ($G_{D1b} = G_{T1} > G_{M1} > G_{D1a}$) and with the glycoprotein component of the TSH
receptor of thyroid membranes. Each protein competes with the other for binding
to thyroid and neural membranes. Despite these similarities in receptors,
tetanus toxin does bind better to neural membranes while TSH binds better to
thyroid. In addition, tetanus toxin appears to stimulate thyroid function in
$vivo$.

Studies of papain digested fragments show that fragment C (a part of the
putative binding subunit) interacts with gangliosides in a manner identical
with the holotoxin, as determined by direct binding to liposomes containing
gangliosides and by the ability of gangliosides to block binding to membrane
preparations. This nontoxic fragment C also appears to undergo retrograde
axonal transport. Thus, neither transport nor binding alone can account for
the toxicity expressed by the holotoxin, nor are these parameters sufficient
to account for differences in the susceptibility of various animal species to
tetanus.

INTRODUCTION

Recent studies in our laboratories have centered on the initial interaction
of tetanus toxin with biological membranes. In this paper we will describe
the structure-function relationships of tetanus toxin as they have been explored
by several approaches:

1. Comparison of the receptors for tetanus toxin and the glycoprotein
    hormone, thyrotropin (TSH).

2. Examination of the structure of tetanus toxin in relation to its ability
    to express biological effects.

3. The relationship of the sensitivity to tetanus toxin of several animal
    species to their competency in various in $vitro$ assays.

The difficulties in studying structure-function relationships in the absence
of satisfactory in $vitro$ assays for function will become apparent.

Receptor studies

The ability of brain tissue to neutralize the *in vivo* toxic effects of tetanus toxin has been known for over 75 years.[1] In 1959, van Heyningen first attributed this neutralizing effect to gangliosides present in neural tissue, and subsequently identified the most efficient gangliosides (in terms of their ability to block toxicity) as G_{D1b} and G_{T1}.[2] The specificity of the tetanus toxin-ganglioside interaction is very similar to that which has subsequently been described for the TSH-ganglioside interaction.[3] Additional studies showed that thyroid membranes could bind[4] and neutralize[5] tetanus toxin and that the toxin and the hormone compete for binding to thyroid membranes.[4] These and other data led us to the concept that the similarities in the receptors for TSH and tetanus toxin could be exploited to provide information on structure-function determinants of the toxin molecule. The receptor for TSH is known to consist of both a glycolipid (ganglioside) and a glycoprotein component; each of these components can be incorporated into liposomes and will bind TSH specifically.[6] Tetanus toxin can also bind to liposomes which contain either this glycoprotein or gangliosides (Fig. 1). There is no binding to control liposomes or to liposomes containing the monosialosyl ganglioside, GM_3. Toxin binding to

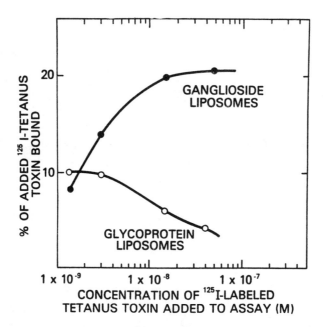

Fig. 1. The binding of ^{125}I-labeled tetanus toxin to dipalmitoyl phosphatidylcholine-cholesterol liposomes containing either mixed brain gangliosides or a preparation of the bovine thyroid glycoprotein receptor for TSH. Binding assays in this and subsequent figures involve determination of the radioactivity retained by Millipore filtration of the protein-membrane mixture.[3,4] From Ref. 7.

ganglioside-containing liposomes exhibits positive cooperativity (Fig. 1), a
phenomenon that is also seen in toxin binding to neural membranes, but not to
thyroid membranes.[7] The accumulation of a radiolabeled tetanus toxin fragment
in rat superior cervical ganglia after intraocular injection also shows positive
cooperativity.[8] These observations may indicate that toxin binding to gangliosides predominates over binding to glycoproteins in neural tissue, since the
glycoprotein binding is not positively cooperative. A glycoprotein receptor
for toxin in neural tissue has not yet been reported.

Tetanus toxin and TSH compete with each other for binding to thyroid membrane
receptors.[4] These two proteins also compete for the binding of ^{125}I-TSH to
brain membranes (Fig. 2A). In contrast, however, pre-incubation of rat brain
membranes with TSH enhances the binding of tetanus toxin (Fig. 2B); several
other hormones have no effect on the binding of toxin or TSH. Therefore, TSH
and tetanus toxin seem to interact in a similar manner with thyroid membranes,
but not with neural tissue. Further study of these effects reveals that toxin
binds better to brain membranes than does TSH, while TSH binds better to
thyroid membranes than does toxin (Fig. 3). Thus, despite all of the apparent
similarities in their receptors, the toxin and the hormone nonetheless exhibit
tissue specificity.

The tissue binding specificity must be considered along with the observation
that tetanus toxin can apparently stimulate thyroid function in mice;[5] it is not
yet known if this stimulation represents a direct effect of toxin on the thyroid
or whether, for example, it might be mediated by the pituitary. Although tetanus
in man is primarily a neuromuscular disease, some clinical observations are
consistent with abnormal thyroid function (summarized in Reference 4).

Structure and properties of tetanus toxin and its subunits

Extracellular tetanus toxin has a molecular weight of about 150,000 and is
composed of two chains; a light chain of about 50,000 MW, and a heavy chain of
approximately 100,000 MW.[9,10] Neither subunit alone is toxic.[10] The heavy
chain appears to recognize and bind to ganglioside receptors.[9,12] Papain
digestion of the holotoxin can cleave the heavy chain to produce a fragment
(fragment C) of about 47,000 MW which retains the binding function of the whole
molecule.[10,11] The remainder of the holotoxin molecule is termed fragment B
and consists of the light subunit and the remaining half of the heavy subunit.
Neither fragment causes the typical signs of tetanus, although fragment B can
kill mice at doses over 5 million times greater than required for the parent
toxin.[13] Matsuda and co-workers[14] have obtained similar fragments after trypsin

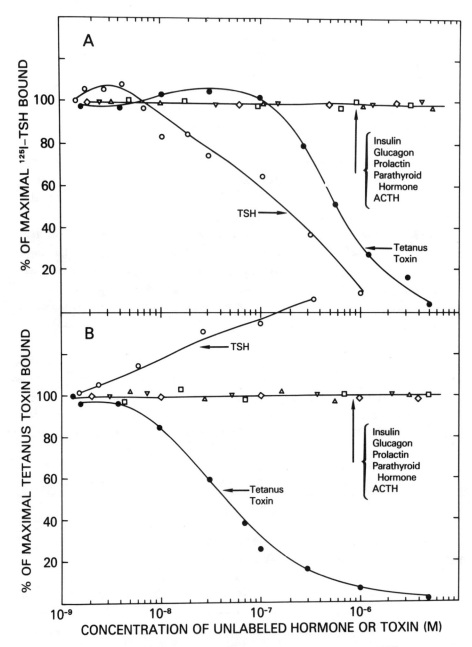

Fig. 2. Effect of unlabeled tetanus toxin or TSH on binding of ^{125}I-labeled TSH (A) or ^{125}I-labeled toxin (B) to rat brain membranes. The unlabeled proteins were incubated with membranes for 30 min prior to the addition of the labeled proteins. From reference 7.

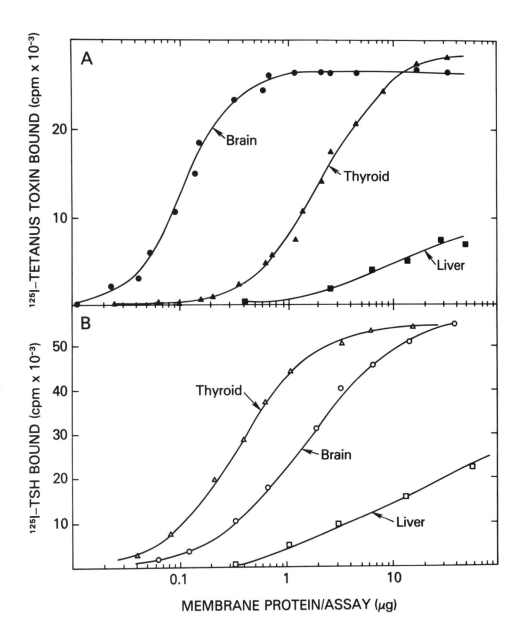

Fig. 3. Binding of ^{125}I-labeled tetanus toxin (A) or ^{125}I-labeled TSH (B) to brain, thyroid, or liver membranes. From Reference 7.

digestion. In the present study, these peptides were kindly supplied by
Dr. Torsten Helting, Behringwerke AG, Marburg/Lahn, Germany.

We have studied the binding of toxin and toxin fragments to membranes by
both direct and indirect means. In direct binding experiments (Fig. 4A, B),
fragment C or native toxin are comparable in their ability to block the binding
of radio-iodinated tetanus toxin to either thyroid or brain membranes. They
are likewise equally able to block the binding of iodo-TSH to thyroid membranes,
although TSH is most effective in this regard (Fig. 4C). Fragment B has no
effect on the binding of either protein. Mixed brain gangliosides are equally
effective as inhibitors of the direct binding of ^{125}I-labeled toxin or ^{125}I-
labeled fragment C to neural membranes (Fig. 5). The salt sensitivity and pH
optimum for binding are also similar for toxin and fragment C.

While tetanus toxin may inhibit synaptic transmission both peripherally and
centrally, the initial signs of the disease result primarily from its effects
in the central nervous system.[15] It is now considered likely that the toxin is
transported centrally via retrograde intra-axonal transport. Retrograde axonal
transport, thus, is a process that is believed to be intimately related to the
final mechanism of action of tetanus toxin *in vivo*. To examine this process in

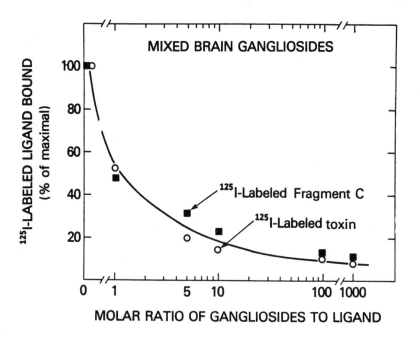

Fig. 5. The ability of mixed brain gangliosides to block the binding of radio-
iodinated tetanus toxin or fragment C to neural membranes. (N. Morris, et al.,
in preparation).

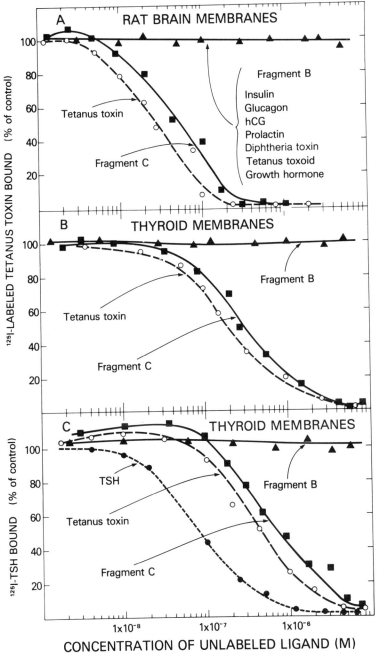

CONCENTRATION OF UNLABELED LIGAND (M)

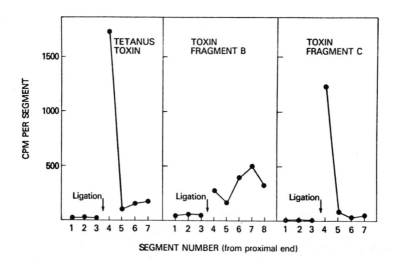

Fig. 6. Retrograde axonal transport of tetanus toxin and toxin fragments. The average of the results obtained using two rats for each protein is presented. The ^{125}I-labeled proteins injected were: Tetanus toxin (1.3 μCi, 4 μg); Fragment B (3.1 μCi, 2 μg); and Fragment C (5 μCi, 2 μg). See text and Reference 7 for further details of method.

rats, the sciatic nerve was ligated just prior to injection of radio-labeled protein into the gastrocnemius (calf muscle) of the same leg. After 18-24 hours, the animal was sacrificed and the sciatic nerve was removed and cut into 3 mm segments both above and below the ligation. Both tetanus toxin and fragment C accumulate markedly on the distal side of the ligation (Fig. 6). The behavior of fragment B is reproducible, but not typical of fast retrograde transport.

Fragment C exhibits binding and transport properties similar, if not identical, to those of the holotoxin; yet fragment C is atoxic. We can conclude that neither binding nor axonal transport is sufficient for toxicity.

Species sensitivity

There are large differences in the susceptibility of various animal species to tetanus. The chicken is an extreme example, being over 10,000 times less sensitive than the mouse.[2] The susceptibility of three common laboratory animals to toxin is shown in Table I; the rat is much more resistant than the

Fig. 4. Effect of unlabeled tetanus toxin, fragment B, or fragment C on ^{125}I-labeled toxin binding to neural membranes (A) or to thyroid membranes (B) and their effect on ^{125}I-labeled TSH binding to thyroid membranes (C). (N. Morris et al., in preparation).

TABLE 1

SPECIES SUSCEPTIBILITY TO TETANUS TOXIN

Species	Animal Weight (g)	Relative Lethal Dose[1]	Relative Susceptibility[2] (weight basis)
Mouse	15	1	1
Rat	300	256	13
Guinea Pig	300	8	0.4

[1]Lethal dose (relative to mouse) determined by injecting twofold dilutions of toxin into two animals per dilution. The toxin preparation contained 10^7 mouse lethal doses per mg of protein.

[2]Relative species susceptibility on an equal weight basis.

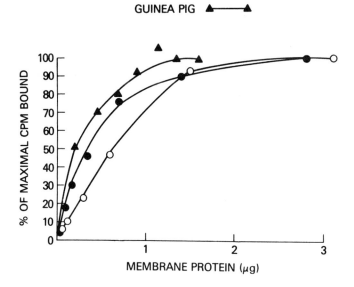

BINDING OF [^{125}I] - TETANUS TOXIN TO BRAIN MEMBRANES

MOUSE ○——○
RAT ●——●
GUINEA PIG ▲——▲

Fig. 7. Binding of ^{125}I-labeled tetanus toxin to brain membranes from mouse, rat, or guinea pig. (N. Morris, unpublished observations).

mouse. Yet, brain membranes from these two species are nearly identical in their ability to bind ^{125}I-tetanus toxin (Fig. 7), perhaps reflecting the nearly identical composition of gangliosides in these two tissues (N. Morris, unpublished observations).

While we have thus far only examined axonal transport in rats, it has been reported that toxin migrates even in chicken neurons.[16] Neither binding nor transport are therefore the limiting factors in the differential species sensitivity to tetanus toxin.

DISCUSSION AND SUMMARY

In the context of structure-function studies, it should be clear that we have not yet found an assay which allows assessment of those structural components of the toxin that are responsible for function. Additional assays are being developed, however, that might provide further information in this area. These include toxin-mediated changes in the permeability of synaptosomes to lipophilic cations[17] or calcium (Grollman et al., in preparation), as well as electrophysiologic[18] or metabolic[19] changes in cultured spinal cord cells.

Several well-studied bacterial toxins (cholera and diptheria) and plant toxins (abrin and ricin) are composed of two subunits, one of which is active and the other responsible for binding.[20] In tetanus, the identity of the binding subunit seems well established. Missing, however, is a functional assay for the active portion of the molecule which does not require binding. By analogy with the previously mentioned toxins, it may be predicted that the nonbinding chain of tetanus toxin will be active in a suitable assay system.

Sequence homologies have been reported between both the A and B subunits of cholera toxin and portions of several of the glycoprotein hormones.[21] When the primary structure of tetanus toxin is determined, it will be of interest to see if there are sequence homologies between TSH and tetanus toxin.

REFERENCES

1. Wasserman, N.A. and Takaki, T. (1898) Berl. Klin. Wochenschr., 35, 5-6.
2. van Heyningen, W.E. and Mellanby, J. (1971) In Microbial Toxins, S. Kadis, T.C. Montie and S.J. Ajl, eds., Vol. IIA, Academic Press, New York, pp. 69-108.
3. Mullin, B.R., Fishman, P.H., et al. (1976) Proc. Natl. Acad. Sci. USA, 73, 842-846.
4. Ledley, F.D., Lee, G., Kohn, L.D., et al. (1977) J. Biol. Chem., 252, 4049-4055.
5. Habig, W.H., Grollman, E.F., et al. (1978) Endocrinology, 102, 844-851.
6. Kohn, L.D. (1978) In Receptors and Recognition, Series A, P. Cuatrecasas and M.F. Greaves, eds., Vol. 5, Chapman and Hall, London, pp. 133-212.
7. Lee, G., Grollman, et al. (1979) J. Biol. Chem., 254, 3826-3832.

8. Dumas, M., Schwab, M.U., and Thoenen, H. (1979) J. Neurobiol., 10, 179-197.
9. Matsuda, M. and Yoneda, M. (1975) Infect. Immun., 12, 1147-1153.
10. Helting, T.B. and Zwisler, O. (1977) J. Biol. Chem., 252, 187-193.
11. Helting, T.B., Zwisler, O., and Wiegandt, H. (1977) J. Biol. Chem., 252, 194-198.
12. van Heyningen, S. (1976) FEBS Ltrs., 68, 4-7.
13. Helting, T., Ronneberger, et al. (1978) J. Biol. Chem., 253, 125-129.
14. Matsuda, M. and Yoneda, M. (1978) Toxicon, Supplement 1, 977-989.
15. Habermann, E. (1978) In Handbook of Clinical Neurology, P.J. Vinker and G.W. Bruyn, eds., Vol. 33, Part 1, Elsevier-North Holland, New York, pp. 491-547.
16. Max, S.R., Schwab, M., Dumas, M. and Thoenen, H. (1978) Brain Res., 159,
17. Ramos, S., Grollman, E.F., et al. (1979) Proc. Natl. Acad. Sci. USA, in press.
18. Macdonald, R.L., Bergey, G.K. and Habig, W.H. (1979) Neurology, 29, 588.
19. Bigalke, H., Dimpfel, W. and Habermann, E. (1978) Naunyn-Schmiedeberg's Arch. Parmacol., 303, 133-138.
20. Gill, D.M. (1978) In Bacterial Toxins and Cell Membranes, J. Jelaszewicz and T. Wadström, eds., Academic Press, New York, pp. 291-332.
21. Ledley, F.D., Mullin, B.R., et al. (1976) Biochem. Biophys. Res. Commun., 69, 852-859.

STRUCTURE-FUNCTION STUDIES ON NERVE GROWTH
FACTOR AND ITS CELL SURFACE RECEPTOR[*]

RALPH A. BRADSHAW, NICHOLAS V. COSTRINI[+], KENNETH A. THOMAS, AND JEFFREY S. RUBIN
Department of Biological Chemistry, Washington University School of Medicine,
St. Louis, Missouri 63110

ABSTRACT

Nerve growth factor (NGF) is a member of a subgroup of trophic factors that share a common ancestral precursor with insulin (proinsulin). As with relaxin and insulin-like growth factor (IGF), it stimulates a variety of metabolic responses by complexation with a receptor molecule on the surface of its responsive cells, i.e., sympathetic and neural crest-derived sensory neurons. Although primary structure comparisons indicate the closest relationship to be between insulin and IGF, the latter substance is more similar to NGF with respect to polypeptide chain structure. That is, unlike insulin and relaxin, neither is cleaved to a two chain active structure, an event that is apparently related to storage. In one of the two sites where high concentrations of NGF are stored, the adult male mouse submaxillary gland, a higher molecular weight complex (7S NGF) is formed. One of the two additional types of polypeptide chains (γ subunit) is an arginine-specific esteropeptidase which shows 41% homology to bovine trypsin based on structural analyses.

The receptor for NGF in adult rabbit superior cervical neurons is an intrinsic membrane protein that is solubilized by Triton X-100. It retains the same binding characteristics as the membrane bound form. Hydrodynamic studies of the receptor-Triton X-100 complex indicate it has a Stokes radius of 71 Å, an $S_{20,w} = 4.3S$, and a $\bar{v} = 0.74$ cc/g. Thus, by calculation, the receptor has a $M_r = 135,000 \pm 15,000$ with an $f/f_o = 1.8$, indicating it is a highly asymmetrical molecule with only minimal amounts of bound detergent. These results contrast with the properties of the nuclear receptor which is not solubilized by detergent. A mechanism that links these two receptors to the same temporal basis as the biologic responses of the hormone has been suggested.

[*] The work described originating in the authors' laboratory was supported by U.S. Public Health Service Research Grant NS 10229.
[+] Present address: Department of Medicine, Section of Gastroenterology, Medical College of Wisconsin, Veterans Administration Hospital, 5000 W. National Avenue, Milwaukee, Wisconsin 53193.

Among the many functions of protein molecules in living organisms, one of the most important is to act as messengers to transfer information from one cell type to another. Unlike enzymes, such functions are not exclusively carried out by proteinoid substances; however, they do contribute a substantial number covering a variety of biological activities such as hormones, neurotransmitters, lectins, immunoglobulins, bacterial and plant toxins, etc. One such group containing substances that can be broadly classified together is the growth factors,[1] which represent a wide diversity of biological functions such as the maintenance of the sympathetic nervous system, the development of the skeleton and connective tissues and the preparation of the birth canal prior to the delivery of the fetus.[2-4] Despite such wide variance in the nature of the functional response at the organ level, it has been suggested that at the molecular level, a much greater similarity might exist.[5] This is particularly likely among the substances of this class which show evidence of evolutionary relatedness to a common ancestral gene. In this regard, the subset of growth factors or hormone-like substances that are related to insulin, which include nerve growth factor (NGF), insulin-like growth factor (IGF), and relaxin, appears to provide a particularly good example of this phenomenon.[5,6] Thus, it might be concluded that detailed knowledge of the structure, function and mechanism of action of one of these substances would provide important clues in understanding the action of the others. While this "mechanistic" homology remains to be established, the rapidly developing knowledge of the structure-function relationships of NGF[7] should soon provide the opportunity to test such a hypothesis in detail.

As with other polypeptide hormone or hormone-like systems, the growth factors exert their mechanism of action, at least initially, through interaction with responsive target cells at the plasma membrane. This interaction appears to invariably occur through specific complexation with a protein molecule generally termed a receptor, in a highly selective process analogous to the interaction of an enzyme with its substrate or an antibody with its antigen. As in all such two component systems, a complete description of the biological activities that are induced requires detailed knowledge of both participants. This report will describe the present knowledge of the chemistry of NGF and its cell surface receptor.

The chemistry of nerve growth factor (NGF)

Sources of NGF. Unlike most other hormones of higher vertebrates, NGF is not elaborated by only a single tissue but seems to be the product of a variety of organs and cells in the body.[7] While this observation was originally confusing, it now appears simply to be a manifestation of the morphologically expansive target tissues of the hormone, namely the sympathetic nervous system and many sensory neurons derived from the neural crest. Since the tissues that are innervated by these neurons produce and supply NGF to them during all stages of development of the organism, this growth factor is produced by a wide variety of tissues, albeit, in virtually all cases in extremely small quantities, insufficient for molecular characterization.[8] There are three known exceptions to this situation: the venom of the three families of poisonous snakes,[9] the adult male submaxillary gland of the mouse[10] and, as demonstrated quite recently, the prostate gland of the guinea pig.[11] There is no detailed characterization of the last NGF, and only sparing information is available about the NGF from a number of the snake venoms.[12] In contrast, mouse NGF has been studied in detail and provides, at this juncture, virtually all of our knowledge of the chemistry of this hormone.

Mouse NGF. Nerve growth factor is found in high concentrations in the submaxillary gland of male mice who have passed puberty. Considerable amounts are also found in the saliva of these animals which can be further induced by α-adrenergic agonists.[13] The reason for the high concentrations of the factor in either the gland or its secretions are not presently appreciated. However, the fact that it is now found in high concentrations in the prostate,[11] another exocrine gland, suggests the possibility that NGF may exert functions in addition to its role as a regulator of the growth, development and maintenance of sympathetic and selected sensory neurons.

The mouse submaxillary protein is found both in the glands and the saliva as a high molecular weight complex referred to as 7S NGF.[14,15] This sedimentation coefficient is indicative of its molecular weight of approximately 130,000. There are two copies of each of the three unique polypeptide sequences present in the complex. The β subunit, which can independently bind each of the other two types of polypeptide chains, contains all of the biological activity associated with maintaining neuronal function.[16,17] The β subunit is actually composed of two identical polypeptide chains, each of 13,250 molecular weight.[18,19] As with all the subunits of the 7S complex, the two polypeptides are associated by noncovalent forces. Each polypeptide of the β subunit is capable of binding one α and one γ subunit.[17] In addition, each γ subunit appears to bind one zinc ion.[20] Both the α and γ subunits have molecular weights of approximately 26,000.[21] As described below, the γ subunit is an arginine esteropeptidase of the well-studied "serine" proteinase family.[17,22] However, there is no known biological function for the α subunit other than its specificity to form complexes with the β subunit.

The biological role of the 7S NGF seems to be unrelated to the neuronal

Fig. 1. Schematic representation of the amino acid sequence of the primary subunit of 2.5S NGF from mouse submaxillary glands. (Taken from Ref. 19.)

stimulation activity of the β subunit. As a complex, it is inactive in all
biological tests associated with neuronal tissue. It has been proposed that it
may serve as a storage function in the submaxillary gland.[21] Alternatively, the
high molecular weight complex may be important for exocrine functions, and it
will thus be of interest to ascertain the quaternary structure of the prostate
hormone.

The amino acid sequence of the polypeptide chain of the β subunit is shown
in Fig. 1. The longest form of this chain, as isolated from the gland, contains 118 amino acids in a continuous polypeptide chain.[19] Three intrachain
disulfide bonds are also present. Depending on the methods of preparation,[23]
proteolytic processing can occur, resulting in the maximum removal of eight
residues from the amino terminal and one residue from the carboxyl terminal of
either or both polypeptides.[24,25] This processing seems to occur primarily
during preparation and is without effect on either biological or immunological
activity.

The γ subunit of NGF is an enzyme of high specificity for the peptide or ester bonds formed by the carboxyl group of arginine residues.[22] This specificity, along with the presence of a carboxyl terminal arginine residue in the β poly-peptide chain (See Fig. 1), has led to the suggestion that the γ protein may play a role in the processing of a pre-β-subunit.[19] In this regard, Berger and Shooter[26] have demonstrated the synthesis of a 22,000 molecular weight precursor molecule which can be processed by the γ subunit to a polypeptide the same size as β NGF. It still remains curious, however, that an enzyme with this biologi-cal role would be present in stoichiometric amounts.

The γ subunit also undergoes proteolytic processing, which as shown in Fig. 2, gives rise to a total of four constituent fragments, A, B1, B2 and C, depend-ing on whether there are one or two cleavages in the polypeptide backbone. All four fragments can be obtained in homogenous form by the two chromatographic steps depicted in the lower half of Fig. 2. These fragments have been extremely useful in the amino acid sequence analysis of this protein. The structural data is summarized in Fig. 3 in a comparative fashion to bovine trypsin. The rela-tionship of this murine serine protease to those of other species is summarized in Table 1. As can be seen, it shows 40%, 41% and 38% identities, respectively, to the trypsins of dogfish, cow and pig. It is somewhat less closely related to chymotrypsin A and B of cow and even less so to the blood-clotting enzymes factor X and thrombin. It is least related to porcine elastase and the rat group specific protease.[27] Thus, it seems most like, species differences not-withstanding, the trypsins, enzymes that share a similar specificity. However, the lower homology seen with thrombin and factor X, which also show a high degree of specificity for arginine residues, argues that substrate specificity is not particularly manifested in overall sequence homology. Thus, there seems

TABLE 1

GAMMA SEQUENCE HOMOLOGIES

Trypsin (Dogfish)	46%
Trypsin (Cow)	41%
Trypsin (Pig)	38%
Chymotrypsin A (Cow)	34%
Chymotrypsin B (Cow)	33%
Factor X (Cow)	29%
Thrombin (Cow)	29%
Elastase (Pig)	30%
Group Specific Protease (Rat)	27%

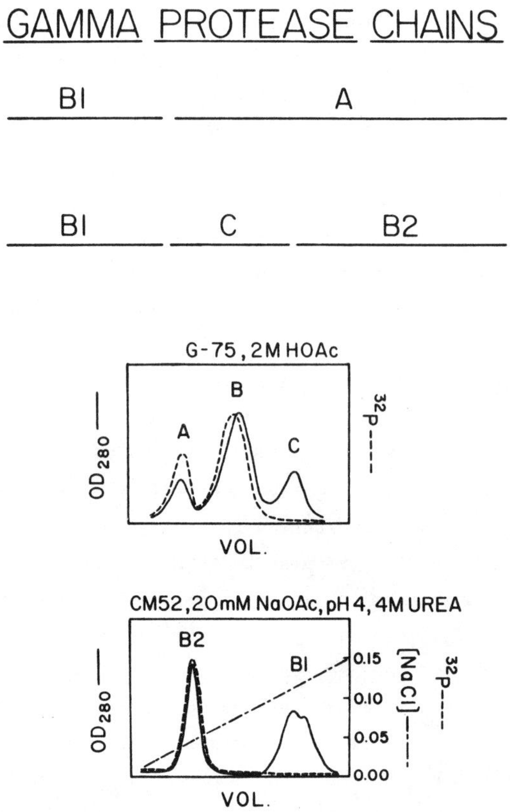

GAMMA PROTEASE CHAINS

Fig. 2. The γ-subunit of mouse NGF is a mixture of two and three polypeptide chain species (upper diagram). After reduction and carboxymethylation, the four fragments (A, B1, B2 and C) can be separated on Sephadex G-75 (middle panel) into A, B and C chains. The B1 and B2 chains are separated by CM-cellulose (CM-52) column chromatography (lower panel). After labeling with ^{32}P-diisopropyl fluorophosphate (DFP), both the A and B2 fragments contain the radioactive label.

to be little about the evolutionary origin of the γ subunit that provides any clues about its possible biological function in either the 7S complex or the submaxillary gland.

There is essentially no structural information available about the α subunit of the complex. Electrophoretic characterization suggests that it too can undergo proteolytic processing resulting in one or more internal cleavages which, as with the γ subunit, can produce multiple, electrophoretically distinct forms of the protein.[21] It may be hoped that the sequence analysis of this subunit, presently underway, will be useful in providing insight into the functional properties of this polypeptide.

453

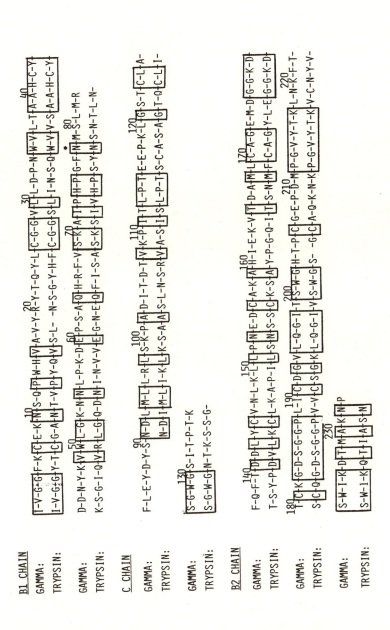

Fig. 3. The amino acid sequence of mouse NGF γ subunit. Identical residues with bovine trypsin[54] are enclosed in boxes. The asparagine at postion 78 (*) is glycosylated. The one letter code is defined in Fig. 4.

454

The insulin-related subset. Many proteins with similar but distinct functions have been shown to be related sequentially, suggesting evolution from a common precursor. Statistically significant homology in protein families such as NAD-dependent dehydrogenases, the pituitary gonadotropins and several types of cytochromes, to name a few, have been shown.[28] Recently, a new such group, a family of growth factors or growth regulatory substances, that are related to the well known hormone insulin, has emerged.[5,6] Nerve growth factor was the first of these molecules to be so identified,[29] followed more recently by the addition of three other substances, i.e., insulin-related growth factor I and II[30] and the ovarian hormone relaxin.[31,32] The insulin-like growth factors are isolated from human serum, and as the name implies, show a considerable spectrum of insulin-like functions. The similarity is also manifested in primary structure relatedness which, in the regions of the molecules corresponding to the A and B chains of insulin, is approximately 50%.[30] However, unlike insulin, they are not processed to a two chain structure, but exist as a continuous polypeptide chain. In addition, the C-bridge region has been foreshortened to about one-third the length found in proinsulin, and there are several additional residues at the carboxyl terminus. In contrast, relaxin shares a chain structure more similar to that of insulin in that it is processed from a precursor molecule to A and B chains forming a covalently linked dimer of approximately equal size to that of insulin. Like both IGF molecules, the disulfide structure of relaxin exactly mimics that found in the insulin molecule.[30,32] However, the extent of homology is somewhat less being only about 25%.

Functionally, IGF-I meets the criteria of a somatomedin,[33] that is, it is a mediator of the action of growth hormone on cartilage tissue. It promotes the incorporation of radioactive sulfate into the proteoglycans of cartilage, its serum concentration is regulated by growth hormone and it exerts insulin-like activities on extraskeletal tissues.[4] It is also interesting to note that the molecular responses to somatomedin with respect to insulin-like activity are similar to those of NGF. On the other hand, the responses to relaxin are quite different at the organ level. At present, the biological response of relaxin appears to be manifested through changes in the pubic symphysis at the termination of pregnancy as well as responses of the cervix, uterus and possibly the mammary gland.[34] There is little information available about the actions of relaxin at the molecular level. However, it is without effect in either somatomedin or NGF assays (M. Niall, J. Jacobs, I.K. Mariz, W.H. Daughaday and R.A. Bradshaw, unpublished observations).

The structural comparison of the A and B chains of insulin, IGF-I, NGF and relaxin are shown in Fig. 4. Those residues which are identical in at least

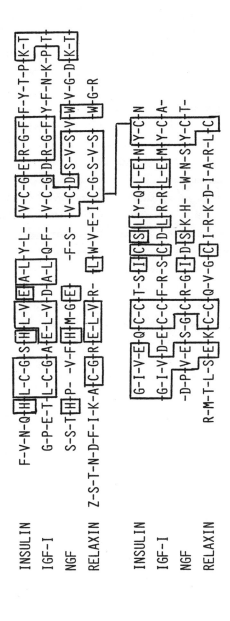

Fig. 4. The comparison of the amino acid sequences of the B (upper) and A (lower) chains of human insulin,[55] insulin-like growth factor-I,[30] nerve growth factor (NGF)[19] and relaxin.[31] Solid boxes indicate residues identical in at least two of the polypeptides. Deletions were inserted arbitrarily to increase identities. The connecting line between the half-cystinyl (C) residues represents the one disulfide bond conserved in all four hormones. The one letter code used: A, Ala; C, Cys; D, Asp; E, Glu; F, Phe; G, Gly; H, His; I, Ile; K, Lys; L, Leu; M, Met; N, Asn; P, Pro; Q, Gln; R, Arg; S, Ser; T, Thr; V, Val; W, Trp; Y, Tyr; Z, PCA.

two of the molecules are enclosed in solid boxes. It is noteworthy that one disulfide bond is conserved in all four molecules, as indicated by the solid line. Based only on the number of identities present, insulin and IGF-I appear more closely related than either relaxin or NGF are to the pancreatic hormone. In addition, particularly in the region of the B chain surrounding the conserved disulfide, relaxin and NGF show greater similarities. However, on the basis polypeptide chain structure, insulin and relaxin can be grouped together as can IGF and NGF, which are continuous polypeptide chains. The removal of an intervening peptide fragment (C-bridge) in two of the molecules but not in the other two is perhaps suggestive of important functional differences. However, it seems more likely that this difference may reflect alterations in the biosynthetic preparation of the hormones to meet different demands in their biological function.[5,35] That is to say, both insulin and relaxin are processed to a granular storage form in their respective sites of synthesis, the pancreas and the corpus luteum. In turn, they are released in bolus amounts into the bloodstream on receipt of the proper stimulii. In contrast, NGF and IGF do not appear to be storage hormones with respect to tissue synthesis and endocrine function. That is, both NGF and IGF-I appear to be synthesized and immediately released to carry out their biological functions. Although, the site of synthesis of IGF-I remains controversial, it does appear clear that there is no storage depot. As noted in the introductory remarks, there are two tissues in which NGF is stored in higher vertebrates (the submaxillary gland of mice and the prostate of the guinea pig and perhaps other species). However, the role of the high concentrations of NGF in these tissues seems unrelated to its function as a regulator of growth and development of the nervous system, and therefore, the storage in these tissues may be viewed as being under separate control.[7] In the sympathetic end organs which provide the NGF for the maintenance of neurons, there is no comparable storage. This is also emphasized by the fact that females of all species of higher vertebrates have an equal requirement for NGF in maintenance of sympathetic and selected sensory neurons and do not show the high concentration in these tissues. Thus, the processing to remove an intervening sequence to produce the two-chain structure in insulin and relaxin may be viewed as an obligatory step in the formation of the storage granules, which is not required for IGF-I and NGF. An acquirement of the dibasic residues to provide the sites of cleavage in insulin and relaxin or, alternatively, a loss of such sites in the development of IGF-I and NGF, presumably occurred via normal evolutionary events.[35]

The nerve growth factor receptor

As with other molecules of similar activity, NGF initiates its complex series of cellular and metabolic responses in its target cell by first complexing with a specific receptor molecule located in the plasma membrane.[7] The specific binding properties of the receptors for NGF from a number of tissues have been examined by equilibrium and kinetic binding methodology.[36-38] In each case, NGF has been specifically marked with an ^{125}I nuclide, introduced into the molecule by a variety of techniques. Although early studies used tracers of this kind, which showed significant amounts of nonspecific binding as well as low specific activity, recent advances have provided labeled NGF molecules essentially devoid of these undesirable properties. The exact nature of the binding properties to these receptors has been somewhat controversial with varying results from different laboratories. There is, however, common agreement on the presence of a high affinity receptor, the occupation of which appears to be proportional to those concentrations required to initiate biological responses. The presence of apparent low affinity receptors resulting from negatively cooperative interactions in the receptors has been proposed by Frazier et al.[38] In contrast, Sutter et al.[39] have observed two classes of receptors present in responsive neurons with no evidence of negatively cooperative interactions. These discrepancies remain to be clarified.

The NGF receptor from dorsal root or superior cervical neurons is readily solubilized by nonionic detergents but not by simple extractions with salt solutions.[40,41] This indicates it to be an intrinsic membrane protein embedded in or transversing the lipid bilayer of the plasma membrane. As initially reported by Banerjee et al.,[40] the solubilized receptor from superior cervical neurons retains its high affinity binding properties. Costrini and Bradshaw[41] have examined the receptor from adult rabbit superior cervical ganglia as solubilized by Triton X-100 and have demonstrated it to be a molecule with a hydrodynamic radius of 71 Å as judged by gel filtration on columns of Sepharose 6B in the presence of the detergent. This value, which is similar to that reported for the insulin receptor from IM-9 lymphocytes,[42] is unaffected by the presence of sucrose. As a result, it was possible to calculate the sedimentation coefficients and partial specific volume in sucrose density gradient centrifugations both in H_2O and in D_2O. These parameters are summarized in Table 2. The partial specific volume of 0.74 cc/g is indicative of a small amount of bound detergent in the soluble complex. This has been confirmed by independent measurements which indicate less than 0.25 mg detergent bound per mg protein. The molecular weight calculated from these values is 135,000 $\pm$ 15,000, and it is not unsimilar to molecular weight determinations of the insulin receptors.[43,45]

TABLE 2

PHYSICAL CHARACTERISTICS OF THE TRITON X-100 SOLUBILIZED

SOLUBILIZED NERVE GROWTH FACTOR RECEPTOR

Parameter	Value[a]
Stokes radius ($\mathring{A}$)	71 ± 2.2 (5)
Sedimentation coefficient ($s_{^\circ 20,w}$)	4.3 ± 0.1 (11)
Partial specific volume (cc/g)	0.74 ± 0.03 (2)[b]
Molecular weight (M_r)	$135,000 \pm 15,000$
Detergent bound (mg/mg protein)	< 0.25
Frictional ratio, f/f_o	1.8[c]

[a]Values are mean $\pm$ SE for the number of determinations shown in parentheses.

[b]Two separate experiments each of which consisted of parallel sucrose density gradients prepared in H_2O and D_2O. Mean value based on eleven gradient analyses.

[c]The value of the solvation factor was taken to be 0.2 g of solvent/g of protein.

The other parameter reported in Table 2 is that of the frictional ratio f/f_o. The value of 1.8 is quite high, and indicates that the solubilized receptor for NGF is a highly asymmetrical molecule. While the frictional ratio is too general to draw any specific conclusions about the shape or the orientation of the receptor molecule relative to the membrane it is not inconsistent with a dumbbell-shaped molecule in which large amounts of the protein are found on both the internal and external faces with a connecting segment spanning the lipid bilayer. It should be emphasized, however, that such a model is not presently supported by any direct observations. It should also be noted that the molecular weight value calculated in this study represents the minimal unit solubilized by the detergent. It provides no information about a possible subunit structure composed of more than one identical or dissimilar polypeptide chains.

Concluding remarks

As presently appreciated, NGF is a hormone-like substance required for the maintenance and development of the sympathetic and portions of the sensory nervous system that has evolved from a common precursor which has also yielded relaxin, insulin-like growth factor and insulin. It is not inconsistent with our present knowledge that this evolutionary relatedness is also paralleled by functional similarities, at least at the molecular level for all four substances. This is certainly true insofar as the initial interaction of each molecule is with a plasma membrane receptor of the responsive target cells. In this regard, it is noteworthy that the molecular characterization of the NGF receptor appears to be quite similar to values reported for that of the insulin receptor.

However, the receptor interaction which marks the initial event in the NGF mechanism is not sufficient for the complete manifestation of the biological response. It has been established by Hendry, Thoenen, and their colleagues that following this complex formation, an endocytotic event, which results in the uptake of the hormone as well as presumably its receptor, occurs.[46-48] This is followed in mature neurons by retrograde axonal transport of the vesicle containing the hormone to the cell body. In immature neurons lacking processes and axonal elements, such a step is not required. Upon reaching the cell bodies, the fate of the NGF appears to be diverse. Whereas a considerable amount of material is found subsequently in lysosomes, which presumably mediate degradation, it is also found in vesicles fused with the smooth endoplasmic reticulum[49] and, as determined by subcellular distribution experiments, a portion is found in the nuclear fraction.[50] This last observation is consistent with the discovery of specific receptor sites in the nuclei of responsive cells.[51] These sites of interaction appear distinct from those found in the plasma membranes in that they are not solubilized by nonionic detergents. While occupation of these receptors has not been clearly demonstrated to produce further biological responses, the well known regulation of transcriptional events by the hormone producing such changes as the specific induction of tyrosine hydroxylase and dopamine-β-hydroxylase[52] may well be associated with such events. This two-stage mechanism of action readily explains the temporal basis of the NGF response in terms of both short and long-term effects and provides a rational explanation for the various biochemical and morphological changes induced by the hormone.[7] It conceptually differs from the mechanism of action of other hormones such as glucagon, epinephrine and parathyroid hormone, which induce production of intracellular cyclic AMP by the stimulation of membrane bound adenylcyclase.[5,6] In the case of NGF, and perhaps with the evolutionarily related members of the insulin subset, it acts as its own intracellular second messenger. The recently reported observations for the receptor-mediated uptake of many classes of hormone and hormone-like molecules[53] has strengthened the notion that the internalization of such molecules is mechanistically important and suggests that the proposed mechanism for NGF[7] may have broad applicability.

460

REFERENCES

1. Gospodarowicz, D., and Moran, J.S. (1976) Ann. Rev. Biochem., 45, 531-558.
2. Levi-Montalcini, R., and Angeletti, P.U. (1968) Physiol. Revs., 48, 534-569.
3. Fevold, H.L., Hisaw, F.L. and Meyer, R.K. (1930) J. Am. Chem. Soc., 52, 3340-3348.
4. Daughaday, W.H., Hall, K., et al. (1972) Nature, 235, 107.
5. Bradshaw, R.A. and Niall, H.D. (1978) Proc. 6th Asia and Oceania Cong. of Endocrinology, 1, 358-364.
6. Bradshaw, R.A. and Niall, H.D. (1978) Trends in Biochem. Sci., 3, 274-276.
7. Bradshaw, R.A. (1978) Ann. Rev. Biochem., 47, 191-216.
8. Young, M., Murphy, R.A., et al. (1976) in Surface Membrane Receptors: Interface Between Cells and Their Environment, Bradshaw, R.A., Frazier, W.A., et al. eds., Plenum Pub. Corp., New York, 247-267.
9. Cohen, S. (1959) J. Biol. Chem., 234, 1129-1137.
10. Cohen, S. (1960) Proc. Natl. Acad. Sci. U.S.A., 46, 303-311.
11. Harper, G.P., Barde, Y.A., et al. (1979) Nature, 279, 160-162.
12. Hogue-Angeletti, R.A. and Bradshaw, R.A. (1979) in Snake Venoms, Lee, C.Y., ed., Handbook of Experimental Pharmacology, Springer-Verlag, Berlin, 52, 276-294.
13. Wallace, L.J., Partlow, L.M., and Wardell, L.J. (1977) Trans. Amer. Soc. Neurochem., 8, 135.
14. Varon, S., Nomura, J., and Shooter, E.M. (1968) Biochemistry, 7, 1296-1303.
15. Burton, L.E., Wilson, W.H., and Shooter, E.M. (1978) J. Biol. Chem., 253, 7807-7812.
16. Smith, A.P., Varon, S., and Shooter, E.M. (1968) Biochemistry, 7, 3259-3268.
17. Silverman, R.E. (1978) Ph.D. Thesis, Washington University, St. Louis.
18. Angeletti, R.H., Bradshaw, R.A. and Wade, R.D. (1971) Biochemistry, 10, 463-469.
19. Angeletti, R.H., and Bradshaw, R.A. (1971) Proc. Natl. Acad. Sci. U.S.A., 68, 2417-2420.
20. Pattison, S.E., and Dunn, M.R. (1975) Biochemistry, 14, 2733-2739.
21. Server, A.C., and Shooter, E.M. (1977) Adv. Prot. Chem., 31, 339-409.
22. Greene, L.A., Shooter, E.M., and Varon. S. (1968) Proc. Natl. Acad. Sci. U.S.A., 60, 1383-1388.
23. Jeng, I.M., and Bradshaw, R.W. (1978) in Research Methods in Neurochemistry, Marks, N., and Rodnight, R., eds., Plenum Pub. Corp., New York, 4, 265-288.
24. Moore, J.B., Jr., Mobley, W.C., and Shooter, E.M. (1974) Biochemistry, 13, 833-840.
25. Mobley, W.C., Schenker, A., and Shooter, E.M. (1976) Biochemistry, 15, 5543-5552.
26. Berger, E.A., and Shooter, E.M. (1977) Proc. Natl. Acad. Sci. U.S.A., 74, 3647-3651.
27. Woodbury, R.G., Katunuma, N., et al. (1978) Biochemistry, 17, 811-819.
28. Dayhoff, M.O., (1977) ed., Atlas of Protein Sequence and Structure, National Biomedical Research Foundation, Silver Spring, Maryland, 5.
29. Frazier, W.A., Angeletti, R.H., and Bradshaw, R.A. (1972) Science, 176, 482-488.
30. Rinderknecht, E., and Humbel, R.E. (1978) J. Biol. Chem., 253, 2769-2776.
31. James, R., Niall, H., et al. (1977) Nature, 267, 544-546.
32. Schwabe, C.J., and McDonald, K.J. (1977) Science, 197, 914-915.
33. Megyesi, K., Kahn, R.C., et al. (1975) J. Biol. Chem., 250, 8990-8996.
34. Niall, H.D., Bradshaw, R.A., and Bryant-Greenwood, G.D. (1979) in Progress in Clinical and Biological Research, Bitensky, M., Collier, R.J., et al., eds., Vol. 31, Alan R. Liss. Inc., N.Y., 651, 658.
35. Dodson. G.G., Isaacs N., et al. (1979) Proc. of Int. Symp. on Insulin. Aachen, in press.

461

36. Banerjee, S.P., Snyder, S.H., et al. (1973) Proc. Natl. Acad. Sci. U.S.A., 70, 2519-2523.
37. Herrup, K., and Shooter, E.M. (1973) Proc. Natl. Acad. Sci. U.S.A., 70, 3884-3888.
38. Frazier, W.A., Boyd, L.F., and Bradshaw, R.A. (1974) J. Biol. Chem., 249, 5513-5519.
39. Sutter, A., Riopelle, R.J., et al. (1979) J. Biol. Chem., in press.
40. Banerjee, S.P., Cuatrecasas, P., and Snyder, S.H. (1976) J. Biol. Chem., 251, 5680-5685.
41. Costrini, N.V. and Bradshaw, R.A. (1979) Proc. Natl. Acad. Sci. U.S.A., in press.
42. Ginsberg, B.H., Kahn, C.R., et al. (1976) Biochem. Biophys. Res. Commun., 73, 1068-1074.
43. Jacobs, B., Schecter, Y., et al. (1977) Biochem. Biophys. Res. Comm., 77, 981-988.
44. Yip, C.C., Yeung, C.W.T., and Moule, M.L. (1978) J. Bio. Chem., 253, 1743-1745.
45. Pilch, P.F. and Czech, M.P. (1979) J. Biol. Chem., 254, 3375-3381.
46. Hendry, I., Stoeckle, K., et al. (1979) Brain Res., 68, 103-121.
47. Stoeckel, K., Paravicini, U., and Thoenen, H. (1974) Brain Res., 76, 413-421.
48. Paravicini, U., Stoeckel, K., and Thoenen, H. (1975) Brain Res., 84, 279-291.
49. Schwab, M., and Thoenen, H. (1977) Brain Res., 122, 459-474.
50. Johnson, E.M., Jr., Andres, R.Y., and Bradshaw, R.A. (1978) Brain Res., 150, 319-331.
51. Andres, R.Y., Jeng, I., and Bradshaw, R.A. (1977) Proc. Natl. Acad. Sci. U.S.A., 74, 2785-2789.
52. Thoenen, H., Angeletti, P.U., et al. (1971) Proc. Natl. Acad. Sci. U.S.A., 68, 1598-1602.
53. Goldstein, J.L., Anderson, R.G.W., and Brown, M.S. (1979) Nature, 279, 679-685.
54. Walsh, K., and Neurath, H. (1969) Proc. Natl. Acad. Sci. U.S.A., 52, 884-889.
55. Nicol. D.S.H.W., and Smith, L.F. (1960) Nature, 187, 483-485.

VI
Other Topics

STRUCTURE AND FUNCTION OF C-REACTIVE PROTEINS

EDUARDO B. OLIVEIRA[*†], EMIL C. GOTSCHLICH[+] and TEH-YUNG LIU[#]
*Instituto de Ciências Biológicas, Universidade Federal de Minas Gerais, Belo
Horizonte 30 000 Brazil; +The Rockefeller University, New York, New York 10021;
and #Division of Bacterial Products, Bureau of Biologics, Food and Drug
Administration, Bethesda, Maryland 20205

SUMMARY

The complete amino acid sequence has been derived for human C-reactive protein (CRP).[1] The protein yielded a unique sequence containing 187 amino acids in a single polypeptide chain. The two half cystines residues at positions 36 and 78 are involved in a disulfide bond. Based on sequence data, a minimal molecular weight of 20,946 has been calculated for CRP.

Computer analysis showed no significant repeating sequences within the CRP molecule; distant homologies were noted to the C_H2 domain of immunoglobulin G and to C_3a anaphylatoxin. The homologies noted were insufficient to support a common evolutionary origin of these proteins. The binding properties of human (CRP) and rabbit (CxRP) C-reactive proteins were investigated by studying the reactions with pneumococcal C-polysaccharide and three derivatives of bovine serum albumin (BSA). One derivative (BSA-PC) contained phosphodiester linked phosphorylcholine, the other two contained phosphorylethanolamine linked to BSA through the amino nitrogen, either as a secondary amine (BSA-I-PEA-A), or through an amide bond (BSA-I-PEA-B). Like the pneumococcal C-polysaccharide, these BSA derivatives formed a precipitate upon reaction with CRP or CxRP in the presence of Ca^{+2} ions. The results of the quantitative precipitin reactions and their inhibitions by phosphorylcholine and α-glycerolphosphate indicated that the binding site of C-reactive proteins consist of two loci; a primary locus responsible for the calcium dependent binding of the phosphorylester, and a secondary locus for the binding of the cationic group. A major difference that was observed in the binding properties of human CRP and rabbit CxRP is that for precipitin formation with ligands, binding by the two loci in CRP is essential, whereas with CxRP only the binding of phosphoryl groups through calcium is obligatory.

INTRODUCTION

During the acute phase of infectious diseases and various other inflammatory

---

[†]Fellow from Fundacão de Amparo à pesquisa do Estado de São Paulo, Brazil

processes there appears in the serum of man a protein not normally detected. This substance is called C-reactive protein (CRP) because of its property of precipitating with C-polysaccharide derived from the cell wall of *Streptococcus Pneumonia*.[1] Proteins analogous to CRP have been isolated from sera of other species.[2,3]

C-reactive protein mediates, *in vitro*, many reactions that are also mediated by immunoglobulins. C-reactive protein causes the precipitation of bacterial cells and certain polymers[4] and acts as an opsonin.[5] It **differs**, however, from immunoglobulins in some respects; for instance, many different stimuli can cause the appearance of the same CRP, which is both antigenically and structurally[6] distinct from immunoglobulins.

C-reactive protein may participate in nonspecific resistance to infections,[7] but, as yet, the definitive mechanism of its biological function has not been established.

The results presented in this paper are concerned with two basic features of the CRP molecule, the determination of its chemical structure and the delineation of some of its binding properties.

METHODS

Preparation of C-reactive protein and C-polysaccharide. The human CRP was prepared by a modification[8] of the method of Wood et al.[9] from pooled pleural fluid of patients with various neoplastic diseases. Mouse and rabbit CRPs were obtained as described in the text. Pneumococcal C-polysaccharide and Cx-polysaccharide were prepared as described previously.[2]

Amino acid sequence determination. The human CRP was freed of salt by dialysis against water, lyophilized, and subjected to cyanogen bromide cleavage[10] in 70% formic acid at 37°C for 18 h. The fragments were separated by gel filtration on Sephadex G-100 in 30% acetic acid. Proteolytic digestions of the protein and its cyanogen bromide fragments were carried out with trypsin, α-chymotrypsin, staphylococcal protease, thermolysin, α-lytic protease, and papain. The resulting peptides were separated by a combination of gel filtration and ion-exchange chromatography on cation and/or anion exchanger resins. Peptides in the eluent fractions were analyzed with o-phthalaldehyde[11] after alkaline hydrolysis. The amino acid sequences of the purified peptides were determined by subtractive Edman degradation[12] and hydrolysis with carboxypeptidases. The phenylthiohydantoin derivative released at each step was also identified as the amino acid after acid hydrolysis.[13] Peptides were hydrolyzed with 4N methanesulfonic acid,[14] and amino acid analysis was performed according to Spackman et al.[15] with a Beckman 121-M analyzer. The detailed conditions for these procedures were published elsewhere.[16]

Preparation of sepharose-phosphorylcholine resin. Aminohexyl Sepharose 4B
(Pharmacia Fine Chemicals AB, Uppsala, Sweden) substituted with caproylphos-
phorylcholine was kindly given to us by Drs. J.J. Moore and D.M. Segal of the
National Institutes of Health, Bethesda, Maryland. It was acetylated by treatment with 0.1 M 1-cyclohexyl-3-(2-morpholinoethyl)-carbodiimide metho-p-toluene
sulfonate in 0.4 M acetate buffer, pH 4.75 for 24 h at room temperature to
block the remaining free amino groups on the resin.

Synthesis of bovine serum albumin (BSA) derivatives. BSA-phosphorylcholine
derivative (BSA-PC) was prepared by allowing BSA (48 mg) to react with phosphorylcholine-caproyl-p-nitrophenyl ester (60 mg) in 0.48 ml of 0.05 M sodium
acetate, pH 7.0. The reaction mixture was magnetically stirred at room temperature while 10 µl aliquots of 0.05 N NaOH were added until a strong yellow
color developed. The derivatized BSA was desalted on a Sephadex G-25 column
(2 x 20 cm) equilibrated with water and lyophilized. Quantitative analysis of
phosphorus in BSA-PC gave an estimate of the degree of derivatization.

Bovine serum albumin-phosphorylethanolamine derivative A (Fig. 2, BSA-I-PEA-
A) was prepared by first converting the amino groups of BSA-I into bromoacetyl-
groups by the dropwise addition of chilled bromoacetylbromide (200 mg/ml in
acetonitrile) into a solution of BSA-I (50 mg) in 3.0 ml of guanidinium hydrochloride maintained at pH 8.0 with the addition of 10% Na_2CO_3. Upon completion
of the reaction, phosphorylethanolamine (282 mg) was added and the pH of the
reaction mixture raised to 9.5 with 1.0 N NaOH to allow the amino groups of
phosphorylethanolamine to react with the bromoacetyl groups of BSA-I over a
period of 18 h at room temperature. The unreacted bromoacetyl groups in BSA-I
were blocked by reacting with 2-mercaptoethanol (100 µl). The derivative was
desalted and the degree of derivatization estimated as described for BSA-PC.

Bovine serum albumin-I-phosphorylethanolamine derivative B (Fig. 2, BSA-I-
PEA-B) was prepared in the following manner. To 3.0 ml of a solution of phosphorylethanolamine (14 mg/ml in 0.1 N $NaHCO_3$) was added, dropwise, 0.2 ml of
bromoacetylbromide (200 mg in 1.0 ml of chilled acetonitrile) at room temperature while the pH was maintained at 8.0 by addition of 10% Na_2CO_3. Upon completion of the reaction, the reaction mixture was extracted with chloroform. To
the aqueous phase, containing the N-bromoacetyl-ethanolamine phosphate, was
added guanidinium hydrochloride to a final concentration of 6 M. The pH of the
solution was raised to 9.3 with solid carbonate and BSA-I (10 mg) was added to
allow the bromoacetyl groups of the derivatized ethanolamine phosphate to react
with the amino groups of BSA-I for 18 h at room hemperature. The protein was
desalted and the extent of the derivatization measured as described for BSA-PC.

Precipitin assay. Quantitative precipitin assays of CRP and CxRP with C-polysaccharide and BSA-conjugates were performed in 0.02 M Tris buffer, pH 7.0 containing 0.05 M NaCl and 0.001 M $CaCl_2$. Crystalline CRP or CxRP were equilibrated with the buffer by gel filtration. Aliquots (0.2 ml) of the protein solutions adjusted to 0.1 mg/ml were added to 0.2 ml of C-polysaccharide or BSA-conjugate solutions (0 to 930 µg/ml). The tubes were incubated for 1 h at 35°C and overnight at 5°C. The precipitates were recovered by centrifugation, washed twice with 0.5 ml of cold buffer and redissolved in 0.4 ml of 0.1 N NaOH. The protein content was estimated by absorbance at 280 nm.

Inhibition assay of the precipitin reaction. The inhibitory effect of phosphorylcholine and α-glycerolphosphate on the precipitin reaction was tested at pH 7.0 using the same buffer system and protein solution described in the previous section. C-reactive protein and CxRP solutions (0.2 ml) were mixed with increasing amounts of one of the inhibitors in 0.15 ml of the buffer and incubated for 10 min at room temperature. The quantity of C-polysaccharide, Cx-polysaccharide or BSA-conjugate required for maximal precipitation in the absence of inhibitor was added in 0.05 ml of buffer, and the reaction was continued for 1 h at 35°C and overnight at 5°C. C-reactive protein, with only the inhibitor served as control, and C-polysaccharide or BSA-conjugates alone were used as blanks. The amount of precipitate formed was measured as described in the previous section.

pH-dependence of the precipitin reaction. Precipitin reaction of human CRP with BSA-PC and BSA-I-PEA-A were performed over pH ranges of 5.0 to 9.5 in buffers consisting of 0.05 M Tris, 0.05 M NaCl, 0.05 M N-2-hydroxyethylpiperazine-N-2-ethanesulfonic acid, 0.5 M acetic acid, 0.001 M $CaCl_2$, and the pH adjusted appropriately with 2 N NaOH or HCl. C-reactive protein solutions (0.1 mg in 0.1 ml of the buffer) were mixed with solutions of BSA-PC (20 µg in 0.15 ml of buffer) or BSA-I-PEA-A (40 µg in 0.15 ml of buffer). After 1 h incubation at 35°C, the tubes were kept at 5°C overnight. Controls consisted of CRP or BSA-conjugates alone in the buffers at different pHs. The amount of precipitate formed was measured as described in the previous section.

Preparation of rabbit and mouse C-reactive proteins. Acute phase rabbit serum was obtained 24 h after intramuscular injection of 0.5 ml turpentine into the animals. The experimental conditions used for the purification of the CxRP are given in the legend of Fig. 6. Crystallization of the protein was achieved by the described method.[2]

Mouse CRP was prepared from acute phase serum essentially as described for rabbit CxRP using the Sepharose-phosphorylcholine affinity column.

Computer methods for comparing protein sequences. Two types of computer comparisons were made for CRP with proteins of known sequences. The programs used were SEARCH, for hints of evolutionary similarities with other sequences, and RELATE. These programs are described in detail by Dayhoff.[17]

RESULTS

Amino acid sequence of human C-reactive protein. The complete covalent structure of human CRP is shown in Fig. 1. Data in support of the proposed primary structure are summarized as follows:

Isolation and alignment of the cyanogen bromide peptides--CRP contains two residues of methionine per mole, which would give rise to three CNBr fragments. The expected three fragments were isolated and their sequences determined as indicated below. The cyanogen bromide fragments were aligned as shown in Fig. 1 by the isolation and characterization of two methionine containing peptides.

Purification and sequencing of peptides--Further cleavage of the cyanogen bromide fragments was accomplished by enzymic hydrolysis. In all, 65 peptides were purified by a combination of gel filtration on Sephadex columns and anion and/or cation exchange chromatography. Peptides of suitable size were then subjected to subtractive manual Edman degradation. The description of each peptide was given elsewhere.[16]

The disulfide bridge--CRP contains a pair of half-cystines per mol. Because

PCA-THR-ASP-MET-SER-ARG-LYS-ALA-PHE-VAL-PHE-PRO-LYS-GLU-SER-ASP-THR-SER-TYR-VAL- 20
SER-LEU-LYS-ALA-PRO-LEU-THR-LYS-PRO-LEU-LYS-ALA-PHE-THR-VAL-(CYS)-LEU-HIS-PHE-TYR- 40
THR-GLU-LEU-SER-SER-THR-ARG-GLY-TYR-SER-ILE-PHE-SER-TYR-ALA-THR-LYS-ARG-GLN-ASP- 60
ASN-GLU-ILE-LEU-PHE-GLU-VAL-PRO-GLU-VAL-THR-VAL-ALA-PRO-VAL-HIS-ILE-(CYS)-THR-SER- 80
TRP-GLU-SER-ALA-SER-GLY-ILE-VAL-GLU-PHE-TRP-VAL-ASP-GLY-LYS-PRO-ARG-VAL-ARG-LYS- 100
SER-LEU-LYS-LYS-GLY-TYR-THR-VAL-GLY-ALA-GLU-ALA-SER-ILE-ILE-LEU-GLY-GLN-GLU-GLN- 120
ASP-SER-PHE-GLY-GLY-ASN-PHE-GLU-GLY-SER-GLN-SER-LEU-VAL-GLY-ASP-ILE-GLY-ASN-VAL- 140
ASN-MET-TRP-ASP-PHE-VAL-LEU-SER-PRO-ASP-GLU-ILE-ASN-THR-ILE-TYR-LEU-GLY-GLY-PRO- 160
PHE-SER-PRO-ASN-VAL-LEU-ASN-TRP-ARG-ALA-LEU-LYS-TYR-GLU-VAL-GLN-GLY-GLU-VAL-PHE- 180
THR-LYS-PRO-GLN-LEU-TRP-PRO(COOH)

Fig. 1. The covalent structure of human C-reactive protein. PCA, pyrrolidine-carboxylic acid.

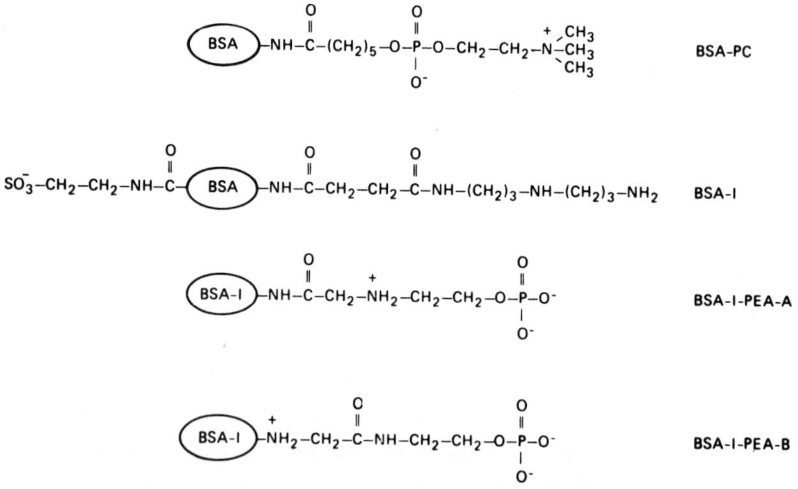

Fig. 2. The proposed structures of the bovine serum albumin conjugates; BSA-PC, BSA-I-PEA-A and BSA-I-PEA-B. BSA-I is an intermediate in the synthesis of the two phosphorylethanolamine conjugates. Note that the only difference between BSA-I-PEA-A and BSA-I-PEA-B is the relative position of the fragment linking BSA-I to phosphorylethanolamine.

no sulfhydryl group is detected either by titration with 5,5'-dithiobis(2-nitrobenzoic acid) in 8 M urea or alkylation with iodoacetamide in 6 M guanidinium hydrochloride,[18] the two half-cystine residues must exist in a disulfide bond.

Molecular weight of CRP based on sequence data--The protein yielded a unique sequence containing 187 amino acids in a single polypeptide chain. Based on the amino acid composition derived from sequence data, a minimal molecular weight of 20,946 has been calculated for human CRP. This value agrees well with the molecular weight of 21,500 established by gel filtration in 5 M guanidinium HCl.[18]

Bovine serum albumin (BSA) derivatives. Reaction of BSA with phosphorylcholine-caproyl-p-nitrophenyl ester yielded a product with an average number of 45 phosphorylcholine residues attached covalently to the protein (Fig. 2, BSA-PC). The degree of derivatization was based on the results of the amino acid and phosphorus[19] analyses.

To prepare the two phosphorylethanolamine derivatives (Fig. 2, BSA-I-PEA-A and BSA-I-PEA-B) a modified BSA was used as the carrier macromolecule (Fig. 2, BSA-I). Amino acid analysis of BSA-I indicated that 79% of the total number of free carboxylic groups had been converted into $-SO_3^-$ groups. Bromoacetyl bromide

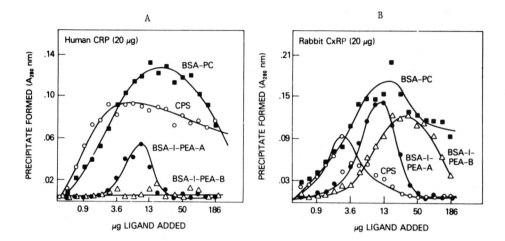

Fig. 3. Quantitative precipitin reactions of a) Human CRP, left, and b) CxRP, right, with pneumococcal C-polysaccharide (o), BSA-PC (▢), BSA-I-PEA-A (●), and BSA-I-PEA-B (△). The reactions were carried out in 0.02 M Tris buffer pH 7.0 containing 0.05 M NaCl and 0.001 M $CaCl_2$ by adding 200 µl of CRP or CxRP solution (0.1 mg/ml) to tubes containing the indicated amounts of polysaccharide or BSA conjugate in 200 µl of buffer. After incubation for 1 h at 35°C and for overnight at 5°C, the amount of precipitates formed were estimated by absorbance measurement at 280 nm of the corresponding solutions in 0.4 ml of 0.1 N NaOH.

treatment followed by reaction with phosphorylethanolamine gave rise to the derivative BSA-I-PEA-A. The results of phosphorus analysis showed that 27 moles of phosphorylethanolamine had been introduced per mol of BSA-I. The derivative BSA-I-PEA-B was obtained by first reacting phosphorylethanolamine with bromo-acetyl bromide followed by reaction with BSA-I. Analysis for phosphorus revealed that 17 phosphorylethanolamine residues were incorporated per mol of BSA-I.

The structural differences between the two derivatives of BSA-I is obvious from the formulas shown in Fig. 2.

Binding studies. In Figure 3 a and b are shown the quantitative precipitin assays of CRP and CxRP with C-polysaccharide and the three BSA-conjugates. Differences in reactivities of CRP and CxRP with four ligands were noted. The most significant difference was that while CxRP reacted and precipitated with BSA-I-PEA-B, CRP failed to do so with this ligand.

In Figure 4 are shown the results of inhibition studies on the reaction of CxRP with four ligands by phosphorylcholine and α-glycerolphosphate. The amounts of each inhibitor required for 50% inhibition is indicated in the figure.

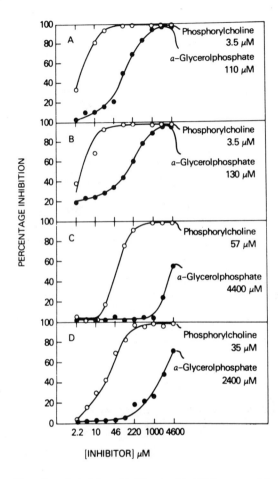

Fig. 4. Inhibition of the precipitin reaction of 50 µg of CxRP with (A) 25 µg BSA-I-PEA-B; (B) 5 µg C-polysaccharide; (C) 12.5 µg BSA-PC; and (D) 5 µg Cx-polysaccharide, by increasing concentrations of phosphorylcholine and α-glycerolphosphate. The quantities of BSA conjugates and polysaccharides used in each reaction were those required for maximal precipitation in the absence of inhibitor. The concentrations of phosphorylcholine and α-glycerolphosphate required for 50% inhibition are listed in the figure for each individual reaction.

Phosphorylcholine is 30- to 80-fold more effective as an inhibitor compared with α-glycerolphosphate. Similar results were obtained for the human CRP; the concentration of phosphorylcholine required for 50% inhibition of CRP/C-polysaccharide and CRP/BSA-PC precipitin reactions were both 75 µM, respectively.

pH dependence of the CRP reaction. The pH dependence of the precipitin reaction of these studies are summarized in Fig. 5. The ratio of the precipitate formed with CRP/BSA-I-PEA-A to that of CRP/BSA-PC over the pH range examined showed two inflection points, one at pH 6.5, the other at pH 9. The results reflect the different effect imposed by changes in pH on the binding properties of phosphorylcholine and phosphorylethanolamine attached to the BSA in their reactions with the CRP molecule.

Purification of C-reactive proteins from mice and rabbits. C-reactive proteins from mice and rabbits were purified by affinity chromatography using a

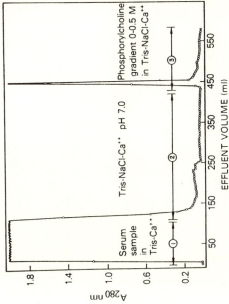

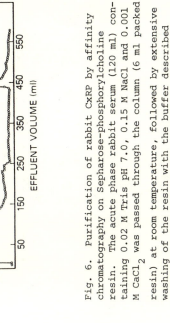

Fig. 5. Effect of pH on the precipitation of C-reactive protein with BSA-PC and BSA-I-PEA-A. Human CRP (100 μg) was incubated with BSA-PC (20 μg) or BSA-I-PEA-A (40 μg) in 0.4 ml of Tris-HEPES-Acetate buffer containing 0.001 M $CaCl_2$ at the indicated pH value. The amount of precipitate formed in each tube after reaction for 1 h at 35°C and storage overnight at 5°C was estimated by absorbance measurement at 280 nm of the corresponding precipitate dissolved in 0.4 ml of 0.1 N NaOH. The ratio of the precipitates formed at each pH value is indicated in the lower part of the figure.

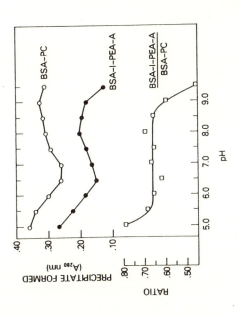

Fig. 6. Purification of rabbit CxRP by affinity chromatography on Sepharose-phosphorylcholine resin. The acute phase rabbit serum (120 ml) containing 0.02 M Tris pH 7.0, 0.15 M NaCl and 0.001 M $CaCl_2$ was passed through the column (6 ml packed resin) at room temperature, followed by extensive washing of the resin with the buffer described above. A gradient of phosphorylcholine up to 0.5 M prepared in that buffer was used to elute the bound CxRP, which emerged at the beginning of the gradient. Yield: 14.7 mg.

TABLE 1

AMINO ACID COMPOSITION OF RABBIT AND MOUSE C-REACTIVE PROTEINS[a]

Amino acid	Rabbit		Mouse	
Tryptophan	2.0	(2)[e]	4.7	(5)
Lysine	12.0	(12)	11.8	(12)
Histidine	2.6	(3)	2.1	(2)
Arginine	5.4	(5)	5.4	(5)
Half-cystine[b]	2.0	(2)	2.0	(2)
Aspartic acid	19.5	(20)	16.8	(17)
Threonine[c]	9.1	(9)	8.0	(8)
Serine[c]	17.9	(18)	16.3	(16)
Glutamic acid	19.5	(20)	17.3	(17)
Proline	10.6	(11)	10.2	(10)
Glycine	17.2	(17)	14.5	(15)
Alanine	10.3	(10)	9.3	(9)
Valine	15.2	(15)	11.9	(12)
Methionine[d]	3.8	(4)	2.8	(3)
Isoleucine	9.4	(9)	7.8	(8)
Leucine	17.4	(17)	14.2	(14)
Tyrosine	10.3	(10)	8.6	(9)
Phenylalanine	11.9	(12)	9.4	(9)
Total residues	196		173	

[a] Based on two residues of half-cystine per mol. Hydrolysis period 22 h at 115°C.

[b] Determined as cysteic acid. The values were corrected for the theoretical value of 85%.[29]

[c] The values were not corrected for destruction during hydrolysis.

[d] Determined as methionine sulfone.

[e] Number in parentheses indicates the nearest integer.

Sepharose-phosphorylcholine column. A typical experiment is shown in Fig. 6. The two proteins were apparently homogenous when analyzed by gel electrophoresis. The amino acid compositions of these two proteins are presented in Table 1.

DISCUSSION

The primary structure of CRP has been examined for internal homology and compared to all known proteins whose primary structures were published before

April, 1978. Two computer programs[17] were used for these comparisons; in the first instance, segments of CRP containing 27 amino acids were compared to all 1059 protein sequences in the data collection. In the searches, each 27-amino acid segment of CRP picked up a different selection of proteins among the highest scoring matches. In no case did the C_H2 domain of IgG contribute the highest scoring segment retrieved. In addition, using a different program, the entire sequence of CRP was compared against all the constant homology regions of the human gamma chain, against the C3 homology regions of α, μ, and ε chains, against the constant region of κ and λ chains, against three representatives of the human V region sequences, against 2-microglobulin, and against human C3a anaphylatoxin. The C_H2 region of IgG scored the best, but the score was not statistically significant (2.3 SD for 300 random runs). The sequence stretches in which this distant similarity is found correspond to the residues 1-27 and 44-88 of CRP and residues 233-277 of γ-chain. It should be noted that the segments mentioned above are not those noted by Osmand et al.[20] No relationship is evident to other homology regions of IgG or to the same homology region in other immunoglobulin chains. In addition, the most highly preserved residues which are common to most immunoglobulin homology regions such as the half-cystine residues and certain proline, valine, serine, and tryptophan residues are not found in human CRP. Thus, it seems unlikely that the distant homology of CRP to the C_H2 region of IgG has evolutionary significance. The comparisons carried out by the computer also showed no repeating sequences within the CRP molecule.

A distant homology which was not statistically significant (SD 2.14) of residues 1-57 of CRP and residues 24-77 of C3a anaphylatoxin[21] was also observed. C3a is a fragment of the α-chain of C3, the third component of complement. It remains to be seen whether the sequence of the remainder of C3 α-chain bears more extensive homology to CRP.

In sum, the homologies noted cannot be regarded as sufficient evidence to support a common evolutionary origin of these proteins. At this point in time, it is preferable to assign human CRP, rabbit CxRP, and a protein known as 9.5 S α-glycoprotein, P component of amyloid, and C_1t to a new super family unrelated to any other proteins investigated. The homology between these proteins have been convincingly demonstrated by Osmand et al.[22]

The discovery of the homology between CRP and the protein C_1t is enormously interesting and is further strengthened by the data presented here indicating that residues 2 and 3 (Thr-Asp) of both proteins are identical. One is then left with the problem that CRP, a protein of unknown function, has now been demonstrated to be closely related to another protein whose function is equally mysterious.

There are two differences between the sequence of human CRP and that reported for rabbit CxRP which are noteworthy. Whereas CRP has a half-cystine at position 36 and a histidine at position 38, the respective residues in the rabbit sequence have been reported to be serine and tyrosine. As a result of these substitutions, the similarity between the sequence of human CRP and HLA-B7 is limited to a single tripeptide (Phe-Tyr-Thr) which is not shared by HLA-A2 antigen.[23] In addition, it is indeed surprising that two proteins as closely related as CRP and CxRP should have as major a difference as a change in the position of a disulfide loop. In regard to the histidine at position 38, it is of interest to note that the disulfide loop brings this residue into juxtaposition with the second histidine residue in the molecule at position 76. The limited information available on metalloproteins and metal-protein complexes indicates a prominent role for histidine residues.[24,25] C-reactive protein has been shown to bind calcium ions and, in addition, exhibits its various binding properties only in the presence of divalent cations.

The CRP for structure determination was purified from pooled starting materials and usually represented product derived from five or more individuals. The fact that a unique sequence was obtained shows that variable sequence regions are not present. In addition, if genetic polymorphism is appreciable, then one gene product is far more commonly expressed.

The three BSA-derivatives (Fig. 2) synthesized for binding studies with CRPs have structural differences designed to elucidate the possible binding site characteristics of CRPs. Bovine serum albumin-PC contains phosphodiester-linked phosphorylcholine with the positively-charged quaternary ammonium ion at the terminus. The other two contain phosphorylethanolamine linked to the protein through the amino nitrogen, either as a secondary amine (BSA-I-PEA-A) or through an amide bond (BSA-I-PEA-B), with the negatively-charged phosphoryl group at the terminus. At pH below 9, the secondary amine in BSA-I-PEA-A assumes a positive charge 5 atoms from the terminal negatively charged phosphoryl group. The distance between the positive and negative charges on the ligands BSA-PC and BSA-I-PEA-A are thus identical. In BSA-I-PEA-B, however, the positive charge is 7 atoms away from the phosphoryl group.

Pneumococcal C-polysaccharide prepared by autolysis is a heteropolymer consisting of N-acetyl-galactosamine phosphate, murein,[26,27] ribitol phosphate,[28] choline[29] and diaminotrideoxyhexose.[30] An analogous Cx-polysaccharide is prepared from the same strain of pneumococcus by rapid lysis with deoxycholate.[2] Cx-polysaccharide has a composition similar, if not identical, to the C-polysaccharide (Liu, T.-Y., and Egan, W., unpublished observation).

The results of quantitative precipitin tests of CRP and CxRP with BSA-PC and their quantitative inhibition assays by phosphorylcholine and α-glycerol-phosphate strongly argue for the presence of two loci in the binding of CRP with phosphorylcholine ligand in BSA-PC. Such binding site in CRPs may consist of a primary locus for the calcium mediated binding of the phosphoryl group[18] and a secondary locus for the binding of the cationic group of phosphorylcholine. Similar conclusions can be drawn from the binding studies of CRP and CxRP with BSA-I-PEA-A. An unexpected finding came from the observation that CxRP but not CRP reacted and precipitated with BSA-I-PEA-B. This observation points to a major difference in the binding properties of CRP and CxRP. It suggests that in order to form stable precipitating complexes with human CRP, ligands must bind both the primary phosphate binding site and the secondary cation binding locus, whereas with rabbit CxRP only calcium mediated binding of the phosphoryl group is obligatory since CxRP, but not CRP, reacts and precipitates with BSA-I-PEA-B. Alternatively, it may indicate that the CxRP is more tolerant in regard to the distance which separates the negative and the positive charges, which, in BSA-I-PEA-B is 7 atoms. These observations together with the fact that any phosphomonoester is an inhibitor of the precipitin reaction of CRP with C-polysaccharide while choline is completely inactive,[26] we propose that the calcium-mediated phosphoryl binding locus in both proteins is the "primary" one. Notwithstanding that the reaction of CxRP with BSA-I-PEA-B results in precipitating complex formation with the ligand, much stronger complexes are formed with BSA-I-PEA-A where both binding loci are correctly engaged. This is demonstrated by the fact that a much higher concentration of inhibitors is needed to inhibit the complex formation (Fig. 4).

It is instructive to compare the pH dependence of the precipitin curves of CRP/BSA-PC with that of CRP/BSA-I-PEA-A. Examination of the structure shown in Fig. 2 reveals that phosphorylcholine, linked through phosphodiester bond to the BSA, would have no titratable groups in the pH range studied. Thus, the precipitin curve obtained would reflect titration of ionizable groups in CRP that are involved in binding with BSA-PC, or it would simply reflect the solubility of the complex formed. On the other hand, the structure of BSA-I-PEA-A possesses two titratable groups; the -OH group of the phosphomonoester with a pK around 6, and the secondary amine with a pK around 9.5. The curve obtained by plotting pH vs the ratio of the two precipitin curves BSA-I-PEA-A/BSA-PC (Fig. 5) showed that as the pH reaches 9.5, where the secondary amine of phosphorylethanolamine loses its positive charge, the complex formation between CRP and BSA-I-PEA-A decreases rapidly. The results are consistent with the view that the positive charge in the ligand on BSA is of importance in the complex formation with CRP.

Comparison of the results shown in Fig. 4 A,B,C and D, further suggests that the interaction of CxRP with C-polysaccharide most likely involves only the primary binding locus of CxRP. The concentration of phosphorylcholine and α-glycerolphosphate required for 50% inhibition of the precipitin reaction of CxRP with C-polysaccharide and that of CxRP with BSA-I-PEA-B were similar; 3.5 µM for phosphorylcholine and about 110 µM for α-glycerolphosphate, whereas for the reaction of CxRP with BSA-PC a much higher concentration of inhibitors was needed; 57 µM for phosphorylcholine and 4400 µM for α-glycerolphosphate.

For the human CRP, the profile of the quantitative inhibition curve for the two systems, CRP/C-polysaccharide and CRP/BSA-PC, were similar, suggesting that both loci in CRP are involved in the complex formation with the C-polysaccharide.

The difference in reactivity observed between Cx-polysaccharide and C-polysaccharide toward CxRP is due most likely to their conformational difference, since the two polysaccharides, including phosphorylcholine content, consist of identical chemical constituents. Since C-polysaccharide contains phosphorylcholine, it is logical to assume that the binding of CRP to C-polysaccharide is on the basis of its reactivity with phosphorylcholine residues. While the results obtained in this study with artificial ligands leave no doubt concerning the ability of phosphorylcholine or related determinants to cause calcium dependent binding to CRP, this may, however, not be the determinant in the C-polysaccharide primarily responsible for precipitation with CRP. The finding that compels caution is the ability of CRP to bind and precipitate with pneumococcal type IV polysaccharide once this material has had pyruvyl groups removed by mild acid hydrolysis.[27] Approximately five-fold greater quantities of this material are required to produce equivalent precipitation with CRP. In the current series of experiments, we have confirmed this finding and demonstrated by nuclear magnetic resonance measurements that the content of phosphorus, or of quaternary amine (choline) in type IV polysaccharide used for these studies was too low to account for the observed reactivity (Liu, T.-Y., and Egan, W., unpublished results). The only component common to the two polysaccharides is N-acetyl galactosamine. It is therefore possible that CRP may contain a site capable of interacting with structures bearing a similarity to N-acetyl galactosamine. This site is either contiguous or modulated by events in the phosphorylcholine, divalent cation binding site, since the precipitation with C-polysaccharide or with depyruvylated type IV polysaccharide is calcium dependent and readily inhibited by phosphorylcholine.

Proteins analogous to human CRP have been isolated and purified by affinity chromatography on Sepharose-phosphorylcholine. These proteins share some similarities with human CRP in that they cross-react with human CRP when assayed with goat anti-human CRP and that they bind to pneumococcal C-polysaccharide, to BSA-PC and Sepharose-phosphorylcholine in the presence of calcium ions.

The Sepharose-phosphorylcholine affinity column enabled us to demonstrate that in the acute phase, serum of both rabbits and mice obtained 24 h after turpentine injection, CRP was the major (> 95%) protein that bound specifically and reversibly to the resin; only a small amount of protein, comprising at least 15 different species as revealed by polyacrylamide gel electrophoresis, was eluted from the column with 6 M guanidinium hydrochloride following the phosphorylcholine gradient (Fig. 6).

Similarity was also noted in the amino acid compositions and subunit molecular weight of the rabbit ($\sim$ 22,000) and mouse ($\sim$19,500) CRPs with the human CRP ($\sim$ 21,000). The two histidine residues at the proposed binding site of divalent cations in human CRP[6] are present in rabbit and mouse CRPs; so are the pair of half-cystines that form a loop to bring the two histidine residues into juxtaposition. Further evaluation of the structural similarities that may exist between CRPs from different species must await detailed primary sequence analysis of each protein. Rabbit CxRP has been reported to be devoided of half-cystine.[28] The results presented in this paper clearly indicate that there are two half-cystines present per mole of rabbit CxRP.

ACKNOWLEDGMENT

This study was supported in part by the Pan American Health Organization, by the Wellcome Trust and by the Institut Merieux.

REFERENCES

1. Tillet, W.S., and Francis, T., Jr. (1930) J. Exp. Med., 52, 561-571.
2. Anderson, H.C., and McCarty, M. (1951) J. Exp. Med., 93, 25-36.
3. Bodmes, B., and Siboo, R. (1977) J. Immunol., 118, 1086-1089.
4. Patterson, L.T., and Higginbotham, R.D. (1965) J. Bact., 90, 1520-1524.
5. Kindmark, C.-O. (1972) Clin. Exp. Immunol., 11, 215-221.
6. Oliveira, E.B., Gotschlich, E.C., and Liu, T.-Y. (1977) Proc. Natl. Acad. Sci. USA, 74, 3148-3151.
7. Kindmark, C.-O. (1971) Clin. Exp. Immunol., 8, 941-948.
8. Gotschlich, E.C., and Edelman, G.M. (1965) Proc. Natl. Acad. Sci. USA, 54, 558-566.
9. Wood, H.F., McCarty, M., and Slater, R.J. (1954) J. Exp. Med., 100, 71-79.
10. Gross, E., and Witkop, B. (1962) J. Biol. Chem., 237, 1856-1860.
11. Roth, M. (1971) Anal. Chem., 43, 880-882.
12. Konigsberg, W. (1972) Methods Enzymol., 11, 351-361.
13. Kortt, A.A., Wysocky, J.R., and Liu, T.-Y. (1976) J. Biol. Chem., 251, 1941-1947.
14. Simpson, R.J., Neuberger, M.R., and Liu, T.-Y. (1976) J. Biol. Chem., 251, 1936-1940.
15. Spackman, D.H., Stein, W.H., and Moore, S. (1958) Anal. Chem., 30, 1190-1206.

16. Oliveira, E.B., Gotschlich, E.C., and Liu, T.-Y. (1979) J. Biol. Chem., 254, 489-502.
17. Dayhoff, M.O. (1976) Atlas of Protein Sequence and Structure, National Biomedical Research Foundation, Washington, D.C., Vol. 5, Suppl. 2.
18. Gotschlich, E.C., and Edelman, G.M. (1967) Proc. Natl. Acad. Sci. USA, 57, 706-712.
19. Chen, P.S., Jr., Toribara, T.Y., and Warner, H. (1956) Anal. Chem., 28, 1756-1761.
20. Hugli, J.E. (1975) J. Biol. Chem., 250, 8293-8301.
21. Osmand, A.P., Friedenson, B., Gewurz, H., et al. (1977) Proc. Natl. Acad. Sci. USA, 74, 739-743.
22. Osmand, A.P., Gewurz, H., and Friedenson, B. (1977) Proc. Natl. Acad. Sci. USA, 74, 1214-1218.
23. Terhorst, C., Parham, P., et al. (1976) Proc. Natl. Acad. Sci. USA, 73, 910-914.
24. Mildvan, A.S. (1970) in The Enzymes, Boyer, P., ed., Academic Press, New York, Vol. II, 445-536.
25. Liu, T.-Y. (1977) in The Proteins, Neurath, H., and Hill, R.H., eds., Academic Press, New York, 3d Ed., Vol. 3, 240-402.
26. Volanakis, J.E., and Kaplan, M.H. (1971) Proc. Soc. Exp. Biol. Med., 136, 612-614.
27. Heildelberger, M., Gotschlich, E.C., and Higginbotham, J.D. (1972) Carbohydr. Res., 22, 1-4.
28. Bach, A.B., Gewurz, H., and Osmand, A.P. (1977) Immunochem., 14, 215-219.
29. Hirs, C.H.W. (1971) Meth. Enzymol., 11, 197-199.

Published 1980 by Elsevier North Holland, Inc.
Liu/Mamiya/Yasunobu, eds. Frontiers in Protein Chemistry

481

PURIFICATION AND PROPERTIES OF THE *LIMULUS* CLOTTING ENZYME

ROBERT C. SEID, JR.,[*] AND TEH-YUNG LIU
Division of Bacterial Products, Bureau of Biologics, Food and Drug Administration, Bethesda, Maryland 20205

SUMMARY

The clotting enzyme(s) from the *Limulus* amoebocyte lysate has been purified to apparent homogeneity as judged by sodium dodecyl sulfate-gel electrophoresis. The enzyme(s) was found to be stabilized by the detergent Tween-20. Without the detergent, enzymatic activity decreased rapidly. Two active forms of the clotting enzyme were demonstrated in the *Limulus* amoebocyte lysate. The enzymatic material existing as higher molecular weight aggregates was purified to a single protein band on sodium dodecyl sulfate-polyacrylamide gel electrophoresis with an estimated molecular weight of 80,000 $\pm$ 5,000. The enzymatic material eluting in the lower molecular weight region of the gel-permeation column yielded a protein with a molecular weight of about 40,000 $\pm$ 5,000. The amino acid compositions of the two proteins were similar indicating a "monomer-dimer" relationship but attempts to interconvert the 80,000 protein into the 40,000 protein have not been successful. Both enzymes have the same Ca^{2+} and endotoxin requirement for exhibiting maximal clotting activity and the same specificity toward synthetic substrates.

INTRODUCTION

In 1964, Levin and Bang discovered that gram-negative bacterial endotoxin could rapidly induce gelation of *Limulus* amoebocyte lysate.[1] Because of this unique property and its extreme sensitivity, *Limulus* lysate has been used to detect endotoxin in a variety of biological fluids.[2] For the last three years, our laboratory has been engaged in the study of *Limulus polyphemus* amoebocyte lysate coagulating system to further understand the chemical and the physiological principles underlying the clotting processes. We have purified a pro-clotting enzyme,[3] a coagulogen,[4] and more recently, an endotoxin binding protein[5] from the *Limulus* amoebocytes. A transglutaminase activity has also been demonstrated[6] in freshly prepared *Limulus* lysate and its possible involvement in stabilizing the coagulin polymer has been investigated.

---

*Present address: The Walter Reed Institute of Research, Department of Bacterial Diseases, Washington, D.C. 20012.

In this manuscript we report the purification of the clotting enzyme(s) from the *Limulus* lysate.

EXPERIMENTAL PROCEDURES

Materials

Endotoxin (*E. coli* 026:B6) was purchased from Difco Laboratories, Detroit, Michigan. Tritiated *E. coli* 026:B6 endotoxin (specific activity, 3.58 mCi/mg) was prepared by New England Nuclear Co. BioGel A-5m and BioGel hydroxylapatite were purchased from BioRad Laboratories; Octyl-Sepharose CL-4B from Pharmacia; Tween-20 from Fisher Scientific Co.; deoxycholic acid from Sigma Co.; S-2222 (Benzoyl-Ile-Glu-(γ-OCH$_3$)-Gly-p-nitroanilide) chromogenic substrate from Ortho Diagnostic, Inc.; pyrogen-free water from Abbott Laboratory. *Limulus* coagulogen, used for the clotting assay, was prepared by the acetic acid extraction method described by Tai et al.[4]

Methods

Assay for clotting activity. Assay of *Limulus* clotting enzyme was performed by reacting 0.1 ml of enzyme test solution with 0.1 ml of coagulogen (5 mg/ml in 0.05 M Tris-HCl, 0.1 M NaCl, 0.005 M CaCl$_2$, pH 7.5) in a 10 x 75 mm test tube at 37° for 60 min. Solid gel formation indicated the presence of clotting enzyme.

Assay of clotting enzyme amidase activity. The use of the S-2222 chromogenic substrate to assay the clotting enzyme in *Limulus* lysate from the horseshoe crab (*Tachypleus* and *Limulus*) has been described earlier by Nakamura et al.[7] For our assay, 10 μl of clotting enzyme solution was added to a mixture containing 0.8 ml of 0.05 M Tris-HCl buffer, pH 8.0, in 0.1 M NaCl and 0.2 ml of 0.001 M solution of S-2222 substrate. After a 15 min incubation period at 37°, the reaction was stopped by the addition of 0.2 ml of glacial acetic acid and the released p-nitroaniline was estimated from its absorbance at 405 nm. When necessary, proper dilution of the enzyme fraction was made in order that absorbance remained within the previously determined linear range. An enzyme unit was arbitrarily defined as an optical density change of 0.01 at 405 nm using a 1.0-cm cell. Specific activity was expressed as enzyme unit per mg protein.

The K_m values of the Peak I (Fig. 2A) and Peak II (Fig. 2B) enzymes were determined by following the rate production of p-nitroaniline (ξ 10600 M^{-1}cm^{-1}) at 405 nm with the Cary 14 CM recording spectrophotometer. The concentrations of the S-2222 substrate were between 0.005 M to 0.05 M prepared in 0.05 M Tris-HCl buffer, pH 7.5, containing 0.1 M NaCl and 0.005 M CaCl$_2$. Reaction was initiated by adding 0.010 ml of enzyme solution to 0.990 ml of substrate solution in a 1.0-cm cell and the recording was started after stirring for 1-2 seconds with a glass rod. Reactions were followed to completion at 20°. Values for k_{cat} were determined from the zero order part of the spectrophometric trace where $S_o \gg K_m$.

Assay for protein concentration. Protein concentrations were measured by the method of Lowry et al.[8] Standard curves of bovine serum albumin were always run in the same medium, including detergent, in which the protein concentration was being measured.

Amino acid analysis. The purified clotting enzyme (10 to 20 µg) was hydrolyzed in 50 µl of 4 N methanesulfonic acid containing 0.2% tryptamine[9] and analyzed on a Beckman 121 M analyzer. Performic acid oxidation was carried out by the method of Moore.[10]

Gel electrophoresis. Sodium dodecyl sulfate-gel electrophoresis was performed essentially as described by Weber and Osborn.[11] Protein standards used for molecular weight estimation were catalase, aldolase, serum albumin, ovalabumin, chymotrypsinogen A, and cytochrome C. The molecular weight of clotting enzyme was calculated on the basis of a linear relationship of the log molecular weight to the distance migrated in the sodium dodecyl sulfate gel. For detergent-protein samples, gel electrophoresis was performed according to the methods described by Nielsen and Reynolds.[12]

Preparation of enzyme extract from Limulus lysate clot. Limuli, obtained from the beaches of Long Island, N.Y., were bled by the cardiac puncture method described by Jorgensen and Smith.[13] The blood, collected in ice cold 250 ml siliconized bottles, was centrifuged at 1000 x g for 15 min at 4°. After removal of the blue plasma by decantation, the amoebocytes (5-10 ml) were washed twice with pyrogen-free 3% NaCl. Pyrogen-free water in a 1:3 ratio was added and the amoebocytes were lysed by vortexing at 4°. After centrifugation (1000 x g, 30 min) to remove cellular debris, the supernatant was induced to clot by the addition of 1 ml of endotoxin solution (0.5 mg/ml) and 0.1 ml of 1 M $CaCl_2$. After a 4 hr incubation at 37°, the solid gel formed was broken with a stirring rod and centrifuged at 5,000 x g for 30 min, 4°. The resulting supernatant (15 ml), containing the clotting enzyme, was recovered and the precipitate discarded. An additional amount of endotoxin (1.1 mg) was added prior to adding Tween-20 and NaCl to a final concentration of 0.1% w/v and 0.5 M, respectively. This mixture (10 ml) was immersed in an Ultramet III sonic bath for 30 sec and applied to a BioGel A-5m column equilibrated with 0.05 M Tris-HCl buffer containing 0.2% Tween-20 and 0.1 M NaCl, pH 7.5 (Fig. 1)

Gel permeation chromatography. A 2.5 x 88 cm BioGel A-5m column was used. Blue dextran and tritiated water were used as void and total volume markers. The column was equilibrated with 0.05 M Tris-HCl, pH 7.5, buffer containing 0.2% Tween-20 and 0.1 M NaCl. Fractions of 4.3 ml were collected.

A similar BioGel column was used in the gel filtration experiment with sodium deoxycholate. The equilibrating buffer was 0.05 M Tris-HCl, pH 8.0, containing 0.2% deoxycholate and 0.1 M NaCl.

Endotoxin coating of hydroxylapatite resin. An experiment designed to show that $E.$ $coli$ 026:B6 endotoxin binds to hydroxylapatite was performed. A sonicated, tritiated endotoxin solution (5 mg in 10 ml of 0.005 M potassium phosphate buffer) was added to 20 ml of an hydroxylapatite slurry previously equilibrated with the same buffer. The resulting suspension was stirred for 30 min at room temperature before pouring into a column. The packed column (0.9 x 20 cm) was washed extensively with the starting buffer. Radioactivity determination of the wash solution indicated that 40% (2 mg) of the endotoxin remained bound to the resin. A subsequent wash with 0.03 M phosphate buffer resulted in the elution of about 0.1% (20 µg) of the bound endotoxin from the hydroxylapatite column, whereas a stronger buffer, 0.5 M in phosphate released about 75% of the bound endotoxin.

A nonradioactive endotoxin-coated hydroxylapatite column was prepared under similar conditions as described above. The samples, pooled Peak I, 86 ml, and Peak II, 43 ml, from Fig. 1 were concentrated to 5.3 ml and 7.0 ml, respectively, by ultrafiltration with an Amicon 52 stirred cell with a PM-10 membrane. They were then dialyzed four times at 4° against 4,000 ml of starting phosphate buffer to remove most of the Tween-20 detergent. No loss in enzymatic activity occurred during dialysis. These treated samples were applied to the endotoxin-coated column.

Inhibition studies. The effect of various inhibitors, diisopropyl fluorophosphate (DFP), phenylmethylsulfonyl fluoride, iodoacetamide, and N-ethylmaleimide on the clotting enzyme activity was measured by the two assay procedures described above. The enzyme solution (25 µg/100 µl) was preincubated with the inhibitor (0.005 M) at 37° for 45 min at pH 7.5 in 0.01 M Tris-HCl buffer containing 0.005 M $CaCl_2$. This was followed by the addition of one of the substrates, the coagulogen solution or the S-2222 chromogenic peptide solution, and the resulting mixture was further incubated at 37° for 60 min.

RESULTS

Gel-filtration using Tween-20

Figure 1 shows a typical chromatogram of an extract from the clotted $Limulus$

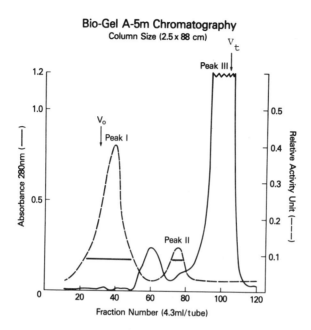

Fig. 1. Gel permeation chromatography of *Limulus* clotting enzyme extract from lysate clot. The arrows indicate elution positions of markers for void volume (V_o) and total volume (V_t). Dashed line and solid bar represent amidase and clotting activities, respectively. See Experimental Procedures for assay methods.

lysate on BioGel A-5 m column equilibrated and eluted with a Tris-HCl buffer containing 0.2% Tween-20. The amidase and the clotting activities both appeared in two regions, Peak I and Peak II. Based on its elution volume, the enzymatic material in Peak I existed presumably as high molecular weight aggregates and exhibited very low absorbance at 280 nm. The material in Peak II showed a lower specific enzymatic activity due to contamination with other proteins.

The total amidase activity recovered from Peak I and Peak II represented a 140-fold increase compared to the original extract (Table 1). The specific activities of the Peak I and Peak II enzymatic materials were 2857 and 353 amidase units/mg protein, respectively, which represented a 607 and 75-fold purification compared to the specific activity of the original extract.

Affinity chromatography on endotoxin-coated hydroxylapatite

Peak I and Peak II enzymatic materials were exhaustively dialyzed and concentrated as described in the Experimental Procedures and subjected to further

TABLE 1

PURIFICATION OF *LIMULUS* CLOTTING ENZYME

Purification Step	Volumn (ml)	Protein mg/ml	Total Units	Specific Activity Units/mg	Fold Purification
Extract of lysate clot	10	4.2	197	4.7	1.0
Peak I	5.3	1.55	23470[b]	2857	607
Peak II	7.0	1.7	4200[b]	353	75
Hydroxylapatite					
Peak I	2.0	0.40[a]	18512	23140	4923
Peak II	1.0	0.73[a]	10000[c]	13700	2914

[a] In these cases, the protein concentrations were determined by total amino acid analysis.

[b] A total activity enhancement after BioGel filtration with Tween-20. See text for explanation.

[c] An activity enhancement is again observed after Peak II was chromatographed on the endotoxin-coated resin.

purification on a column of endotoxin-coated hydroxylapatite (Figs. 2A and 2B). In both instances, amidase and clotting activities eluted as a single peak with the 0.03 M phosphate buffer. The specific activities expressed as amidase units were 23,140 and 13,700 for the Peak I and Peak II materials, representing a 4900 and 2900-fold purification, respectively, when compared to the original extract (Table 1).

On sodium dodecyl sulfate-polyacrylamide gel electrophoresis, the Peak I material yielded a single protein band (Fig. 3) corresponding to a molecular weight of about 80,000. The Peak II material showed one major band (estimated to be 95% of total protein content on gel) with a molecular weight of about 40,000 (Fig. 3). Several minor components were observed.

The amino acid compositions of the two enzyme preparations are given in Table 2).

When a nonendotoxin coated hydroxylapatite column was used, more than 75% of the enzymatic activities eluted in the initial buffer wash together with other proteins. The remaining activity eluted spuriously with the subsequent buffers.

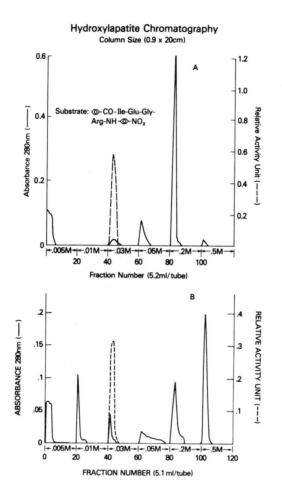

Fig. 2. Affinity chromatography on endotoxin-coated hydroxylapatite resin.
(A) chromatogram of Peak I material obtained from BioGel A-5 m filtration (Fig.
1). (B) chromatogram of Peak II material. A stepwise increase in potassium
phosphate was used to elute the clotting enzyme. Clotting activity was found
to coelute with the amidase activity represented by the dashed line (see
Experimental Procedures for assay methods).

Gel-filtration in sodium deoxycholate medium

Gel filtration in deoxycholate medium has been used to separate proteins
from their bound lipids.[14] In some cases, the delipidated proteins thus
obtained were enzymatically inactive, but activities could be restored when
specific phospholipids or detergents were added.[14-18]

A BioGel A-5 m column (2.5 x 88) equilibrated with a 0.05 M Tris-HCl buffer,
pH 8.0, containing 0.5% deoxycholate and 0.1 M NaCl was charged with 10 ml of

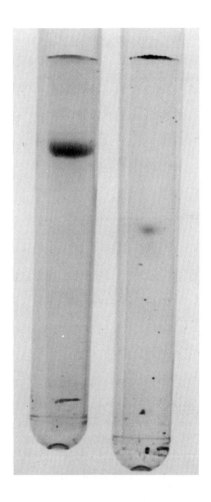

Fig. 3. Acrylamide gel electrophoresis in 0.1% sodium dodecyl sulfate, 0.1 mM dithiothreitol, 0.1 M pH 7.2 phosphate buffer. Gel 1 represents 0.02 M clotting enzyme fraction (Fig. 2A) obtained from affinity chromatography; Gel 2 represents 0.03 M enzyme fraction of Fig. 2B.

the extract from the clotted *Limulus* lysate prepared as described above. Aliquots of 0.1 ml were withdrawn from each fraction (5.0 ml) and added to 0.1 ml of buffer solution with and without endotoxin (0.40 mg) and calcium chloride (0.22 mg). A third tube supplemented with 0.1 ml of 1% Tween-20 solution was also used. After a 30 min incubation period at 37°, 0.5 ml of the S-2222

TABLE 2

AMINO ACID COMPOSITION OF *LIMULUS* CLOTTING ENZYME

Amino Acid	Peak I g/1000g Protein	Peak II g/1000g Protein
Lys	42.9	44.0
His	33.4	33.6
Arg	30.5	32.6
Asp	100.0	96.2
Thr	47.2	42.2
Ser	162.3	164.5
Glu	196.1	196.6
Pro	34.9	32.1
Gly	97.7	103.5
Ala	52.6	51.5
Val	49.2	51.5
Met	10.6	9.7
Phe	27.6	29.2
Leu	42.4	39.5
Tyr	34.8	35.9
Phe	28.4	28.9
Cys	9.4	8.9

chromogenic substrate solution was added to each tube. The p-nitroaniline released was estimated after an additional 30 min incubation period. The result of this experiment is shown in Fig. 4. Amidase activity was observed for those tube supplemented with both endotoxin and calcium. Endotoxin alone was minimally effective in restoring enzymatic activity, while Tween-20 was ineffective (Fig. 4).

Hydrophobic interaction chromatography

An Octyl-Sepharose 4B-CL column (0.9 x 10 cm) was used to show the presence of hydrophobic regions on the *Limulus* clotting enzyme. Both the crude extract of the clotted lysate and the purified clotting enzyme were applied to the hydrophobic matrix previously equilibrated with a high ionic strength buffer, 0.05 M Tris-HCl, pH 7.5, in 2 M NaCl. Following an initial wash with the 2 M NaCl solution, a series of decreasing ionic strength buffers (1 M, 0.5 M, 0.1 M, and 0 M NaCl in 0.05 M Tris-HCl, pH 7.5) were passed through the column in a

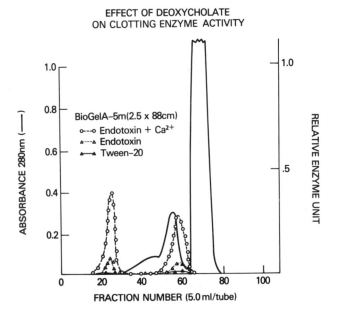

Fig. 4. Stimulation by endotoxin and calcium of amidase activity in enzyme
fractions eluted from a BioGel A column with 0.5% deoxycholate buffered with
50 mM Tris-HCl, pH 8.0. Experimental details are described in text, O assays
on fractions supplemented with both endotoxin and calcium; Δ , assays on
fractions supplemented with endotoxin; ▲ , assays on fractions supplemented with
Tween-20.

stepwise manner. Tris buffer containing Tween-20 was used as a final eluent.
In both instances no measurable enzymatic activity could be detected in the
aqueous buffers, even though proteins were being eluted as evidenced by their
absorbances at 280 nm. Enzymatic activities were, however, recovered with the
Tween-20 buffer. Table 3 indicates a 60-fold increase in amidase activity from
the original extract of the lysate clot, whereas full activity was recovered
from the purified clotting enzyme.

The requirement for Tween-20 to elute the clotting enzyme from the gel bed
suggests strong hydrophobic interaction between the protein molecule and the
gel matrix. In general, water soluble proteins can usually be eluted from an
hydrophobic matrix with low ionic strength buffers, whereas the more hydrophobic
proteins, such as membrane proteins, require detergent buffer.[19]

Specificity and inhibition of the clotting enzyme
The specificity of the purified clotting enzyme was examined with the aid of

TABLE 3

RECOVERIES OF *LIMULUS* CLOTTING ENZYME ACTIVITY AFTER HYDROPHOBIC
INTERACTION CHROMATOGRAPHY ON OCTYL-SEPHAROSE 4B-CL

Sample Applied	Volume (ml)	Protein mg/ml	Initial Activity Units[a]	Final Activity Units
Extract of lysate clot	5.0	4	120	7200
Clotting enzyme	1.0	0.050	1250	1100

[a]Amidase activity was assayed with the S-2222 chromogenic substrate as described

a number of synthetic substrates. At a 0.002 M substrate concentration in 0.05
M Tris-HCl buffer, pH 7.5, in 0.1 M NaCl and 0.005 M $CaCl_2$, benzoyl-Ile-Glu-
(γ-OCH$_3$)-Gly-Arg-p-nitroanilide was hydrolyzed 30 times faster than benzoyl-Phe-
Val-Arg-p-nitroanilide, whereas benzoyl-Arg-p-nitroanilide and benzyloxyl-
carbonyl-Gly-Pro-Arg-p-nitroanilide were not hydrolyzed. The specificity of
the *Limulus* clotting enzyme is different from that of trypsin although both
enzymes cleaved the same -Arg-Lys- peptide bond in *Limulus* coagulogen for clot
formation.[4]

The K_m values, determined from the Lineweaver-Burk reciprocal plot of $1/v$ vs
$1/s$, for the Peak I and Peak II enzymes were 2.6 µM and 2.3 µM, respectively.
The k_{cat} values were 1811 sec^{-1} and 1820 sec^{-1}.

Both the amidase and the clotting activities of the enzyme were totally
inhibited by DFP and phenylmethylsulfonyl fluoride under the conditions
employed as described in the Experimental Procedures. Sulfhydryl reagents,
such as iodoacetamide and N-ethylmaleimide, has no effect on enzymatic activities. These results suggest that the *Limulus* clotting enzyme is a serine
rather than a sulfhydryl protease.

DISCUSSION

A preliminary note on the isolation and partial characterization of the
Limulus clotting enzyme was reported by Sullivan and Watson.[20] They showed
that the clotting enzyme exists as a dimeric protein with a molecular weight of
84,000 which dissociates into its subunit structure with a molecular weight of
43,000 on sodium dodecyl sulfate-polyacrylamide gel electrophoresis. In our
initial attempts to purify the clotting enzyme without Tween-20, we encountered

a consistent problem of protein aggregation and precipitation with concomitant loss of enzymatic activity. It was found that Tween-20 added to the buffer medium stibilized the clotting enzyme, thus, providing a suitable means for purification. The detergent stabilized enzyme showed no loss in both amidase and clotting activities over a period of a year at 4° and was stable to heat treatment at 50° for 24 hrs. It is possible that Tween-20 interacts with the hydrophobic portion of the enzyme, thus, rendering a more stable conformation. Detergent binding to proteins and stabilization of enzymatic activity have been reported.[21-23]

The *Limulus* clotting enzyme displays properties resembling other membrane-bounded proteins. The clotting enzyme is stabilized by Tween-20. It absorbs on an Octyl-Sepharose column and can be desorbed with a detergent containing buffer. It binds to an endotoxin-coated hydroxylapatite column. Finally, deoxycholate inactivates the clotting enzyme and endotoxin restores enzymatic activity in the presence of calcium.

Previously, we have shown that both endotoxin and calcium are required for the activation of the pro-clotting enzyme in *Limulus* lysate.[3] The results of the present experiment with deoxycholate suggest that both components are needed to restore the active conformation of the deoxycholate-inactivated enzyme.

The 140-fold increase in total recovery of amidase activity is probably due to the separation of an enzyme inhibitor(s) during gel filtration of the lysate clot extract with the Tween-20 buffer. Sullivan and Watson reported the removal of a clotting enzyme inhibitor(s) from *Limulus* lysate.[24] Its removal, effected by a chloroform extraction of the lysate, improved the lysate sensitivity to bacterial endotoxin a hundred-fold; however, the ability of the lysate to form a firm gel was impaired. In the gel filtration experiment with Tween-20, the recovered clotting enzyme remained capable of producing a firm clot when added to a solution of purified coagulogen. The inhibitor(s) most likely is a low molecular weight substance(s) as evidenced by the finding that addition of detergent-free Peak III material in Fig. 1 to the purified clotting enzyme diminished enzymatic activity. Attempts are being made to purify and identify the chemical nature of this "inhibitor".

Two active forms of the clotting enzyme with molecular weights of 80,000 and 40,000 have been demonstrated to exist in the extract of the *Limulus* lysate clot. Attempts to interconvert the two proteins were not successful. The amino acid analyses of the two enzymes, shown in Table 2, indicate similar compositions. Both enzymes induce the clotting of purified coagulogen and exhibited similar specificity toward synthetic substrates. They both have the same calcium and endotoxin requirement for expressing maximal enzymatic activity and are inhibited by the serine protease inhibitors such as DFP and phenylmethylsulfonyl fluoride. Moreover, both enzyme exhibited approximately the same K_m and k_{cat} values, suggesting similar binding properties for the S-2222 substrate. Their rather low K_m values indicate an apparent high affinity of the tetrapeptide for both enzymes. These results suggest that the 80,000 and 40,000 proteins are the "dimeric" and the "monomeric" forms of the clotting enzymes. Failure to dissociate the "dimeric" form to the "monomeric" form could be due to strong noncovalent interaction between the subunit components. It is not possible at this time to eliminate entirely the possible existence of an intercovalent linkage, other than a disulfide bond, as an alternate explanation.

REFERENCES

1. Levin, J., and Bang, F.B. (1964) Bull. Johns Hopkins Hosp., 115, 337-345.
2. Cooper, J.F., Hochstein, H.D., and Seligmann, E.F., Jr. (1972) Bull. Parenter. Drug Assoc., 26, 153-162.
3. Tai, J.Y., and Liu, T.-Y. (1977) J. Biol. Chem., 252, 2178-2181.
4. Tai, J.Y., Seid, R.C., Jr., et al. (1977) J. Biol. Chem., 252, 4773-4776.
5. Liang, S.-M., Sakmar, T.P., and Liu, T.-Y. (1980) J. Biol. Chem., in press.
6. Chung, S.I., Seid, R.C., Jr., and Liu, T.-Y. (1977) Thromb. and Haemostasis, 38, 182.
7. Nakamura, S., Morita, T., et al. (1977) J. Biochem., 81, 1567-1569.
8. Lowry, O.H., Rosebrough, N.J., et al. (1951) J. Biol. Chem., 193, 265-275.
9. Simpson, R.J., Neuberger, M.R., and Liu, T.-Y. (1976) J. Biol. Chem., 251, 1936-1940.
10. Moore, S. (1963) J. Biol. Chem., 238, 235-237.
11. Weber, K., and Osborn, M. (1969) J. Biol. Chem., 244, 4406-4412.
12. Nielsen, T.B., and Reynolds, J.A. (1978) Methods Enzymol., 47, 3-10.
13. Jorgensen, J.H., and Smith, R.F. (1973) Appl. Microbiol., 26, 43-48.
14. Helenius, A., and Simons, K. (1975) Biochim. Biophys. Acta., 415, 29-79.
15. Hong, K., and Hubbell, W.L. (1973) Biochemistry, 12, 4517-4523.
16. Tanaka, Y., et al. (1976) J. Biochem., 80, 821-830.
17. Agnew, W.S., and Popjak, G. (1978) J. Biol. Chem., 253, 4574-4583.
18. Dean, W.L., and Tanford, C. (1977) J. Biol. Chem., 252, 3551-3553.
19. Hjerten, S. (1973) J. Chromatogr., 87, 325-331.
20. Sullivan, J.D., Jr., and Watson, S.W. (1975) Biochem. Biophys. Res. Commun., 66, 848-855.
21. Robinson, N.C., and Tanford, C. (1975) Biochemistry, 14, 369-378.
22. Osborn, H.B., et al. (1974) Eur. J. Biochem., 44, 383-390.
23. Marchesi, V.J., and Palade, G.E. (1967) Proc. Natl. Acad. Sci. U.S., 58, 991-995.
24. Sullivan, J.D., Jr., and Watson, S.W. (1975) Appl. Microbiol., 28, 1023-1026.

HORSESHOE CRAB COAGULOGENS: THEIR STRUCTURES AND GELATION MECHANISM

S. NAKAMURA, F. SHISHIKURA[*], S. IWANAGA[+], T. TAKAGI[**],
K. TAKAHASHI, M. NIWA[++], AND K. SEKIGUCHI[*]
Primate Research Institute, Kyoto University, Inuyama, Aichi 484; [*]Institute of
Biology, The University of Tsukuba, Ibaragi 300-31; [+]Department of Biology,
Faculty of Science, Kyushu University, Fukuoka 812; [**]Biological Institute,
Faculty of Science, Tohoku University, Sendai 982; and [++]Department of Bacteri-
ology, Osaka City University Medical School, Osaka 545, Japan

SUMMARY

A clottable protein, named coagulogen, was isolated, respectively, from the
hemocyte lysates of four species of horseshoe crabs (Tachypleus (T) tridentatus,
T. gigas, Carcinoscorpius (C) rotundicauda, and Limulus (L) polyphemus). The
isolated coagulogens all consisted of a single basic polypeptide chain with a
molecular weight of 20,000, and they formed a gel by the action of a clotting
enzyme purified from the respective hemocyte lysates. The gelation involved
limited proteolysis at the position of the Arg-46–Gly-47 linkages in common and
the Arg-18-Thr-19 (T. tridentatus), Arg-18-Asn-19 (C. rotundicauda) or Arg-18-
Lys-19 (T. gigas and L. polyphemus) all located in the NH_2-terminal portions
of each coagulogen, liberating peptide C from its inner portion. The resulting
gel protein consisted of two chains of A and B, bridged by disulfide bonds. The
complete amino acid sequence of T. tridentatus was established and the partial
sequences of the others were also elucidated. These sequences had partial
homology with that of mammalian fibrinogen and platelet Factor 4, suggesting
that coagulogen is a common ancestor of these proteins. The sequence of L.
polyphemus coagulogen differed from those of other three species, and the
structural difference was also suggested by the comparison of their immunological properties.

INTRODUCTION

The hemolymph of horseshoe crab contains only one type of cell, called
amoebacytes, which participates both in hemostasis and in the defence against
invading microorganisms.[1] The amoebacytes carry a clottable protein, named
coagulogen, and a high-molecular-weight proenzyme, which after activation by
endotoxin (lipopolysaccharides) reacts with coagulogen to produce a gel.[2,3]
A gelation test using the amoebacyte lysate has been widely employed as a simple
assay method for bacterial endotoxins.[4] The sequence leading to the coagulation of Limulus amoebacyte lysate has recently been discussed by Young et al.,[5]
and Niwa and Waguri.[6] They suggested that phenomenon is very similar to

mammalian blood coagulation, which involves sequential transformation of several proenzymes to enzymes, accompanied by a limited proteolysis. In this respect, it is of great interest to investigate the overall molecular events of the coagulation system in *Limulus* amoebacytes, in comparison with those of mammalia, because the former may be a prototype of the latter.

Our attention today concerns mainly the chemical structure and gelation mechanism of coagulogen, which was isolated from the amoebacyte lysate of Japanese horseshoe crab, *Tachypleus tridentatus* (Fig. 1). First of all, we would like to talk about the biochemical properties of the isolated coagulogen and its structural change during gelation. Secondly, we will describe the complete amino acid sequence of *Tachypleus* coagulogen. Thirdly, we will discuss the sequence homology of the coagulogen with other related proteins. Finally, we shall show the structural and immunological differences between Asian *Tachypleus* coagulogens and American *Limulus* coagulogen.

Fig. 1. Japanese horseshoe crabs (*Tachypleus tridentatus*), which inhabit the Inland Sea of Japan, called "Seto Sea" (taken from Kabutogani Jiten (in Japanese), edited by Nishii, H., 1973).

MATERIALS AND METHODS

Preparation of amoebacytes lysates. A hemolymph of *T. tridentatus* (adult and both sexes shown in Fig. 1) was collected by cardiac puncture. Fifty to 150 ml per individual (1-3 kg) was drawn into a bottle which contained 3% NaCl containing 0.1 M caffeine amounting to one-tenth of the original blood volume of hemolymph. Amoebacytes were obtained by centrifuging pooled hemolymph at 2,500 rpm for 10 min and washed twice with 3% NaCl. The cells were dispersed in one-tenth of the original volume of a lysate buffer consisting of 0.05 M Tris-HCl, 0.001 M $CaCl_2$ and 0.15 M NaCl, pH 7.2, and the suspension was frozen and thawed. The thawed cells were homogenized by a mechanical blender and centrifuged at 2,500 rpm for 15 min. The sediment, which still contained unbroken amoebacyte, was homogenized again with one-thirtieth or more of the original blood volume of lysate buffer and centrifuged. The combined supernate was spun again at 5,000 rpm for 30 min to remove cell debris. The lysate thus obtained contained 15-20 mg of protein per ml.[7] The other lysates of *C. rotundicauda*, *T. gigas*, and *L. polyphemus* were prepared with the similar procedures to those for the *T. tridentatus* lysate.

Isolation of coagulogens. The amoebocyte lysate of *T. tridentatus* was gel-filtrated on an endotoxin-free Sepharose Cl-6B column (2.9 x 135 cm), equilibrated with pyrogen-free saline. The elution profile is shown in Fig. 2. The third peak (Fraction No. 117-132) exhibited a potent clotting activity after activation with *Salmonella minnesota* R595 endotoxin.[8] A coagulogen detected by gel formation with trypsin was found in the fourth peak (Fraction No. 148-163), designated Fraction B. These fractions were collected and concentrated by adding $(NH_4)_2SO_4$ to give 40% saturation. The precipitate was dissolved in small amount of 0.05 M acetate buffer, pH 5.6 and then dialyzed overnight against the same buffer at 4°C. The dialyzate was applied to a Sephadex G-50 column (1.7 x 143 cm), equilibrated with 0.1 M $(NH_4)HCO_3$ buffer, pH 8.1. Through these procedures, about 30 mg of pure coagulogen was obtained from 10 ml of the lysate. For purification of coagulogens from other species, it was extracted from each amoebacyte with 10% acetic acid. The extract was neutralized by adding 0.2 N NH_4OH and concentrated with $(NH_4)_2SO_4$ to give 40% saturation. The precipitate was dissolved in the same acetate buffer used above, and the solution was gel-filtrated on a Sephadex G-50 column under the same conditions as described for the purification of *Tachypleus* coagulogen.[8]

Preparation of clotting enzyme. A *Tachypleus* clotting enzyme from the hemocyte lysate was separated by gel-filtration on Sepharose Cl-6B[9] and partially purified by DEAE-cellulose chromatography after activation with an endotoxin from *Salmonella minnesota* R595.[10-12] The enzymes from other species were also prepared by the same procedures as those of *Tachypleus* clotting enzyme.[9]

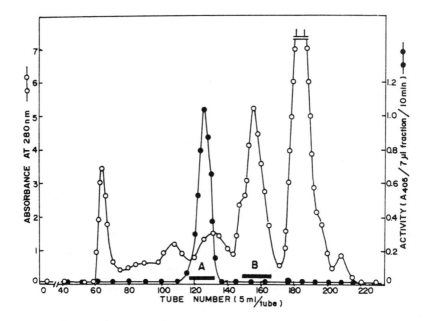

Fig. 2. Gel filtration of hemocyte lysate prepared from Japanese horseshoe crab (T. *tridentatus*) on a pyrogen-free Sepharose Cl-6B column.[9,10] The lyophilized lysate (about one gram protein) was dissolved in 15 ml of a sterilized water and gel-filtrated on the column (2.9 x 135 cm), equilibrated with pyrogen-free saline. Fractions of 5 ml were collected at a flow rate of 50 ml per hr at 4°C. Fraction A contained proclotting enzyme and exhibited an amidase activity after activation with endotoxins, as determined by using specific chromogenic sub-strate.[10-12] Fraction B contained coagulogen detected by a clot formation with trypsin. These fractions A and B were used for further purification.

RESULTS AND DISCUSSION

Biochemical properties of T. *tridentatus* coagulogen. The purified coagulogen gave a single protein band on 7.5% gel electrophoresis at pH 3.2 in the presence of 6 M urea, as shown in Fig. 3. The band moved very rapidly to the cathode, suggesting a basic protein nature. Therefore, on electrophoresis at pH 8.3, coagulogen did not penetrate the gel and was retained at the top of the gel. Sodium dodecyl sulfate (SDS)-gel electrophoresis of the samples also showed a single band in the presence or absence of 2-mercaptoethanol, suggesting that it consists of a single polypeptide chain. Column 5 shows gel electrofocusing of purified coagulogen, using 5% polyacrylamide gel. The isoelectric point was estimated to be approximately pH 10.

The physical and chemical properties of T. *tridentatus* coagulogen are shown in Table 1. The sedimentation coefficient of this protein was calculated to be

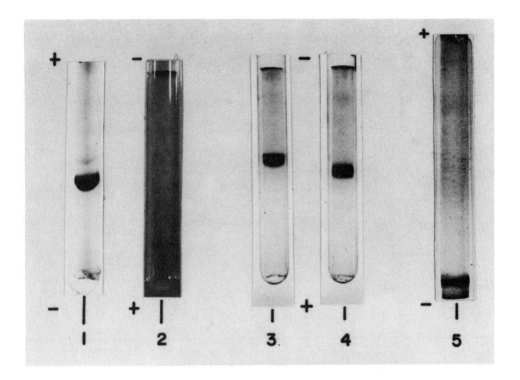

Fig. 3. Polyacrylamide gel electrophoresis of purified coagulogen (T. *triden-tatus*).[7] 1, at pH 3.2 in the presence of 6 M urea; 2, at pH 8.3 in the absence of 6 M urea; 3, SDS-gel without 2-mercaptoethanol; 4, SDS-gel with the reducing agent; and 5, gel electrofocusing using 5% polyacrylamide gel, which contains an Ampholine mixture of pH 5.7-7, 7-10 and 9-11.

2.6 S. The minimal molecular weight was estimated to be 19,500 by SDS-gel electrophoresis and calculated to be 19,956 from the amino acid composition, as described later. Coagulogen had an UV absorption maximum at 276 nm, and showed small shoulders at 253 nm, 257 nm, and 265 nm due to the high content of phenylalanine residues. Its amino-terminal sequence was Alanyl-Aspartyl-Threonine, and the carboxy-terminal end was phenylalanine.[7] It is of interest that this amino-terminal sequence is homologous with the first three amino-terminal residues of Aα-chain of primate fibrinogen.

Structural change of T. *tridentatus* coagulogen during gel formation. Figure 4 shows SDS-gel electrophoretogram of the reaction products of coagulogen with an endotoxin-activated clotting enzyme. The upper figure was electrophoretogram

TABLE 1

PHYSICAL AND CHEMICAL PROPERTIES OF *TACHYPLEUS* COAGULOGEN

$S^{\circ}_{20,W}$	2.6 S
Molecular weight	15,300 $\pm$ 400 (sedimentation equilibrium)
	19,956 (chemical analyses)
	19,500 $\pm$ 1,000 (SDS-gel electrophoresis)
Amino acid composition	175 residues
Isoelectric point	10.0 $\pm$ 0.2
UV absorption spectrum	λ_{max}: 276 nm
	Shoulder: 253, 257 and 265 nm
$A^{1\%}_{280\ nm}$	8.52 (pH 6.5)
	9.80 (0.2 N NaOH-8 M urea)
N-terminal sequence	Ala-Asp-Thr
C-terminal residue	Phe

of the reduced sample and the lower figure was that of the unreduced sample.
At time zero, a single protein band was seen. On incubation for 5 to 10 min,
the coagulogen gradually disappeared, and a gel protein accumulated at the
top of the SDS-gels. On incubation for more than 40 min, a fragment, named
peptide C, appeared on the gel. On the other hand, on incubation for 5 to 10
min, three new bands, named F-C·B, F-B, and F-A·C, appeared on the reduced gel.
Of these fragments, two major fragments, F-B and peptide C, seemed to be accumulating in the reaction mixture. These results suggest that fragment F-A·C links
with the large fragment F-B by a disulfide bridge.[8] To estimate the relative
amounts of these fragments, the bands on gels were scanned with a densitometer,
and the amount of gel protein formed was measured separately. The result is
shown in Fig. 5. The amount of F-A·C increased in the initial stage of gel
formation and its decrease seemed to be roughly proportional to increase in the
amount of peptide C. The amount of gel protein also increased in parallel with
release of peptide C. For separation of the fragments which appeared on SDS-gel
electrophoresis, 7 mg of coagulogen was incubated at 37°C with *Tachypleus* clotting enzyme for 3 hr. Then, the reaction mixture was reduced and S-carboxymethylated, and the resulting material was applied to a Sephadex G-50 column

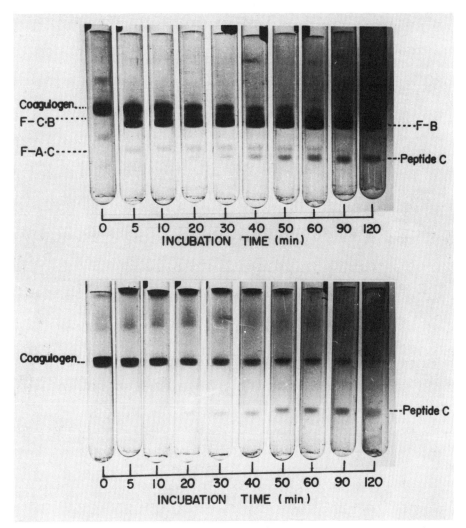

Fig. 4. SDS-gel electrophoresis of the reaction product of coagulogen with *Tachypleus* clotting enzyme.[8] Electrophoresis was performed by the method of Weber and Osborn[26] on 10% polyacrylamide gel, and the gels were stained with Coomassie brilliant blue R-250 at 50°C for 30 min. The upper figure was on the reduced samples and the lower on unreduced samples.[8]

(1.6 x 110 cm), equilibrated with 10% acetic acid. The elution pattern is shown in Fig. 6. We pooled each of the peaks and determined their amino acid compositions as well as their amino- and carboxyl-terminal residues. The results are given in Table 2. A-chain (F-A in Fig. 6) consisted of 18 residues with amino-

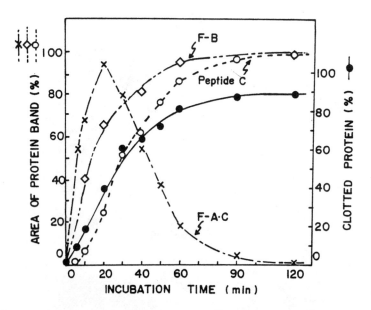

Fig. 5. Densitometric analysis of the fragments obtained from coagulogen during proteolysis and estimation of gel protein accumulated in the reaction mixture.[8,9] The stained gels (reduced) in Fig. 4 were scanned with a densitometer, model CS-900, Shimadzu Seisakusho, Ltd., and the relative intensities of the bands were calculated, taking the maximum intensity of each fragment as 100%.[8] Gel protein was estimated by the method of Miller.[16]

terminal alanine and carboxyl-terminal arginine, and the subsequent NH_2-terminal sequence was determined to be Ala-Asx-Thr, which is the same as that of intact coagulogen. Peptide C was 28 residues long and its NH_2-terminus and COOH-terminus were threonine and arginine, respectively. Fragment F-AC was estimated to contain 46 residues which agrees with the sum of total residues in A-chain and peptide C. The large fragment B-chain (F-B) consisted of 129 residues with NH_2-terminal glycine and its COOH-terminal phenylalanine was identical with that of intact coagulogen. From these results, it was concluded that coagulogen consists of three polypeptide segments, A-chain, peptide C and B-chain, and has a total of 175 amino acid residues.

Figure 7 shows alignment of the peptide segments and structural changes of coagulogen during gel formation. The clotting enzyme seems to cleave the Arg-46-Gly-47 linkage and Arg-18-Thr-19 linkage, both located in the NH_2-terminal region of coagulogen. It is interesting that the Arg-Gly linkage cleaved by the clotting enzyme is the same type as that cleaved by α-thrombin in the conversion of mammalian fibrinogen to fibrin.

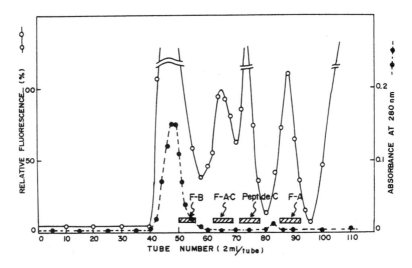

Fig. 6. Fractionation of the peptide fragments produced from coagulogen during enzymatic gel formation.[14] The reaction mixture containing 7 mg of coagulogen was reduced, S-carboxymethylated and gel-filtrated on a Sephadex G-50 column (1.6 x 110 cm), equilibrated with 10% acetic acid. Elution was carried out with the same solution, and the fragments were detected by the O-phthalaldehyde method.[17]

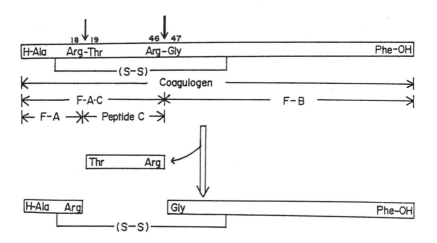

Fig. 7. Alignment of the peptide segments and structural change of *T. tridentatus* coagulogen during gel formation.[8] The arrows indicate the sites cleaved by *Tachypleus* clotting enzyme.

TABLE 2

AMINO ACID COMPOSITIONS AND N- AND C-TERMINAL RESIDUES OF
POLYPEPTIDE FRAGMENTS FROM *TACHYPLEUS TRIDENTATUS* COAGULOGEN

Amino Acid	A-chain	Peptide C		F-AC	B-chain
Aspartic acid	3.1 (3)	1.1	(1)	4.1	6.5 (6)
Threonine	1.0 (1)	2.9	(3)	4.1	6.9 (7)[a]
Serine	0	1.9	(2)	2.1	9.1 (9)[a]
Glutamic acid	1.1 (1)	6.1	(6)	6.8	13.4 (13)
Proline	1.9 (2)	0		2.0	8.0 (8)
Glycine	2.1 (2)	2.1	(2)	4.0	10.4 (10)
Alanine	2.0 (2)	3.0	(3)	4.8	3.3 (3)
CM-cystine	2.1 (2)	0		1.4	14.8 (14)
Valine	1.1 (1)	3.9	(4)	4.7	10.2 (11)
Isoleucine	1.0 (1)	2.8	(3)	3.9	2.0 (2)
Leucine	2.1 (2)	0		2.1	4.0 (4)
Tyrosine	0	0		0	5.0 (5)
Phenylalanine	0	0		0	13.1 (13)
Lysine	0	2.8	(3)	3.0	4.3 (4)
Histidine	0	0		0	5.2 (5)
Arginine	1.1 (1)	0.9	(1)	1.9	13.8 (14)
Tryptophan	0	0		0	1.3 (1)[b]
Total	18	28		46	129
N-terminal	Ala	Thr		Ala	Gly
C-terminal	Arg	Arg		Arg	Phe

Figures represent residues per mole calculated from values for 24 h hydrolysate.
Values in parentheses were obtained from sequence analyses. a: Extrapolated
value, b: Hydrolysed with 4 M methanesulfonic acid.

Complete amino acid sequence of T. *tridentatus* coagulogen. The complete
amino acid sequences of peptide C and A-chain, which constitutes the NH_2-terminal
portion of coagulogen, are shown in Fig. 8. The COOH-terminal octapeptide
sequences of these peptides exhibited great homology for each other. This suggests that a specific oligopeptide sequence immediately preceding the bond to
be cleaved is required for the clotting enzyme to split the Arg-Gly and Arg-Thr
linkages so as to initiate gel formation. The most interesting finding, however,
is a remarkable sequence homology of peptide C and A-chain with primate fibrino-

Tachypleus coagulogen

F - A - - - - - - - - - Asp-Glu-Pro-Gly-Val-Leu-Gly-Arg

Peptide C - - - - - - Gln-Glu-Ser-Gly-Val-Ser-Gly-Arg

Fibrinogen

Rhesus monky
fibrinopeptide B - - - - - Glu-Glu-Ser-Pro-Phe-Ser-Gly-Arg

Human
fibrinopeptide B - - - - - Glu-Gly-Gly-Phe-Phe-Ser-Ala-Arg

Factor X III

Bovine
"activation peptide"- - - - - Glu-Leu-Gln-Gly-Leu-Val-Pro-Arg

Human
"activation peptide"- - - - - Glu-Leu-Gln-Gly-   -Val-Pro-Arg

Fig. 8. Sequence homologies among *T. tridentatus* coagulogen fragments, primate fibrinopeptides B[18,19] and Factor XIII "activation peptides".[20,21]

peptide B,[19-21] as shown in Fig. 8. Especially, the last eight COOH-terminal residues of peptide C are very similar to those of the *Rhesus* monkey fibrino-peptide B.[18]

Figure 9 shows the amino acid sequence of B-chain.[14] For the sequence determination, tryptic, chymotryptic, thermolytic digestions on the S-carboxymethylated B-chain were used. Upon tryptic digestion, all of the 17 theoretically expected sensitive peptide bonds were cleaved, and the 19 major peptides were isolated. The sequence analysis of these peptides was performed by direct Edman degradation and dansyl-Edman methods. The alignment of the tryptic peptides was confirmed with the 18 chymotryptic peptides and 39 thermolytic peptides, as shown in Fig. 9. These experiments made it possible to establish the complete sequence of *Tachypleus* coagulogen (Fig. 10). *Tachypleus* coagulogen consisted of 175 residues. It contained a total of 16 half-cystines, which are linked with disulfide bonds, and five of them are clustered in the COOH-terminal region. The arrangements of these disulfide linkages are now under investigation. Comparing the amino acid sequence of the coagulogen with that of human fibrinogen, there was no convincing sequence homology. However, the most interesting finding was a partial sequence homology of coagulogen with platelet factor 4, which is known as antiheparin.

As shown in Fig. 11, when the sequences are aligned for maximum homology, coagulogen was found to contain a number of similar partial sequences compared with platelet Factor 4. Especially, the first ten NH_2-terminal residues of coagulogen are very similar to those of platelet Factor 4.[22,23] Moreover, a

506

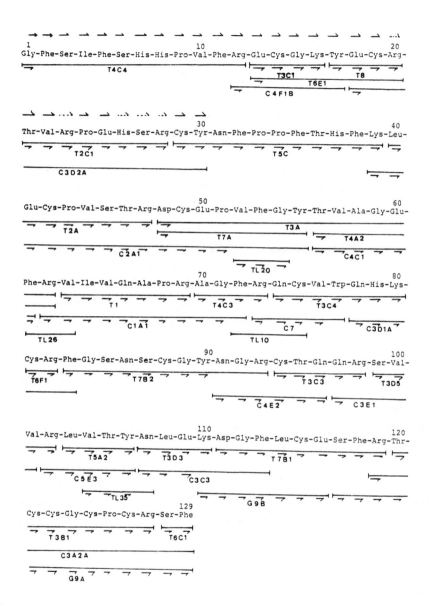

Fig. 9. The amino acid sequence of B chain (F-B) isolated from *T. tridentatus* coagulogen.[14] T, trypsin; C, α-chymotrypsin; G, Staphylococcal protease; TL, thermolysin.

similar sequence with platelet Factor 4 was found in between the residues of 110 and 122. These facts suggest that coagulogen and platelet Factor 4 are derived

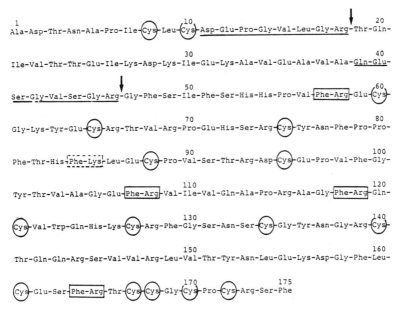

Fig. 10. The complete amino acid sequence of *T. tridentatus* coagulogen.[14] The arrows indicate the sites cleaved by *Tachypleus* clotting enzyme.

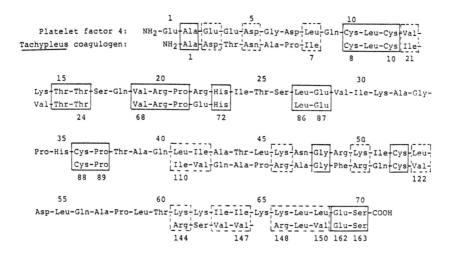

Fig. 11. Comparison of the amino acid sequences of platelet Factor 4 and *T. tridentatus* coagulogen.[24] The whole sequence of the former was taken from the data of Begg et al.[23] Solid boxes enclose identical residues and dotted boxes enclose the residues with chemical similarities.

TABLE 3

MOLECULAR WEIGHTS AND N- AND C-TERMINAL RESIDUES OF *T. TRIDENTATUS*,
C. ROTUNDICAUDA, *T. GIGAS*, and *L. POLYPHEMUS COAGULOGENS*

	T. triden-tatus	*C. rotundi-cauda*	*T. gigas*	*L. polyphemus*
Molecular weight				
SDS-gel electrophoresis (reduced sample)	20,000	20,000	20,000	20,000
SDS-gel electrophoresis (unreduced sample)	21,000	21,000	21,000	24,000
Gel filtration	15,500	15,500	15,500	19,500
N-terminal sequence	Ala-Asp-Thr	Ala-Asp-Thr	Asp-Asp-Thr	Gly-Asp-Pro
C-terminal residue	Phe	Tyr	Phe	Ser

from a common ancestor or that the coagulogen is not only a prototype of fibrinogen[13] but also platelet Factor 4.[24]

Biochemical characteristics of coagulogens isolated from other species, *T. gigas, C. rotundicauda* and *L. polyphemus*. Recently, we have succeeded in isolating a clottable protein from the hemocyte lysates of *Carcinoscorpius rotundicauda* and *Tachypleus gigas*, which inhabit the area of southeast Asia, and also from that of American *Limulus polyphemus*.[15] Table 3 shows the molecular weights and the amino and carboxyl terminal residues of the isolated four coagulogens for comparison. Their molecular weights, determined by SDS-gel electrophoretic method, were estimated in common to be about 20,000. Moreover, the first three NH_2-terminal residues and COOH-terminal end of these materials were similar to each other, except for the *Limulus polyphemus* coagulogen.

Table 4 shows the amino acid compositions of three different coagulogens from *Carcinoscorpius*, *Tachypleus gigas*, and *Limulus polyphemus*. The composition of *Tachypleus tridentatus* previously determined is also listed.[7] All the coagulogens consisted of 173 to 175 residues and their compositions were very similar, such as no methionine, high proportions of basic amino acids, and high contents of cystine, valine and phenylalanine.[7] These analytical data for the amino acids in the coagulogens corresponded well to integral molar ratios. This may be taken as strong evidence for the homogeneities of these proteins.

Figure 12 shows the SDS-gel electrophoretogram of the reaction product of *Limulus polyphemus* coagulogen with a *Limulus* clotting enzyme. This pattern was

TABLE 4

AMINO ACID COMPOSITIONS OF *T. TRIDENTATUS*, *C. ROTUNDICAUDA*, *T. GIGAS*, AND *L. POLYPHEMUS COAGULOGENS*

Amino acid	*T. tridentatus*		*C. rotundicauda*		*T. gigas*		*L. polyphemus*	
	Residues per mole							
Aspartic acid	10.9	(11)	12.2	(12)	13.4	(13)	11.9	(12)
Threonine	11.4	(11)	9.5	(10)	9.8	(10)	12.9	(13)
Serine	10.9	(11)	9.8	(10)	9.8	(10)	11.2	(11)
Glutamic acid	19.1	(19)	21.1	(21)	19.2	(19)	21.0	(21)
Proline	9.3	(10)	11.4	(11)	10.1	(10)	9.5	(10)
Glycine	13.5	(14)	13.9	(14)	14.7	(15)	12.5	(13)
Alanine	7.8	(8)	8.1	(8)	8.0	(8)	7.0	(7)
CM-cysteine	16.1	(16)	16.4	(16)	16.1	(16)	16.2	(16)
Valine	15.2	(15)	16.4	(16)	13.2	(13)	14.1	(14)
Methionine	0		0		0		0	
Isoleucine	5.8	(6)	3.2	(3)	4.7	(5)	6.4	(6)
Leucine	5.9	(6)	8.2	(8)	7.3	(7)	6.0	(6)
Tyrosine	5.1	(5)	6.4	(6)	3.9	(4)	5.4	(5)
Phenylalanine	12.8	(13)	11.9	(12)	13.6	(14)	11.6	(12)
Lysine	6.9	(7)	7.3	(7)	7.9	(8)	8.6	(9)
Histidine	4.9	(5)	5.3	(5)	5.0	(5)	4.1	(4)
Arginine	15.7	(16)	15.3	(15)	14.6	(15)	12.8	(13)
Tryptophan[a]	1.3	(1)	0.9	(1)	1.1	(1)	0.8	(1)
Total residues	172.6	(174)	174.7	(174)	172.4	(173)	172.1	(172)

Values of 24, 48 and 72 hr hydrolysis.   a. Determined spectrophotometrically.

510

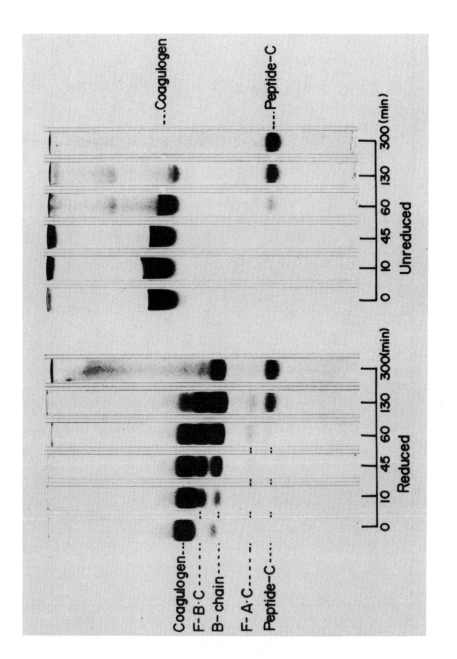

Fig. 12. SDS-gel electrophoresis of the reaction product of *Limulus polyphemus* coagulogen with a *Limulus* clotting enzyme. Electrophoresis was performed as described in Fig. 4.

the same as that of *Tachypleus* coagulogen shown in Fig. 4. The other coagulogens from two Asian species, *Tachypleus gigas* and *Carcinoscorpius*, also showed the identical patterns of structural changes during the gel formation. These results suggest that upon gelation of all the coagulogens by clotting enzyme, a large peptide, peptide C, is released in common from the inner portion of the parent molecules (Fig. 12). The resulting gel proteins consist of two chains, named A and B. Therefore, as shown in Fig. 13, the enzymatic formation of gel involved limited proteolysis of the Arg-X and Arg-Gly linkages located in the NH_2-terminal portions of all the coagulogens, liberating peptide C.

To establish the NH_2-terminal structures of the other newly isolated three coagulogens, the partial sequence studies on the A-chain, peptide C and B-chain were performed. The results are shown in Fig. 14. First of all, we would like to point out that the NH_2-terminal sequences up to 55 residues in the three coagulogens, *Tachypleus tridentatus*, *Tachypleus gigas* and *Carcinoscorpius*, were very similar, showing 80% to 87% homology with the *T. tridentatus* coagulogen. There were no deletions or insertions of residues up to 55. Secondly, on comparing American *Limulus* coagulogen with *Tachypleus* coagulogen, 22 amino acid

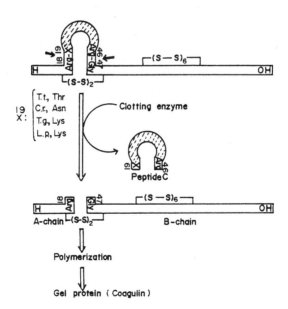

Fig. 13. Structural changes of four kinds of coagulogens during gel formation. The arrows indicate the sites cleaved by the clotting enzyme prepared, respectively, from the hemocyte lysates of *T. tridentatus* (T.t), *T. gigas* (T.g.), *C. rotundicauda* (C.r) and *L. polyphemus* (L.p).

A-chain
```
              1         5              10             15        18
T.t.   H -Ala-Asp-Thr-Asn-Ala-Pro-Ile-Cys-Leu-Cys-Asp-Glu-Pro-Gly-Val-Leu-Gly-Arg-OH
C.r.   H -Ala-Asp-Thr-Asn-Ala-Pro-Leu-Cys-Leu-Cys-Asp-Glu-Pro-Gly-Ile-Leu-Gly-Arg-OH
T.g.   H -Asp-Asp-Thr-Asn-Ala-Pro-Ile-Cys-Leu-Cys-Asp-Glu-Pro-Gly-Val-Leu-Gly-Arg-OH
L.p.   H -Gly-Asp-Pro-Asn-Val-Pro-Thr-Cys-Leu-Cys-Glu-Glu-Pro-Thr-Leu-Leu-Gly-Arg-OH
```

Peptide-C
```
              20              25              30              35
T.t.   H -Thr-Gln-Ile-Val-Thr-Thr-Glu-Ile-Lys-Asp-Lys-Ile-Glu-Lys-Ala-Val-Glu-Ala-
C.r.   H -Asn-Gln-Leu-Val-Thr-Pro-Glu-Val-Lys-Glu-Lys-Ile-Glu-Lys-Ala-Val-Glu-Ala-
T.g.   H -Lys-Glu-Phe-Val-Ser-Asp-Ala-Thr-Lys-Thr-Ile-Ile-Glu-Lys-Ala-Val-Glu-Glu-
L.p.   H -Lys-Val-Ile-Val-Ser-Gln-Glu-Thr-Lys-Asp-Lys-Ile-Glu-Glu-Ala-Val-Gln-Ala-

              40           46
T.t.      Val-Ala-Gln-Glu-Ser-Gly-Val-Ser-Gly-Arg-OH
C.r.      Val-Ala-Glu-Glu-Ser-Gly-Val-Ser-Gly-Arg-OH
T.g.      Val-Ala-Lys-Glu-Gly-Gly-Val-Ser-Gly-Arg-OH
L.p.      Ile-Thr-Asx-Lys-Asp-Glu-Ile-Ser-Gly-Arg-OH
```

B-chain
```
              50              55                              175
T.t.   H  Gly-Phe-Ser-Ile-Phe-Ser-His-His-Pro-----------------------Phe-OH
C.r.   H  Gly-Phe-Ser-Ile-Phe-Ser-His-His-Pro-----------------------Tyr-OH
T.g.   H  Gly-Phe-Ser-Ile-Phe-Ser-His-His-Pro-----------------------Phe-OH
L.p.   H  Gly-Phe-Ser-Ile-Phe-Gly-His-His-Pro-----------------------Ser-OH
```

Fig. 14. Comparisons of amino acid sequences of A-chain, peptide C and B-chain derived from four kinds of coagulogens. T.t = *T. tridentatus*, C.r. = *C. rotundicauda*, T.g. = *T. gigas*, L.p. = *L. polyphemus*.

replacements in their sequences can be found. This significant difference between *Limulus* and *Tachypleus* seems reasonable, because they belong to different genere on the basis of classical taxonomy. Thus, it seems that *Limulus polyphemus* has evolved independently during the larger part of evolution of the Arthropoda. This view is supported by classical taxonomy, in which *Limulus polyphemus* is an early offshoot in the phylogenetic tree of the *Limulus*. Thirdly, because the sequence homologies around the NH_2-terminal portions of B-chains among four coagulogens were extremely high, as shown in Fig. 14, one of the polymerization sites, which contribute to gel formation, may be involved in this region.

Figure 15 shows immunological comparison with four kinds of coagulogens.[25] In the left panel, anti-*Tachypleus tridentatus* coagulogen antibody was placed in the center well, and in the right panel, anti-*Limulus polyphemus* coagulogen antibody was placed in the center well. Anti-*Tachypleus* coagulogen antibody did not form any precipitin line against anti-*Limulus* coagulogen antibody, whereas

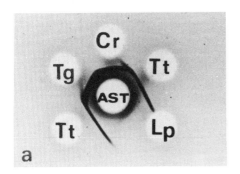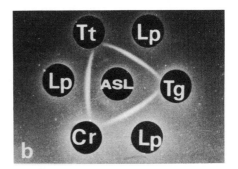

Fig. 15. Immunological comparison with four kinds of coagulogens. In the panel a, anti-*T. tridentatus* coagulogen antibody was placed in the center well, and in the panel b, anti-*Limulus* coagulogen antibody was placed in the center well.[25] T.t.= *T. tridentatus*, T.g. = *T. gigas*, C.r. = *C. rotundicauda*, L.p. = *L. polyphemus*.

the other Asian coagulogens formed a single fused precipitin line for each other on immunodiffusion. On the other hand, anti-*Limulus* coagulogen antibody gave a single fused precipitin line only against American *Limulus* coagulogen. These results indicate that all the Asian coagulogens are immunologically indistinguishable, whereas American *Limulus* coagulogen differs immunologically from *Tachypleus* coagulogens.

In conclusion, we have established the complete amino acid sequence of *T. tridentatus* coagulogen. It contained a total of 175 amino acid residues, and a part of the sequences were found to have similarity with primate fibrinopeptide B and platelet Factor 4. Upon gelation of coagulogen, a large peptide, peptide C, which contains 28 residues, was released, and the resulting gel protein consisted of A and B chains, bridged by two disulfide linkages. This structural change during gel formation was in common with all the coagulogens isolated from other species, including American *Limulus* coagulogen. The coagulogens isolated from two Asian species had extremely higher sequence homology with *T. tridentatus* coagulogen in their NH_2-terminal portions. However, the NH_2-terminal sequence of *Limulus* coagulogens. In good agreementwith the structural studies, an immunological difference of *Limulus* coagulogen from the *Tachypleus* coagulo-gens was also found, whereas three Asian species were immunologically similar.

ACKNOWLEDGMENTS

We wish to express our thanks to Messrs. Y. Hokama and T. Miyata and Dr. T. Morita, for their help in the sequence analyses. The assistance of Miss K. Fukunishi in performing the amino acid analysis is appreciated. Thanks are also due to Dr. Y. Sumita, for his kind assistance in the bleeding of *Tachypleus tridentatus*.

REFERENCES

1. Bank, F.B. (1956) Bull. Johns Hopkins Hosp., 98, 325-351.
2. Levin, J. and Bang, F.B. (1968) Thromb. Diath. Haemorrh., 19, 18j-197.
3. Solum, N.O. (1973) Thrombosis Res., 2, 55-70.
4. Yin, E.T., Galanos, C., et al. (1972) Biochim. Biophys. Acta, 261, 284-289.
5. Young, N.S., Levin, J. and Predergast, P.A. (1972) J. Clin. Invest., 51, 1790-1797.
6. Niwa, M. and Waguri, O. (1975) Seikagaku (in Japanese), 47, 1-13.
7. Nakamura, S., Iwanaga, S., et al. (1976) J. Biochem., 80, 1011-1021.
8. Nakamura, S., Takagi, T., et al. (1976) J. Biochem., 80, 649-652.
9. Nakamura, S. and Iwanaga, S. (1978) Protein Nucleic Acid Enzyme (in Japanese), 23, 277-290.
10. Harada, T., Morita, T. and Iwanaga, S. (1978) J. Med. Enzymol. (in Japanese), 3, 43-60.
11. Nakamura, S., Morita, T., et al. (1977) J. Biochem., 81, 1567-1569.
12. Iwanaga, S., Morita, T., et al. (1978) Haemostasis, 7, 183-188.
13. Nakamura, S., Takagi, et al. (1976) Biochem. Biophys. Res. Communs., 72, 902-908.
14. Takagi, T., Hokama, Y., et al. (1979) Proceedings of the Symposium on Bio-medical Application of *Limulus polyphemus* (Horseshoe Crab), Cohen, E., et al., eds., Alan R. Liss, Inc., New York, in press.
15. Tai, J.Y., Seid, R.D. Jr., et al. (1977) J. Biol. Chem., 252, 4773-4776.
16. Miller, G.L. (1959) Anal. Chem., 31, 964-966.
17. Benson, J.R. and Hare, P.E. (1975) Proc. Natl. Acad. Sci., USA, 72, 619-622.
18. Wooding, G.L., and Doolittle, R.F. (1972) J. Human Evol., 1, 553-563.
19. Blombäck, B., Blombäck, M., et al. (1966) Biochim. Biophys. Acta, 115, 371-382.
20. Nakamura, S., Iwanaga, S., et al. (1974) Biochem. Biophys. Res. Communs., 58, 250-256.
21. Takagi, T. and Doolittle, R.F. (1974) Biochemistry, 13, 750-756.
22. Deuel, T., Keim, P.S., et al. (1977) Proc. Natl. Acad. Sci., USA, 74, 2256-2258.
23. Begg, C.S., Pepper, D.S., et al. (1978) Biochemistry, 17, 1739-1744.
24. Takagi, T., Nakamura, S., et al. (1979) Thrombosis and Haemostasis, 42, 272.
25. Shishikura, F. and Sekiguchi, K. (1978) J. Exp. Zool., 206, 241-246.
26. Weber, K. and Osborn, M. (1969) J. Biol. Chem., 244, 4406-4412.

FUNCTIONAL ROLE AND IDENTIFICATION OF AN ESSENTIAL HISTIDYL RESIDUE

IN CYTOSOLIC ASPARTATE AMINOTRANSFERASE FROM PIG HEART

YOSHIMASA MORINO, SUMIO TANASE AND MASAYUKI MIYAWAKI
The Department of Biochemistry, Kumamoto University Medical School, 2-2-1, Honjo,
Kumamoto 860 Japan

ABSTRACT

Methylene blue-sensitized photooxidation of cytosolic aspartate aminotransferase from pig heart produced a preparation which was almost inactive in transamination with a natural dicarboxylic substrate such as L-aspartate or L-
glutamate but fully active with L-alanine or 3-chloro-L-alanine which lacks the
distal carboxyl group. Upon extensive photooxidation, the enzyme was rendered
catalytically more competent for these monocarboxylic substrates and behaved as
if the enzyme lost its apparently strict substrate specificity for dicarboxylic
substrates. Photoinactivation was accompanied by loss of two out of eight
histidyl residues per monomeric unit of this enzyme. Correlation of the rate
of oxidation of these two histidyl residues with the rate of inactivation
revealed that the second, slower oxidizable histidyl residue was essential to
the enzyme activity toward natural dicarboxylic substrates whereas the first,
rapidly oxidized histidyl residue was not essential to the enzymic catalysis.
Quantitative analysis of histidine contents in the peptide fragments obtained
by cleavage of a photooxidized preparation with cyanogen bromide revealed that
the rate of oxidation of His-317 was much greater than that of the enzyme
inactivation while the rate of oxidation of His-405 was in parallel with that
of inactivation. Thus, HIS-405 was identified to be the residue that is intimately involved in the interaction with the distal carboxylate and hence
determines the substrate specificity of this enzyme for dicarboxylic amino
acids and keto acids.

INTRODUCTION

Aspartate aminotransferase (EC 2.6.1.1) catalyzes transamination between L-
aspartate and 2-oxoglutarate to form oxalacetate and L-glutamate. Cytosolic
aspartate aminotransferase from pig heart is composed of two identical polypeptide chains, each of which possesses a molecular weight of 46,600 and binds
one molecule of pyridoxal 5'-phosphate as the coenzyme.[1] The complete amino
acid sequence was determined by Ovchinnikov et al.[2] and Doonan et al.[3] This
enzyme is highly specific for dicarboxylic substrates. L-Alanine, which lacks

ω-carboxyl group and hence is a poor substrate for this enzyme, reacts at a rate 0.1% that for a natural dicarboxylic substrate such as L-aspartate. However, the presence of a high concentration of formate ion enhances markedly the rate of reaction with L-alanine.[4] Similar accelerating effect of formate ion was observed on the α,β-elimination reaction with 3-chloro-L-alanine.[4] These findings were explained as follows. Formate binds to a specific site which normally binds the distal carboxylate of a natural substrate. When the substrate is L-alanine, the enzyme active site recognizes the formate structure as the distal carboxylate of a natural dicarboxylic substrate. Formate ion thus bound to a discrete subsite exerts a stimulatory effect on the catalytic efficiency of the enzyme in these reactions. A common intermediary step in both transamination and α,β-elimination reactions should be the withdrawal of the α-hydrogen atom of a bound substrate.[5,6] Then the effect of formate ion must be on this step. It was previously proposed[7] that the proton-abstracting base would be provided by the ε-amino group of the lysyl residue (Lys-258) involved in the formation of an almidime bond with pyridoxal 5'-phosphate and formate might act in enhancing the reactivity of this lysyl side chain. When this argument is carried over to the normal transamination reaction, one may conceive that the distal carboxylate not only assists in binding the substrate to the enzyme but also participates in the actual catalytic process by enhancing the reactivity of the ε-amino group of Lys-258 as a base to abstract the α-hydrogen atom of a bound substrate.

The presence of two histidyl residues in or near the active site of this enzyme had been suggested by Martinez-Carrion et al.[8] Although these workers proposed that the histidyl residues might act in the withdrawal of the α-hydrogen of a substrate,[9] we postulated the histidyl residue(s) as the formate-, or distal carboxylate-binding site[10] and investigated the alteration of aspartate aminotransferase-catalyzed reactions during methylene blue-sensitized photooxidation of histidyl residues in the cytosolic enzyme. The result was that the activity of transamination with the natural dicarboxylic substrates decreased progressively as reported previously[8] whereas the activity for transamination with L-alanine as well as that for α,β-elimination of 3-chloro-L-alanine remained unaltered during photooxidation.[11] This finding is consistent with the functional role postulated for a histidyl residue of this enzyme as the binding site for the distal carboxylate of a natural substrate.

The present paper describes the striking alterations in catalytic properties induced upon photooxidative destruction of a critical histidyl residue in cytosolic aspartate aminotransferase from pig heart and the identification of this important histidyl residue in the primary structure of this enzyme.

EXPERIMENTAL PROCEDURES

Materials. 3-Chloro-L-alanine was synthesized as described by Fischer and Raske.[12] Cytosolic aspartate aminotransferase was purified from pig heart by a modification[13] of the procedures previously described.[14,15] Lactate dehydrogenase was obtained in crystalline form from pig heart as described.[16] Other reagents were obtained from commercial sources and of reagent grade.

Methods. Activities for transamination between L-aspartate and 2-oxoglutarate,[17] for transamination with L-alanine[4] and for α,β-elimination of 3-chloro-L-alanine[10] were determined by the procedures cited. Detailed conditions have been described in the legends to figures.

Photooxidation was performed in a test tube (1.8 x 15 cm) thermostated at 10 to 15°C by irradiating with a 150-W tungsten lamp installed at a distance of 5 cm from the center of the test tube which contains 15 to 100 mg of the pyridoxal form of cytosolic aspartate aminotransferase, 0.0002 to 0.0005% methylene blue, 0.2 mM dithiothreitol and 20 mM potassium pyrophosphate buffer (pH 8.0) in a total volume of 5 ml. The reaction mixture was constantly stirred during photooxidation. The rate of inactivation was dependent upon the concentration of the enzyme, ranging from 15 minutes (2-3 mg enzyme/ml) to 40 minutes (10-20 mg/ml) in the time of irradiation required for inactivating 50% of the enzyme initially present.

Solutions containing 3 μmol of photooxidized preparations (residual activity 20% and 65%) in 10 ml were passed over Sephadex G-25 columns (4 x 30 cm) previously equilibrated with 50 mM Tris-HCl buffer (pH 7.5) containing 0.1 mM dithiothreitol and 0.1 mM EDTA to remove methylene blue. The effluents were concentrated by adding ammonium sulfate to 80% saturation, followed by dialysis overnight at 4°C against the above buffer. The resulting preparations were treated with $NaBH_4$ (5 mg) to reduce the coenzyme aldimine bond, followed by dialysis against distilled water. The dialyzed solutions were adjusted to 6 M in guanidine HCl, 1 mM in EDTA, and 2.5 mM dithiothreitol, and incubated at 37°C for 4 hours to ensure the unfolding of the enzyme protein. At the end of incubation, 5 mM sodium iodoacetate (pH 6.5) was added and the mixtures were further incubated at 37°C for 15 minutes. Excess reagents were removed by dialysis against distilled water. After dialysis, the heavy precipitates were collected by centrifugation and suspended in 70% formic acid. To the suspensions was added 300 molar excess of cyanogen bromide over the methionine contents of the enzyme preparations and the mixtures were incubated at 0°C for 24 hours under nitrogen gas. The resulting solutions were diluted 10 times with distilled water and shell-dried.

A solution containing 100 mg (2.1 μmol) of cyanogen bromide-cleaved

preparation in 5 ml of 6 M urea in 5% formic acid was passed over a Sephadex
G-75 (2.7 x 140 cm) which was equilibrated with 7 M urea in 5% formic acid.
The effluent fractions were measured for the absorbance at 280 nm and for the
histidine content using 5-diazo-1-H tetrazole.[17] Fractions IV and V (cf.
Table 1) were further purified by repetition of chromatography on Biogel P-10
and P-6 columns. Finally, three peptides containing His-317, -378, and -405,
respectively, were obtained in pure forms.

Amino acid analysis was performed in a Hitachi KLA-5 automatic amino acid
analyzer. Alkaline hydrolysis of the enzyme preparations and their amino acid
analysis was carried out as described.[18] Acid hydrolysis was performed in 5.7 N
HCl at 110°C for 20 hours in evacuated, sealed tubes.

RESULTS

Alteration of aspartate aminotransferase-catalyzed reactions during methylene blue-sensitized photooxidation

Methylene blue-sensitized photooxidation of cytosolic aspartate aminotransferase resulted in a progressive decrease in the activity of transamination
between L-aspartate and 2-oxoglutarate as reported previously.[8,11] The pH
dependence of the photoinactivation rate and the absence of protection by substrates from the inactivation were in accord with the previous result.[8] It was
also previously described that photooxidation did not affect the activity of
α,β-elimination of 3-chloro-L-alanine as well as transamination with L-alanine.[11]
The present study confirmed this. In addition, the activity for these monocarboxylic substrates were found to increase considerably with the progress of
photooxidation (Fig. 1). A preparation having 20% remaining activity of the
normal transamination exhibited the activities for these monocarboxylic substrates 3 to 4 times those exhibited by the untreated native enzyme. The
transamination of L-alanine as well as α,β-elimination as catalyzed by the
native enzyme are markedly enhanced in the presence of formate.[4] However, it
was found that the extent of acceleration of α,β-elimination of 3-chloro-L-
alanine by formate decreased with the progress of photoinactivation and, as
measured in the presence of 3 M formate, the rate of α,β-elimination decreased
in parallel with that of the normal transamination with that of the normal
transamination with dicarboxylic substrates (Fig. 1). All these findings favor
the view that photoinactivation does not result from the destruction of a
catalytic residue responsible for the withdrawal of the α-hydrogen atom of
bound substrates, a common step prerequisite to both of transamination and α,β-
elimination reactions, but results from the destruction of a residue involved
in the interaction with the distal carboxylate of natural dicarboxylic substrates.

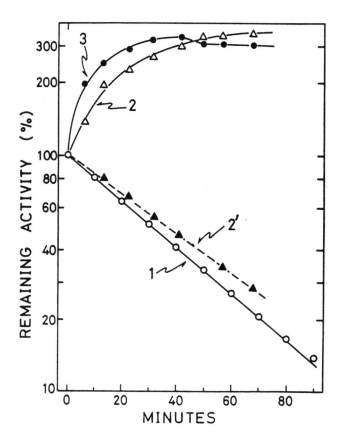

Fig. 1. Alterations of aspartate aminotransferase-catalyzed reactions during methylene blue-sensitized photooxidation. Photooxidation was performed as described
under "Experimental Procedures". At intervals aliquots were withdrawn from the
reaction mixture and passed over Sephadex G-25 columns to remove the dye. Activity of transamination between the natural dicarboxylic substrates (curve 1) was
determined under a standard assay condition where the reaction mixture contained,
in a total volume of 3 ml, 20 mM L-aspartate, 10 mM 2-oxoglutarate and 50 mM
Tris/HCl buffer(pH 8.0) and the reaction was monitored by the absorbance change
at 280 nm. The α, β-elimination reaction with 3-chloro-L-alanine was assayed
with (curve 2) or without 3 M potassium formate (curve 2'). The reaction mixture
contained, in a total volume of 1 ml, 20 mM 3-chloro-L-alanine, 0.1 mM NADH,
5 μg crystalline lactate dehydrogenase, 50 mM Tris/HCl buffer (pH 7.5) and with
or without 3 M formate. The reactions were followed by recording the absorbance
change at 340 nm. Transamination with L-alanine was determined from the rate of
conversion of the pyridoxal form to the pyridoxamine form of the enzyme (curve 3).
The reaction mixture contained, in a total volume of 1 ml, 2 mg of the enzyme in
pyridoxal form and 50 mM Tris/HCl buffer (pH 7.5). The reaction was initiated by
adding 50 μl of 1 M L-alanine and the decrease in absorbance at 370 nm was recorded at 25°C. The absorbance at 370 nm represents the concentration of the
pyridoxal form present. The pseudo first order rate constant for the conversion
was calculated from plots of $\log\left[100(A_o - A_t)/A_o\right]$ versus time, where A_o denotes
the initial value of absorbance at 370 nm and, A_t, that at time, t.

Properties of photooxidized aspartate aminotransferase

Figure 2 compares the reactions of a photooxidized preparation (having 37% activity) and the native enzyme with 3-chloro-L-alanine. The native enzyme was previously shown to be inactivated during α,β-elimination reaction with this haloalanine.[4] The rate constant for the inactivation of the native enzyme was 0.03 min^{-1} as determined from the first order plot of the logarithm of the remaining activity versus the incubation time (Fig. 2) Such a plot for the photooxidized enzyme revealed the presence of at least two phases differing in rates of inactivation (Fig. 2, curve 2). The slower phase (0.03 min^{-1}) was judged as representing the event occurring to the remaining native enzyme since the inactivation rate of the native enzyme was also approximately 0.03 min^{-1}. Then the fast phase must include the event occurring to the photooxidized enzyme. Subtraction of the slower phase from the fast phase of the first order plot yielded a value of 2.3 min^{-1} for the rate constant of the inactivation of the photooxidized enzyme. It was previously demonstrated that Lys-258 of cytosolic aspartate aminotransferase is covalently modified by 3-chloro-L-alanine during α,β-elimination reaction.[7] A photooxidized preparation which had only 20% residual activity of normal transamination but retained a full activity for α,β-elimination was rapidly inactivated by incubating with 3-chloro-L-[U-^{14}C]alanine. Chemical analysis of the resulting labeled enzyme was performed as described previously[7] and revealed that the modified site was Lys-258 as was the case with the native enzyme (data not shown). The reactivity of this important lysyl residue appears to be enhanced some 80-fold upon photooxidation of the enzyme. In addition, the pK value for the ε-amino group of Lys-258 in the apoenzyme prepared from the photooxidized preparation was found to be 8.40 (unpublished data) as compared with the corresponding value of 7.98[19] for the native apoenzyme. It is premature at this stage to give any specific mechanistic significance to this increase in the pK value of Lys-258 in the apoenzyme upon photooxidation. However, this perturbation of the ionization state of the active site lysyl residue, the apparent loss of the substrate specificity and the striking acceleration of the rate of labeling of the active site by 3-chloro-L-alanine, all indicate that photooxidation caused a profound alteration of the structure and function of the active site of this enzyme.

Photooxidative destruction of histidyl residues and its correlation with the inactivation

Cytosolic aspartate aminotransferase from pig heart contains 8 histidyl residues per monomeric unit.[2,3] There has been no indication of the presence of functional cooperativity between two subunits composing this enzyme molecule.[20,21] Thus, one can assume that each subunit acts independently and hence

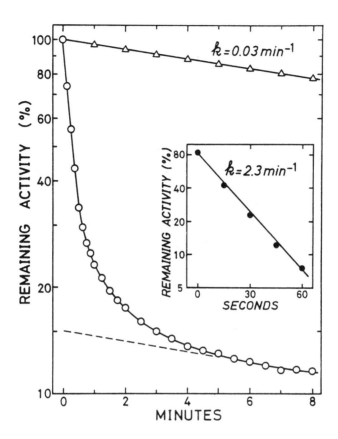

Fig. 2. The rate of inactivation of the native and photooxidized aspartate aminotransferase with 3-chloro-L-alanine. The reaction mixture contained, in a total volume of 1 ml, 20 mM 3-chloro-L-alanine, 0.15 mM NADH, μg lactate edhy-drogenase and 50 mM Tris/HCl buffer (pH 7.5). The reaction was started by adding an enzyme sample and followed by recording the absorbance change at 340 nm at 25°C. Logarithm of [100 x (rate of pyruvate formation at time, t)/(initial rate of pyruvate formation)] was plotted against the time of reaction. Native enzyme, -Δ-; photooxidized enzyme having 37% remaining activity, -o-. Inset shows a secondary plot for the portion of the early rapid phase in the original first order plot.

the environment of each histidyl residue on one subunit polypeptide chain should be identical to that of the corresponding residue on the other subunit.

Upon methylene blue-sensitized photooxidation of the enzyme in the pyridoxal form, the content of histidyl residues decreased progressively. In accord,with the previous report,[8] the amino acid analysis of extensively photooxidized enzyme preparations (residual activity: 25 to 15%) indicated the loss of 2.2 to 2.3 histidine residues per monomeric unit of the enzyme and no appreciable loss of tyrosine, cysteine, methionine and tryptophan residues. The semilogarithmic plot of the remaining histidine contents versus time of photooxidation revealed the presence of two rapidly oxidizable histidyl residues (Fig. 3). This is also in good agreement with the previous result.[8] However, the overall rate constant for the oxidation of the two histidyl residues was 0.050 min^{-1} whereas the rate of inactivation was 0.021 min^{-1}. This is in contrast to the previous report[8] describing the identity of these two rate constants. If there were two histidyl residues oxidizable at two different rates, the first order plot for the overall process of the oxidation of two histidyl residues should in principle exhibit two distinct phases, the faster phase followed by a slower phase. However, the experimental errors inherent in such analytical procedures did not allow us to be aware of these two different processes only by inspection of the first order plot (Fig. 3, inset). Thus, as a working hypothesis, we assumed the presence of an essential histidyl residue which should be oxidized at a rate identical with that of the inactivation (k = 0.021 min^{-1}).

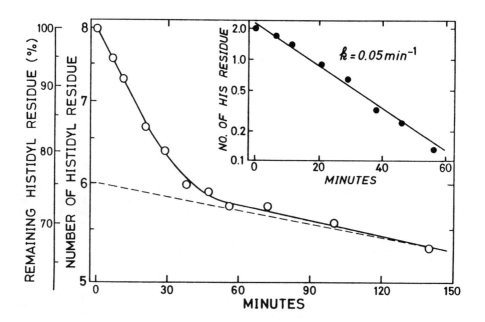

Fig. 3. Rate of destruction of histidyl residues in cytosolic aspartate aminotransferase during photooxidation. Conditions for photooxidation were the same as in Fig. 1. Total number of histidine residues per monomer of this enzyme was assumed to be eight from the sequence.[2] The inset shows a secondary plot for the destruction of two rapidly oxidized histidyl residues as assessed from the original first order plot.

Since the overall rate for the oxidation of the two histidyl residues was
0.050 min^{-1}, the other histidyl residue would be oxidized at a faster rate
(calculated to be 0.079 min^{-1}) and should not be essential to enzymic catalysis.
To prove this assumption, it is imperative to identify these two histidine
residues in the primary structure of this enzyme and to correlate quantitatively
the extent of destruction of a particular histidyl residue to that of the enzyme
inactivation.

Positioning of photooxidizable histidyl residues in the primary structure and identification of an essential histidyl residue

The identification of photooxidizable histidine residues in the primary
structure of this enzyme was attempted by chemical analysis of the peptide fragments obtained by cleavage with cyanogen bromide. Both of the native and photooxidized preparations were subjected to fragmentation by treatment with cyanogen
bromide. Comparison of chromatographic patterns of the fragments from the
native enzyme and an extensively photooxidized preparation (residual activity:
20%) on Sephadex G-75 columns in 6 M urea[22] indicated that the histidine contents decreased markedly in Fraction IV and significantly in Fraction V.
Fraction I contained the peptide [1-212]. Since this peptide is large and contains as many as four histidyl residues (His-68, His-143, His-189, and His-193),
the difference in histidine contents of Fraction I between the native and
photooxidized preparations was not apparent from the inspection of the chromatographic patterns. The corresponding peptides from the native and photooxidized
enzymes were further purified by chromatography on Sephadex G-100 (2.7 x 140 cm)
equilibrated with 6 M urea in 5% formic acid. The amino acid compositions of
the resulting pure peptides did not show any significant difference between the
native and photooxidized preparations. Thus, it is evident that these four
histidyl residues were all inaccessible to photooxidation under the present
experimental conditions. Fraction IV contained the peptide [288-326] containing
His-317 and Fraction V appeared to contain the peptides [360-389] and [390-412],
containing His-378 and His-405, respectively. These peptides were further purified by repetition of chromatography on Biogel P-6 and P-10 columns equilibrated
with 30% acetic acid. Table 1 summarizes the results of amino acid analyses of
these purified peptides. Each peptide was obtained in fairly good yields. The
histidine content in the peptide [390-412] isolated from the photooxidized preparation having 65% residual activity was 66% as compared with that in the corresponding peptide isolated from the native enzyme. This finding clearly indicates that His-405 was photooxidized at a rate identical to that of inactivation
of the enzyme. In contrast, the histidine content of the peptide [288-326]

TABLE 1

BASIC AMINO ACID COMPOSITIONS OF CYANOGEN-BROMIDE PEPTIDES FROM NATIVE
AND PHOTOOXIDIZED PREPARATIONS OF CYTOSOLIC ASPARTATE AMINOTRANSFERASE

Amino acid residues	Cyanogen-Bromide Fragments[a]						
	I [1-212][c]	IV [288-336]		V-1[b] [360-389]		V-2[b] [390-412]	
	P-20[d]	P-65[e]	N[f]	P-65	N	P-65	N
	number of residues						
Lysine	8.00(8)[f]	2.00	2.00(2)	2.00	2.00(2)	2.00	2.00(2)
Histidine	3.90(4)	0.22	0.99(1)	0.97	0.98(1)	0.66	0.97(1)
Arginine	12.70(13)	2.73	2.85(3)	1.17	1.25(1)	0	0 (0)
Yield (%)	63	33	24	16	18	25	31

[a] Cyanogen-bromide fragments obtained by chromatography on a Sephadex G-75 column as described under "Experimental Procedures" were numbered from I to V in the order of their appearance from the column (see reference 22).

[b] Peptides V-1 and V-2 were obtained by further purification of Fraction V on Biogel P-6 and P-10 columns.

[c] Positions in the whole sequence.[2]

[d] Photooxidized enzyme having 20% residual activity.

[e] Photooxidized enzyme having 65% residual activity.

[f] Native enzyme.

[g] Numbers in parenthesis denote the number of residues expected from the sequence of this enzyme.[2]

decreased to 22%. This value is very close to that (23%) expected from a histidyl residue which was oxidized at a rate of 0.079 min^{-1}. Thus, His-317 should be the one which was photooxidized most rapidly and was assumed to be nonessential to enzymic catalysis. The amino acid compositions of the peptides [360-389] from both the native and photooxidized enzymes were identical, indicating that His-378 was not accessible to photooxidation. The total amount (1.2 mole) of decomposed histidyl residues in this particular preparation of photooxidized enzyme was fully accounted for by the sum of the decreased amount of His-405 (0.34 mole) and His-317 (0.78 mole).[*] These results are consistent with the earlier assumption that the most rapidly oxidized histidyl residue was His-317 and the second rapidly photooxidized residue was His-405 which is essential to enzymic catalysis.

[*] We have not isolated the peptide [334-359] containing His-352. However, it is reasonable to assume that this histidyl residue would not significantly be oxidized under the present conditions since the sum of the decomposed amounts of His-317 and His-405 accounted for the total amount of photooxidized residues in the enzyme.

DISCUSSION

The present investigation demonstrates unequivocally that a profound alteration in the catalytic properties of cytosolic aspartate aminotransferase occurs
upon photooxidative destruction of HIS-405. All aspects of the catalytic properties exhibited by the photooxidized enzyme were consistent with the view that
HIS-405 is the important residue intimately related to the interaction of the
enzyme with the distal carboxyl group of a natural dicarboxylic substrate.
Cytosolic aspartate aminotransferase from pig heart exhibits a very low affinity
(Km = 250 mM) for L-alanine. However, photooxidation resulted in a striking
decrease in the Km value to 30 mM (unpublished observation). Thus, photooxidation appears to lead to loss of the substrate specificity of this enzyme
and His-405 plays an important role in determining the substrate specificity of
this enzyme. The reported absence of interaction of α-methylaspartate and some
dicarboxylic analogs with the photooxidized enzyme[23] is consistent with this
view. The result of ^{1}H - n.m.r. spectroscopy of this enzyme in the histidine
region[24] seems to be also consistent with the above view. Cytosolic aspartate
aminotransferase in the pyridoxal form exhibits four discrete resonances in the
low field region (7.7 to 9 ppm). One of these resonances was assigned to the
coenzyme internal aldimine proton on the carbon atom. Thus, with this enzyme,
only three out of eight histidyl residues per monomer appear as distinct peaks
on the n.m.r. spectra. The resonance at the most down field chemical shift
position (8.9 ppm down field from sodium 2, 2-dimethyl-2-silapentane-5-sulfonate)
did not titrate with pH in the range of pH 5.6 to 9.0 whereas the other two
resonances titrated as histidyl residues having pK 6.8 and 9.1. The resonance
at the most low field position was found to broaden upon the addition of saturating concentration of α-methylaspartate reflects probably His-405.

Erythro-β-hydroxyaspartate interacts with aspartate aminotransferase and
gives rise to a characteristic absorption species at 492 nm which has been
attributed to a quinonoid intermediate.[25] In accord with the previous observation,[9] a photooxidized preparation showing 20% residual activity gave rise to
but a trace of the 492 nm species upon the addition of erythro-β-hydroxyaspartate.
This fact may reflect that, first, the analog does not bind to the enzyme;
secondly, even if it is bound, the subsequent deprotonation step is impaired;
thirdly, the rate determining step is altered upon photooxidation such that the
transamination reaction with this analog proceeds without appreciable accumulation of the 492 nm-intermediate. The third possibility was ruled out by confirming that the formation of the keto acid expected from this analog was not
detectable in the reaction mixture containing a photoinactivated enzyme, hydroxy-
aspartate and either of 2-oxoglutarate or pyruvate as an amino group acceptor.

The second possibility may not directly be ruled out. However, this possibility appears to be remote since the catalytic competence of the enzyme for α,β-elimination of 3-chloro-L-alanine and transamination of L-alanine was fully retained after extensive photooxidation and the deprotonation at the α-position of these substrates was reasonably assumed[10,11] to be performed by the same catalytic residue as that for the reactions with erythro-β-hydroxyaspartate or natural dicarboxylic substrates. Thus, the first possibility seems to be most plausible. This may also be extended to a possibility that the dicarboxylic substrates bind to the enzyme active site in an abortive fashion. This point is currently under investigation.

It may be premature to visualize the catalytic role of His-405 and Lys-258 as the structural constituents of the active site of this enzyme. Although there is no evidence that the distal carboxylate of dicarboxylic substrates is directly interacting with His-405, one of the mechanism may be that the binding of the distal carboxyl group to this imidazolyl residue would induce the proton transfer from the ε-amino group of Lys-258 to the distal carboxylate of a substrate via this imidazolyl group, thus enhancing the reactivity of Lys-258 as a base to abstract the α-hydrogen atom of the bound substrate.

Mitochondrial aspartate aminotransferase is the isozymic counterpart of the cytosolic enzyme.[1] Both isoenzymes are structurally quite distinct from each other.[26] The complete amino acid sequence of 401 residues in the monomeric unit has recently been established and has been found to be 11 residues shorter than that of the cytosolic isoenzyme.[27] Sequence identity between the two isoenzymes was 48%. The sequences around Lys-258 (cytosolic enzyme) and Lys-250 (mitochondrial enzyme) were not identical but highly homologous. The sequence at the COOH-terminal region containing His-405 are also homologous to that around His-397 at the COOH-terminal region of the mitochondrial isoenzyme. In contrast, there is no histidyl residue in the mitochondrial isoenzyme at the position corresponding to His-317 of the cytosolic isoenzyme. Thus, the comparison of the positions of these histidyl residues in the amino acid sequences of both isoenzymes is consistent with the fact that His-405 is essential but His-317 is not. As estimated by the procedure for the prediction of the secondary structures,[28] the COOH-terminal regions in both isoenzymes are characteristic of β-structure. A preliminary study indicates that the COOH-terminal region is resistant to hydrolysis by carboxpeptidase Y even in 4 M urea at pH 5.5, suggesting a fairly rigid structure of this region.

ACKNOWLEDGMENTS

This work was supported in part by a Grant-in-Aid from the Ministry of Education, Science and Culture of Japan, and by the Yamada Science Foundation.

REFERENCES

1. Braunstein, A.E. (1973) in Enzymes, Boyer, P.D., ed., Vol. 9, Academic Press, New York, 379-481.
2. Ovchinnikov, Yu., Egorov, C.A., et al. (1973) FEBS Lett., 29, 31-34.
3. Doonan, S., Doonan, A.J., et al. (1975) Biochem. J., 149, 497-506.
4. Morino, Y., and Okamoto, M. (1972) Biochem. Biophys. Res. Commun., 47, 498-504.
5. Metzler, D.E., and Snell, E.E. (1954) J. Am. Chem. Soc., 26, 648-652.
6. Braunstein, A.E., and Shemyakin, M.M. (1953) Biokhimiya, 18, 393-411.
7. Morino, Y., and Okamoto, M. (1973) Biochem. Biophys. Res. Commun., 50, 1061-1067.
8. Martinez-Carrion, M., Turano, C., et al. (1967) J. Biol. Chem., 242, 1426-1430.
9. Peterson, D.L., and Martinez-Carrion, M. (1970) J. Biol. Chem., 245, 806-813.
10. Morino, Y., Osman, A.M., and Okamoto, M. (1974) J. Biol. Chem., 249, 6684-6692.
11. Yamasaki, M., Tanase, S., and Morino, Y. (1975) Biochem. Biophys. Res. Commun., 65, 652-657.
12. Fischer, E., and Raske, E. (1907) Chem. Ber., 40, 3717-3724.
13. Morino, Y., Tanase, S., Watanabe T., et al. (1977) J. Biochem. (Tokyo), 82, 847-852.
14. Jenkins, W.T., Yphantis, D.A., and Sizer, I.W. (1958) J. Biol. Chem., 234, 51-57.
15. Martinez-Carrion, M., Turano, C., et al. (1967) J. Biol. Chem., 242, 2397-2409.
16. Reeves, W.J., and Fimognari, G.M. (1966) Methods Enzymol., 9, 288-294.
17. Okamoto, M., and Morino, Y. (1973) J. Biol. Chem., 248, 82-90.
18. Glazer, A.N., Delange, R.J., and Sigman, D.S. (1975) Chemical Modification of Proteins, Chapter 2, North-Holland Publishing Company, Amsterdam, 13-67.
19. Slebe, J.C., and Martinez-Carrion, M. (1976) J. Biol. Chem., 251, 5663-5669.
20. Boettcher, B., and Martinez-Carrion, M. (1976) Biochemistry, 15, 5657-5664.
21. Schlegel, H., Zaoralek, P.E., and Christen, P. (1976) J. Biol. Chem., 251, 1853-1858.
22. Osman, A.M., Yamano, T., and Morino, Y. (1976) Biochem. Biophys. Res. Commun., 70, 153-159.
23. Cheng, S., and Martinez-Carrion, M. (1972) J. Biol. Chem., 247, 6597-6602.
24. Yamasaki, M., Taura, Y., et al. (1976) Seikagaku, 48, 456.
25. Jenkins, W.T. (1961) J. Biol. Chem., 234, 1121-1125.
26. Wada, H., and Morino, Y. (1964) Vitamins and Hormones, 22, 411-444.
27. Kagamiyama, H., Sakakibara, R., et al. (1977) J. Biochem. (Tokyo), 82, 291-294.
28. Chou, P.Y., and Fasman, G.D. (1974) Biochemistry, 13, 222-245.

Published 1980 by Elsevier North Holland, Inc.
Liu/Mamiya/Yasunobu, eds. Frontiers in Protein Chemistry

RHODANESE–ASSOCIATED SYNTHESIS OF IRON–SULFUR CENTERS

MARGUERITE VOLINI, KATHLEEN OGATA AND DEBORAH CRAVEN
Department of Biochemistry and Biophysics, University of Hawaii, Honolulu,
Hawaii 96822

ABSTRACT

Evidence is accumulating that the sulfurtransferase, rhodanese, is multifunctional in the biological environment and serves as a focal point of sulfane
metabolism. The present study indicates that in one of its physiological roles
the enzyme is involved in the synthesis of protein-bound iron-sulfur centers.

A partially purified mitochondrial extract containing rhodanese activity
equivalent to 36 mg of pure enzyme has been resolved on Sephadex G-100. In
addition to activity being observed in the fraction usually expected to contain
rhodanese, a small amount of active enzyme ($\sim 2\%$ of the total activity) has also
been detected in association with a higher molecular-weight component. In the
presence of iron ions, mercaptoethanol and the rhodanese substrate, thiosulfate,
subfractions of both components showed time-dependent formation of iron-sulfur
centers. Essentially the same rate constant was estimated from measurements of
iron binding, "labile" sulfur binding and increasing absorbance at 410 nm.
However, the rates with different subfractions varied as much as four fold. The
amount of iron bound ranged from 37-210 nmoles per mg of protein and the recovery
of labile sulfur was close to 1.0 mole per mole of iron. The relative recoveries
for the different subfractions indicate that there are at least two proteins
present of different molecular weight that bind iron-sulfur centers. It has
been shown previously that rhodanese of full specific activity does not bind
iron-sulfur centers under the same conditions. The identity of the iron-binding
proteins is currently under investigation.

INTRODUCTION

The sulfanes are acidic inorganic sulfur compounds which consist of linear
chains of divalent sulfur atoms bonded only to each other or to chain-terminal
hydrogen atoms.[1,2] The rhodanese substrates, thiosulfate ion, $O \diagdown S - \underline{S}^{(-2)}$, and

cysteine persulfide $\underset{O}{\overset{\ominus O \diagdown}{\diagup}} C - \overset{\overset{+}{N}H_3}{\underset{|}{C}}H - CH_2 - S - \underline{S}^{\ominus}$ are examples of common

metabolites that contain sulfane-sulfur atoms.

Reactions involving the transfer of sulfane-sulfur atoms occur in virtually all living organisms but the full scope and diversity of these reactions are not yet known.[2,3]

The construct of a metabolic pool of sulfane sulfur,[4,5] provides a framework for speculating on the network of these reactions. Sulfane sulfur may be contributed to the pool through the mediation of enzymes that catalyze reactions involving sulfane compounds as products. At present, there are four enzymes known that may serve in this role: rhodanese [thiosulfate:cyanide sulfur-transferase (E.C.2.8.1.1)], 3-mercaptopyruvate sulfurtransferase (E.C.2.8.1.2), cystathionase [cystathionine γ-lyase (E.C.4.4.1.1)] and thiosulfate reductase (E.C. unassigned). Together these enzymes are capable of supplying sulfane sulfur from both organic and inorganic sources.

Some of these enzymes also have the potential to serve in critical physio-logical roles as users of the sulfane pool. Metabolic processes that appear to make use of the pool include: (1) the biosynthesis of iron-sulfur centers of proteins that are ubiquitous in biology,[6-9] (2) the activation of enzymes involved in oxidative processes,[10-13] (3) the detoxication of cyanide especially in organisms that synthesize or assimilate cyanogenic glycosides,[14-16] (4) the detoxication of sulfide arising from the catabolism of organic sulfur compounds pounds,[17-20] (5) the thiolation of pyrimidine bases in transfer RNAs,[21-26] and (6) some pathways of cysteine biosynthesis.[4,27] Other uses, less well under-stood, are suggested by studies on the fate of sulfane sulfur in animals[2,4,28-30] and by studies on bacterial genera including *Desulfovibrio* and *Thiobacillus*.[2,3] In plasma, serum albumin serves as a carrier for much of the sulfane sulfur derived from labeled thiosulfate.[4] Polythionates and unidentified protein persulfides are also formed.

The sulfurtransferase rhodanese appears to be of central importance in sulfane metabolism. This enzyme has the potential to serve both as a supplier and user of the sulfane pool since it transfers sulfane sulfur directly. It was originally thought to function in the detoxication of cyanide exclusively.[14,15] However, its broad substrate specificity and wide phylogenetic distribution as well as the large amounts of activity that are found in biological systems have led several investigators[4-6,8,11,17,27-32] to propose additional physiological roles for the enzyme.

Recent work in this laboratory provides new evidence that rhodanese is multifunctional in the physiological environment. Different forms of the enzyme have been identified in mammalian liver mitochondria, one of which appears to be involved in the synthesis of iron-sulfur centers.[9,33] The present study extends our observations on the formation of iron-sulfur centers in mitochondrial extracts containing endogenous rhodanese.

EXPERIMENTAL

The mitochondrial suspension was prepared free of rhodanese activity from fresh bovine liver and tested for purity by assay of marker enzyme activities. The latent rhodanese activity was liberated by repeated freezing and thawing and partially purified by pH and ammonium sulfate fractionation as previously described.[9] The extract was further fractionated by molecular exclusion chromatography on Sephadex G-100. The ammonium sulfate precipitate, pH 4.3, was dissolved in Tris acetate buffer, 0.05 I, pH 8.6. 5.4 ml of this solution was added to a column of G-100, 95.9 cm × 2.5 cm, which had been preequilibrated with the same buffer. ‹Effluent fraction volumes ranged from 3.0-5.0 ml.

For generation of protein-bound iron-sulfur centers, 0.59 ml of the G-100 fractions were incubated with 7 µl of 2.5 M sodium thiosulfate, 16 µl of 17.9 mM ferric chloride and 8 µl of 2-mercaptoethanol. Following incubation at 7°C, 0.5 ml aliquots were chromatographed on separate columns of Sephadex G-25, ∼20 cm × 0.7 cm, preequilibrated with Tris acetate buffer, 0.05 I, pH 8.6. Effluent fractions, <1.0 ml, were collected in tared, plastic vials. Fraction volumes were calculated assuming a specific gravity of 1.0 for the solutions. Spectra were recorded on a Cary Model 15 spectrophotometer either in the wavelength region from 600 nm to 250 nm for the G-100 fractions or from 750 nm to 250 nm for the G-25 fractions.

Iron content was measured using a Perkin-Elmer Model 290b atomic absorption spectrophotometer. Standards were prepared in the same buffer used for chromatography. "Labile" sulfur was measured by the procedure of Chen and Mortenson[34] but final volumes of the reaction mixtures were either 1.65 ml or 5.0 ml. Both systems were tested for linearity with fresh sodium sulfide solutions standardized by assay with DTNB (dithio-bis-nitrobenzoic acid). The extinction coefficient used for calculations of recovered sulfide was 3.05×10^4 $M^{-1}cm^{-1}$ as given by Chen and Mortenson.[34]

The protein concentrations of the G-100 fractions were measured by the biuret method of Zamenhof[35] following exchange of the Tris acetate eluent buffer for 0.05 M KPO_4, pH 7.0 by chromatography on columns of Sephadex G-25. Rhodanese activity was measured as previously described.[36] Glucose-6-phosphatase and arylsulfatase activities were determined by the methods of Swanson[37] and Breslow and Sloan,[38] respectively.

RESULTS

The mitochondrial preparations used for these analyses contained only minimal amounts of other subcellular fractions as shown by measurement of their glucose-6-phosphatase and arylsulfatase activities. The latter two enzymes are

532

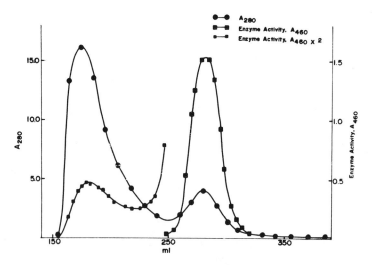

Fig. 1. Molecular exclusion chromatography of the mitochondrial extract on
Sephadex G-100. In the case of the higher molecular-weight component enzyme,
activities are plotted as 2 x the O.D.$_{460}$ values actually observed using 20 μl
aliquots of the fractions in the standard assay system.[36] For the greater
activities associated with the lower molecular-weight component, the values are
reported for 1 μl aliquots. For other details, see text.

localized in the microsomal and lysosomal fractions, respectively, whereas

rhodanese activity has been found exclusively in the mitochondria of liver

cells.[39]

Resolution of the mitochondrial extract

It was shown previously that a rhodanese-containing component purified from

mitochondrial extracts by ammonium sulfate fractionation and chromatography on

Sephadex G-100 binds iron-sulfur centers.[9] In the present studies we have

extended these analyses to larger scale preparations. For the chromatogram

shown in Fig. 1, the extract contained rhodanese activity equivalent to 36 mg of

pure enzyme. Absorbance measurements at 280 nm indicated that the extract

proteins were resolved into two major components with different molecular-

weight ranges. The lower molecular-weight component contained most of the

rhodanese activity as expected. However, a small amount of active enzyme (~2%

of the total activity) was also detected in association with the higher

molecular-weight component. SDS-gel electrophoresis of subfractions from

smaller columns indicate that both components give rise to a band with a

mobility similar to that observed for crystalline rhodanese,[9] also suggesting

association of the enzyme with a higher molecular-weight component.

The experimental points have been averaged for clarity in Fig. 1. However, over fifty fractions were collected and analyzed for absorption spectra and activities in this volume range. In the region from 270 ml to 300 ml where the low molecular-weight component is eluted, ten successive fractions exhibited nearly constant specific activity as judged from their activity-absorbance ratios. This observation, in agreement with earlier results,[9] suggests that a rhodanese-apoprotein complex may be formed.

Time-dependent formation of iron-sulfur centers

Subfractions from each of the G-100 components with near maximal absorbances at 280 nm were selected for intensive study of their iron- and sulfur-binding efficiencies. Aliquots of each fraction were incubated at 7°C with iron ion, mercaptoethanol, and the rhodanese substrate, thiosulfate ion. At the incubation times given in Figs. 2 and 3, individual aliquots were chromatographed on Sephadex G-25 to stop the reaction and remove excess iron ions, mercaptoethanol and thiosulfate or its conversion product, bisulfite ion. Both visible and ultraviolet absorption spectra were recorded on the resulting protein fractions.

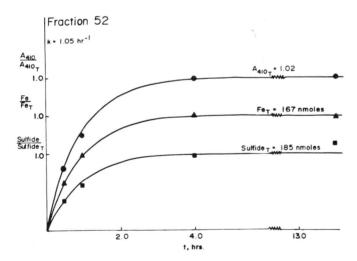

Fig. 2. The time course of iron binding, labile sulfur binding, and increasing absorbance at 410 nm for a subfraction of the higher molecular-weight component resolved on Sephadex G-100. The midpoint elution volume of subfraction 52 from the G-100 column was 193 ml. Aliquots were incubated with 28 mM thiosulfate ion, 0.46 mM iron ions, and 0.18 M mercaptoethanol and chromatographed on Sephadex G-25 at the incubation times specified by the experimental points. The experimental values are superposed on theoretical first order curves (see text).

534

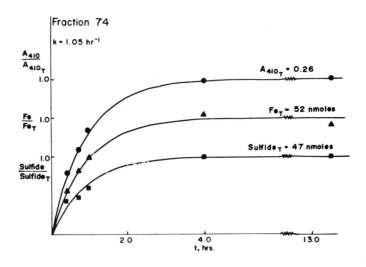

Fig. 3. The time course of iron binding, labile sulfur binding and increasing absorbance at 410 nm for a subfraction of the lower molecular-weight component resolved on Sephadex G-100. The midpoint elution volume of subfraction 74 from the G-100 column was 289 ml. Aliquots were incubated with 28 mM thiosulfate ion, 0.46 mM iron ions, and 0.18 M mercaptoethanol and chromatographed on Sephadex G-25 at the incubation times specified by the experimental points. The experimental values are superposed on theoretical first order curves (see text).

Their iron content was measured by atomic absorption spectrophotometry and their "labile" sulfur content was determined by a modification of the procedure of Chen and Mortenson.[34] Since known sample volumes were added to the G-25 columns and recoveries were quantitated by collecting the fractions in tared vials, the time course data shown in Figs. 2 and 3 could be calculated.

In both figures, the experimental values are superposed on theoretical curves generated for a rate constant of 1.05 hr^{-1}. The limiting values estimated at infinite time for the increase in absorbance at 410 nm, the iron content and the labile sulfur content are shown above their respective curves for each subfraction. The observations that: (1) the same rate constant is obtained for all three parameters, (2) the iron-sulfur ratio is close to 1.0, and (3) the extinction coefficient at 410 nm is close to 5000 M^{-1}cm^{-1} per mole of bound iron indicate that nonheme iron-sulfur centers are formed.

Though the rate constant was essentially the same for both subfractions, the rate of formation of iron-sulfur centers in the higher molecular-weight subfraction (52-Fig. 2) was more than three-fold the rate observed with the lower molecular-weight subfraction (74-Fig. 3). Furthermore, though the total number

TABLE 1

IRON-BINDING EFFICIENCIES, EXTINCTION COEFFICIENTS AND

SULFIDE-IRON RATIOS FOR INCUBATED SUBFRACTIONS FROM SEPHADEX G-100

Fraction	Fe, $\dfrac{\text{nmoles}}{\text{mg protein}}$	ε_M, $M^{-1}cm^{-1}$	$\dfrac{\text{mole } S^=}{\text{mole Fe}}$
52	37	6100	1.1
58	110	5800	1.2
74	52	5000	0.9
78	210	5200	1.0

of centers bound was greater in the high molecular-weight subfraction, there were, in fact, fewer centers bound per milligram of protein in this fraction (Table 1). The protein concentrations of subfractions 52 and 74 as determined by biuret analyses were 9.0 mg ml^{-1} and 2.0 mg ml^{-1}, respectively. In Table 1 the limiting values of bound iron per milligram of protein are listed for subfractions 52 and 74 and for two additional subfractions as well. The relative values indicate that there are at least two proteins present in the extract that bind iron-sulfur centers. The ratio of "labile" sulfur to iron recovered in all of the fractions was close to 1.0 supporting this conclusion. It was shown previously that crystalline rhodanese does not bind iron-sulfur centers under the same experimental conditions.

The features of the spectrum shown in Fig. 4 were typical of those observed for the various incubated subfractions although the percentage of iron-sulfur protein varied giving differences in the ratio of absorbances at 280 nm and 410 nm. Since 410 nm was near a maximum in the visible spectrum, it proved to be a convenient wavelength for measuring the increased absorbance resulting from new iron-sulfur centers. Some of the subfractions particularly those of the higher molecular-weight component exhibited absorbance at 410 nm prior to incubation and a low iron content. These values did not exceed 20% of the total following incubation in subfraction 52 or 10% in subfraction 74. The average extinction coefficient for the higher molecular-weight component was near 6000 M^{-1}cm^{-1} per mole of bound iron. The average value was near 5000 M^{-1}cm^{-1} per mole of bound iron for the lower molecular-weight component (Table 1).

DISCUSSION

In an early investigation into the origin of the "labile" sulfur atoms,

536

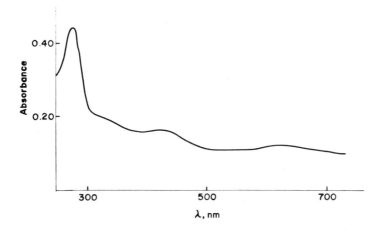

Fig. 4. The absorption spectrum of a subfraction of the lower molecular-weight component following the binding of iron-sulfur centers. Subfraction 78 was incubated for 15 hrs. with 28 mM thiosulfate, 0.46 mM iron ions, and 0.18 M mercaptoethanol and chromatographed on Sephadex G-25.

Suzuki and Kimura[40] showed that a compound with the properties of a nonheme iron protein could be formed by incubation of bovine serum albumin with iron ions and mercaptoethanol. Lovenberg and McCarthy[41] confirmed this observation but these authors also demonstrated that the inorganic sulfide of the iron protein did not arise from cysteinyl residues and that it was necessary to add a source of sulfide to the reaction mixtures. Subsequently, Finnazzi-Agrò and coworkers[6] reported that rhodanese, in the presence of iron ions, mercaptoethanol, and the substrate thiosulfate, could aid in the reconstitution of ferredoxin from apoferredoxin as judged from increasing absorbance at 422 nm. Using similar criteria, Taniguchi and Kimura[7] reported that adrenal ferredoxin could be regenerated from its apoprotein in the presence of the enzyme, β-mercaptopyruvate sulfurtransferase together with iron ions, a reducing agent, and the substrate, β-mercaptopyruvate. Since both aproproteins can be regenerated to their active forms under nonenzymic conditions, the question is raised as to whether the sulfur-transferases serve in a direct donor capacity in these systems or only to supply the necessary sulfide ions from their respective substrates.

Our studies indicate that in soluble mitochondrial extracts containing endogenous rhodanese at least two proteins bind iron-sulfur centers. The nearly constant specific activity observed in our experiments over a large part of the lower mulecular-weight component suggests a complex may be formed between the enzyme and the iron-binding protein. Complex formation with an iron-binding protein in the higher molecular-weight component is also suggested by the observation of a small amount of rhodanese activity associated with this fraction. Bonomi and coworkers[10] have reported that the interaction of rhodanese with succinate dehydrobenase greatly reduces the catalytic activity of the enzyme with respect to thiocyanate formation. In our system, a similar interaction with an iron-binding protein in the higher molecular-weight component could account for the presence of the rhodanese-like band visualized by SDS-gel electrophoresis.[9]

The variation in rates observed in our experiments for formation of new iron-sulfur centers in the two molecular-weight components allows the conclusion that sulfide production is not the rate-limiting step in this process.

The conditions used for partial purification of rhodanese in the mito-chondrial extract prior to molecular exclusion chromatography are those expected to release iron-sulfur centers from proteins (precipitation by high salt con-centrations at acidic pH). Examination of the supernatant from this purifica-tion step by atomic absorption spectrophotometry showed that it contained a large quantity of iron (more than 2000 nanomoles). This finding could explain the fact that mainly apoprotein is present in the rhodanese-containing precipi-tate. Similar purification steps are used in the isolation of the pure enzyme from whole liver extracts. Successful crystallization of the enzyme in the latter preparations appears to be related to their content of nonheme iron protein.[9]

Solubilization of nonheme iron proteins by sonication of mitochondrial suspensions has been reported by different groups of investigators.[42-44] Recently, Ruzicka and Beinert have identified one of these, a high-potential iron protein, as the enzyme aconitase.[45] A soluble iron-protein of the adrenodoxin type has also been reported in mitochondrial extracts by Onashi and Omura.[46] The identity and characteristics of the iron-binding proteins observed in the present study are currently under investigation.

ACKNOWLEDGMENT

This work was supported by a grant (PCM 75-23299) from the National Science Foundation.

REFERENCES

1. Burton, K.W.C. and Machmer, P. (1968) in Inorganic Sulphur Chemistry, Nickless, G., ed., Chap. 10, Elsevier Publishing Company, Amsterdam, pp. 335-366.
2. Roy, A.B. and Trudinger, P.A. (1970) The Biochemistry of Inorganic Com-pounds of Sulphur, Cambridge University Press, London, pp. x-xiii.

538

3. Siegel, L.M. (1975) Metab. Pathways, 3d Ed., 7, 217-286.
4. Schneider, J.F. and Westley, J. (1969) J. Biol. Chem., 244, 5735-5744.
5. Westley, J. (1977) in Bioorg. Chem., van Tamelen, E.E., ed., Vol. 1, Chap. 15, Academic Press, Inc., New York, pp. 371-390.
6. Finazzi-Agró, A., et al. (1971) FEBS Lett., 16, 172-174.
7. Taniguchi, T. and Kimura, T. (1974) Biochim. Biophys. Acta, 364, 284-295.
8. Pagani, S., et al. (1975) FEBS Lett., 51, 112-115.
9. Volini, M., Craven, D. and Ogata, K. (1977) Biochem. Biophys. Res. Commun., 79, 890-896.
10. Bonomi, F., et al. (1977) Eur. J. Biochem., 72, 17-24.
11. Massey, V. (1973) in Iron-Sulfur Proteins, Lovenberg, W., ed., Vol. I, Chap. 10, Academic Press, New York, pp. 301-360.
12. Finazzi-Agró, A., et al. (1976) Biochem. Biophys. Res. Commun., 68, 553-560.
13. Tomati, U., et al. (1976) Phytochemistry, 15, 597-598.
14. Lang, K. (1933) Biochem. Z., 259, 243-256.
15. Sorbo, B. (1975) Metab. Pathways, 3d Ed., 7, 433-456.
16. Conn, E.E. (1969) J. Agr. Food Chem., 17, 519-526.
17. Szczepkowski, T.W. and Wood, J.L. (1967) Biochim. Biophys. Acta, 139, 469-478.
18. Koj, A. and Frendo, J. (1962) Acta Biochim. Polon., 9, 373-379.
19. Borysiewicz, J., Frendo, J. and Koj, A. (1962) Folia Biol., 10, 169-177.
20. Koj, A., Frendo, J. and Borysiewicz, J. (1964) Acta Med. Polon., 5, 109-115.
21. Wong, T.W., et al. (1970) Biochem., 9, 2376-2386.
22. Wong, T.W., Harris, M.A. and Jankowicz, C.A. (1974) Biochem., 13, 2805-2812.
23. Wong, T.W., Harris, M.A. and Morris, H.P. (1975) Biochem. Biophys. Res. Commun., 65, 1137-1145.
24. Abrell, J.W., Kaufman, E.E. and Lipsett, M.N. (1971) J. Biol. Chem., 246, 294-301.
25. Harris, C.L., Kerns, F.T. and St. Clair, W. (1975) Cancer Res., 35, 3608-3610.
26. Harris, C.L. (1978) Nucleic Acids Res., 5, 599-613.
27. Schneider, J.F. and Westley, J. (1963) J. Biol. Chem., 238, PC3516-PC3517.
28. Koj, A. (1968) Acta Biochim. Polon., 15, 161-169.
29. Sido, B. and Koj, A. (1972) Acta Biologica Crac., 15, 97-103.
30. Koj, A., Frendo, J. and Janik, Z. (1967) Biochem. J., 103, 791-795.
31. Wang, S.-F. and Volini, M. (1968) J. Biol. Chem., 243, 5465-5470.
32. Schievelbein, H., Baumeister, R. and Vogel, R. (1969) Naturwissenschaften, 56, 416-417.
33. Volini, M., Craven, D. and Ogata, K. (1978) J. Biol. Chem., 253, 7591-7594.
34. Chen, J. and Mortenson, L. (1977) Anal. Biochem., 79, 157-165.
35. Zamenhof, S. (1955) in Methods in Enzymology, Colowick, S.P. and Kaplan, N.O., eds., Vol. 3, Academic Press, New York, p. 702.
36. Wang, S.-F. and Volini, M. (1973) J. Biol. Chem., 248, 7376-7385.
37. Swanson, M. (1955) in Methods in Enzymology, Colowick, S.P. and Kaplan, N.O., eds., Vol. 2, Academic Press, New York, pp. 541-543.
38. Breslow, J.L. and Sloan, H.R. (1972) in Methods in Enzymology, Ginsburg, V., ed., Vol. 28, Academic Press, New York, pp. 880-884.
39. Koj, A., Frendo, J. and Wojtczak, L. (1975) FEBS Lett., 57, 42-46.
40. Suzuki, K. and Kimura, T. (1967) Biochem. Biophys. Res. Commun., 28, 514-518.
41. Lovenberg, W. and McCarthy, K. (1968) Biochem. Biophys. Res. Commun., 30, 453-458.
42. Ruzicka, F. and Beinert, H. (1974) Biochem. Biophys. Res. Commun., 58, 555-563.
43. Albracht, A. and Heidrich, H. (1975) Biochim. Biophys. Acta, 347, 231-236.
44. Backström, D., et al. (1973) Biochem. Biophys. Res. Commun., 53, 596-602.
45. Ruzicka, F. and Beinert, H. (1978) J. Biol. Chem., 253, 2514-2517.
46. Onashi, M. and Omura, T. (1978) J. Biochem., 83, 249-260.

Liu/Mamiya/Yasunobu, eds. Frontiers in Protein Chemistry

EFFECTS OF CHEMICAL MODIFICATION OF ANTHOPLEURIN A, A PEPTIDE HEART STIMULANT

ROBERT NEWCOMB, KERRY T. YASUNOBU, DAVID SERIGUCHI[+] AND TED R. NORTON[+]
Department of Biochemistry-Biophysics, University of Hawaii School of Medicine,
Honolulu, Hawaii 96822 and [+]Department of Pharmacology, University of Hawaii
School of Medicine, Honolulu, Hawaii 96822

SUMMARY

Anthropleurin A, a heart stimulant isolated from the sea anemone *Anthropleura xanthogrammica*, shows great potential as a clinically usable heart stimulant, and as a tool to investigate the mechanism of myocardial contractibility. As a prelude to the chemical synthesis, and for purposes of studying structure-function relationships, chemical modifications of Anthopleurin A were performed, and the activity of the derivatives as heart stimulants on the isolated cardiac muscle was investigated. It was concluded that arginine, lysine, histidine and tyrosine modifications did not alter heart stimulant activity. Modification of 1-2 carboxyl groups and reduction of the 3 cystine residues followed by reaction with iodoacetic acid resulted in inactivation of heart stimulant activity. Modification of all the tryptophan residues with 2-nitrophenylsulfenyl chloride yielded a derivative which showed full or partial activity. Trinitrobenzene-sulfonic acid inactivated Anthopleurin A, but reductive alkylation with $NaBH_4$-acetone yielded an active product and thus ruled out the essentiality of the epsilon amino side chains. The results are discussed in terms of possible molecular conformation and essential residues of Anthopleurin A.

Anthopleurin A (AP-A) is one of the three cardiac muscle specific stimulating agents which have been isolated from sea anemones of the genus Anthopleura.[1-3] These peptides have been demonstrated to have pharmacological properties superior to those agents currently in commercial use as heart stimulants.[4,5] The sequences of the Anthopleurin stimulants AP-A[6] and AP-C[*] have been determined, and they show sequence homology with the toxin isolated from the sea anemone *Anemonia sulcata*. This toxin, which has been labeled toxin II,[7] has been shown to stimulate Ca^{++}, and to a lesser degree, Na^+ uptake in cultured embryonic heart cells. The mechanism(s) of action of the Anthopleurin stimulants have yet to be determined, but AP-A does not appear to have identical properties to those reported for toxin II, which also is isolated from sea anemone.[8-10]

---

[*]Unpublished results of Carl D. Bennett, Merck Institute for Therapeutic Research, West Point, Pennsylvania (see Reference 3).

At present, the clinical usefulness of Anthopleurin A may be limited by the supply of anemones and by its antigenicity. The object of the present study was, by the technique of chemical modification, to determine those residues essential for the biological activity of the molecule. The information gained may be used to aid in the synthesis of peptide analogues with similar biological activity. A further object of this study was to interpret the biological activity of AP-A in terms of the conformation of the molecule, and in terms of the sidechain functional groups either essential or nonessential for activity.

MATERIALS AND METHODS

Materials. 2, 4, 6-Trinitrobenzene-sulfonic acid, diethyl pyrocarbonate, phenylglyoxal, o-nitrophenylsulfonyl chloride, quanidine HCl and NaBH$_4$ were purchased from Sigma Chemical Company. Carboxypeptidase Y, 1-ethyl-3(3-dimethyl-aminopropyl), carbodiimide and 5, 5'-dithio-bis(2-nitrobenzoic acid) were from Pierce Chemical Company. Dithiothreitol and 2-hydroxy-5-nitrobenzyl bromide were from Calbiochem. Cyclohexanedione and 3-nitro-L-tyrosine were from Aldrich Chemical Company. Aminopeptidase M was from Rotten, tetranitromethane from K and K Division of I.C.N. Chemicals, and L-taurine from Nutritional Biochemicals. All of the above reagents were used without further purification. Iodoaceta-mide was recrystallized before use.

General Methods. In cases where a large amount of material was not required for quantitation, time courses were first run in order to determine the optimal conditions for modification. The activity was checked using at least two dif-ferent reaction products of a given modification type, with bioassays performed with and without separation of the derivative from the modifying reagents in most cases. A Beckman Model 121-MB analyzer was used for amino acid analysis.[11] AP-A was isolated by previously published procedures.[3] Several preparations of AP-A were used in the study, and varied from 90-99.5% purity. Separation from reagents, unless otherwise indicated, was on 1.5 x 50 cm columns of Sephadex G-25 coarse equilibrated with 0.05% propionic acid. Concentrations for the bio-assay were determined based on amino acid analyses, with hydrolyses, unless otherwise specified, performed for 24 hr in constant boiling 6 N HCl.

Carboxyl Modifications. Carboxyl groups were modified by a variation of the procedure of Hoare and Koshland.[12] For modification of the native molecule, 0.5 ml of 0.4 M 1-ethyl-3(3-dimethyl aminopropyl) carbodiimide in water was added to 0.15-5 mg of AP-A. The solution was stirred for 10 minutes and then 0.5 ml to 2 M glycine ethyl ester in water, or 0.5 ml of water saturated with taurine, was added. The pH was then maintained between 4.5 and 5.0 for 2.5 hr. For modification in the presence of guanidine hydrochloride (GuHCl), the above

solutions were made 6 M in GuHCl at pH 5.0, and care was taken not to let the
pH fall below 4.5 during the initial stage of the reaction. The derivative was
passed through Sephadex as described above, lyophilized, and then passed again
through a freshly prepared column. The extent of modification was quantitated
by amino acid analysis with nonhydrolyzed samples being applied to the analyzer
to monitor for free taurine or glycine ethyl ester (not observed). Sequence
analysis was performed in the Beckman 890 protein sequencer as described in an
earlier paper.[13] A sample of AP-A which had been S-carbamidomethylated and then
treated in an identical fashion served as a standard in determining the number
of carboxyl groups modified. This derivative was stored in 0.05% proprionic
acid until tested for activity (storage in dilute acid has no effect on the
activity of AP-A).

In order to show that the inactivation produced by the above treatment was
due to carboxyl modification, several controls were performed. Treatment of
AP-A as described above, but without added free amine, resulted in an active
product. Separation from reagents on a Sephadex G-25 column equilibrated with
0.1 M NH$_2$OH – HCl pH 7, followed by incubation at room temperature for 8 hours
and storage in frozen state for several days in the same NH$_2$OH solution resulted
in an inactive product. Bioassay immediately after reaction, without separation (but with dilution from 10^{-4} M AP-A to 10^{-9} M) from reagents, with and
without incubation for 1/2 hour in 0.1 M NH$_2$OH pH 7, resulted in an inactive
product. Incubation in 6 M GuHCl, pH 5 for 2.5 hours followed by testing without separation (but with dilution) from the GuHCl, resulted in an active product.

Arginine Modifications. Arginine was selectively modified according to the
procedure of Takahashi.[14] For determining the time course of the reaction, and
for initial testing of biological activity, 0.6 ml of a solution containing
0.75% phenylglyoxal in 0.2 N N-methylmorpholine-acetic acid, pH 8.0 was added
to 1.1 mg of AP-A and the mixture was then stirred at room temperature. After
20, 60 and 95 minutes of reaction, 0.2 ml aliquots were removed for chromatography on Sephadex G-25 as described above. The derivative was stored frozen in
0.05% proponic acid until tested for activity, with no regeneration of arginine
being noted following several weeks of storage in this manner.

The activity of phenylglyoxal modified AP-A was also checked immediately
after reaction under the above conditions for 80 min, with no separation from
reagents, but with the 10^5-fold dilution required for the assay. The extent of
arginine modified for this bioassay was quantitated by amino acid analysis.

Tyrosine Modifications. The procedure of Sokolovsky et al.[15] was used to
selectively modify the tyrosine residue of AP-A. For determining the time course
of reaction, 100 µl of a solution of AP-A at 0.5 mg/100 µl in H$_2$O was added to

200 µl of 0.1 M Tris pH 8.0. To this was added 1 µl of tetranitromethane, and
the mixture was stirred at room temperature. Aliquots of 50 µl were removed at
various times and placed in hydrolysis tubes containing a large excess (∼20 mg)
of phenol in 6 N HCl. The extent of reaction was then quantitated by amino
acid analysis. The biological activity of the derivative was tested on a sample
which had been treated as above for 85 min and then separated from reagents as
described above. The derivative was also tested for activity immediately following 90 min of the above treatment, with no separation from reagents (but with
dilution), and with the extent of reaction being quantitated as described above.

Tryptophan modifications. The procedures of Scoffone et al.[16] and Koshland
et al.[17] were used to selectively modify the tryptophan residues in AP-A. Modification of 3.1 of 3 tryptophan residues with o-nitrophenylsulfenyl chloride
(o-NPS-Cl) was accomplished as follows: AP-A (0.3 mg) was dissolved in 400 µl
of a 17 mg/ml solution of o-NPS-Cl in acetic acid was added, slowly, and with
stirring. The stirring was then continued at room temperature for 2 hr. The
derivative was separated from reagents on Sephadex G-25, lyophilized, and then
passed again through a column of Sephadex G-25. The absorbance in 80% acetic
acid at 360 nm was used to quantitate the extent of reaction, utilizing an
extinction coefficient of 4,000 M^{-1} cm^{-1},[15] with concentrations being determined
by amino acid analysis. The time course of the reaction was followed by removing aliquots and treating them as described above. The reaction was over after
30 min. The derivative was stored frozen in 2% propionic acid until tested for
activity. The bioassay was also performed on this derivative immediately following reaction for 1 and 2 hrs with o-NPS-Cl.

For modification of 1 to 2 of 3 tryptophan residues with 2-hydroxy-5-nitro-
benzylbromide (φ'Br), 20 µl of a 0.015 M solution of φ'Br in acetone was added
to 0.5 mg AP-A in 0.4 ml 1% propionic acid. The derivative was twice passed
through Sephadex and lyophilized, then dissolved in H_2O. An aliquot of this was
removed and brought to 0.25 N in KOH for determining the extent of modification.
A molar absorbancy index at 410 nm of 19,200 M^{-1} cm^{-1}[18] was used for quantitation, with the concentration of AP-A being determined by amino acid analysis.
The derivative was stored frozen in water for several days before it was tested
for biological activity.

Amino group modifications. Procedures described by Means and Feeney,[19] and
Habeeb,[20] were used to selectively modify the free amino groups of AP-A. For
reductive isopropylation, 1 mg of AP-A was dissolved in 1 ml in borate buffer
(10% v/v in acetone, 0.15 M, pH 9.0). One mg of $NaBH_4$ was added and the mixture
was incubated overnight at 0°C, with 1 mg of $NaBH_4$ and 50 µl acetone being added
at hourly intervals for the first four hours of incubation. The reaction was

quenched by dilution with 3-fold v/v excess of 0.5% propionic acid, and the
derivative was stored in the resultant solution several days before testing.
The extent of reaction was quantitated by amino acid analysis.

For determining the time course of the reaction of AP-A with trinitrobenzene-
sulfonic acid (TNBS) at 40°, 0.7 mg of AP-A was dissolved in 1.4 ml of a solution containing 2% $NaHCO_3$ and 0.1% TNBS, and then incubated in the dark. At
various times, 200 µl aliquots were added to 200 µl of 6 N HCl, and the extent
of reaction was quantitated by amino group analysis and by an E_{340} of 1.4 x 10^4
M^{-1} cm^{-1}.[20] A similar procedure was used for determining the extent of reaction
at 25°. However, the extent of modification in this case was quantitated only
by amino acid analysis.

For the bioassay, AP-A was modified as described above to the extent of 2 (of
2) lysine residues per mole at 40° and 25°, and to the extent of 1 lysine residue per mole of AP-A at 25°. These derivatives were tested for activity without
separation (but with dilution) from reagents, with the extent of modification
being determined by amino acid analysis. As a control, a sample of AP-A was
heated to 40° for 3 hrs in 2% $NaHCO_3$ and then tested for activity.

Histidine modifications. Iodoacetamide[21] and diethyl pyrocarbonate (EFA)[22]
were used to modify the histidine residues present in AP-A. For modification
with iodoacetamide, 0.5 mg of AP-A was incubated in the dark at room temperature
in a solution containing 200 µl, 0.2 M potassium phosphate buffer, pH 7.2, 1 M
iodoacetamide and 10 µl of toluene. After 65 days of reaction, 100 µl of the
reaction mixture was removed. The derivative was desalted above, and stored for
several days frozen in 0.05% propionic acid, with the extent of modification
being quantitated by amino acid analysis.

For modification with EFA, 2 µl of EFA was added to 1 ml of a 2 x 10^{-5} M
solution of AP-A in potassium phosphate buffer, 0.1 M, pH 6.2. The extent of
reaction was followed at 242 nm assuming an E_{242} of 3200 M^{-1} cm^{-1}, with concentrations being determined by amino acid analysis. Following complete modification (25 min), aliquots were either diluted for the bioassay into 150 ml of
physiological saline pH 7 containing 2 µl of EFA, or were first incubated in
0.5 M NH_2OH pH 7.0 for 30 min before being assayed for activity.

Reduction of the disulfide bonds and S-β-carboxymethyl-cysteinyl derivative
of AP-A. For the experiment, about 3 mg of AP-A in 1.0 ml of 0.1 M Tris buffer,
pH 8.6, was reduced with a slight excess of 2-mercaptoethanol and was flushed
with nitrogen for 4 hours. A 5% molar excess of recrystallized iodoacetic acid
was added over the amount of mercaptoethanol present and the reaction was
allowed to proceed for 20 minutes.[23] The reaction mixture was passed through a
Sephadex G-25 column and then the proteins were detected from the absorbance at
280 nm, pooled and lyophilized.

Enzymatic digests. For digestion with carboxypeptidase Y, 0.5 mg of AP-A was incubated with either 0.05 or 0.005 mg of Cpase Y in 0.3 ml pyridine-acetate 0.1 N pH 6.0 at room temperature. At various times, 50 μl aliquots were removed from the digest and from a blank containing only Cpase Y, and were then applied to the amino acid analyzer at pH 2.3 without hydrolysis.

For digestion with the aminopeptidase M, 1.1 mg of aminopeptidase M was dissolved in 0.5 ml 0.01 M potassium phosphate buffer, pH 7.2 and 0.4 ml of this was added to 0.2 ml of a solution containing AP-A at 0.5 mg/100 μl in water. The digest was incubated at 37° for 24 hours, with aliquots being removed at various times for amino acid analysis. The digest was then applied to Sephadex G-24 in 5% propionic acid (the elution of the aminopeptidase is severely retarded under these conditions, and it may easily be separated from the peptide). The recovered AP-A was lyophilized, and the digestion and separation were repeated. Amino acid analysis and NH_2-terminal analysis by the dansyl chloride method[24] were used to determine the effect of the aminopeptidase on AP-A.

Pharmacological assay. The spontaneously beating right atria from rat (200-300 g) of either sex were used for test of cardiotonic activity. The whole atria were dissected and mounted vertically in a 50 ml tissue bath containing Krebs-Ringer bicarbonate solution of the following composition (mM): NaCl, 120; KCl, 4.8; $MgSO_4$, 7; H_2O, 1.3: KH_2PO_4, 1.2; $CaCl_2$, 1.2; glucose, 5.8; and $NaHCO_3$, 25.2; pH 7.4, bubbled with a gas mixture of 95% O_2 and 5% CO_2 and maintained at 30°C, pH 7.4. One end of the tissue was connected to the force transducer by a silk ligature and the other end was secured to the portion of the Plexiglas tissue holder containing the platinum stimulating electrode. Tension was recorded isometrically through a force displacement transducer (Grass FT. 03) and displayed on Grass 7 polygraph.

In the bioassay described above, unmodified AP-A will give a cardiotonic effect of concentrations between 3×10^{-10} and 3×10^{-9} M. To draw worthwhile conclusions from the assay, it was thus necessary that modifications be nearly quantitative (97-100% modified). A modification was considered not to observably affect the cardiotonic activity of AP-A, if the extent of modification was between 97 and 100%, and if the derivative was active between 3×10^{-10} and 3×10^{-9} M. A modification was considered to effect the cardiotonic activity of AP-A if no activity was observed below 3×10^{-8} M. Two derivatives were active between 3×10^{-10} and 3×10^{-9} and 3×10^{-8} M. Conclusions regarding these specific cases are dealt with in the Results section.

RESULTS

Carboxyl group modifications. Possible side reactions in the carbodiimide mediated coupling of free amines to protein carboxyl groups include possible intermolecular crosslinking with lysine amino groups,[25] and the formation of O-arylureas with tyrosine residues.[26] Treatment with carbodiimide and free amine, as described in the Methods section, completely inactivated AP-A, and it was then necessary to investigate these possible side reactions. Incubation of AP-A with carbodiimide, but without free amine, does not inactivate AP-A, and so intermolecular crosslinking can be ruled as an inactivating side reaction. Incubation of taurine coupled AP-A in NH_2OH does not restore activity, and thus the tyrosine side reaction can be ruled out as the sole inactivating factor.

Under conditions reported to give quantitative modification of free carboxyl groups, treatment of AP-A with carbodiimide and nucleophile (taurine) resulted in modification of 2.4 of 3 carboxyl groups. For verification of this, a sample of AP-A was first carbamidomethylated and then coupled to taurine. This treatment yielded a value of 3 of 3 carboxyl groups modified. The inactivation of AP-A by carbodiimide mediated coupling of carboxyl groups to taurine may thus be said due to modification of either 1 or 2 carboxyl groups per mole. If the inactivation was due to modification of the third carboxyl group, then activity would have been observed at the higher concentrations of AP-A in the bioassay.

Arginine modifications. Nearly quantitative reaction of the arginine residue of AP-A with phenylglyoxal results in a derivative which still possesses full biological activity. Due to the bulky nature of the adduct formed on reaction of phenylglyoxal with guanidine groups, it is unlikely that the arginine residue of AP-A is either directly essential for activity, or is in a conformationally sensitive region of the molecule.

Tyrosine modifications. Quantitative reaction of the tyrosine residue of AP-A with tetranitromethane produced a derivative retaining full activity. The major product, identified as 3-nitrotyrosine by amino acid analysis, has a lower pK than unmodified tyrosine,[15] but the adduct formed is not particularly bulky. Thus, it is unlikely that the tyrosine residue of AP-A is essential in a charge sense, but other interactions related to activity may have gone undetected.

Tryptophan modifications. Less than quantitative modification of the 3 tryptophan residues of AP-A with either o-nitrophenylsulfenyl chloride (o-NPS-Cl) or 2-hydroxy-5-nitrobenzyl bromide (φ'Br) resulted in an active product. Modification with o-NPS-Cl to the extent of 3.0 residues per mole resulted in a product which displayed either full or partial activity.

Amino group modifications. Nearly quantitative modification of lysine amino groups with $NaBH_4$ and acetone, and with TNBS resulted in active and inactive products, respectively. These results might be explained by the more bulky nature of the adduct formed upon reaction with TNBS, and indicate that one or both lysine residues may be in conformationally sensitive regions of the molecule, but that lysine amino groups are not directly essential for the activity of AP-A

Histidine modifications. Treatment with iodoacetamide and ethoxyformic anhydride resulted in modification of both the histidine residues of AP-A. In the case of iodoacetamide, reaction of one residue of lysine per mole of AP-A was also observed. This product was active at 3×10^{-9} or 10^{-8} M (pure AP-A has an ED-50 *ca.* 2×10^{-9} M). However, due to quantitative modification of histidine observed with iodoacetamide, and also to the full activity observed after modification of 2 histidine residues per mole with EFA, it is doubtful that either of the histidine residues of AP-A are directly essential for the activity of AP-A. The two histidine residues of AP-A do not appear to show widely differing reactivities to EFA as shown in Fig. 1.

Reduction of disulfide bonds. Amino acid analyses of the S-β-carboxymethyl-cysteine derivative showed the presence of 6 carboxymethylcysteine residues per mole of AP-A. The samples when assayed on the heart showed no heart stimulant activity at a concentration of 10^{-7} M.

Effect of exopeptidase and summary of the chemical modifications. Amino-peptidase M failed to liberate detectable amounts of amino acids from AP-A, despite the drastic conditions used. Carboxypeptidase Y did release the COOH terminal glutamine and lysine in about 10% yield after prolonged (36 hrs)

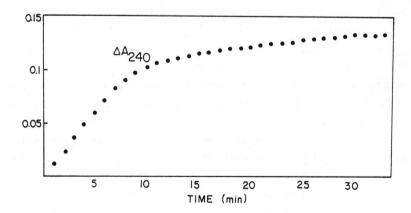

Fig. 1. Reaction of AP-A with ethoxyformic acid anhydride. About 2 µl of EFA was added to 1 ml of 2×10^{-5} M solution of the peptide in potassium phosphate buffer, 0.1 M, pH 6.2. The extent of the reaction was followed at 252 nm assuming an E_{242} of 3200 M^{-1} cm^{-1}.

TABLE 1

CHEMICAL MODIFICATION OF ANTHOPLEURIN A

Amino		Reagent	Residues present	Residues altered	Activity[a]
1. Histidine	a)	Iodoacetamide	2	2.0	+[d]
	b)	Ethoxyformic acid	2	2.0[c]	+
2. Lysine	a)	Trinitrobenzene-sulfonic acid	2	2.0	−
	b)	Reduction alkylation	2	1.95	+
3. Arginine	a)	Phenylgloxal	1	0.95	+
4. Tyrosine	a)	Tetranitromethane	1	1.0	+
5. Tryptophan	a)	2-Hydroxy-5-nitro-benzyl-Bromide	3	1-2[c]	+
	b)	o-Nps-Cl	3	1-2[c]	+[d]
	c)	o-Nps-Cl	3	3.0	+
6. COOH	a)	1-Ethyl-3-(3-dimethyl-aminopropyl) carbodiimide	3	2.4[b]	−
7. Cystine	a)	Reduction-iodoacetic acid	3	3.0	−

[a]Activity (+) defined as positive inotropic effect between 3×10^{-10} and 3×10^{-9} M. Inactivity (−) defined as no positive inotropic effect below 3×10^{-8} M.

[b]Determined by analyses of the acid hydrolyzates on the automatic amino acid analyzer.

[c]Determined spectrophotometrically. See Experimental section.

[d]Active at 10^{-8} M after quantitative modification.

incubation, but this yield was not sufficient for study of the structure activity relationships. Carboxypeptidase Y also released glycine from the glycine ethyl ester derivative of the native AP-A in about a 2:1 ratio over the lysine released. However, the release was at the same slow rate. The results obtained for all the chemical modification experiments of AP-A are summarized in Table 1.

DISCUSSION

Anthopleurin A is a very interesting peptide heart stimulant which is extremely potent at 10^{-9} M concentration, and at this concentration it has few side effects. However, the mode of action of the heart stimulant remains unknown. The molecular approach towards ellucidating the structure-function relationship of the heart stimulant requires that the structure of the heart stimulant be ellucidate. Tanaka et al.[6] have determined the amino acid sequence of the peptide and shown it to be a single polypeptide chain peptide which consists of 49 amino acids. The heart stimulant was also shown to contain three cysteine residues but the pairing of the component cysteine residues were not determined.

Studies in progress (unpublished results of Tanaka, Ishizaki, McKay and Yasunobu) show that cysteine 6 is -SS- bonded to cysteine 36. The presence of the -SS-bond causes the molecule to form a loop like structure, and it follows that cysteine-4 must be bonded in cysteine-46 or 47, and also that cysteine-29 is bonded to either cysteine-46 or 47 which inserts two additional loops. Thus, the Anthopleurin A has a complex conformation despite its rather low molecular weight.

Besides a knowledge of the sequence of the heart stimulant, it is desirable to know what amino acid side chains are essential for activity. The present investigation has demonstrated the lysine residues-37 and 48 are unessential for activity. However, it is interesting to note that the reagent trinitrobenzene-sulfonic acid which reacted with both of the ε-NH$_2$ groups of lysine residues resulted in an inactive derivative. However, reductive isopropylation of both lysine residues led to an active derivative. Histidine residues-34 and 39 are probably nonessential since ethoxyformic acid anhydride fully modified the two residues and yet the derivative was fully active. Iodoacetamide quantitatively modified both histidine residues and one of two lysine residues, yet the derivative maintained sufficient activity such that histidine could not be considered essential for activity. The partial loss of activity is interesting, and could be due either to the histidine modification, or to lysine modification. A single arginine residue is present at position 14 from the NH$_2$-terminal position of AP-A and modification of it with phenylglyoxal yielded an active derivative showing the unessential nature of this residue. Therefore, despite the basic isoelectric point of AP-A, the basic sidechains are unessential for activity. Substitution of tyrosine residue-25 with tetranitromethane did not inactivate AP-A.

Substitution of 2-3 of the three carboxylic groups (Asp at positions 7 and 9, and the COOH-terminal Gln) resulted in the synthesis of an inactive derivative. Activity was lost with either the small highly charged taurine, or with the relatively bulky uncharged glycine ethyl ester. Activity was lost with or without guanidine in the reaction mixture; however, in the presence of guani-dine, two treatments with carbodiimide and nucleophile were often required to affect fully loss of activity. Sequence studies with the taurine derivative of native AP-A are in progress to more fully determine the extent of modification at positions 7, 9 and the COOH terminus.

When 2-nitrophenylsulfenyl chloride was used to modify the 3 tryptophan residues, full or partial activity was observed, which would appear to rule out an essential role for the tryptophan residues. The reduction of the 3 cysteine bonds and carboxymethylation of cysteine residues 4, 6, 29, 36, 46 and 47 resulted in an inactive derivative. Aminopeptidase M, despite being used at rather high concentrations, failed to release any amino acids from AP-A. After prolonged incubation, Cpase Y did release glutamine and lysine in 10% yield.

The three-dimensional structure and mode of action of AP-A must be ellucidated before definite conclusions concerning the role of the essential residues and of the aspartate carboxyls of AP-A have normal pK values, while one of the aspartate carboxyls has an unusually low pK around 2. This carboxyl may be involved in an intramolecular salt linkage, possibly with one of the lysine ε-amino groups. The inactivating effects of carboxyl modification may be due either to conformational side effects (of breaking the salt linkage?) or to blockage of a site necessary for interaction of AP-A with metal ions or heart macromolecules. The fact that the small negatively charged taurine adduct is able to inactivate AP-A argues that its activity is sensitive to very small structural changes. This is supported by the opposite effects of lysine modification by reductive alkylation or with trinitrobenzenesulfonic acid. Here, the bulky adduct destroys activity while the less bulky group does not. All of the aspartate carboxyls and the lysine ε-amino groups are located within several residues of the 3 cystine bonds of AP-A. The AP-A molecule appears to have a highly ordered secondary structure,[28] quite possibly consisting of some β-bonds, and the cysteine bonds have been shown by lasar raman techniques[27] to have an ordered gauche-gauche-gauche conformation. It is perhaps not surprising that chemical modification in the region of the cysteine bonds, and also reductive carboxymethylation of the cysteine residues, destroys the activity of AP-A. Further studies are needed to determine the role of carboxyl groups in this interesting and potentially useful molecule.

ACKNOWLEDGMENTS

The authors wish to thank Ms. Priscilla Searl for some of the bioassay experiments. Thanks are also due to Dr. M. Tanaka for providing the S-β-carboxymethyl-cysteine derivative of AP-A and to Neil Reimer for help with the sequencing work.

This research was supported in part by Grant 76-752 from the American Heart Association, Hawaii Heart Association, Grants GM 22556 and HL 14991 from the National Institutes of Health, and Grants GB 43448 and SER77-06923 from the National Science Foundation.

REFERENCES

1. Norton, T.R., Shibata, S., et al. (1976) J. Pharm. Sci., 65, 1368.
2. Shibata, S., Dunn, D.F., et al. (1974) J. Pharm. Sci., 63, 1332.
3. Norton, T.R., Shibata, S. and Kashiwagi, M. (1978) Proceedings of 5th Drugs-Food From the Sea Conference (Norman, Oklahoma).

550

4. Shibata, S., Norton, T.R., et al. (1976) J. Pharmacol. Exp. and Ther., 199, 298.
5. Scriabine, A., Van Arman, C.G., et al. (1977) Fed. Proc., 36, 973.
6. Tanaka, M., Haniu, M., et al. (1977) Biochemistry, 16, 204-208.
7. Wunderer, G., Macheleidt, W. and Wachter, E. (1976) Hoppe-Seyler's Z. Physiol. Chem., 357, 239-240.
8. Shimizu, T., Iwamura, N., et al. (1977) Fed. Proc., 36, 415.
9. Shibata, S., Izumi, T., et al. (1978) J. Pharmacol. Exp. and Ther., 204.
10. Shibata, S., Seriguchi, D.G., et al. (1977) Fed. Proc., 36, 946.
11. Spackman, D.H., Moore, S. and Stein, W.H. (1958) Anal. Chem., 30, 1190.
12. Hoare, D.G. and Koshland, D.E. (1967) J. Biol. Chem., 242, 2447.
13. Tanaka, M., Haniu, M., et al. (1977) Biochemistry, 11, 3525.
14. Takahashi, K. (1958) J. Biol. Chem., 243, 6171.
15. Sokolovsky, M., Riordan, J.F. and Vallee, B.L. (1966) Biochemistry, 5.
16. Scoffone, E., Fontana, A. and Rocchi, R. (1968) Biochemistry, 7, 971.
17. Koshland, D.E., Karkhanis, Y.D. and Latham, H.G. (1964) J. Am. Chem. Soc., 86, 1448.
18. Horton, H.R. and Koshland, D.E. (1967) Methods Enzymol., 11, 556.
19. Means, G.E. and Feeney, R.E. (1968) Biochemistry, 7, 2192.
20. Habeeb, A.F.S.A. (1966) Anal. Biochem., 14, 328.
21. Hartzell, C.R., Hardman, K.D., et al. (1967) J. Biol. Chem., 242, 47.
22. Rosen, C.G. and Fedorcsak, I. (1966) Biochim. Biophys. Acta, 130, 401.
23. Crestfield, A.M., Moore, S. and Stein, W.H. (1963) J. Biol. Chem., 238, 618.
24. Woods, K.R. and Wang, K.-T. (1967) Biochim. Biophys. Acta, 133, 369.
25. Kurzur, F. and Douraghi-Zadeh, K. (1967) Chem. Rev., 67, 107.
26. Carraway, K.L. and Triplett, R.B. (1970) Biochim. Biophys. Acta, 200, 564.
27. Norton, R.S. and Norton, T.R. (1979) J. Biol. Chem., in press.
28. Ishizaki, H., Yasunobu, K.-T., et al. (1979) J. Biol. Chem., 254, 9651.

ISOLATION OF PURE BOVINE PLASMA AMINE OXIDASE A AND THE EFFECT OF EVOLUTION
ON THE CHEMICAL AND ENZYMATIC PROPERTIES OF CU-AMINE OXIDASES

KAZUHO WATANABE, HIROYUKI ISHIZAKI, VALERIE FUJITA AND KERRY YASUNOBU
Department of Biochemistry and Biophysics, University of Hawaii School of
Medicine, Honolulu, Hawaii 96822

ABSTRACT

Bovine plasma has been shown to contain more than one form of the enzyme. A
procedure is described for isolating two forms of the enzyme in pure forms. They
have been designated enzyme-A and enzyme-B. It is suggested that the two forms
of the enzyme present are due to the mixing of the blood plasma from two different species of steer. The effect of evolution on the structure and enzymatic
activities of the bovine, mold and plant Cu-amine oxidase were investigated.
The protomer of the animal and plant enzyme consists of two subunits which are
linked by disulfide bonds. The *Aspergillus niger* protomer is also probably
similar to the animal and plant enzymes but the two subunits are not linked by
disulfide bonds. The preliminary data suggests that the NH_2-group of the N-
terminal amino acid is blocked and the site remains as the potential location
of the covalently bonded organic cofactor. The physiocochemical properties of
the Cu-amine oxidases investigated to date suggest that all of the Cu-amine
oxidases investigated here have evolved from the same common ancestor. However,
the Cu-amine oxidases gene is rapidly undergoing mutational events as evidenced
from the lack of the formation of insoluble antigen-antibody complexes in the
tests with the mold and plant enzymes. Enzyme inhibition by anti-bovine enzyme-B
sera indicated the formation of enzyme-inhibitor complexes in the case of the
bovine and A. *niger* Cu-amine oxidases but no such interaction of inhibition was
observed with the plant enzyme. The changes in the structures of the Cu-amine
oxidases during the process of evolution are compared with the changes observed
in the case of the FAD-amine oxidases.

INTRODUCTION

Two basic forms of amine oxidases are found in nature. These types include
the FAD-amine oxidases and the Cu-amine oxidases.[1] Highly purified Cu-amine
oxidases have been isolated from animal plasma,[2-4] animal kidney,[5,6] pea seedling[7] and the mold.[8,9] Due to the availability of some of the Cu-amine oxidases
in pure forms, our laboratory decided to take a preliminary look on the effect
of chemical evolution on the structures and activities of the Cu-amine oxidases
available to us. The details of these investigations are presented in this
report.

MATERIALS AND METHODS

Materials. The isolation procedure for preparing pure bovine plasma amine oxidase B has been published previously.[2] The isolation of the A-form is described in the Results section. Partially purified pea seedling Cu-amine oxidase was a gift from J. Hill (Harpen-den, Hertz, England) and its isolation has been reported earlier.[7] Crystalline preparation of the *Aspergillus niger* Cu-amine oxidase was a generous gift from Dr. O. Adachi (Kyoto University). All chemicals used were of reagent grade and were obtained from either the standard chemical sources or from sources previously reported.[2,8] The Sepharose 4B was purchased from Pharmacia Fine Chemicals. Cyanogen bromide was a product of J.T. Baker. Immunodiffusion plates were purchased from Cappel Labs., Inc.

Methods. The procedures for the measurements of activity and specific activity have been described in previous reports from our laboratory.[2,8] The methods for the determination of amino acid compositions and the end groups of the enzymes have also been reported previously.[9,10] Amino acid analyses were determined in the Beckman Model 121MB automatic amino acid analyzer as described by Spackman et al.[11]

Molecular weights of the enzymes were determined by the SDS-disc electrophoretic procedure of Weber and Osborne[12] both in the presence and absence of 2-mercaptoethanol. Concanavalin A-Sepharose 4B affinity support was prepared essentially as reported by Cuatracasas and Anfinsen[13] and the diamine-hexane-Sepharose 4B support by the procedure reported by Toraya et al.[14]

Anti-bovine plasma amine oxidase B was obtained as follows. Four milligrams of the plasma amine oxidase B in one milliliter of Freund's adjuvant was injected into the rabbit. One month later, the same dose was again injected. Two weeks after the final injection, the rabit was bled through the ear consecutively every two weeks for three months. Thirty milliliter samples were removed per bleeding.

RESULTS

Isolation of pure plasma amine oxidase A and B. Our laboratory has earlier reported on the isolation procedure for obtaining pure bovine plasma amine oxidase[2] and noted more than one form of the enzyme. However, only one form was isolated in a pure form and the reason for the presence of more than one form of the enzyme was not examined. In the present study, it was noted that not all batches of plasma contained the A-form of the enzyme but all preparations contained the B-form. It appears that different species of steer contain chemically distinct forms of the enzyme and since the plasma was collected from 4-5 steers, occasionally more than one form of Cu-amine oxidase was present in the

TABLE 1

SUMMARY OF PURIFICATION OF BOVINE PLASMA AMINE OXIDASES-A AND -B

Step	Vol. (ml)	Activity (units/ml)	Protein (mg/ml)	Total units	Total protein (mg)	Yield (%)	Specific activity	Purification
1. Plasma	20,000	80	39.0	1600×10^3	780,000	100	2.1	1
2. 0.35-0.55 AmSO$_4$ ppt	2,000	740	36.0	1480×10^3	72,000	93	20.6	98
3. A) 1st DEAE-cellulose eluate	2,000	720	12.9	1440×10^3	25,800	90	55.8	26.6
B) After AS fractionation	445	3000	20.2	1335×10^3	8,989	83	148.5	70.7
4. 2nd DEAE-cellulose step after AS fractionation								
A) 0.03 M eluate	(250)	(2600)	(18.5)	(650×10^3)	(4,625)	(41)	(140.5)	(66.9)
B) 0.07 M eluate	100	4200	10.8	420×10^3	1,080	26	338.9	161.4
5. Con A Sepharose								
A) Eluate of A-form	(958)	(205)	(0.59)	(196×10^3)	(565)	(12)	(347.5)	(165.5)
B) Eluate of B-form	755	390	0.32	294×10^3	242	28	1218.7	580.3
6. Diamino Hexane-Sepharose eluate of A-form	(290)	(280)	(0.39)	(81×10^3)	(113)	(5)	(717.9)	(341.9)

pooled plasma. There are other possible explanations for the presence of two forms of the enzyme such as nicking by plasma proteolytic enzymes but the molecular weight and chemical properties of the two forms of the enzyme argue against this possibility. Also, both forms of the enzyme interact with Concanavalin A-Sepharose 4B suggesting that the carbohydrate moiety is present in both forms of the enzyme. As documented below, the A-form of the enzyme is eluted from the DEAE-cellulose column by 0.03 M potassium phosphate buffer, pH 7.0. Table 1 summarizes the data obtained from the purification procedure which led to pure bovine plasma amine oxidases-A and -B from 50 L of blood. The B-form of the enzyme is much easier to purify than the A-form of the enzyme and an additional diaminohexane-Sepharose 4B, a substrate affinity chromatography step, was essential before the enzyme-A was pure. From an enzymatic standpoint, the enzyme purified by the substrate affinity column chromatography appeared unstable because possibly the substrate liberated sulfhydryl groups failed to reoxidize properly after elution of the enzyme from the column by the substrate (butylamine).

Subunit molecular weights of the various enzymes. The two forms of the bovine plasma enzyme, the A. *niger* enzyme and the pea seedling Cu-amine oxidases were all examined by the SDS-disc electrophoretic procedure in the presence and absence of 2-mercaptoethanol. The electrophoretic patterns are shown in Fig. 1. The results are in agreement that, all except for the A. *niger* enzyme, the Cu-amine oxidases under investigation contain two subunits which are covalently bonded by cystine linkages. Also, in the case of the A. *niger* enzyme, when the disc electrophoretic experiments were conducted in the presence of SDS but absence of the reducing agent 2-mercaptoethanol, the components detectable after staining showed molecular weights of 77,000 (monomer), 165,000 (protomer) and 245,000 (trimer). The molecular weight of the subunits of these Cu-amine oxidases are shown in Fig. 2. The molecular weight of the bovine enzyme has been carefully determined by ultracentrifugal analysis and was determined to be 170,000[15] but was shown to form dimers and trimers which accounted for the earlier reported value of about 261,000.[16] It is quite possible the A. *niger* enzyme has a protomer molecular weight of about 165,000 but due to the presence of polymers (dimers and trimers), the average molecular weight of the monomer, dimer and trimer mixture was 251,000. However, the $s_{20,w}$ value for the bovine enzyme was about 9.3 S.[15] Obviously, additional studies are needed to settle the molecular weight question.

Amino acid composition of the various Cu-amine oxidases. The amino acid compositions of bovine enzymes-A and -B were determined from 24 hour hydrolyzates (Table 2). In addition, the B-form of the enzyme was hydrolyzed for 24, 48 and

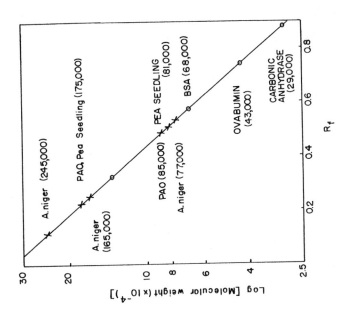

Fig. 2. Subunit molecular weights of some Cu-amine oxidases. The molecular weights were estimated from the disc electrophoretic experiments described in the legend in Fig. 1. The abbreviation PAO stands for bovine plasma amine oxidase.

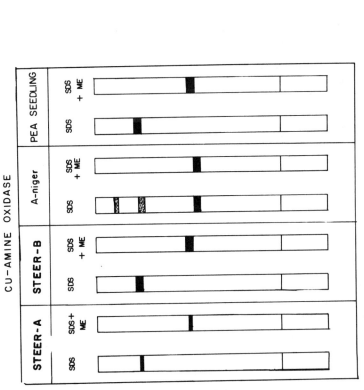

Fig. 1. SDS-Disc electrophoretic patterns of some Cu-amine oxidases. Five percent polyacrylamide gels containing 0.1% SDS and in the designated cases contained samples which were first reduced with a 1% SDS-1% 2-mercaptoethanol solution. The buffer used was 0.1 M sodium phosphate buffer, pH 7.2. Five mA was applied to each tube for 6 hours. The proteins were tracked with Coomassie blue.

556

TABLE 2

AMINO ACID COMPOSITIONS OF SOME CU-AMINE OXIDASES

Amino Acid	Moles per mole		
	Bovine Enzyme A	Bovine Enzyme B[a]	A. *niger* Enzyme[b]
Lysine	46	38	72
Histidine	47	43	47
Arginine	75	68	97
Aspartic Acid	116	116	161
Threonine	72	76	72
Serine	104	78	78
Glutamic Acid	160	167	156
Proline	116	118	110
Glycine	135	126	104
Alanine	80	108	90
Half Cystine	21	23	20
Valine	106	114	124
Methionine	20	21	20
Isoleucine	38	39	71
Leucine	126	121	103
Tyrosine	56	58	45
Phenylalanine	78	81	75
Tryptophan	37	35	17
Total Residues	1455	1430	1462
Acidic/Basic Residues	1.64	1.90	1.47

[a] Assumed mol. wt. of 170,000 of which 152,000 is due to protein.
[b] Assumed mol. wt. of 158,000.

72 hours and the amino acid composition determined. The data for the A. *niger* amine oxidase was also determined from the 24, 48 and 72 hour hydrolyzates. In order to calculate the number of residues, the molecular weight data was used and the total residues were made to add up to the proper molecular weight after correction for the presence of carbohydrate and prosthetic groups. Although the amino acid composition data is only approximate, it was concluded that all of these enzymes have very similar amino acid compositions. In agreement with iso-electric focusing data (not shown as well as the affinity of these enzymes for DEAE-cellulose, one is led to the conclusion that they are all acidic proteins, the A. *niger* enzyme being the most acidic. From the DEAE-cellulose chromatography elution pattern, the plasma amine oxidase B would be expected to be more acidic than the A-form of the enzyme. Although the amide contents have not been determined, the ratio of acidic/basic amino acid residues are in agreement with the elution of the two forms of the enzyme from the anion exchange resin. The other point of interest is that the bovine plasma enzymes A and B show

considerable differences in their amino acid compositions especially with
respect to their serine and alanine contents.

  End group analyses of the Cu-amine oxidases. Dansylation of the bovine
plasma amine oxidases A and B, and the A. *niger* amine oxidase disclosed no definite NH_2-terminal residue. These are preliminary results which need to be further checked. However, it is certain that the bovine plasma amine oxidase has
an N-terminal residue with a blocked $-NH_2$ group. When the S-β-carboxymethyl-
cysteine-derivative was analyzed in the Beckman 890 Protein Sequencer, no Pth-
amino acid could be detected in stoichiometric yields. In addition, a peptide
liberated from the phenylhydrazine derivative of the enzyme has led to the isolation of yellow peptide which supposedly contained the organic cofactor
attached to the peptide via the amino group. This has led us to speculate that
the organic cofactor is attached to the NH_2-group of the N-terminal amino acid
of the enzyme.[17] Therefore, the Cu-amine oxidases may possibly all contain the
organic cofactor acylated to the N-terminal amino acid. Carboxypeptidase A was
used to determine the C-terminal amino acid of the various Cu-amine oxidases.
It has been reported previously that the C-terminal amino acid of bovine plasma
amine oxidase B was tyrosine.[1] Carboxypeptidase A digestion of the bovine Cu-
amine oxidase A was also found to yield tyrosine. The C-terminal amino acid of
the A. *niger* enzyme was difficult to identify since valine, alanine, leucine,
isoleucine and threonine were rapidly liberated by carboxypeptidase A. Although
additional investigations are necessary, the preliminary data indicate the Cu-
amine oxidases under investigation contain N-terminal amino acids with blocked
NH_2-groups but the C-terminal amino acids can be liberated by carboxypeptidase
A.

  Immunological similarity of the various amine oxidases. Rabbit antibody to
bovine plasma amine oxidase B was used to check the immunochemical similarities
of the bovine, mold and plant enzymes and results are shown in Fig. 3, which

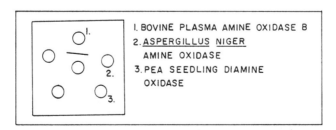

1. BOVINE PLASMA AMINE OXIDASE B
2. ASPERGILLUS NIGER
   AMINE OXIDASE
3. PEA SEEDLING DIAMINE
   OXIDASE

Fig. 3. Ouchterlony double immunodiffusion analysis of some Cu-amine oxidases.
Rabbit antisera to the bovine plasma amine oxidase B was added to the center well.

shows the Ouchterlony immunodiffusion tests. No precipitin lines were observed with the A. *niger* or the pea seedling Cu-amine oxidases, only a single precipitin line with the bovine enzyme B, as expected.

Effect of rabbit antisera to the bovine plasma enzyme-B on the activity of the various Cu-amine oxidases. The enzyme was assayed by the spectrophotometric assay of Tabor et al.[18] in which the oxidation of benzylamine to benzaldehyde is monitored at 250 nm. Various amounts of antisera were added to the mixture which contained either the bovine plasma amine oxidase B, the A. *niger* amine oxidase or the pea seedling amine oxidase. The results are shown in Fig. 4 and it can be seen that the antibody to the bovine plasma amine oxidase B did not inhibit the pea seedling enzyme, inhibited the A. *niger* amine oxidase to an intermediate extent and inhibited the bovine plasma amine oxidase B to the greatest extent.

Specific activities of various amine oxidases with benzylamine as substrate. All of the enzymes listed in Table 3 oxidize benzylamine. The rates of oxidation are dependent upon pH and the values shown in the table should be considered

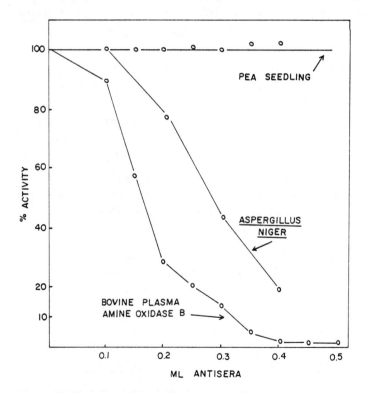

Fig. 4. Inhibition of some Cu-amine oxidases by antisera to the bovine enzyme B.

TABLE 3

SPECIFIC ACTIVITIES OF SOME CU-AMINE OXIDASES WITH BENZYLAMINE AS THE SUBSTRATE

Source of Enzyme	Specific Activity	Expt'l Conditions	K_m
Cu-amine Oxidases			
1. Bovine enzyme A	>700	pH 7.2, 25°	--
2. Bovine enzyme B	1,200	pH 7.2, 25°	2×10^{-3}
3. Pea Seedling enzyme	>3,000	pH 7 , 25°	--
4. A. *niger* enzyme[a]	9,800	pH 7 , 25°	1.7×10^{-5}
FAD-amine Oxidase			
1. Bovine Liver Monoamine Oxidase enzyme	11,000	pH 7.4, 25°	2.2×10^{-4}

[a]Adachi.[8]

approximate. Among the Cu-amine oxidases, the mold enzyme has the highest
specific activity while the bovine enzyme has the lowest value. The bovine
liver FAD-amine oxidase has the highest specific activity when benzylamine is
used as the substrate, being more active than any Cu-amine oxidase. In the case
of the bovine enzyme-B, kinetic investigations have shown that $K_m = K_s$.[19] If
the kinetics for the other Cu-amine oxidases is the same as reported for the
bovine amine oxidase, the smaller K_m value for the A. *niger* amine oxidase indicates that it has a higher affinity for benzylamine than the bovine enzyme.

DISCUSSION

Cu-amine oxidases probably all contain 2 g atoms of copper per protomer and
contain an unknown organic cofactor. The chemical identification of the covalently bonded organic cofactor is a major problem left to be solved. In
addition, we have noted multiple forms of the bovine plasma amine oxidase and
now the two forms have been isolated in pure forms and we have designated them
as the A-form and the B-form of the enzyme. The A-form is more basic than the
B-form and are probably present because different species of steer contain different forms of the enzyme. However, of the numerous pooled steer plasma
studied, the B-form was always present but the A-form was occasionally lacking.
It is interesting to note that the pig, for example, has only one form of the
enzyme. The A-form and the B-form of the plasma enzymes are not degraded forms
of one another since the molecular weights and the end groups appear to be the
same. Also, both forms contain carbohydrate since each is bound by Concanavalin

A-Sepharose 4B affinity column. A more detailed study of the properties of the
bovine plasma amine oxidase A were not carried out since the diaminohexane
affinity chromatography step of enzyme purification led to a slightly altered,
unstable form of the enzyme. However, the B-form of the enzyme, which was purified without the substrate-Sepharose 4B chromatography step, was isolated in a
pure active state. Therefore, the procedure described for the purification of
the bovine plasma amine oxidase represents an improvement over the previously
described method.[2]

The availability of the pure forms of the bovine enzyme plus the availability
of the A. *niger* and pea seedling enzymes prompted us to look at the chemical
differences in the properties of these enzymes, especially from the standpoint
of chemical evolution. The following conclusions were derived from the present
investigation. The physicochemical properties of the various Cu-amine oxidases
indicated the various enzymes under study have all evolved from the same common
ancestor. Possibly the ancient enzyme precursor had a molecular weight equal
to the present day subunit (70,000-85,000) and contained one gram atom of copper.
Changes possibly occurred in the sequence of this precursor and the subunit-
subunit binding region developed. Further sequence changes occurred in the subunit-subunit binding region and the cysteine residues were incorporated into
this region and eventually disulfide bridges between the two subunits were
formed. This concept is summarized in Fig. 5 and also shows the changes observed

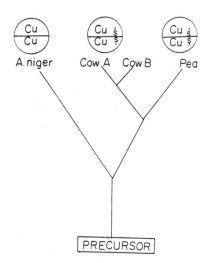

Fig. 5. Changes in the subunit interactions of some Cu-amine oxidases during
the process of evolution.

in the subunit structure of the Cu-amine oxidases during the process of evolution. The figure also points out the mold enzyme is more ancient than the plant or animal enzymes and also that the steers have either the A- or the B-form of the enzyme. However, it should be emphasized that since the pig contains only one form of the plasma amine oxidase, not all animals contain two different forms of the enzyme. The immunological studies in which antiplasma amine oxidase B rabbit sera was used, surprisingly showed that no precipitin lines were formed between the antibody and the plant or mold enzymes. However, the rabbit antisera was able to inhibit the oxidation of substrate by the A. *niger* enzyme but not the plant enzyme. This result was not expected since the mold is more ancient than the plant, and the plant and the animals are more phylogenetically related. Chemical data appears to be more suitable for creating phylogenetic trees than immunological methods. Since no precipitin line was formed when the A. *niger* enzyme and antisere were tested by the Ouchterlony double immunodiffusion test and in view of the inhibition of activity by the antibody observed, a soluble antibody-mold enzyme complex may have been formed. However, the plant amine oxidase since the antiplasma amine oxidase B did not inhibit the plant enzyme and no precipitin line was formed either.

Nature has evolved two different enzymes which oxidize benzylamine. The second type (FAD-amine oxidase) is a better catalyst than the Cu-amine oxidases on a mole-per-mole basis. the FAD-type of amine oxidase has been isolated from animals and bacteria.[20-22] The bovine liver enzyme has been reported to consist of two subunits, each of molecular weight 52,000. There are no disulfide linkages between the subunits.[23] The FAD in the enzyme was reported to be covalently attached to the enzyme as 8-α-cysteinyl-FAD.[25] Two FAD-amine amine oxidases have been isolated from bacteria, i.e., tyramine oxidase from *Sacrina lutea*[20] and putrescine oxidase[21,22] from *Micrococcus rubrens*. The FAD is dissociable from the protein in the case of the bacterial FAD-amine oxidases. The possible subunit structure of these bacterial amine oxidases and the possible phylogenetic relationship of the cow and the bacteria which contain these enzymes are summarized in Fig. 6. In this figure, the viewpoint was adopted that all of the FAD-enzymes have evolved from the same common ancestor but the scheme is very speculative since detailed physiocochemical studies of these enzymes have not yet been carried out.

ACKNOWLEDGMENTS

The research project was partially supported by grants SER 77-06923 from the National Science Foundation and MH 21539 from the National Institutes of Health.

562

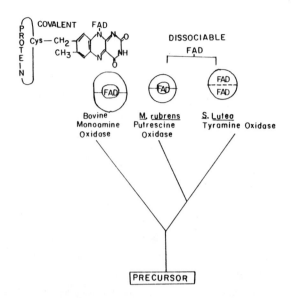

Fig. 6. Changes in the subunit structure and the binding of FAD in some FAD-amine oxidases during the process of evolution.

REFERENCES

1. Yasunobu, K.T., Ishizaki, H. and Minamiura, N. (1976) Mol. Cell Biochem., 13, 3-29.
2. Yamada, H. and Yasunobu, K.T. (1962) J. Biol. Chem., 237, 1511-1516.
3. Buffoni, F. and Blaschko, H. (1964) Proc. Roy. Soc. B., 161, 153-167.
4. Lindstrom, A., Ollson, B. and Petterson, G. (1974) Eur. J. Biochem., 48, 237-243.
5. Yamada, H., Kumagai, H., et al. (1967) Biochem. Biophys. Res. Commun., 29, 723-727.
6. Mondovi, B., Rotilio, G., et al. (1967) J. Biol. Chem., 242, 1160-1167.
7. Hill, J.M. (1972) Methods in Enzymol., 17B, 730-735.
8. Adachi, O. (1969) Ph.D. Thesis, Kyoto University, Japan.
9. Yamada, H., Adachi, O. and Ogata, K. (1965) Agricult. Biol. Chem. (Tokyo), 29, 649-654.
10. Tanaka, M., Haniu, M., et al. (1977) Biochemistry, 16, 3525-3537.
11. Spackman, D.H., Moore, S. and Stein, W.H. (1958) Anal. Chem., 30, 1190-1194.
12. Weber, K. and Osborne, M. (1969) J. Biol. Chem., 224, 4406-4412.
13. Cuatrecasas, P. and Anfinsen, C.B. (1971) Methods in Enzymol., 22, 345-378.
14. Toraya, T., Fujimura, M., et al. (1976) Biochim. Biophys. Acta, 420, 316-322.
15. Achee, F.M., Chervenka, C.H., et al. (1968) Biochemistry, 12, 4329-4335.
16. Yamada, H., Gee, P., et al. (1964) Biochim. Biophys. Acta, 81, 165-171.
17. Watanabe, K., Smith, R.A., et al. (1972) Advances in Biochem. Psychopharmacol., 5, 107-117.
18. Tabor, C.W., Tabor, H. and Rosenthal, S.M. (1954) J. Biol. Chem., 208, 645-661.
19. Oi, S., Inamasu, M. Yasunobu, K.T. (1970) Biochemistry, 9, 3378-3382.
20. Yamada, H., Kumagai, H., et al. (1967) Agricult. Biol. Chem. (Tokyo), 31, 897-901.
21. Yamada, H., Adachi, O. and Ogata, K. (1965) Agricult. Biol. Chem. (Tokyo), 29, 1148-1149.
22. DeSa, R.J. (1972) J. Biol. Chem., 247, 5527-5534.

Index